Microwave Wireless Communications

Microwave Wireless Communications

From Transistor to System Level

Antonio Raffo
University of Ferrara, Ferrara, Italy

Giovanni Crupi
University of Messina, Messina, Italy

ELSEVIER

AMSTERDAM • BOSTON • HEIDELBERG • LONDON
NEW YORK • OXFORD • PARIS • SAN DIEGO
SAN FRANCISCO • SINGAPORE • SYDNEY • TOKYO
Academic Press is an imprint of Elsevier

AP

Academic Press is an imprint of Elsevier
32 Jamestown Road, London NW1 7BY, UK
525 B Street, Suite 1800, San Diego, CA 92101-4495, USA
50 Hampshire Street, 5th Floor, Cambridge, MA 02139, USA
The Boulevard, Langford Lane, Kidlington, Oxford OX5 1GB, UK

British Library Cataloguing in Publication Data
A catalogue record for this book is available from the British Library

Library of Congress Cataloging-in-Publication Data
A catalog record for this book is available from the Library of Congress

ISBN: 978-0-12-803894-9

Publisher: Joe Hayton
Acquisition Editor: Tim Pitts
Editorial Project Manager: Charlie Kent
Production Project Manager: Jason Mitchell
Designer: Victoria Pearson

Ad Alessandro, l'Amore di papà
Antonio Raffo

Ai miei cari genitori Carmelo e Teresa
Ai miei fratelli Vincenzo, Felice, Isodiana e Giuseppe
Ed a mia moglie Gabriella
Giovanni Crupi

Contents

Contributors

G. Avolio
KU Leuven, Leuven, Belgium

K. Burger
Qualcomm, San Diego, CA, United States

A. Caddemi
University of Messina, Messina, Italy

V. Camarchia
Politecnico di Torino, Torino, Italy

W. Chen
Tsinghua University, Beijing, China

X. Chen
Tsinghua University, Beijing, China

E. Cipriani
University of Rome Tor Vergata, Rome, Italy

P. Colantonio
University of Rome Tor Vergata, Rome, Italy

G. Crupi
University of Messina, Messina, Italy

S. Farsi
KU Leuven, Leuven, Belgium; Qualcomm Inc., San Diego, CA, United States

A. Ferrero
Keysight Technologies, Santa Rosa, CA, United States

F. Giannini
University of Rome Tor Vergata, Rome, Italy

E. McCune
RF Communications Consulting, Santa Clara, CA, United States

B. Moser
Qorvo, Inc, Greensboro, NC, United States

M. Pirola
Politecnico di Torino, Torino, Italy

A. Raffo
University of Ferrara, Ferrara, Italy

F. Ramirez
University of Cantabria, Santander, Spain

S. Sancho
University of Cantabria, Santander, Spain

D.M.M.-P. Schreurs
KU Leuven, Leuven, Belgium

A. Suarez
University of Cantabria, Santander, Spain

G. Vannini
University of Ferrara, Ferrara, Italy

Z. Wang
Microsoft Corp., Beijing, China

P.J. Zampardi
Qorvo, Inc, Newbury Park, CA, United States

S. Zhang
Smarter Micro Inc., Shanghai, China

About the Editors

Antonio Raffo is a research associate at the University of Ferrara, Italy, where he teaches courses in semiconductor devices and electronic instrumentation and measurement. His research activity is mainly oriented to nonlinear electron device characterization and modeling and circuit-design techniques for nonlinear microwave and millimeter-wave applications. Antonio has coauthored over 100 publications in international journals and conferences. He is a member of the Technical Programme Committee of the IEEE INMMiC conference and serves as an associate editor *of International Journal of Numerical Modelling: Electronic Networks, Devices and Fields*. Antonio is a member of the IEEE Microwave Measurement Technical Committee (MTT-11).

Giovanni Crupi is a tenure track assistant professor at the University of Messina, Italy, where he teaches microwave electronics, laboratory of wireless technologies, bioengineering, and optoelectronics. Since 2005, he has been a repeat visiting scientist with KU Leuven and IMEC, Leuven, Belgium. Giovanni's main research interests include small and large signal modeling of advanced microwave devices. He is a member of the Technical Programme Committee of the IEEE INMMiC and TELSIKS conferences and serves as an associate editor of *International Journal of Numerical Modelling: Electronic Networks, Devices and Fields*. Giovanni is the chair of the IEEE Microwave Theory and Techniques Society (MTT-S) Fellowship program.

Foreword by Charles F. Campbell

Mobile communications and connectivity have driven a multitude of advancements in RF and microwave semiconductor and circuit technology. While these advancements have enabled the mobile communication revolution, they also present circuit and system designers with many different options and approaches for product realization. The days are long gone when designers simply optimized a Gallium Arsenide integrated circuit for saturated output power and efficiency. Today's circuit and systems engineers must be far more knowledgeable in the overall product development process. Proficiency in device physics, modeling and characterization, circuit design, architecture and application, signal modulation formats, measurements, and industry standards are all required to successfully design modern receive and transmit components. Many engineering and design decisions must be considered to balance cost, performance, and cycle time while simultaneously meeting challenging product specifications. Specific examples would be selecting an optimum semiconductor technology, device characterization and modeling, circuit architecture, linearization strategy, and overall system-level considerations. Although there are many excellent books and articles that cover these topics in depth, it is far more difficult to find a comprehensive source of information. In this book, "Microwave Wireless Communications: From Transistor to System Level," a team of well-known industry and academic experts has been assembled to provide a thorough treatment of these topics. The authors do not simply repeat standard explanations of these subjects, but present unique approaches, perspectives, and alternate points of view that greatly enhance one's understanding of these complex ideas.

This book begins with an introduction to transistor modeling and extraction, highlighting device characteristics and behavior. Sections on linear, nonlinear, and noise modeling are included along with modeling results of actual devices. Chapter 2 focuses on test and measurement systems for device characterization, including discussions of nonlinear vector network analyzers and active load pull. Block diagrams of these measurement instruments are provided and explained in detail.

Chapter 3 is the start of the circuit design portion of the book, presenting a comprehensive discussion of oscillators. This highly theoretical discussion is introduced by way of simple idealized oscillator circuits—to bring the reader up to speed with basic principles. This is followed by a more rigorous analysis, applicable to real oscillators, along with design examples using a commercially available device. Chapter 4 covers the topic of power amplifiers, and starts with parameter definitions and figures of merit. This is followed by an intuitive discussion of load line analysis and power amplifier modes operation. Circuit architectures, implementation, and amplifier linearity are also described in this section.

For Chapter 5, the book reviews how system-level constraints impact circuit design decisions. Aspects such as battery life, probability functions for transmit power, amplifier linearization techniques, and power amplifier module architectures are discussed at a high level. This is followed by a discussion of available devices, passives and board technologies, and how each relates to performance, cost and schedule. Chapter 6 covers power amplifier design, including architecture and linearization techniques. The Doherty amplifier concept is thoroughly reviewed as well as broadband and digital implementations. This is followed by a presentation of popular linearization approaches, including envelope methods (envelope tracking (ET) and envelope elimination and restoration (EER)) and linear amplification with nonlinear components (LINC) approaches (Chireix, multilevel and mode-multiplexing).

The final two chapters focus on the analysis of power amplifier linearization and efficiency enhancement methods. The author of Chapter 7 takes an interesting approach to investigating the effectiveness of efficiency enhancement methods. Using transistor power compression curves, three operational modes are defined (L, P, and C modes) based on the how deep the device is in gain compression. Simple circuit models for each mode are developed which, along with the transistor IV-curves, are used to analyze output power and efficiency of a device subjected to envelope tracking and polar modulation. Chapter 8 covers techniques for the characterization and analysis of nonlinear circuits. Nonlinear circuit transfer functions and Volterra series are reviewed, including systems with and without memory. Then, standard 2-tone characterization techniques are expanded to more sophisticated multisine and offset multisine approaches. The results are applied to the analysis and modeling of an actual power amplifier to illustrate various nonlinear effects.

Hopefully I have successfully highlighted the content of this book and convinced you to examine it personally. Practicing mobile communication engineers, as well as those who simply seek a better understanding of this complex subject, will benefit from the content of this work. The authors and editors should be congratulated for sharing their knowledge and perspective of this topic with the engineering community.

Charles F. Campbell, Ph.D.
Qorvo, Richardson, Texas, United States

Foreword by Ramesh K. Gupta

Systems engineering enables the realization of complex system functions through careful choice, and a combination of interacting system elements assembled to achieve specific and defined objectives. RF and microwave systems are designed for applications such as radio astronomy and space exploration, navigation, radars, wireless terrestrial and satellite communications, remote sensing, RF identification, electronic warfare, medical imaging, monitoring and sensing, automotive collision avoidance systems, etc. Typically, these RF systems are integrated with various subsystems and elements, such as low-noise amplifiers and front ends, frequency sources, frequency converters, intermediate and high power amplifiers with transmit and/or receive antennas, and radiating elements. Realization of these RF subsystems often requires high frequency RF transistors and devices that are embedded and assembled with passive circuits of microstrip lines printed on a substrate, and/or coaxial or waveguide elements, used in the design and particular wireless communication application.

Effective RF and microwave systems engineering requires an in-depth understanding of interactions between devices, circuits, and systems for developing solutions that can be manufactured cost-effectively, and with required performance, repeatability, and reliability. This is particularly important for high volume deployment and production of wireless systems and user equipment, including hand-held and wearable devices (eg, smart phones, tablets, smart watches, etc.) where consumers demand the highest performance, functionality, and repeatable performance at the lowest possible price. It has been a challenge for practicing wireless circuit, device design, and system engineers to find the requisite information for their design and development efforts. Therefore, I believe that Dr. Antonio Raffo and Dr. Giovanni Crupi, the editors of the book "*Microwave Wireless Communications: From Transistor to System Level*," have done a great service to the microwave and RF community by putting together important and relevant information on systems, circuits, and devices in a single volume with contributions from authors that have extensive research experience and expertise in the specific technical areas and topics included in each of the eight chapters.

Going through the table of contents for this book sent me on a nostalgic journey through my own career, which spanned work in RF device and circuit modeling, satellite and wireless subsystem development and system design, and modeling, as well as the deployment and operations of these systems and devices. I started my career with the investigation of two-terminal solid-state devices, such as Gunn and IMPATT diodes, for system applications. The issues of RF parameter de-embedding, linear and nonlinear circuit analysis and modeling, as well as their noise behavior, were relevant then for their system insertion. The two-terminal RF devices were quite constrained because of their "noisy behavior" and signal distortion characteristics during large signal conditions. Emergence of three-terminal Gallium Arsenide metal-semiconductor field-effect transistors (MESFETs) and the many derivatives—including high electron mobility transistors (HEMTs) and pseudomorphic HEMTs (p-HEMTs), etc.—and their integration onto monolithic microwave integrated circuits (MMICs) on GaAs substrates, as well as development of silicon based RF integrated circuits (RFICs), have had a revolutionary impact on the miniaturization of RF systems. Some examples of these systems include the high performance radars using T/R modules, satellite systems with flexible and steerable beams, and wireless systems with smaller size, broader bandwidth, and repeatable performance with higher reliability. By effective system design, these systems can be manufactured and deployed at a significantly reduced cost. During my work at COMSAT Laboratories, using the device, circuit, system, measurement, and calibration techniques similar to ones documented in all the chapters of this book, we were able to develop miniaturized receivers, power amplifiers and transmitters, frequency converters, and switch matrices for dynamic satellite beam switching. We also successfully developed proof-of-concept C-, X-, and Ku-band phased array antennas for satellite beam shaping and steering for on-board and ground satellite applications. It is gratifying to see many of these subsystems and systems and their derivatives now being components of the majority of modern satellite and wireless network architectures.

Given the state of the RF and microwave technology today, I remain very enthusiastic on emergence of new innovative wireless systems. During the coming years and decades, many new applications are likely to become available, including wearable healthcare data sensing and monitoring systems, self-energizing implantable medical systems, autonomous and interconnected vehicles, collision avoidance systems, nondestructive medical imaging using terahertz (THz) frequencies, contactless payment systems, etc. It is my hope that students, engineers, and researchers involved in RF and microwave practice will be able to enhance their understanding of device, circuit, and system interactions through concepts and techniques presented in this book, and help create exciting system applications and solutions for the benefit of mankind.

Ramesh K. Gupta, Ph.D.
Bethesda, Maryland, United States

Preface

The wireless communication market has witnessed tremendous growth in recent years. To the consumer, this growth manifests itself via the proliferation of mobile telephones, tablets, gaming consoles, and soon, the Internet of Things. The design of these wireless communication systems requires in-depth understanding of their microwave functionality, starting from the theoretical underpinnings of their inner workings to their practical system-level deployment.

The design of microwave wireless systems traverses three different stages, oftentimes dubbed the device, circuit, and system levels. In spite of the vast archival literature dealing with each of these design levels, thus far no book has detailed the links and interactions among them. Not surprisingly, microwave engineers often acquire deep expertise in one level, but lack understanding of other levels, especially their interactions, which complicates the collaboration and project handover between design teams, and stifles innovation. As an example, engineers working on device modeling oftentimes experience difficulties understanding the different operating modes of transmitters based on dynamic power supply power amplifiers (DPS-PA), despite the fact that DPS-PA operating modes are inherently and simply related to transistor operation.

The goal of this book is to elucidate links between the different levels in microwave wireless system design by describing and unraveling their interactions. The book's multiple expert authors offer distinct and complementary viewpoints of the subject material. Together, their two forewords and eight chapters provide comprehensive coverage of the vast topic area. While individual contributions focus primarily on one of the three different design levels, altogether they also touch on interactions between levels, and address the presently siloed knowledge of the subject field. For example, the reader will learn that design links between the device and the circuit levels are not so different when dealing with amplifiers, mixers, or oscillators, and that limitations arising from transistor linear and nonlinear parasitic elements are largely the same, even though they impact different quantities (noise, power added efficiency, conversion gain, resonance conditions).

We hope this book will become a valued resource, not only for researchers, engineers, and teachers wishing to extend their knowledge of microwave wireless systems beyond the presently stratified literature, but to all those seeking a better understanding of the linkages and dependencies between different design levels.

We very much hope that the reader will enjoy her or his journey through the three different levels of the microwave wireless system design!

Antonio Raffo and Giovanni Crupi

CHAPTER

Microwave transistor modeling

1

G. Crupi*, A. Raffo†, G. Avolio‡, A. Caddemi*, D.M.M.-P. Schreurs‡ and G. Vannini†

University of Messina, Messina, Italy University of Ferrara, Ferrara, Italy†*
KU Leuven, Leuven, Belgium‡

CHAPTER OUTLINE

1.1 INTRODUCTION

Microwave transistor modeling is an evergreen research field of paramount importance. The significant interest around this topic comes from the fact that the transistor is the key component in high-frequency circuits that are at the heart of modern wireless communication systems, such as mobile telephony. To meet the more and more stringent requirements of wireless communication systems, transistor technologies are incessantly progressing. Hence, the need for exploiting emerging technologies at their best makes microwave transistor modeling an open research field in continuous evolution.

This first chapter is meant to give a comprehensive overview of the fundamentals, state-of-the-art, challenges, and future trends in the field of microwave transistor modeling, going from linear (also noise) to nonlinear operation.

The chapter is structured into six main sections, including this introduction. The next section is devoted to the microwave transistor technologies. Of the various microwave transistors, the analysis of the present chapter is focused on the high-electron-mobility transistor (HEMT). Hence, the physical structure and the operation principles of this microwave active solid state device are concisely reviewed.

Microwave Wireless Communications. http://dx.doi.org/10.1016/B978-0-12-803894-9.00001-9

In particular, transistors fabricated using both gallium-arsenide and gallium-nitride materials are theoretically and experimentally investigated to highlight pros and cons of each technology. The third section is dedicated to the transistor modeling at high frequencies. The attention is focused on the extraction of equivalent circuit models, which represent a good compromise between physical and behavioral models. In fact, the equivalent circuit model provides indispensable feedback to device technologists for advancing the transistor fabrication process and is a valuable tool to circuit designers for optimizing circuit performance. The bottom-up approach for equivalent-circuit-based modeling is discussed, starting with a description of how to extract a small-signal model in the fourth section, before moving on to its use as a cornerstone to develop both noise and large-signal models in the fifth and sixth sections, respectively.

1.2 MICROWAVE TRANSISTOR TECHNOLOGIES

The first demonstration of a field-effect transistor (FET) using a two-dimensional electron gas (2DEG) in a potential quantum well as the active channel was made in 1980 [1,2]. This type of transistor was named HEMT by the Japanese researchers at Fujitsu Laboratories. Although HEMT is the most popular name, other terms describing its basic operation principle have been proposed by other research groups working on developing this new type of transistor. Examples are modulation doped FET (MODFET, University of Illinois and Rockwell, United States), two-dimensional electron gas FET (TEGFET, Thomson CSF, France), and selectively doped heterojunction transistor (SDHT, Bell Labs, United States). Since its inception, the HEMT has received worldwide attention for its attractive advantages compared to its predecessor, the metal-semiconductor field-effect transistor (MESFET): larger gain, higher operating frequency, and lower noise figure (NF).

The basic structure of a HEMT consists of a heterojunction composed of a doped wide-bandgap semiconductor and an undoped narrow-bandgap semiconductor. The earlier HEMTs were based on the aluminum-gallium-arsenide/gallium-arsenide (AlGaAs/GaAs) heterostructure. Commonly, n-channel GaAs FETs are considered, because electrons have much higher mobility than holes. As illustrated in Figure 1.1, electrons diffuse from the doped wide-bandgap semiconductor to the undoped narrow-band one. Due to the conduction band discontinuity and the accumulation of electrons leaving positive charged ionized dopants, a potential barrier is formed at the interface between the two semiconductors. This barrier prevents the electrons from going back to the doped semiconductor. Therefore, a 2DEG is confined in a triangular potential quantum well. These electrons drift in a pure crystalline material from the source to the drain and the spatial separation of carriers from donors is typically improved by introducing a thin spacer layer of undoped AlGaAs. Such a separation allows for achieving a significant reduction of Coulomb scattering and carriers frozen at ionized donors, which are two mechanisms responsible for degrading the carrier transport properties at low temperature. A considerable improvement

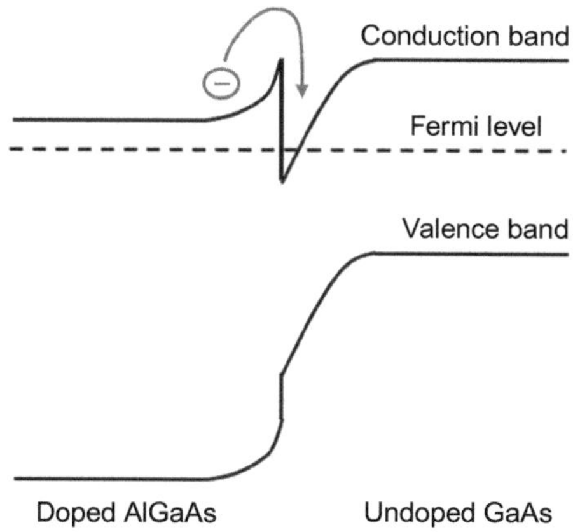

FIGURE 1.1

The energy band diagram for an AlGaAs/GaAs heterostructure.

of the performance of the HEMT is thus achieved at cryogenic temperatures, because the reduction of the phonon scattering processes at lower temperatures allows for obtaining a very high electron mobility [3,4]. Hence, electrons in a 2DEG exhibit a much higher mobility than those in the doped-channel MESFET, especially at low temperatures.

The basic operation principle of a HEMT consists of controlling the output drain current through the gate modulation of the carrier density in the 2DEG. If the gate bias is below the pinch-off voltage, the 2DEG channel is depleted. As the gate voltage is increased, the 2DEG density increases, resulting in a higher output drain current. Typically, a heavy forward gate bias is avoided in order to prevent device degradation from the associated excessive conduction current flowing through the gate Schottky junction. Furthermore, a high V_{GS} may lead to the occurrence of the phenomenon known as "parasitic MESFET," which consists of a transconductance reduction due to a parallel conduction through the undepleted low mobility AlGaAs donor layer [3]. In fact, the increase of V_{GS} lowers the potential energy barrier and then, by increasing V_{DS}, the electrons of the 2DEG can gain enough kinetic energy from the accelerating electric field in the channel to surmount the AlGaAs/GaAs energy barrier.

To improve the performance of the conventional AlGaAs/GaAs HEMT, an undoped layer of narrow bandgap indium-gallium-arsenide (InGaAs) can be introduced between the AlGaAs and GaAs layers. In the case of the AlGaAs/InGaAs/GaAs structure, the carriers flow in the InGaAs quantum well that is square-shaped. As the InGaAs is not lattice matched to its adjacent layers, the InGaAs channel layer has to be thin enough so that it stretches to fit the other two semiconductors. Compared to the conventional HEMT, this type of transistor, known as pseudomorphic HEMT (pHEMT), exhibits improved carrier transport properties due to the presence of indium, and a better carrier confinement due to a larger bandgap difference.

Although silicon (Si) is much cheaper than GaAs because of the lower processing complexity, larger wafer diameter, and better industry infrastructure, GaAs has traditionally been widely used for microwave applications. A fundamental benefit of using GaAs consists of higher electron mobility and saturation velocity compared with those of Si. Furthermore, GaAs being intrinsically a very poor conductor offers semi-insulating substrates, while Si substrates are much more conductive. Although silicon-on-insulator (SOI) technology with a high resistivity substrate allows for minimizing the substrate losses, the SOI wafers have the drawbacks of having higher wafer cost, lower thermal conductivity, and higher defect density than bulk Si wafers. Hence, GaAs is very useful for microwave and millimeter-wave integrated circuits (MMICs), where active and passive components have to be fabricated on the same semiconductor substrate. This is because the GaAs semi-insulating substrate allows for achieving electrical isolation among the various components, while effective shielding solutions are needed for Si-based radio-frequency integrated circuits (RFICs).

MMICs have significant advantages over hybrid circuits such as smaller size, lighter weight, improved reproducibility and reliability, broader band performance, and lower cost for mass production. Although the interest in military and space communication systems has driven and continues to drive the development of MMIC technology, MMICs are now also used in civil applications, such as mobile telephony.

To meet the demanding requirements for wireless communication systems such as those for base stations, a lot of attention is currently devoted to the HEMT based on the aluminum-gallium-nitride/gallium-nitride (AlGaN/GaN) heterostructure. The GaN-based HEMT is undoubtedly the leading technology for microwave high-power applications [5–19] because the wide bandgap of the nitride semiconductors implies a higher breakdown voltage. In the case of GaN-based HEMTs, the physics behind the formation of 2DEG is substantially different from the one illustrated in Figure 1.1 for conventional GaAs-based HEMTs. Owing to the large piezoelectric and spontaneous polarization effects and the large conduction band discontinuity between the AlGaN barrier and the GaN channel layers, an extremely high 2DEG density can be achieved even without intentionally doping the AlGaN layer. The electric field associated with the piezoelectric and spontaneous polarizations can be high enough to ionize electrons and force them to drift towards the heterointerface where they fall into the quantum well, forming the 2DEG. In the case of unintentionally introduced dopant impurities, the 2DEG consists of electrons resulting from covalent electrons, impurities that happen to be present in the materials, and loosely bonded surface electrons.

GaN HEMTs are typically grown on silicon carbide (SiC), silicon, or sapphire substrates. Silicon is available in larger wafer sizes and is cheaper than the other two substrates. On the other hand, silicon carbide is the most expensive solution, but its superior thermal conductivity with respect to silicon and sapphire makes it very suitable for high-power and high-temperature applications. In fact, high-power devices need a substrate material with high thermal conductivity to spread heat efficiently.

To present experimental results as an illustrative example of the SiC technology, Figure 1.2 shows the behavior of I_D versus V_{DS} and V_{GS} for an AlGaN/GaN HEMT

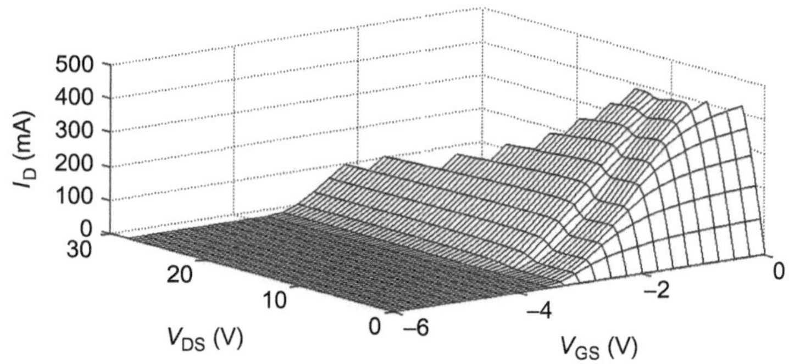

FIGURE 1.2

Behavior of $I_D(V_{DS}, V_{GS})$ for a 0.7 × 800 μm^2 GaN HEMT at 20°C.

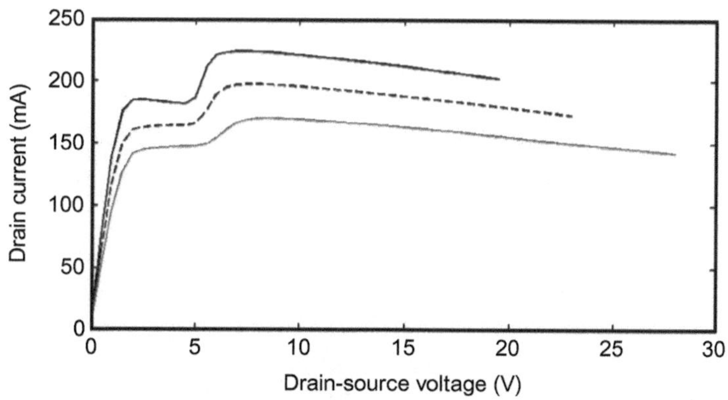

FIGURE 1.3

Output characteristics $I_D(V_{DS})$ for a 0.7 × 800 μm^2 GaN HEMT at $V_{GS} = -2$ V under three different ambient temperatures: 20°C *(solid black line)*, 50°C *(dashed black line)*, and 80°C *(solid gray line)*.

on SiC substrate with a gate length of 0.7 μm [20]. To avoid device degradation, the maximum V_{DS} and the maximum dissipated power P_{max} have been limited to 28 V and 5 W/mm, respectively. As can be observed, the device exhibits a negative slope of $I_D(V_{DS})$ under highly dissipated power conditions because of the self-heating phenomenon.

Figure 1.3 reports $I_D(V_{DS})$ at $V_{GS} = -2$ V under three different ambient temperatures: 20°C, 50°C, and 80°C. It is observed that the self-heating effect is enhanced at higher temperatures and dissipated power, due to the associated increase of the phonon scattering processes leading to a degradation of the electron transport properties. Furthermore, Figure 1.4 shows the load lines at 1 dB gain compression at a fundamental frequency f_0 of 2 GHz under the three ambient temperatures.

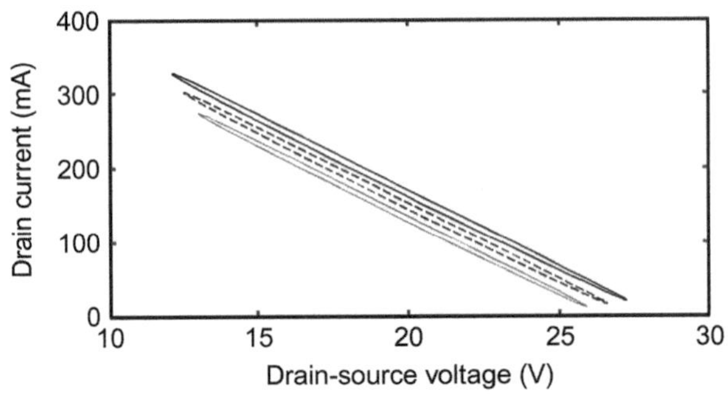

FIGURE 1.4

Load lines at 1 dB compression with a fundamental frequency of 2 GHz for a 0.7×800 μm^2 GaN HEMT at $V_{GS} = -2$ V and $V_{DS} = 19.5$ V under three different ambient temperatures: 20°C *(solid black line)*, 50°C *(dashed black line)*, and 80°C *(solid gray line)*.

FIGURE 1.5

Cryogenic system for microwave transistor characterization.

As expected, the swing of the load line decreases at higher temperatures and this implies a reduction of the output power.

Figure 1.5 shows a picture of an experimental set-up for microwave transistor characterization at cryogenic temperatures. The cryogenic system is based on a closed-circle liquid helium refrigerator and a vacuum chamber having microwave coaxial feedthroughs for connecting the external instrumentation. The temperature is monitored and controlled by using a proportional-integral-differential control loop. To illustrate the performance of a cryogenically cooled HEMT, Figure 1.6 shows the behavior of the magnitude of the forward transmission parameter S_{21} for a packaged AlGaAs/InGaAs/GaAs pHEMT (Mitsubishi MGF4919) versus frequency and V_{GS} at two different temperature conditions: 16.85°C and −243.15°C [3]. By cooling the

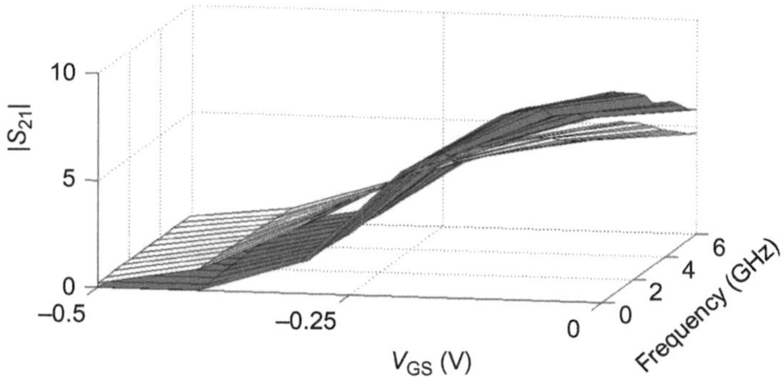

FIGURE 1.6

Magnitude of S_{21} versus frequency and V_{GS} for a GaAs pHEMT (Mitsubishi MGF4919) at $V_{DS}=2$ V under two different ambient temperatures: 16.85°C *(white plot)* and −243.15°C *(gray plot)*.

transistor, the magnitude of S_{21} increases when V_{GS} is far from pinch-off, while it decreases when V_{GS} is near pinch-off. The observed behavior of this figure of merit is due to the fact that, by decreasing V_{GS} towards the pinch-off, the threshold voltage (V_{TH}) shift has a stronger impact on S_{21} than the improvement of the average electron velocity. It should be emphasized that the influence of the threshold voltage shift on S_{21} might be compensated by the DC bias circuitry when the device is biased by imposing I_D rather than V_{GS} [21]. It should be noted that except for this device in a package, all of the transistors analyzed in this chapter are on-wafer devices.

1.3 TRANSISTOR MODELING

What is microwave transistor modeling? It consists of the field of knowledge and problem-solving concerned with how to extract, implement, and validate models able to describe the properties of advanced transistors meant for high-frequency applications. It is worth pointing out that the extracted models have to be properly implementable in a circuit simulator and be accurately validated by means of suitable measurements prior to their release. The importance of this research branch comes from the need for models as helpful feedback to technologists for advancing transistor fabrication and as an effective tool for circuit designers optimizing circuit performance. Although many successful modeling techniques have been developed in the last decades, microwave transistor modeling is an evergreen research field that still attracts extensive interest. This is because innovative techniques are constantly required to model the latest and continuously progressing transistor technologies.

As illustrated in Figure 1.7, transistor models can be divided into three main categories: physical models [22–25], behavioral models [26–29], and equivalent circuit models [30–33]. The physical models (or "transparent boxes") are based on the study of the physical phenomena occurring within the transistor, while the behavioral

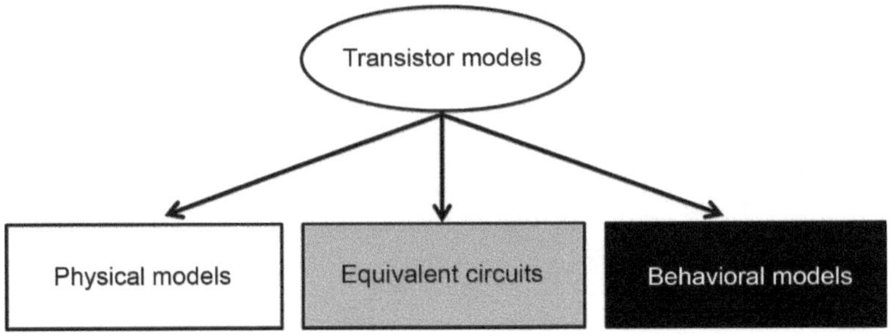

FIGURE 1.7

Classification of the transistor models into physical models (or *transparent boxes*), equivalent circuits (or *gray boxes*), and behavioral models (or *black boxes*).

models (or "black boxes") are based on the experimental characterization of the behavioral input-output descriptions of the transistor. An effective compromise between physical and behavioral models is represented by the equivalent circuits (or "gray boxes"), which are extracted from experimental measurements, yet keeping the connection with the inner physics of the transistor. Equivalent circuit models give the technologists better feedback than black-box models and enable circuit designers to do much faster simulations than with physical models. In fact, physical models are not really suitable for computer-aided design (CAD) environments due to their prohibitive simulation time. On the other hand, behavioral models at transistor level require expensive measurements and they show good predictive capabilities only in the regime of their characterization. It is worth noting that behavioral models at circuit and system levels, where quiescent condition and impedance environment (typically 50 Ω) are a priori known, do not show the same limitations previously discussed, and have been successfully and increasingly used by designers in the last years.

The present chapter focuses on microwave transistor modeling based on the equivalent circuit representation. Equivalent-circuit-based transistor modeling is an interdisciplinary scientific research field that requires expertise from different areas: semiconductor device physics, electromagnetic fields, microwave measurement techniques, circuit network theory, and circuit simulation software packages. Although the microwave performance of a transistor can be fully and straightforwardly characterized by carrying out high-frequency measurements with dedicated instrumentations, it is very useful to extract an equivalent circuit model, as it is a physically meaningful and a very compact representation, also allowing a fast estimation of the scaling of the device performance.

The selection of the appropriate topology of the equivalent circuit for modeling the specific tested transistor is a challenging task, since a higher number of circuit elements may allow for obtaining more accurate results, but the circuit complexity becomes higher as well as the risk of physical inconsistency. Hence, the number of circuit elements should be the right trade-off between model accuracy, physical soundness, and complexity. Typically, the first modeling step consists of extracting a small-signal equivalent circuit that, subsequently, can be used as a cornerstone for building both noise and large-signal models. The remainder of this chapter is dedicated to small-signal, noise, and large-signal modeling. GaAs and GaN HEMTs are used as reference devices to illustrate the application of equivalent circuit modeling techniques with experimental results.

1.4 SMALL-SIGNAL MODELING

Figure 1.8 shows an example of a small-signal equivalent circuit topology for a HEMT. The resistances R_{gsf} and R_{gdf} are often omitted as the conduction gate current is negligible under typical operating bias conditions. Typically, the values of the elements are determined using scattering (S-) parameter measurements, which can be straightforwardly and accurately measured by means of a vector network analyzer (VNA). Given a bias condition, the model extraction is based on defining eight equations that represent the four complex S-parameters in terms of the equivalent circuit elements at each frequency. Several techniques have been developed to solve this ill-conditioned problem and they can be classified into two main categories: numerical optimization [34–36] and analytical technique [33,37,38]. However, it is well known that optimization methods can lead to circuit element values with no physical meaning and by which the results can depend on many factors (eg, the starting element values, local minima, and the optimization method itself). On the other hand, analytical methods allow overcoming these drawbacks by splitting the model extraction in two parts. Basically, the equivalent circuit elements are usually divided into two main groups: the extrinsic elements (ie, C_{pg}, C_{pd}, L_g, L_s, L_d, R_g, R_s, R_d) and intrinsic ones (ie, C_{gs}, C_{gd}, C_{ds}, R_{gsf}, R_{gdf}, R_{gs}, R_{gd}, R_{ds}, g_m, τ). The former are assumed

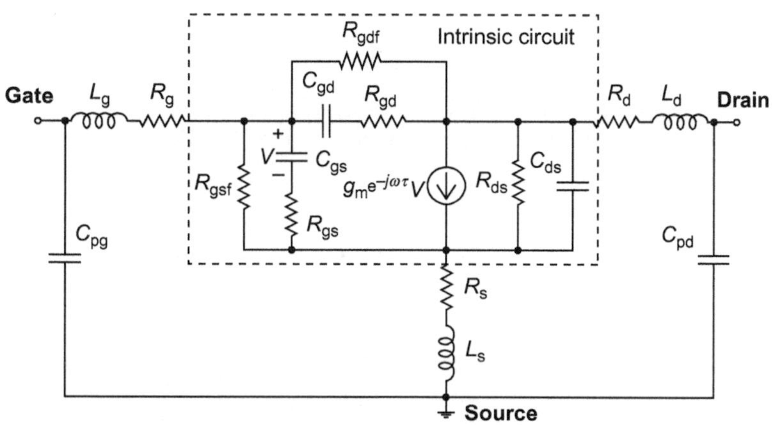

FIGURE 1.8

Small-signal equivalent circuit for a HEMT.

to be bias-independent, whereas the latter are dependent on the bias condition. The analytical techniques are based on determining first the extrinsic elements and successively the intrinsic ones.

Generally, the extrinsic elements are obtained from "cold" S-parameter measurements (ie, $V_{DS}=0$ V) [33,37–48], because the "cold" condition allows a significant simplification of the equivalent circuit model. The term "cold" arises from the fact that the equivalent temperature, representing the carrier average kinetic energy, is cold compared to the typical operating condition and the dissipated power is null. With V_{DS} equal to zero, the voltage controlled current source can be omitted since there are no carriers drifting from source to drain. As a result, two unknowns (ie, g_m and τ) disappear but, at the same time, two equations vanish (ie, $S_{12}=S_{21}$), due to the fact that under cold condition the transistor behaves as a reciprocal network with $[S]=[S]^T$ (ie, not containing any nonreciprocal media such as ferrites or plasmas, or active devices). Additionally, the zero drain-source voltage allows assuming that the gate-source and gate-drain intrinsic circuits are equal. Figure 1.9 shows that the intrinsic cold HEMT can be represented by an RC parallel network (R_y, C_y) modeling the Schottky gate junction and two identical RC parallel networks ($R_{ch}/2$, $2C_{ch}$) modeling the active channel. As the intrinsic resistances decrease at higher V_{GS}, the intrinsic capacitances are negligible under forward condition, whereas the intrinsic resistances are negligible under pinch-off condition [40]. In general, the extrinsic capacitances are extracted from $\text{Im}(Y_{ij})$ of the capacitive network representing the

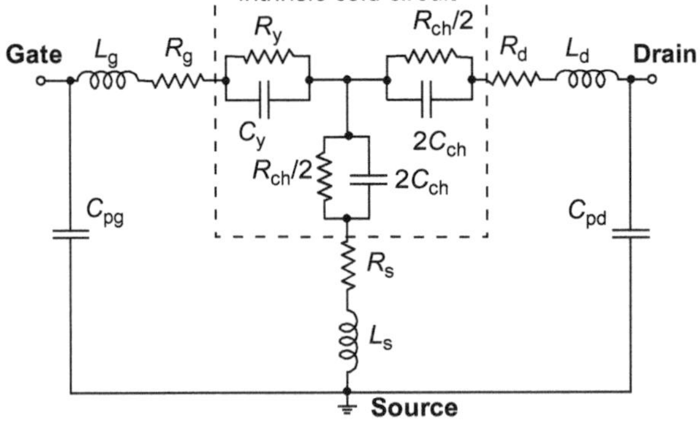

FIGURE 1.9

Small-signal equivalent circuit for a "cold" HEMT.

transistor at low frequencies under pinch-off condition, whereas the extrinsic inductances and resistances are, respectively, obtained from $\text{Re}(Z_{ij})$ and $\text{Im}(Z_{ij})$ of the resistive inductive network representing the transistor at high frequencies under high V_{GS}.

After de-embedding the extracted extrinsic contributions from the S-parameter measurements with simple matrix manipulations [49], the intrinsic elements can be calculated from the intrinsic Y-parameters. In particular, R_{gsf} and R_{gdf} are estimated from the real part of the intrinsic Y_{11} and Y_{12} at low frequencies by assuming that the capacitances behave as open circuits [50]:

$$R_{gsf} = \frac{1}{\text{Re}(Y_{11} + Y_{12})} \tag{1.1}$$

$$R_{gdf} = -\frac{1}{\text{Re}(Y_{12})} \tag{1.2}$$

Therefore, the extracted values of R_{gsf} and R_{gdf} can be subtracted and, finally, the remaining eight intrinsic elements can be calculated from the real and imaginary parts of the four Y-parameters with standard formulas [51]. It should be highlighted that R_{gsf} and R_{gdf} play a significant role only at relatively high V_{GS} when forward biasing the gate Schottky junction. This is confirmed by Figure 1.10 showing the bias

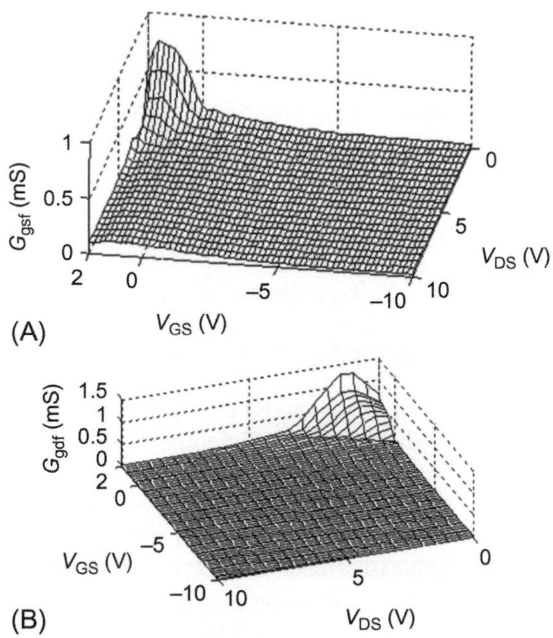

FIGURE 1.10

Behavior of G_{gsf} (A) and G_{gdf} (B) versus V_{DS} and V_{GS} for a $0.2 \times 100\ \mu m^2$ GaN HEMT.

dependence of the two conductances associated to R_{gsf} and R_{gdf} for a GaN HEMT on sapphire substrate with a gate length of 0.2 μm.

It is worth noting that the analysis of the intrinsic Y-parameters versus frequency can suggest to add and/or to neglect some intrinsic circuit elements. In fact, differences among technologies, materials, structures, and studied frequency ranges are reflected in different equivalent circuit topologies. For instance, the present analysis is focused on the intrinsic Y_{22} to highlight the importance of determining the most appropriate circuit topology. The presence of the feedback resistance R_{gd} allows defining this admittance parameter as follows:

$$Y_{22} = g_{ds} + \frac{j\omega C_{gd}}{1 + j\omega R_{gd} C_{gd}} + j\omega C_{ds} \tag{1.3}$$

Then, the real and imaginary parts of the intrinsic Y_{22} can be expressed as follows:

$$\mathrm{Re}(Y_{22}) = g_{ds} + \frac{\omega^2 R_{gd} C_{gd}^2}{1 + \left(\omega R_{gd} C_{gd}\right)^2} \tag{1.4}$$

$$\mathrm{Im}(Y_{22}) = \frac{\omega C_{gd}}{1 + \left(\omega R_{gd} C_{gd}\right)^2} + \omega C_{ds} \tag{1.5}$$

Eq. (1.4) shows that R_{gd} has to be included in the model when Re(Y_{22}) tends to increase with frequency instead of remaining constant and equal to g_{ds}. Contrary to the behavior predicted by Eq. (1.4), Figure 1.11 shows that the intrinsic Re(Y_{22}) decreases at higher frequencies for a GaAs pHEMT with a gate length of 0.25 μm. Consequently, to model this observed behavior, R_{gd} has been omitted and g_{ds} has been replaced by a voltage controlled current source with a time delay τ_{ds}. That solution allows defining Y_{22} as follows [41]:

$$Y_{22} = g_{ds} e^{-j\omega\tau_{ds}} + j\omega C_{gd} + j\omega C_{ds} \tag{1.6}$$

Hence, the real and imaginary parts of the intrinsic Y_{22} are given by:

$$\mathrm{Re}(Y_{22}) = g_{ds} \cos(\omega\tau_{ds}) \tag{1.7}$$

$$\mathrm{Im}(Y_{22}) = \omega\left(C_{gd} + C_{ds}\right) - g_{ds} \sin(\omega\tau_{ds}) \tag{1.8}$$

Eq. (1.7) shows that τ_{ds} is required to model the decrease of Re(Y_{22}) at higher frequencies (see Figure 1.11). When $\omega\tau_{ds}$ is sufficiently small, τ_{ds} can be approximated with an inductance L_{ds} in series with g_{ds}. However, in contrast with the L_{ds} implementation, τ_{ds} allows the simulation also of negative values of Re(Y_{22}) [41].

To verify the accuracy of the extracted equivalent circuit, the model should be able to reproduce the S-parameter measurements. Furthermore, the robustness of the model should be confirmed by its validation under a wide range of operating conditions. In particular, the circuit elements are expected to be physically meaningful and thus their values should be analyzed versus bias condition, temperature, and geometrical dimensions to check if the model meets the physical expectations. As an illustrative example, Figure 1.12 reports the good agreement between measured and simulated

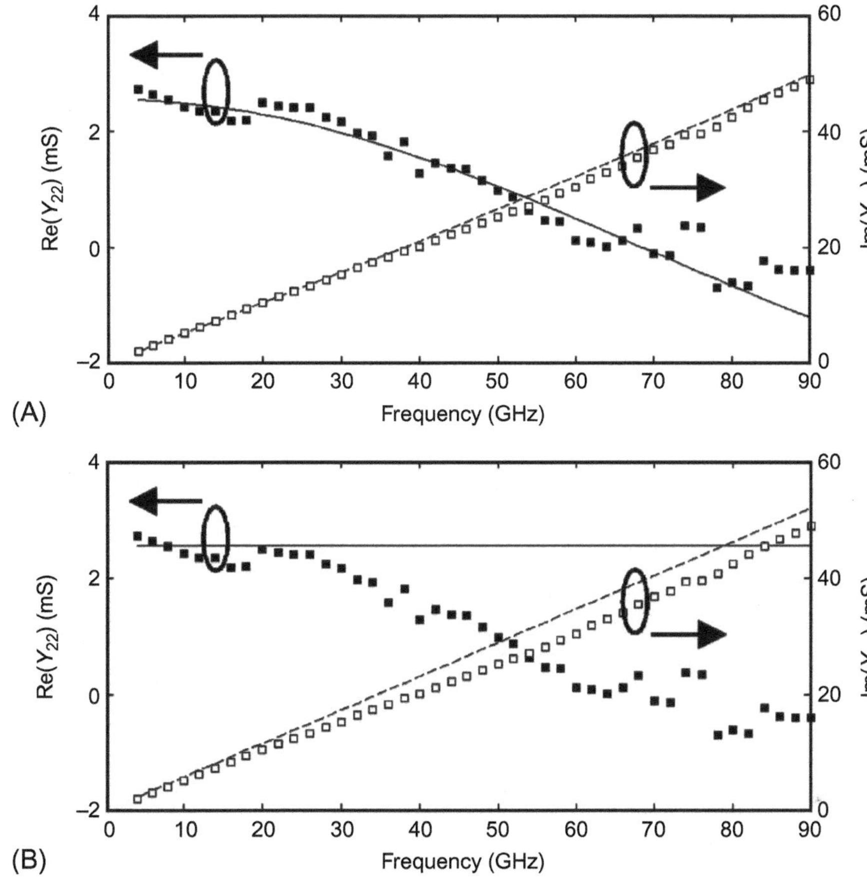

FIGURE 1.11

Comparison between measured *(symbols)* and simulated *(lines)* intrinsic Y_{22} from 4 to 90 GHz for a $0.25 \times 300\ \mu m^2$ GaAs pHEMT at $V_{GS} = -0.65$ V and $V_{DS} = 10$ V. Simulations are obtained with (A) and without (B) accounting for τ_{ds}.

S-parameters for a GaAs pHEMT with a gate length of 0.15 μm [52]. The validation is performed under different bias conditions including negative V_{DS}, because the tested device is meant for cold-FET mixer design. It should be highlighted that the standard formulas for identifying the current source at positive V_{DS} have to be modified for the extraction at negative V_{DS}: the sign of g_m has to be changed and this π phase shift has to be subtracted before calculating τ. The achieved bias dependence of g_m and C_{gs} is reported in Figure 1.13.

Figure 1.14 shows g_m and C_{gs} versus V_{GS} for a GaAs pHEMT at $V_{DS} = 2$ V under two different temperature conditions: 16.85°C and −243.15°C [4]. In agreement with the observed behavior of the magnitude of S_{21} (see Figure 1.6), the tested device shows that the improvement of the carrier transport properties at cryogenic temperature leads to an increase of g_m only when V_{GS} is far from pinch-off. With V_{GS}

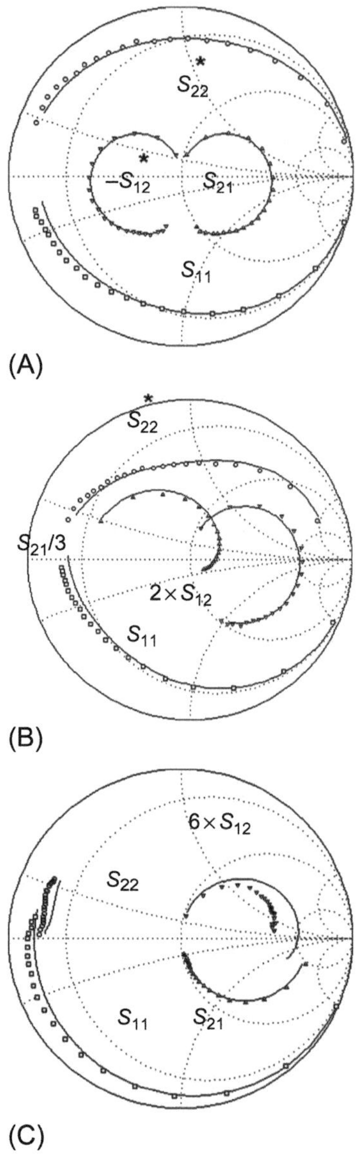

FIGURE 1.12

Comparison between measured *(symbols)* and simulated *(lines)* *S*-parameters from 2.5 to 65 GHz for a $0.15\times 300\ \mu m^2$ GaAs pHEMT at: $V_{DS}=0$ V and $V_{GS}=-1$ V (A), $V_{DS}=0.5$ V and $V_{GS}=-0.8$ V (B), $V_{DS}=-0.5$ V and $V_{GS}=-0.5$ V (C).

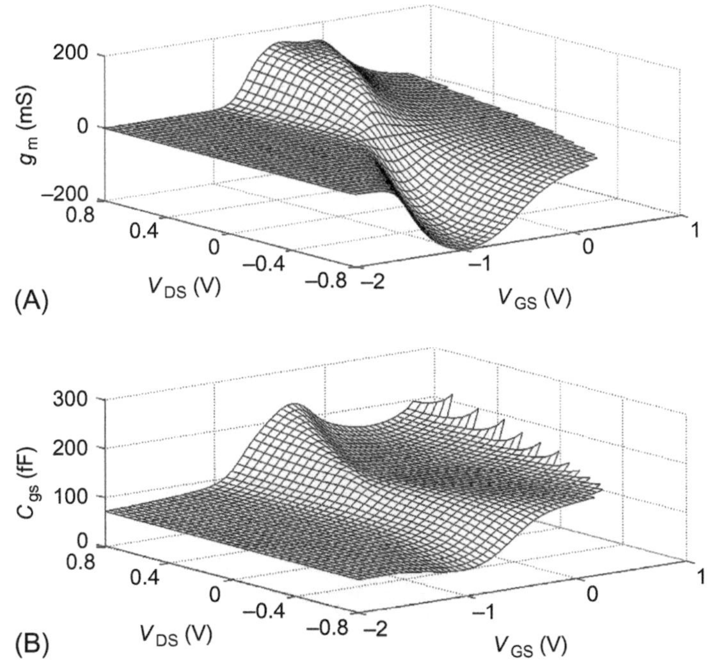

FIGURE 1.13

Behavior of g_m (A) and C_{gs} (B) versus V_{DS} and V_{GS} for a $0.15 \times 300\ \mu m^2$ GaAs pHEMT.

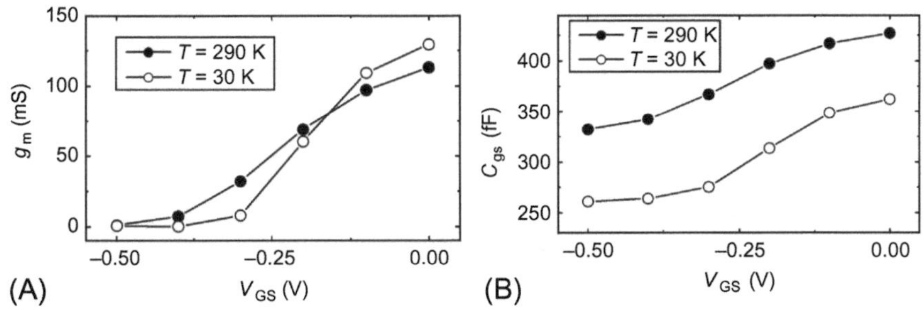

FIGURE 1.14

Behavior of g_m (A) and C_{gs} (B) versus V_{GS} for a GaAs pHEMT (Mitsubishi MGF4919) at $V_{DS} = 2$ V under two different ambient temperatures: 16.85°C *(black symbols)* and −243.15°C *(white symbols)*.

decreasing towards the pinch-off region, the V_{TH} shift leads to a degradation of g_m at lower temperatures. Such a V_{TH} shift can be attributed to the reduction of the net positive charge under the gate, due to a decrease of the thermally activated electron detrapping mechanisms occurring within the AlGaAs donor layer. As a result of this reduction of the detrapping phenomena, a decrease of the "equivalent" doping level can be responsible for the observed reduction of C_{gs} at lower temperatures [4].

The conventional scaling rules are based on the assumption that increasing the gate width can be seen as connecting more intrinsic transistors in parallel. Hence, the intrinsic equivalent circuit elements associated with admittances (ie, capacitances and conductances) and the drain current itself should be proportional to the total gate width, while the intrinsic ones associated with impedances (ie, inductances and resistances) should be inversely proportional to this geometric dimension, and the intrinsic time delay should be constant. Nevertheless, this is an oversimplified assumption that fails to take into account complex phenomena such as thermal effects, border effects, and residual extrinsic element contributions. As illustrative example, Figure 1.15 shows g_m, g_{ds}, C_{gs}, C_{gd}, C_{ds}, and τ versus the gate

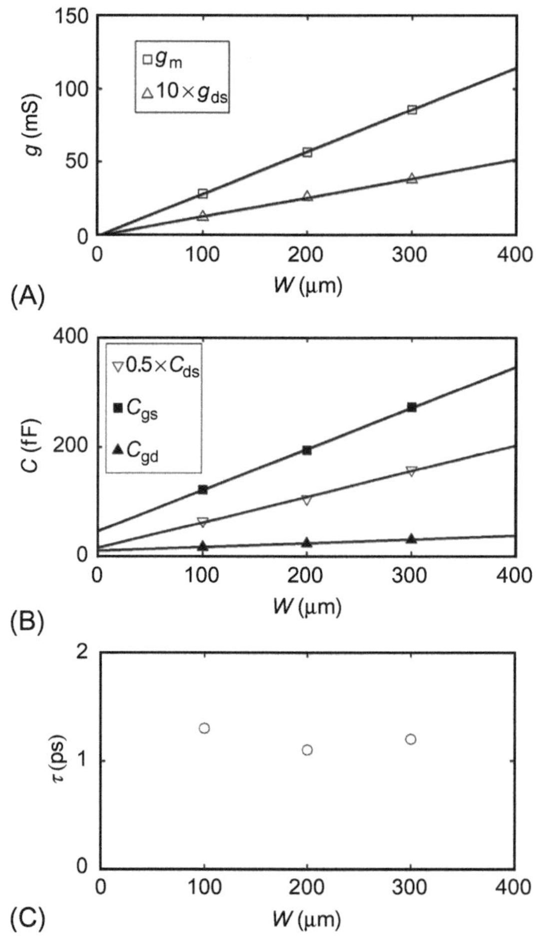

FIGURE 1.15

Behavior of g_m, g_{ds} (A), C_{gs}, C_{gd}, C_{ds} (B), and τ (C) versus the gate width for three GaAs HEMTs ($0.25 \times 100\ \mu m^2$, $0.25 \times 200\ \mu m^2$, and $0.25 \times 300\ \mu m^2$) at $V_{GS} = -0.6$ V and $V_{DS} = 2.5$ V.

width for three AlGaAs/GaAs HEMTs with a gate length of 0.25 μm. As expected, a quite clear linear behavior with respect to the gate width is observed in g_m, g_{ds}, C_{gs}, C_{gd}, and C_{ds}, while τ is almost independent of the gate width. This is because τ can be correlated to the carrier transit time and, therefore, depends mainly on the gate length (in first approximation, proportionally). As can be seen in Figure 1.16A, the line approximating g_m versus the gate width exhibits a positive intercept when V_{GS} is increased to 0 V. This result can be attributed to the fact that the thermal phenomena are enhanced in transistors with larger gate width and at higher V_{GS}, due to the higher dissipated power. To corroborate this observation, g_m is normalized with respect to the gate width in Figure 1.16B and the achieved value is evidently reduced in larger transistors. By considering that the thermal phenomena cannot follow the applied small signal at such high frequencies, the observed degradation of the normalized RF g_m has to be ascribed to the effects of thermal phenomena due to the DC dissipated power.

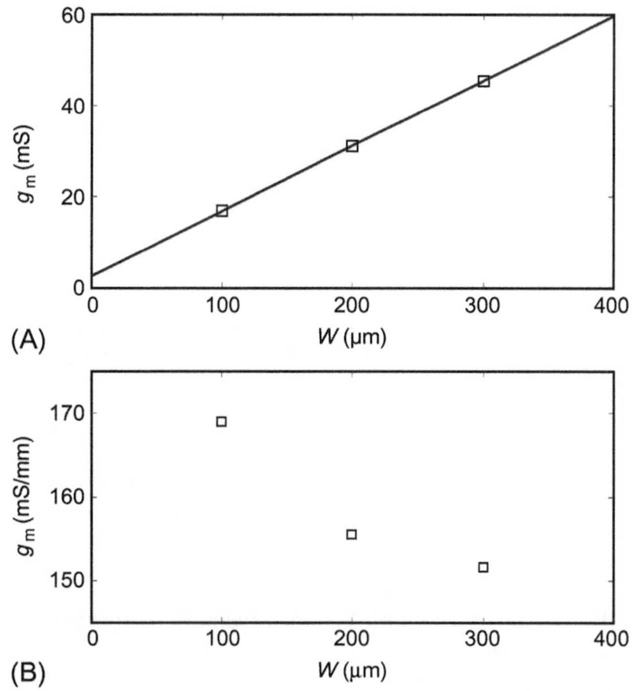

FIGURE 1.16

Behavior of g_m (A) and its normalized value (B) versus the gate width for three GaAs HEMTs ($0.25 \times 100\ \mu m^2$, $0.25 \times 200\ \mu m^2$, and $0.25 \times 300\ \mu m^2$) at $V_{GS}=0$ V and $V_{DS}=2.5$ V.

1.5 NOISE MODELING

High-frequency noise is known to be ubiquitous, because it is inherently present in all active and passive electronic devices, circuits, and systems. The noise behavior of a linear noisy two-port network can be fully represented with a set of four real noise parameters defining the noise factor F (or NF when expressed in decibels) as a function of the source reflection coefficient Γ_s. It is well known that the noise factor is a fundamental figure of merit used to quantify the degradation (or decrease) of the signal-to-noise ratio as the signal passes through the network. In particular, the noise factor is given by the ratio between the available signal-to-noise power ratio at the input and at the output of the network. Although different noise parameter sets can be defined and easily converted to each other, the relationship between F and Γ_s is commonly expressed in terms of the minimum noise factor F_{min}, the noise equivalent resistance R_n, and the complex optimum source reflection coefficient Γ_{opt} [53–58]:

$$F(\Gamma_s) = F_{min} + \frac{4\frac{R_n}{Z_0}\left|\Gamma_s - \Gamma_{opt}\right|^2}{\left|1+\Gamma_{opt}\right|^2\left(1-\left|\Gamma_s\right|^2\right)} \tag{1.9}$$

where the reference impedance Z_0 is typically 50 Ω, F_{min} represents the minimum value of F occurring at $\Gamma_s = \Gamma_{opt}$, and R_n determines how fast F degrades as Γ_s departs from its optimum.

As an illustrative example of the conversion formulas for passing from one representation to another, the following equations show how to transform F_{min}, R_n, and Γ_{opt} into the chain correlation matrix and vice versa [59–62]:

$$[C_A] = 2kT\begin{bmatrix} R_n & \frac{F_{min}-1}{2} - R_n Y_{opt}^* \\ \frac{F_{min}-1}{2} - R_n Y_{opt} & R_n\left|Y_{opt}\right|^2 \end{bmatrix} \tag{1.10}$$

$$R_n = \frac{C_{11A}}{2kT} \tag{1.11}$$

$$F_{min} = 1 + \frac{1}{kT}\left(\mathrm{Re}(C_{12A}) + \sqrt{C_{11A}C_{22A} - (\mathrm{Im}(C_{12A}))^2}\right) \tag{1.12}$$

$$Y_{opt} = \frac{\sqrt{C_{11A}C_{22A} - (\mathrm{Im}(C_{12A}))^2} + j\mathrm{Im}(C_{12A})}{C_{11A}} \tag{1.13}$$

where k is the Boltzmann's constant, T is the absolute temperature in kelvin, Y_{opt} represents the optimum source admittance associated to Γ_{opt}, and the asterisk stands for the complex conjugate.

Eq. (1.9) is particularly useful since it provides direct information to the circuit designer on the optimum source reflection coefficient which must be synthesized to

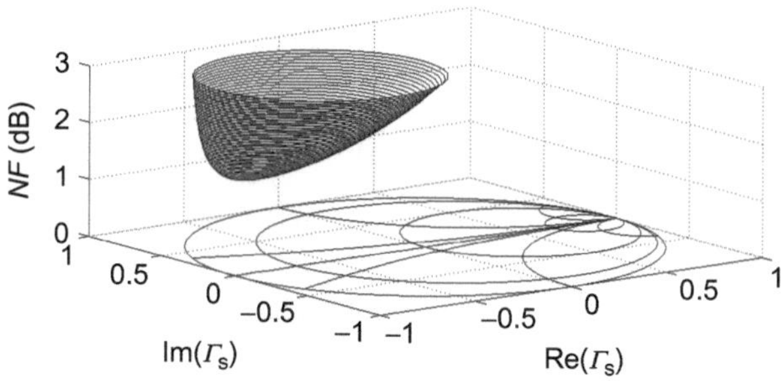

FIGURE 1.17

The noise paraboloid representing NF versus the real and imaginary parts of Γ_s at 10 GHz for a $0.25 \times 400\ \mu m^2$ GaN HEMT at $V_{DS}=10$ V and $I_D=40$ mA.

obtain the minimum NF and an immediate evaluation of the latter. Moreover, four NF measurements, carried out for different source terminations, are theoretically sufficient to determine the noise parameters by means of Eq. (1.9).

To graphically illustrate the meaning of the noise parameters, Figure 1.17 shows the elliptic paraboloid obtained by the 3D representation of the NF as a function of the complex source reflection coefficient for a microwave transistor. The device considered is a GaN HEMT on SiC substrate with a gate length of 0.25 μm. Because the tested device is aimed for X-band applications (ie, 8–12 GHz), the analyzed frequency has been set at 10 GHz. The behavior of NF (Γ_s) has been obtained from Eq. (1.9) by using the four noise parameters, which have been determined with the "source-pull" procedure based on using a noise figure analyzer to measure the NF for at least four different source terminations synthesized by a tuner.

Nowadays, the GaN technology is attracting considerable attention for microwave low-noise applications. This is because GaN-based HEMTs achieve good noise performance that is comparable to those of GaAs devices. Moreover, it is possible to build receivers integrating both the high-power amplifier and the low-noise amplifier in the same chip without requiring any protection or limiting circuitry, due to the high degree of robustness of GaN transistors [63–65]. Motivated by this context, several studies have been aimed at extracting noise models for GaN devices [66–72]. In particular, the attention has been mostly devoted to the noise modeling based on the equivalent circuit representation. With the aim of giving the reader a clear idea how to build an equivalent-circuit-based noise model, the present analysis is focused on reproducing the noise behavior of the GaN HEMT considered in this section. The noise model for the tested device has been obtained by assigning an equivalent noise temperature to each resistor of the small-signal equivalent circuit (see Figure 1.18) [72]. The feedback resistance R_{gd} has been omitted because its contribution has not been found necessary to model the tested device. The value of the temperature T_{ds} associated with the intrinsic output resistance has been determined in order to minimize the difference between simulated and measured 50 Ω noise factor, whereas the other temperatures have been selected to be equal to the ambient temperature,

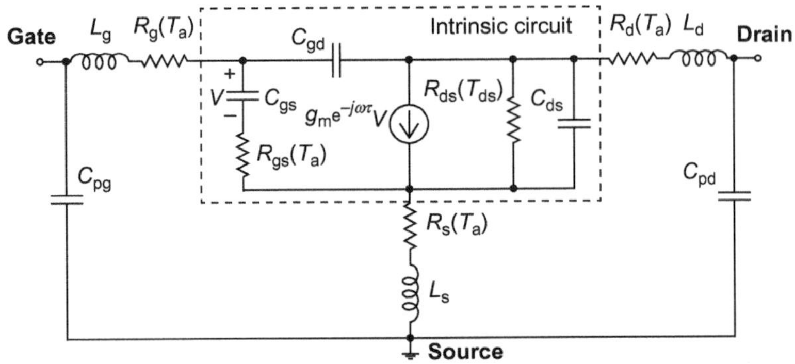

FIGURE 1.18

Noise equivalent circuit for a HEMT. Noise model is implemented by assigning a suitable equivalent noise temperature to each resistor.

coherently with their physics-based contribution to thermal noise. As reported in Figure 1.19, the extracted noise model can accurately predict the measured noise parameters over the investigated frequency range going from 5 to 25 GHz.

1.6 LARGE-SIGNAL MODELING

Large-signal transistor models are mandatory for the design of circuits such as mixers, oscillators, and power amplifiers which represent the core of today's microwave communication systems. A suitable large-signal model must show *acceptable* prediction capability under very different operating conditions, including DC, quasi-linear regime (eg, highly linear amplifiers), and strong nonlinear operation (eg, high-efficiency amplifiers). The word "acceptable" requires a detailed, but necessary, explanation. In fact, one might wonder why a nonlinear model should provide accurate predictions under small-signal regime. The answer appears clear when dealing, for example, with oscillator design (which will be discussed in depth in Chapter 3 [73,74]) where the oscillation start-up condition is rooted in small-signal operation, whereas large-signal operation defines the steady state oscillatory regime. Clearly, a model able to accurately predict both small- and large-signal transistor operations is very powerful to correctly design a similar circuit. Analogous considerations apply also to the stability analysis of microwave amplifier circuits. Moreover, the accuracy of a large-signal model in also reproducing small-signal bias-dependent parameters is a clear indication of its soundness.

The large-signal models oriented to CAD environments can be divided into two categories: equivalent circuit models and behavioral models.

Behavioral models, like for instance those presented in [27, 75–77] are generated from measurement data and have the advantage of potentially guaranteeing very accurate predictions over the measurement sets which they are extracted from.

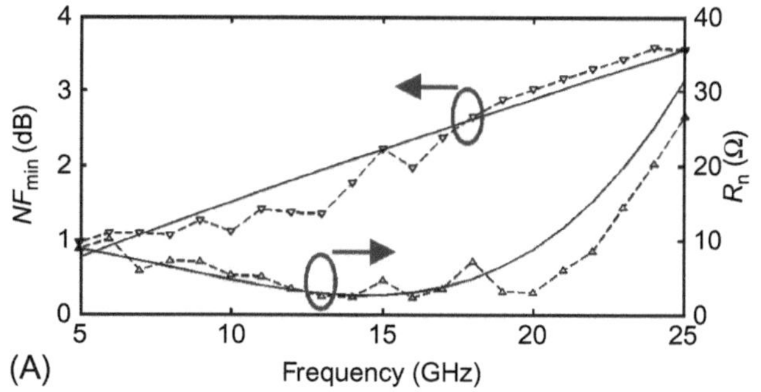

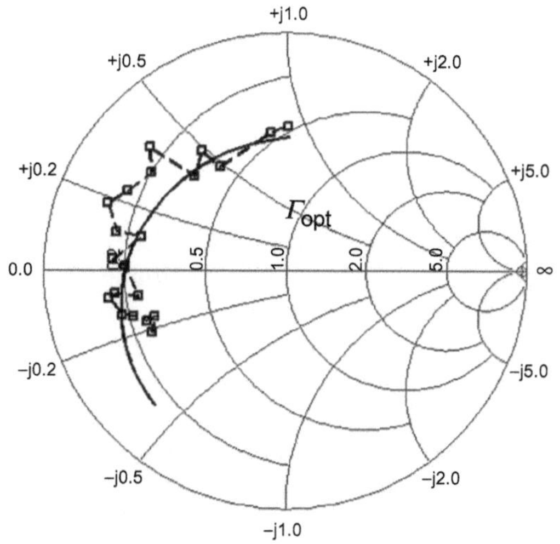

FIGURE 1.19

Measured *(dashed lines with symbols)* and simulated *(solid lines)* noise parameters NF_{min}, R_n (A), and Γ_{opt} (B) from 5 to 25 GHz for a $0.25 \times 400\ \mu m^2$ GaN HEMT at $V_{DS} = 10$ V and $I_D = 40$ mA.

The idea is essentially to measure the device under all the operating conditions of interest and to store these measurements into lookup tables or to map them into analytical functions, as for example artificial neural networks. On the other hand, these models show poor extrapolation capability and do not have any link with the device physics. Additionally, due to limited memory, computational efficiency, limitations of nonlinear measurement setups, it is not possible to measure the device under all the possible operations of interest. Try to imagine the dimension that should have a lookup table including all the possible device operations from small-signal to large-signal, for all the bias conditions, case temperatures, and harmonic terminations of interest. This is clearly unfeasible! Therefore, typically behavioral models represent a good solution when a special purpose model (eg, for a fixed bias-condition, case temperature, and harmonic terminations) is required. Extraction is simple and one can achieve a very high-level of accuracy which, in fact, tends to coincide with the measurement accuracy.

Whenever a general purpose and predictive model is required, equivalent circuit models represent the most appropriate solution, as discussed in Section 1.3. In this case, the large-signal model can be built starting from the multibias small-signal model. As an illustrative example, Figure 1.20 shows a large-signal equivalent circuit with the intrinsic section consisting of nine nonlinear components (ie, R_{gs}, R_{gd}, C_{gs}, C_{gd}, C_{ds}, C_m, I_{dsiR}, I_{gsiR}, I_{gdiR}). This model can be constructed by using the small-signal equivalent circuit in Figure 1.8 as the cornerstone. The effect of τ has been represented by the transcapacitance C_m (ie, $C_m = g_m\tau$), which can be used to account for the first term in the Taylor series expansion of $e^{-j\omega\tau}$ (ie, $e^{-j\omega\tau} = g_m - j\omega C_m$).

The two most widely adopted implementation techniques of equivalent circuit models into commercial circuit simulators are based on analytical functions and lookup tables [78–98]. The analytical approach has the advantage of requiring a limited number of parameters and having better extrapolation capability, while the lookup table approach is technology-independent and allows straightforward implementation without numerical optimization procedures. The equivalent circuit model shown in Figure 1.20 has been used to model the GaAs pHEMT analyzed in Section 1.4. The contributions of I_{dsiR}, I_{gsiR}, I_{gdiR} have been determined from DC measurements. In particular, the forward conduction of the gate Schottky diode has been modeled using dedicated DC measurements at sufficiently high V_{GS}. The approach based on analytical functions has been adopted to represent the gate conduction current, in order to predict more accurately its exponential increase at V_{GS} beyond the investigated bias range. All of the other intrinsic elements of the equivalent circuit have been straightforwardly implemented with the lookup table approach.

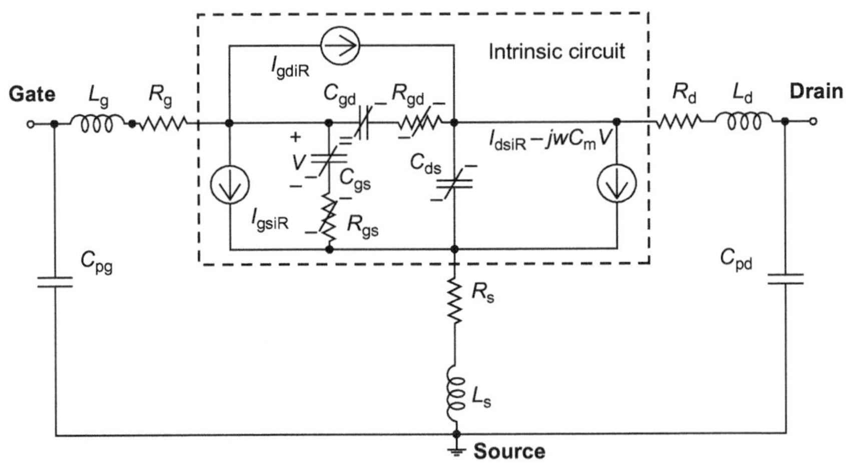

FIGURE 1.20

Large-signal equivalent circuit for a HEMT.

While only DC and small-signal measurements are adopted in the identification phase, nonlinear microwave measurements must be carried out and compared with simulations to validate the large-signal model. As an example of model validation under large-signal operating conditions, Figure 1.21 shows the comparison between simulated and measured current and voltage time-domain waveforms for a GaAs pHEMT at the fundamental frequency of 10 GHz [52]. In addition, a Q-band up-converter has been designed by using the achieved large-signal model. As shown in Figure 1.22, the cold-FET mixer is one of the building blocks of the packaged Q-band up-converter, which integrates on the same MMIC two single balanced cold-FET mixers in I/Q configuration, a frequency quadrupler, which is able to convert the X-band LO input into the Q-band pump signal required for the mixers, and a four-stage RF driver, cascaded to the cold-FET mixers [99]. To better relate the cold-FET model accuracy with the measurements, the comparison is performed between the stand-alone measured mixer and the corresponding simulated circuit. In particular, Figure 1.23A and B shows the good agreement between the simulated conversion gain and image rejection of the designed up-converter and the data obtained from scalar measurements.

As said, the main limitation of lookup table models is related to their nonoptimal extrapolation capability. This appears similar to behavioral descriptions, although in this case the *circuital constraints* imposed by the equivalent circuit topology allow for better prediction outside the measurement grid. The cold-FET mixer represents an excellent example where lookup table implementation gives excellent results. In fact, the limited extension of the region that the transistor experienced under large-signal operation (ie, the triode region) can be fully characterized by means of standard instrumentation such as a DC analyzer and a VNA. A similar condition can be encountered in the design of highly linear amplifiers working under power back-off operation (eg, 0.1 dB compression) [100]. Also in this case, lookup table models based on *S*-parameter measurements carried out on a rectangular grid encompassing the transistor bias point represent a very accurate solution.

Finally, purely analytical models will be dealt with. This is undoubtedly the class of models which is the most used for microwave circuit design. Foundry models typically belong to this class. The main advantage of models based on analytical functions is the optimum trade-off between computational efficiency, accuracy, robust extrapolation, and scaling capability. As will be demonstrated later in this section, analytical models, even if extracted on the basis of only DC and small-signal measurements, can provide already adequate predictions under nonlinear operation. The main drawback of these models is inherent to the simplified analytical functions adopted for describing the complex physics of microwave transistors. Consequently, these models cannot exhibit either the same level of accuracy of behavioral models under nonlinear operation or the same level of accuracy under small-signal operation achievable by lookup table descriptions. Analytical functions are *short blankets* that need to be tailored on the specific application: if the model will be used for the design of power amplifiers, it can be useful to fit the model parameters exploiting nonlinear measurements, whereas when quasi-linear operation is involved, one can adopt

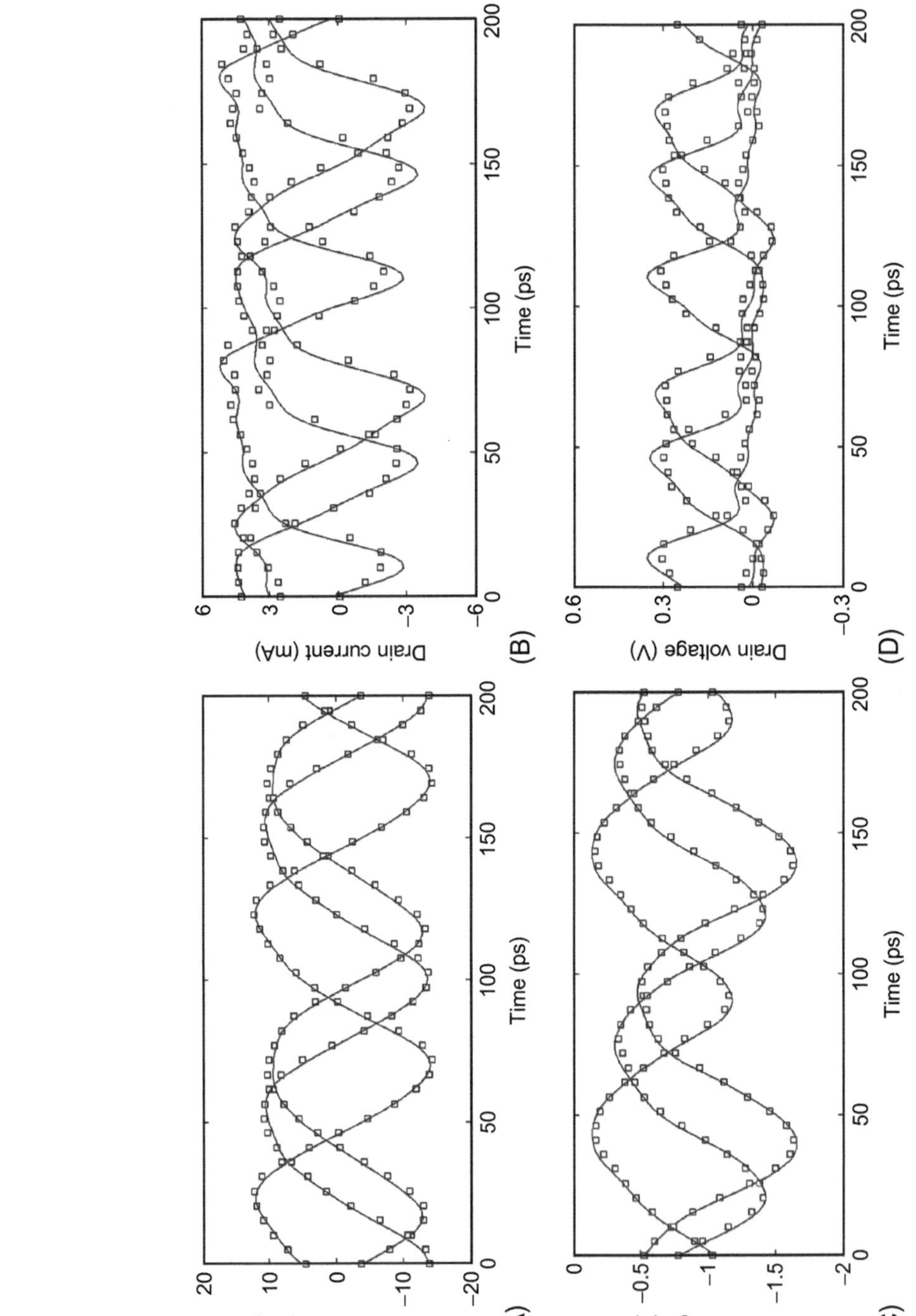

FIGURE 1.21

Comparison between measured *(symbols)* and simulated *(lines)* gate current (A), drain current (B), gate voltage (C), and drain voltage (D) time-domain waveforms for a 0.15 × 300 μm² GaAs pHEMT at f_0 = 10 GHz, V_{DS} = 0 V, and V_{GS} = −1, −0.8, and −0.6 V.

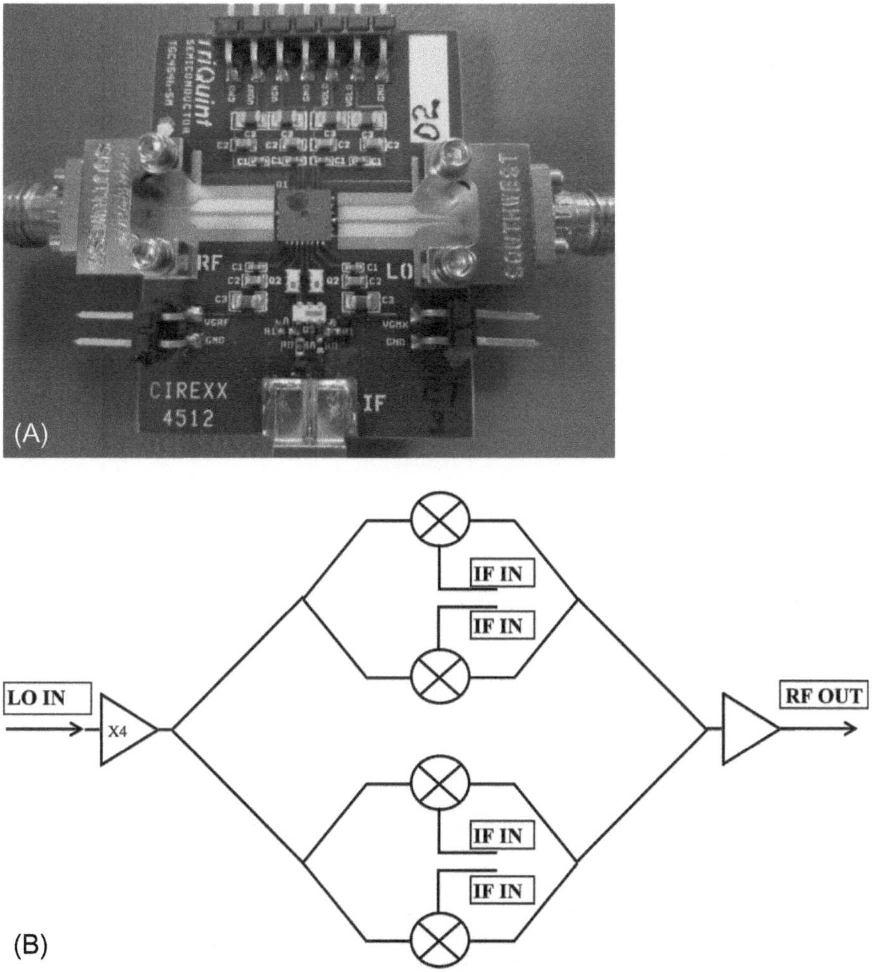

FIGURE 1.22

Photograph (A) and block diagram (B) of a Q-band up-converter.

S-parameter measurements as identification base. Clearly, improving predictions on one side necessarily implies accuracy degradation on the other side. Nevertheless, these models guaranteeing an acceptable accuracy level under very different device operations (eg, temperature and bias condition) have remained over the years the preferred solutions by designers.

On the basis of what has been discussed so far, we can now show and examine the measurement-based extraction procedure of a transistor model available in commercial CAD tools [81,101–103]. The equations of the electro-thermal model representing the nonlinear currents and charges are described thoroughly in Ref. [103] and for clarity

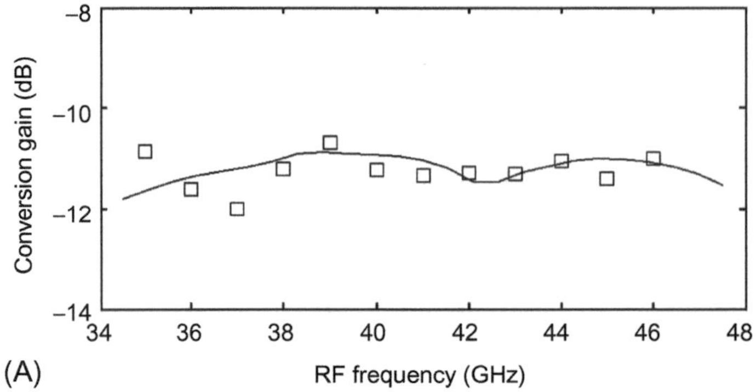

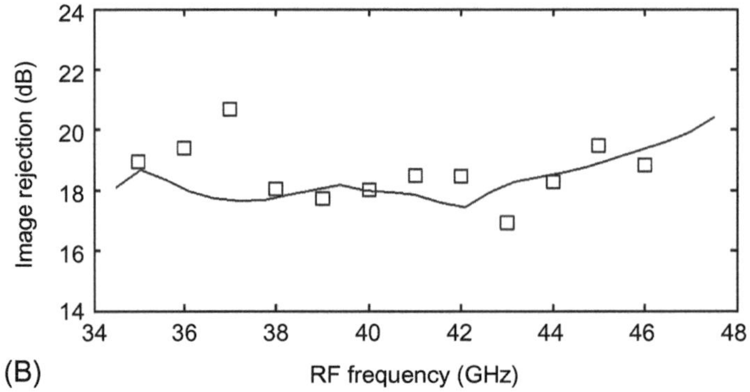

FIGURE 1.23

Comparison between measured *(symbols)* and simulated *(lines)* conversion gain (A) and image rejection (B) of the up-converter in Figure 1.22 with an LO power of 14 dBm and an IF frequency of 2.5 GHz.

reported here also for the resistive drain current (see Eqs. (1.14)–(1.17)) and the gate-source and gate-drain capacitances (see Eqs. (1.18)–(1.23)):

$$I_{dsiR} = I_{pk0}(1 + \tanh(\phi))\tanh(\alpha V_{ds})(1 + \lambda V_{ds}) \tag{1.14}$$

$$\phi = \sum_{i=1}^{3} P_i \times \left(V_{gs} - V_{pkm}\right)^i \tag{1.15}$$

$$\alpha = \alpha_R + \alpha_S \times (1 + \tanh(\phi)) \tag{1.16}$$

$$V_{pkm} = V_{pks} - \Delta V_{pks} + \Delta V_{pks} \times \tanh(\alpha_S V_{ds}) \tag{1.17}$$

$$C_{gs} = C_{GSP} + C_{GS0}(1 + \tanh(\phi_1))(1 + \tanh(\phi_2)) \tag{1.18}$$

$$C_{gd} = C_{GDP} + C_{GD0}(1 + \tanh(\phi_3))(1 + \tanh(\phi_4)) \tag{1.19}$$

$$\phi_1 = P_{10} + P_{11} \times V_{gs} + P_{111} \times V_{ds} \tag{1.20}$$

$$\phi_2 = P_{20} + P_{21} \times V_{ds} \tag{1.21}$$

$$\phi_3 = P_{30} - P_{31} \times V_{ds} \tag{1.22}$$

$$\phi_4 = P_{40} + P_{41} \times V_{ds} \tag{1.23}$$

In the above equations, the parameters I_{pk0}, P_1, C_{GS0}, and C_{GD0} are linearly dependent on the temperature via the thermal parameters T_{cIpk0}, T_{cP1}, T_{ccgs0}, and T_{ccgd0}. For the experimental data that we report, the resistive gate current can be neglected as well as any breakdown current. However, the full model also accounts for them [103].

The device under test is a GaN HEMT on SiC substrate with a gate length of 0.25 μm. The focus here is on the intrinsic part of the model and it is assumed that the extrinsic elements are accurately known. The parameters of the intrinsic model are extracted directly from DC and small-signal measurements [102,103]. DC measurements are used to extract the parameters of the nonlinear current sources and small-signal measurements are used to extract the parameters of the nonlinear capacitances. As compared to behavioral models or equivalent circuit models based on lookup tables, the approach based on analytical functions requires significantly less measurement data for the extraction. Additionally, simulation time is remarkably shorter as compared to lookup table-based models.

In Figure 1.24, the predictions of the model extracted from few DC and *S*-parameter measurements are compared with nonlinear measurements acquired under class AB operation at the fundamental frequency of 5 GHz and sweeping the input power for each load impedance tuned at the fundamental frequency. The chosen bias point is $V_{\text{GS}} = -2.75$ V and $V_{\text{DS}} = 20$ V. The simulations are performed by exciting the model with the measured incident waves at the input and output ports. Figure 1.24 shows the time-domain transistor voltages and currents, the output power as a function of the input power, and the prediction of the load impedances. In the CAD environment, the model is directly excited by the measured "a" waves which therefore coincide with the simulated ones. The simulated "b" waves are instead clearly dependent on the accuracy of the model. By comparing the load impedances, which constitutes important information for the designers, we are indirectly comparing the simulated and measured "b" waves. In Figure 1.25 the measured and simulated *S*-parameters at the bias point under analysis and from 40 MHz to 40 GHz are shown. We used 20 parameters for the nonlinear current sources, including those for the thermal network, and 16 parameters for the capacitive part of the intrinsic model. Some of the parameters are extracted directly from measurements, without any optimization. Other parameters are either set to the default values in the CAD implementation or derived from the foundry datasheet or set to typical values which can be found in literature [103]. The values of the model parameters extracted directly from measurements or set to the default value in the CAD implementation are reported in

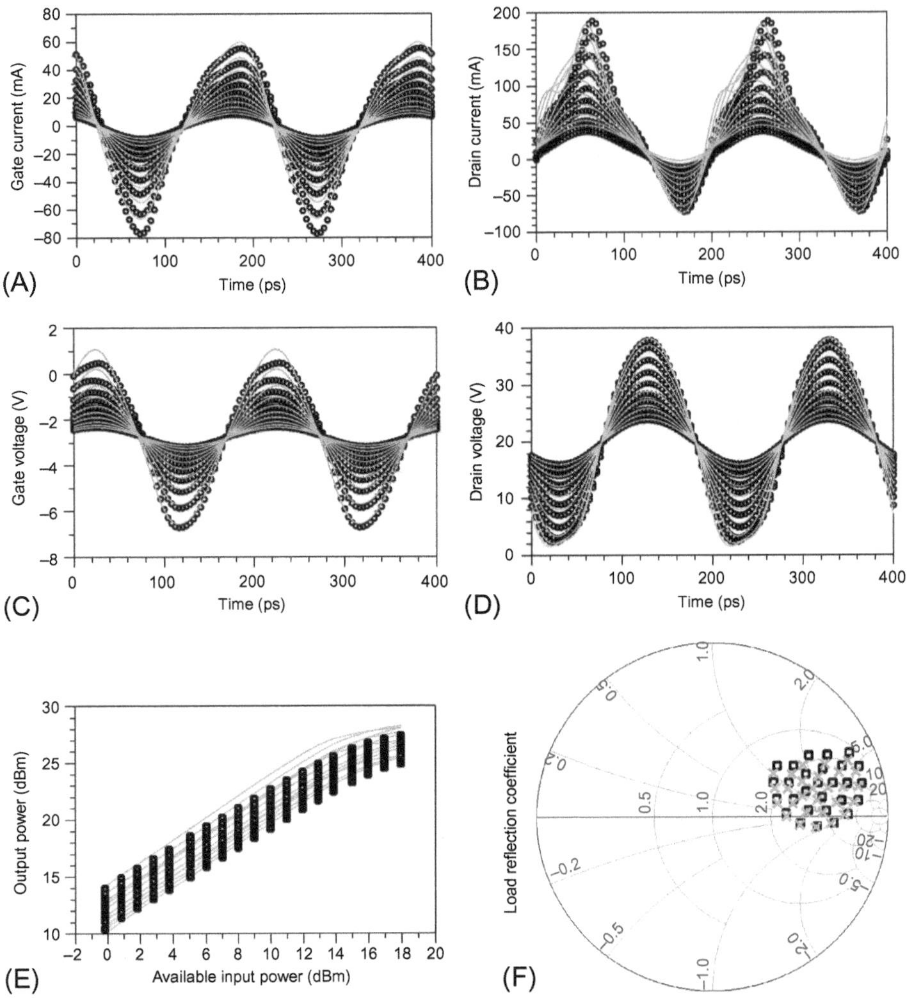

FIGURE 1.24

Fully analytical model extracted directly from DC and *S*-parameter measurements. Measured *(symbols)* and simulated *(lines)* gate current (A), drain current (B), gate voltage (C), and drain voltage (D) time-domain waveforms for a $0.25 \times 200\ \mu m^2$ GaN HEMT at $V_{GS} = -2.75$ V, $V_{DS} = 20$ V, $f_0 = 5$ GHz, $Z_L(f_0) = 91.5 + j144\ \Omega$, and swept input power. Output power at different load impedances (E) and measured *(squares)* and simulated *(crosses)* load impedances (F).

Table 1.1. The parameters of the gate Schottky junction forward current and of the soft breakdown [103] are not reported, because forward conduction and soft breakdown were not reached during the measurements carried out. The comparisons in Figures 1.24 and 1.25 show that this model, which was obtained with the minimal extraction effort, already represents a very good starting point for further

FIGURE 1.25

Fully analytical model extracted directly from DC and *S*-parameter measurements. Measured *(symbols)* and simulated *(lines)* *S*-parameters for a $0.25\times 200\ \mu m^2$ GaN HEMT from 40 MHz to 40 GHz at $V_{GS}=-2.75$ V, $V_{DS}=20$ V.

Table 1.1 Parameters of the Model in Ref. [81] Extracted Directly from DC and *S*-Parameter Measurements or Set to Default Values in the CAD Implementation

I_{pk0} (mA)	V_{pk} (V)	ΔV_{pk} (V)	P_1	λ	α_S	R_{th} (°C/W)
86	−1.375	0.875	0.66	4.5e-6	0.22	5.4
P_2	P_3	α_R	B_1	B_2	T_{clpk0}	T_{cP1}
0	0	0.001	1	3	−0.003	−0.003
C_{GSP} (fF)	C_{GS0} (fF)	C_{GDP} (fF)	C_{GD0} (fF)	C_{DS} (fF)	$P_{11}=P_{41}$	$P_{10}=P_{40}$
240	40	17.5	27	100	2.53	6.96
$P_{21}=P_{31}$	$P_{20}=P_{30}$	P_{111}	T_{ccgs0}	T_{ccgd0}	R_g (Ω)	R_s (Ω)
0.2	0	0.063	0.003	0.003	1.6	0.1
R_d (Ω)	L_g (pH)	L_s (pH)	L_d (pH)	C_{pg} (fF)	C_{pd} (fF)	
3.9	161	1.6	152	26.6	37.6	

optimization and improvements. Clearly, analytical models do not preclude the use of lookup table-based circuit components and vice versa. Actually, one should benefit from the advantages of both approaches to improve the model prediction capability. As an example, one could use an analytical description for the nonlinear current sources (ie, I_{dsiR}, I_{gsiR}, I_{gdiR}), extracted from a few DC measurements, and combine it with a lookup table description of the nonlinear capacitances. As shown in Figures 1.26 and 1.27 in such a way model prediction capability improves, especially when looking at time-domain waveforms, and this is due to the intrinsic higher accuracy of the capacitance lookup table implementation. As before, we did not perform any optimization step in the extraction phase. Obviously, if nonlinear measurements

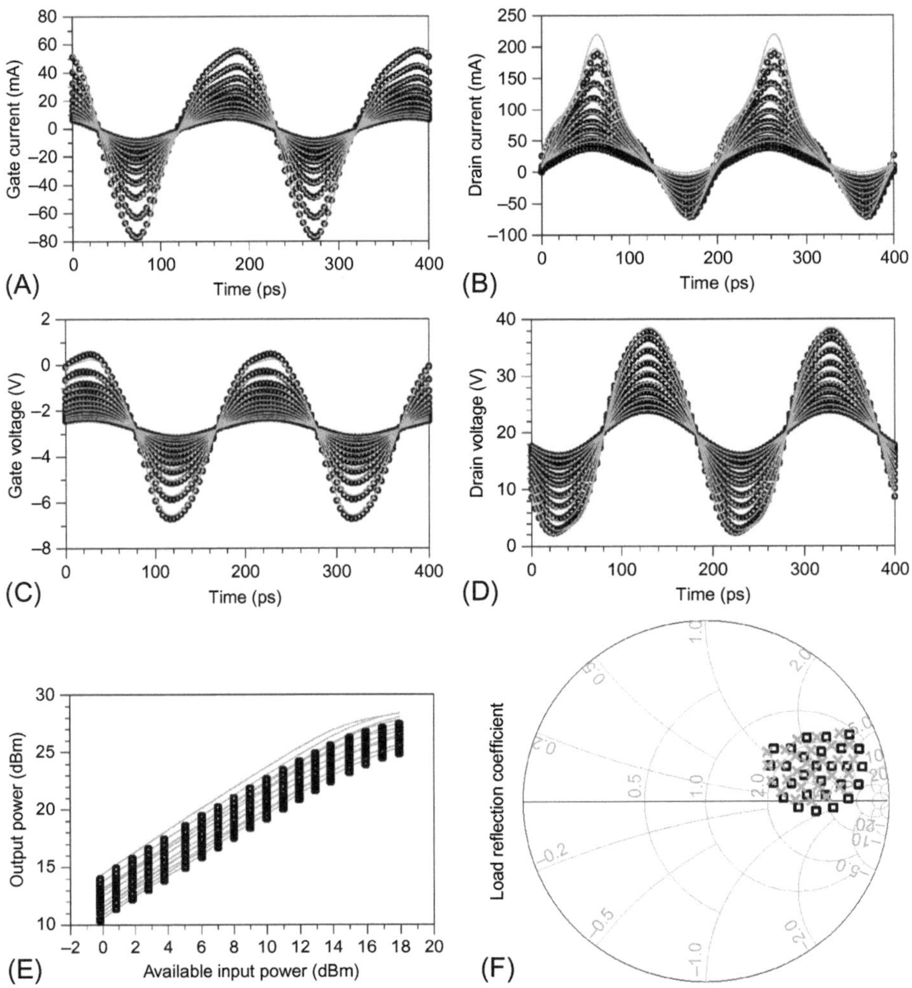

FIGURE 1.26

Model based on an analytical description for the current sources and lookup tables for the nonlinear capacitances. Measured *(symbols)* and simulated *(lines)* gate current (A), drain current (B), gate voltage (C), and drain voltage (D) time-domain waveforms for a $0.25 \times 200\ \mu m^2$ GaN HEMT at $V_{GS} = -2.75$ V, $V_{DS} = 20$ V, $f_0 = 5$ GHz, $Z_L(f_0) = 91.5 + j144\ \Omega$, and swept input power. Output power at different load impedances (E) and measured *(squares)* and simulated *(crosses)* load impedances (F).

FIGURE 1.27

Model based on an analytical description for the current sources and lookup tables for the nonlinear capacitances. Measured *(symbols)* and simulated *(lines)* *S*-parameters for a $0.25\times200\ \mu m^2$ GaN HEMT from 40 MHz to 40 GHz at $V_{GS}=-2.75$ V, $V_{DS}=20$ V.

are available, one should use them to further optimize the model against experimental data which reproduce an operating condition close to the one in the final application. In Figures 1.28 and 1.29 the model predictions, that have been obtained by the fully analytical description, are compared to measurements after optimizing the nonlinear current source parameters and the nonlinear capacitance parameters against low- and high-frequency nonlinear measurements, respectively [104,105]. The numerical optimization improves the overall accuracy under nonlinear operation and an acceptable accuracy is also achieved under small-signal condition. It is worth observing that nonlinear measurements at low frequency are directly linked to the current and voltage across the transistor current source and therefore allow one to capture the so-called "low-frequency dispersion" affecting the transistor dynamic *I-V* characteristics [86,94,105,106]. Dispersive phenomena include thermal effects and trapping effects. When these effects are present, the dynamic *I-V* characteristics of the transistor depend on the bias condition, which in a first-order approximation sets the trapping state, and on the thermal state which is determined by the power dissipated under dynamic operation.

The values of the model parameters that we obtained after numerical optimization are reported in Table 1.2.

These few examples highlight the advantages of analytical models which benefit from fast simulation time, extrapolation capability, robustness in terms of convergence, and the possibility to accurately predict device behavior under different operating conditions with a relatively reduced number of parameters. One can combine direct extraction of the model parameters, based on DC and *S*-parameter measurements, with numerical optimization and better model accuracy can be achieved if nonlinear measurements, which represent the operating

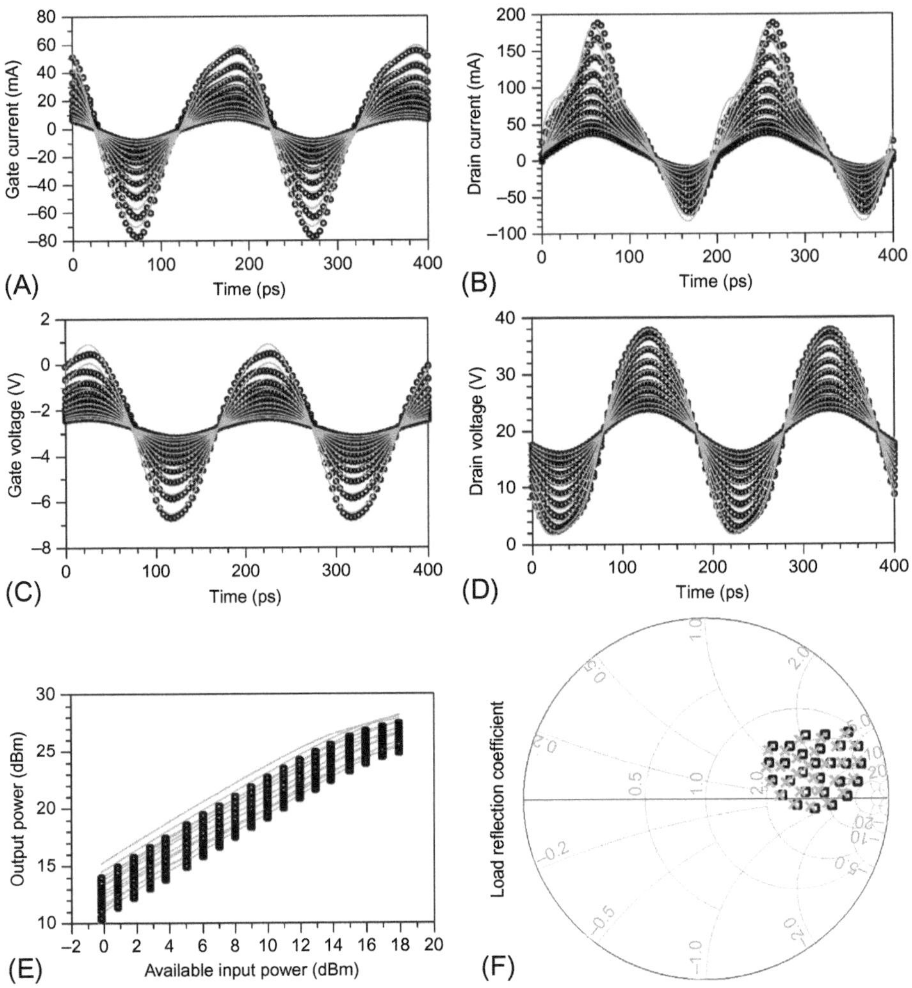

FIGURE 1.28

Fully analytical model extracted after numerical optimization against nonlinear measurements. Measured *(symbols)* and simulated *(lines)* gate current (A), drain current (B), gate voltage (C), and drain voltage (D) time-domain waveforms for a $0.25 \times 200\ \mu m^2$ GaN HEMT at $V_{GS} = -2.75$ V, $V_{DS} = 20$ V, $f_0 = 5$ GHz, $Z_L(f_0) = 91.5 + j144\ \Omega$, and swept input power. Output power at different load impedances (E) and measured *(squares)* and simulated *(crosses)* load impedances (F).

FIGURE 1.29

Fully analytical model extracted after numerical optimization against nonlinear measurements. Measured *(symbols)* and simulated *(lines)* *S*-parameters for a $0.25\times 200\ \mu m^2$ GaN HEMT from 40 MHz to 40 GHz at $V_{GS}=-2.75$ V, $V_{DS}=20$ V.

Table 1.2 Parameters of the Model in Ref. [81] Obtained After Numerical Optimization Against Vector Calibrated Nonlinear Measurements and Using the Parameters in Table 1.1 as Starting Values

I_{pk0} (mA)	V_{pk} (V)	ΔV_{pk} (V)	P_1	λ	α_S	R_{th} (°C/W)
92.5	−1.23	0.466	0.51	6.8e-5	0.165	6
P_2	P_3	α_R	B_1	B_2	T_{clpk0}	T_{cP1}
−0.01	0.05	6.6e-5	1.27	1.65	−0.0016	−0.0055
C_{GSP} (fF)	C_{GS0} (fF)	C_{GDP} (fF)	C_{GD0} (fF)	C_{DS} (fF)	$P_{11}=P_{41}$	$P_{10}=P_{40}$
288	56.3	22.8	25	99.2	3.26	7.1
$P_{21}=P_{31}$	$P_{20}=P_{30}$	P_{111}	T_{ccgs0}	T_{ccgd0}	R_g (Ω)	R_s (Ω)
0.3	0.11	0.053	0.004	0.0025	1.6	0.1
R_d (Ω)	L_g (pH)	L_s (pH)	L_d (pH)	C_{pg} (fF)	C_{pd} (fF)	
3.9	161	1.6	152	26.6	37.6	

condition closer to real life, are included in the extraction/optimization phase. This represents one of the reasons for the huge efforts made in the last decades for the development of microwave instrumentation enabling accurate nonlinear measurements [107–113]. For an extensive treatment of this subject, the reader is referred to the next chapter that is dedicated entirely to linear and nonlinear measurement techniques.

ACKNOWLEDGMENTS

This work was supported by the project PON 01_01322 PANREX with financial support by Italian MIUR, FWO — Vlaanderen (Belgium), Hercules, and EMRP "HF Circuits" project.

REFERENCES

[1] T. Mimura, S. Hiyamizu, T. Fujii, K. Nanbu, A new field-effect transistor with selectively doped GaAs/n $Al_xGa_{1-x}As$ heterojunctions, Jpn. J. Appl. Phys. 19 (5) (1980) L225–L227.

[2] D. Delagebeaudeuf, P. Delescluse, P. Etienne, M. Laviron, J. Chaplart, N.T. Linh, Two-dimensional electron gas MESFET structure, Electron. Lett. 16 (17) (1980) 667–668.

[3] A. Caddemi, G. Crupi, N. Donato, On the soft breakdown phenomenon in AlGaAs/InGaAs HEMT: an experimental study down to cryogenic temperature, Solid State Electron. 49 (6) (2005) 928–934.

[4] A. Caddemi, G. Crupi, N. Donato, Microwave characterization and modeling of packaged HEMTs by a direct extraction procedure down to 30 K, IEEE Trans. Instrum. Meas. 55 (2) (2006) 465–470.

[5] M.A. Khan, A. Bhattarai, J.N. Kuznia, D.T. Olson, High electron mobility transistor based on a GaN-$Al_xGa_{1-x}N$ heterojunction, Appl. Phys. Lett. 63 (9) (1993) 1214–1215.

[6] L.F. Eastman, V. Tilak, J. Smart, B.M. Green, E.M. Chumbes, R. Dimitrov, H. Kim, O. S. Ambacher, N. Weimann, T. Prunty, M. Murphy, W.J. Schaff, J.R. Shealy, Undoped AlGaN/GaN HEMTs for microwave power amplification, IEEE Trans. Electron Devices 48 (3) (2001) 479–483.

[7] P.M. Cabral, J.C. Pedro, N.B. Carvalho, Nonlinear device model of microwave power GaN HEMTs for high power-amplifier design, IEEE Trans. Microw. Theory Techn. 52 (11) (2004) 2585–2592.

[8] R.J. Trew, G.L. Bilbro, W. Kuang, Y. Liu, H. Yin, Microwave AlGaN/GaN HFETs, IEEE Microw. Mag. 6 (1) (2005) 56–66.

[9] P. Colantonio, F. Giannini, R. Giofrè, E. Limiti, A. Serino, M. Peroni, P. Romanini, C. Proietti, A C-band high efficiency second-harmonic-tuned hybrid power amplifier in GaN technology, IEEE Trans. Microw. Theory Techn. 54 (6) (2006) 2713–2722.

[10] P. Colantonio, F. Giannini, R. Giofrè, L. Piazzon, Evaluation of GaN technology in power amplifier design, Microw. Opt. Technol. Lett. 51 (1) (2009) 42–44.

[11] R. Pengelly, S. Wood, J. Milligan, S. Sheppard, W. Pribble, A review of GaN on SiC high electron-mobility power transistors and MMICs, IEEE Trans. Microw. Theory Techn. 60 (6) (2012) 1764–1783.

[12] C.F. Campbell, A. Balistreri, M. Kao, D.C. Dumka, J. Hitt, GaN takes the lead, IEEE Microw. Mag. 13 (6) (2012) 44–53.

[13] G. Ghione, K.J. Chen, T. Egawa, G. Meneghesso, T. Palacios, R. Quay, Guest editorial special issue on GaN electronic devices, IEEE Trans. Electron Devices 60 (10) (2013) 2975–2981.

[14] X. Chen, W. Chen, F.M. Ghannouchi, Z. Feng, Y. Liu, Enhanced analysis and design method of concurrent dual-band power amplifiers with intermodulation impedance tuning, IEEE Trans. Microw. Theory Techn. 61 (12) (2013) 4544–4558.

[15] W. Chen, S. Zhang, Y. Liu, Y. Liu, F.M. Ghannouchi, A concurrent dual-band uneven Doherty power amplifier with frequency-dependent input power, IEEE Trans. Microw. Theory Techn. 61 (2) (2014) 552–561.

[16] H. Jang, P. Roblin, Z. Xie, Model-based nonlinear embedding for power-amplifier design, IEEE Trans. Microw. Theory Techn. 62 (9) (2014) 1986–2002.
[17] C. Florian, F. Mastri, R.P. Paganelli, D. Masotti, A. Costanzo, Theoretical and numerical design of a wireless power transmission link with GaN-based transmitter and adaptive receiver, IEEE Trans. Microw. Theory Techn. 62 (4) (2014) 931–946.
[18] E. McCune, Operating modes of dynamic power supply transmitter amplifiers, IEEE Trans. Microw. Theory Techn. 62 (11) (2014) 2511–2517.
[19] V. Camarchia, M. Pirola, R. Quaglia, S. Jee, Y. Cho, B. Kim, The Doherty power amplifier: review of recent solutions and trends, IEEE Trans. Microw. Theory Techn. 63 (2) (2015) 559–571.
[20] G. Crupi, G. Avolio, A. Raffo, P. Barmuta, D.M.M.-P. Schreurs, A. Caddemi, G. Vannini, Investigation on the thermal behavior for microwave GaN HEMTs, Solid State Electron. 64 (1) (2011) 28–33.
[21] A. Caddemi, G. Crupi, N. Donato, Temperature effects on DC and small signal RF performance of AlGaAs/GaAs HEMTs, Microelectron. Reliab. 46 (1) (2006) 169–173.
[22] F. Bonani, S.D. Guerrieri, G. Ghione, Physics-based simulation techniques for small- and large-signal device noise analysis in RF applications, IEEE Trans. Electron Devices 50 (3) (2003) 633–644.
[23] D. Denis, C.M. Snowden, I.C. Hunter, Coupled electrothermal, electromagnetic, and physical modeling of microwave power FETs, IEEE Trans. Microw. Theory Techn. 54 (6) (2006) 2465–2470.
[24] F.M. Yigletu, S. Khandelwal, T.A. Fjeldly, B. Iniguez, Compact charge-based physical models for current and capacitances in AlGaN/GaN HEMTs, IEEE Trans. Electron Devices 60 (11) (2013) 3746–3752.
[25] W. Grabinski, M. Brinson, P. Nenzi, F. Lannutti, N. Makris, A. Antonopoulos, M. Bucher, Open source circuit simulation tools for RF compact semiconductor device modelling, Int. J. Numer. Modell. Electron. Networks Devices Fields 27 (5–6) (2014) 761–779.
[26] D.M.M.-P. Schreurs, J. Wood, N. Tufillaro, L. Barford, D.E. Root, Construction of behavioural models for microwave devices from time-domain large-signal measurements to speed-up high-level design simulations, Int. J. RF Microwave Comput. Aided Eng. 13 (1) (2003) 54–61.
[27] D.E. Root, J. Verspecht, D. Sharrit, J. Wood, A. Cognata, Broad-band poly-harmonic distortion behavioral models from fast automated simulations and large-signal vectorial network measurements, IEEE Trans. Microw. Theory Techn. 53 (11) (2005) 3656–3664.
[28] M. Koh, J.J. Bell, D. Williams, A. Patterson, J. Lees, D.E. Root, P.J. Tasker, Frequency scalable large signal transistor behavioral model based on admittance domain formulation, in: IEEE MTT-S International Microwave Symposium Digest, Tampa, FL, USA, June, 2014, pp. 1–3.
[29] Z. Marinković, G. Crupi, A. Caddemi, G. Avolio, A. Raffo, V. Marković, G. Vannini, D.M.M.-P. Schreurs, Neural approach for temperature dependent modeling of GaN HEMTs, Int. J. Numer. Modell. Electron. Networks Devices Fields 28 (4) (2015) 359–370.
[30] M.W. Pospieszalski, Modeling of noise parameters of MESFET's and MODFET's and their frequency and temperature dependence, IEEE Trans. Microw. Theory Techn. 37 (9) (1989) 1340–1350.
[31] I. Angelov, H. Zirath, N. Rorsman, A new empirical nonlinear model for HEMT and MESFET devices, IEEE Trans. Microw. Theory Techn. 40 (12) (1992) 2258–2266.

[32] M.F. Barciela, P.J. Tasker, Y. Campos-Roca, M. Demmler, H. Massler, E. Sanchez, C. Currás-Francos, M. Schlechtweg, A simplified broad-band large-signal nonquasi-static table-based FET model, IEEE Trans. Microw. Theory Techn. 48 (3) (2000) 395–405.
[33] G. Dambrine, A. Cappy, F. Heliodore, E. Playez, A new method for determining the FET small-signal equivalent circuit, IEEE Trans. Microw. Theory Techn. 36 (7) (1988) 1151–1159.
[34] W.R. Curtice, R.L. Camisa, Self-consistent GaAs FET models for amplifier design and device diagnostics, IEEE Trans. Microw. Theory Techn. 32 (12) (1984) 1573–1578.
[35] F. Lin, G. Kompa, FET model parameter extraction based on optimization with multi-plane data-fitting and bidirectional search — a new concept, IEEE Trans. Microw. Theory Techn. 42 (7) (1994) 1114–1121.
[36] K. Shirakawa, H. Oikawa, T. Shimura, Y. Kawasaki, Y. Ohashi, T. Saito, Y. Daido, An approach to determining an equivalent circuit for HEMTs, IEEE Trans. Microw. Theory Techn. 43 (3) (1995) 499–503.
[37] R. Tayrani, J.E. Gerber, T. Daniel, R.S. Pengelly, U.L. Rhode, A new and reliable direct parasitic extraction method for MESFETs and HEMTs, in: European Microwave Conference, Madrid, Spain, September, 1993, pp. 451–453.
[38] N. Rorsman, M. Garcia, C. Karlsson, H. Zirath, Accurate small signal modeling of HFET's for millimeter-wave applications, IEEE Trans. Microw. Theory Techn. 44 (3) (1996) 432–437.
[39] J. Wood, D.E. Root, Bias-dependent linear scalable millimeter-wave FET model, IEEE Trans. Microw. Theory Techn. 48 (12) (2000) 2352–2360.
[40] G. Crupi, D. Xiao, D. Schreurs, E. Limiti, A. Caddemi, W. De Raedt, M. Germain, Accurate multibias equivalent circuit extraction for GaN HEMTs, IEEE Trans. Microw. Theory Techn. 54 (10) (2006) 3616–3622.
[41] G. Crupi, D.M.M.-P. Schreurs, A. Raffo, A. Caddemi, G. Vannini, A new millimeter wave small-signal modeling approach for pHEMTs accounting for the output conductance time delay, IEEE Trans. Microw. Theory Techn. 56 (4) (2008) 741–746.
[42] R.G. Brady, H.C.H. Oxley, T.J. Brazil, An improved small-signal parameter-extraction algorithm for GaN HEMT devices, IEEE Trans. Microw. Theory Techn. 56 (7) (2008) 1535–1544.
[43] G. Crupi, D.M.M.-P. Schreurs, A. Caddemi, On the small signal modeling of advanced microwave FETs: a comparative study, Int. J. RF Microwave Comput. Aided Eng. 18 (5) (2008) 417–425.
[44] A. Zarate-de Landa, J.E. Zuniga-Juarez, J.R. Loo-Yau, J.A. Reynoso-Hernandez, M.C. Maya-Sanchez, J.L. del Valle-Padilla, Advances in linear modeling of microwave transistors, IEEE Microw. Mag. 10 (2) (2009) 100, 102–111, 146.
[45] S.-W. Moon, J.-H. Oh, Y.-H. Baek, M. Han, J.-K. Rhee, S.-D. Kim, Millimeter-wave small-signal modeling with optimizing sensitive-parameters for metamorphic high electron mobility transistors, Semicond. Sci. Technol. 25 (8) (2010) 1–7.
[46] R. Alt, D. Marti, C.R. Bolognesi, Transistor modeling: robust small-signal equivalent circuit extraction in various HEMT technologies, IEEE Microw. Mag. 14 (4) (2013) 83–101.
[47] T.T.L. Nguyen, S.D. Kim, A gate-width scalable method of parasitic parameter determination for distributed HEMT small-signal equivalent circuit, IEEE Trans. Microw. Theory Techn. 61 (10) (2013) 3632–3638.

[48] L. Shen, B. Chen, J. Gao, An improved millimeter-wave small-signal modeling approach for HEMTs, Int. J. RF Microwave Comput. Aided Eng. 24 (4) (2014) 417–425.
[49] G. Crupi, D.M.M.-P. Schreurs, Microwave De-embedding: From Theory to Applications, Academic Press, Oxford, 2013.
[50] M. Berroth, R. Bosch, High frequency equivalent circuit of GaAs FET's for large-signal applications, IEEE Trans. Microw. Theory Techn. 39 (2) (1991) 224–229.
[51] A. Caddemi, G. Crupi, A. Macchiarella, On wafer scaled GaAs HEMTs: direct and robust small signal modelling up to 50 GHz, Microw. Opt. Technol. Lett. 51 (8) (2009) 1958–1963.
[52] G. Crupi, A. Raffo, G. Avolio, G. Bosi, G. Sivverini, F. Palomba, A. Caddemi, D.M.M.-P. Schreurs, G. Vannini, Nonlinear modeling of GaAs pHEMTs for millimeter-wave mixer design, Solid State Electron. 104 (2015) 25–32.
[53] A. Caddemi, G. Martines, M. Sannino, HEMT for low noise microwaves: CAD oriented modeling, IEEE Trans. Microw. Theory Techn. 40 (7) (1992) 1441–1445.
[54] J.-M. Collantes, R.D. Pollard, M. Sayed, Effects of DUT mismatch on the noise figure characterization: a comparative analysis of two Y-factor techniques, IEEE Trans. Instrum. Meas. 51 (6) (2002) 1150–1156.
[55] Z. Marinkovic, V. Markovic, Temperature-dependent models of low-noise microwave transistors based on neural networks, Int. J. RF Microwave Comput. Aided Eng. 15 (6) (2005) 567–577.
[56] Fundamentals of RF and microwave noise figure measurements, Agilent application note 57-1, 2010.
[57] W. Wiatr, G. Crupi, A. Caddemi, A. Mercha, D.M.M.-P. Schreurs, Source-pull characterization of FinFET noise, in: International Conference on Mixed Design of Integrated Circuits and Systems, Wrocław, Poland, June, 2010, pp. 425–430.
[58] G. Crupi, A. Caddemi, D.M.M.-P. Schreurs, W. Wiatr, A. Mercha, Microwave noise modeling of FinFETs, Solid State Electron. 56 (1) (2011) 18–22.
[59] H. Hillbrand, P. Russer, An efficient method for computer aided noise analysis of linear amplifier networks, IEEE Trans. Circuits Syst. 23 (4) (1976) 235–238.
[60] C.E. Biber, M.L. Schmatz, T. Morf, U. Lott, E. Morifuji, W. Bachtold, Technology independent degradation of minimum noise figure due to pad parasitics, in: IEEE MTT-S International Microwave Symposium Digest, Baltimore, MD, USA, June, 1998, pp. 145–148.
[61] R.A. Pucel, W. Struble, R. Hallgren, U.L. Rohde, A general noise de-embedding procedure for packaged two-port linear active devices, IEEE Trans. Microw. Theory Techn. 40 (11) (1992) 2013–2024.
[62] S. Colangeli, W. Ciccognani, E. Limiti, A. Caddemi, G. Crupi, G. Salvo, Black-box noise modeling of GaAs HEMTs under illumination, Int. J. Numer. Modell. Electron. Networks Devices Fields. 28 (6) (2015) 698–706.
[63] M. Rudolph, R. Behtash, R. Doerner, K. Hirche, J. Wurfl, W. Heinrich, G. Trankle, Analysis of the survivability of GaN low-noise amplifiers, IEEE Trans. Microw. Theory Techn. 55 (1) (2007) 37–43.
[64] H. Sun, A.R. Alt, H. Benedickter, E. Feltin, J.-F. Carlin, M. Gonschorek, N. Grandjean, C.R. Bolognesi, Low-noise microwave performance of 0.1 μm gate AlInN/GaN HEMTs on SiC, IEEE Microwave Wireless Compon. Lett. 20 (8) (2010) 453–455.
[65] S. Colangeli, A. Bentini, W. Ciccognani, E. Limiti, A. Nanni, GaN-based robust low-noise amplifiers, IEEE Trans. Electron Devices 60 (10) (2013) 3238–3248.

[66] C. Sanabria, H. Xu, T. Palacios, A. Chakraborty, S. Heikman, U. Mishra, R. York, Influence of epitaxial structure in the noise figure of AlGaN/GaN HEMTs, IEEE Trans. Microw. Theory Techn. 53 (2) (2005) 762–769.

[67] S. Lee, K.J. Webb, The influence of transistor nonlinearities on noise properties, IEEE Trans. Microw. Theory Techn. 53 (4) (2005) 1314–1321.

[68] Z.H. Liu, G.I. Ng, S. Arulkumaran, Analytical modeling of high-frequency noise including temperature effects in GaN HEMTs on high-resistivity Si substrates, IEEE Trans. Electron Devices 57 (7) (2010) 1485–1491.

[69] M. Rudolph, R. Doerner, E. Ngnintendem, W. Heinrich, Noise modeling of GaN HEMT devices, in: European Microwave Integrated Circuits Conference, Amsterdam, The Netherlands, October-November, 2012, pp. 159–162.

[70] A. Nalli, A. Raffo, G. Crupi, S. D'Angelo, D. Resca, F. Scappaviva, G. Salvo, A. Caddemi, G. Vannini, GaN HEMT noise model based on electromagnetic simulations, IEEE Trans. Microw. Theory Techn. 63 (8) (2015) 2498–2508.

[71] S. Colangeli, A. Bentini, W. Ciccognani, E. Limiti, Polynomial noise modeling of silicon-based GaN HEMTs, Int. J. Numer. Modell. Electron. Networks Devices Fields 27 (5–6) (2014) 812–821.

[72] G. Crupi, A. Caddemi, A. Raffo, G. Salvo, A. Nalli, G. Vannini, GaN HEMT noise modeling based on 50-ohm noise factor, Microw. Opt. Technol. Lett. 57 (4) (2015) 937–942.

[73] R. Quéré, E. Ngoya, M. Camiade, A. Suarez, M. Hessane, J. Obregon, Large signal design of broadband monolithic microwave frequency dividers and phase-locked oscillators, IEEE Trans. Microw. Theory Techn. 41 (11) (1993) 1928–1938.

[74] S. Sancho, A. Suárez, J. Dominguez, F. Ramirez, Analysis of near-carrier phase-noise spectrum in free-running oscillators in the presence of white and colored noise sources, IEEE Trans. Microw. Theory Techn. 58 (3) (2010) 587–601.

[75] H. Qi, J. Benedikt, P.J. Tasker, Nonlinear data utilization: from direct data lookup to behavioral modeling, IEEE Trans. Microw. Theory Techn. 57 (6) (2009) 1425–1432.

[76] A. Raffo, G. Bosi, V. Vadalà, G. Vannini, Behavioral modeling of GaN FETs: a load-line approach, IEEE Trans. Microw. Theory Techn. 62 (1) (2014) 73–82.

[77] X. Jianjun, R. Jones, S.A. Harris, T. Nielsen, D.E. Root, Dynamic FET model — DynaFET — for GaN transistors from NVNA active source injection measurements, in: IEEE MTT-S IEEE MTT-S International Microwave Symposium Digest, Tampa, FL, USA, June, 2014, pp. 1–3.

[78] W.R. Curtice, A MESFET model for use in the design of GaAs integrated circuits, IEEE Trans. Microw. Theory Techn. 28 (5) (1980) 448–456.

[79] A. Materka, T. Kacprzak, Computer calculation of large-signal GaAs FET amplifier characteristics, IEEE Trans. Microw. Theory Techn. 33 (1985) 129–135.

[80] F. Filicori, G. Vannini, A. Santarelli, A.M. Sanchez, A. Tazon, Y. Newport, Empirical modeling of low-frequency dispersive effects due to traps and thermal phenomena in III-V FET's, IEEE Trans. Microw. Theory Techn. 43 (12) (1995) 2972–2981.

[81] I. Angelov, L. Bengtsson, M. Garcia, Extensions of the Chalmers nonlinear HEMT and MESFET model, IEEE Trans. Microw. Theory Techn. 44 (10) (1996) 1664–1674.

[82] V.I. Cojocaru, T.J. Brazil, A scalable general-purpose model for microwave FET's including DC/AC dispersion effects, IEEE Trans. Microw. Theory Techn. 45 (12) (1997) 2248–2255.

[83] A. Siligaris, G. Dambrine, D.M.M.-P. Schreurs, F. Danneville, A new empirical nonlinear model for sub-250 nm channel MOSFET, IEEE Microw. Wireless Compon. Lett. 13 (10) (2003) 449–451.

[84] C. Fager, J. Pedro, N. Carvalho, H. Zirath, Prediction of IMD in LDMOS transistor amplifiers using a new large-signal model, IEEE Trans. Microw. Theory Techn. 50 (12) (2002) 2834–2842.

[85] A. Jarndal, G. Kompa, Large-signal model for AlGaN/GaN HEMTs accurately predicts trapping- and self-heating-induced dispersion and intermodulation distortion, IEEE Trans. Electron Devices 54 (11) (2007) 2830–2836.

[86] O. Jardel, F. De Groote, T. Reveyrand, J.-C. Jacquet, C. Charbonniaud, J.-P. Teyssier, D. Floriot, R. Quéré, An electrothermal model for AlGaN/GaN power HEMTs including trapping effects to improve large-signal simulation results on high VSWR, IEEE Trans. Microw. Theory Techn. 55 (12) (2007) 2660–2669.

[87] G. Avolio, D.M.M.-P. Schreurs, A. Raffo, G. Crupi, I. Angelov, G. Vannini, B. Nauwelaers, Identification technique of FET model based on vector nonlinear measurements, Electron. Lett. 47 (24) (2011) 1323–1324.

[88] F. Filicori, G. Vannini, Mathematical approach to large-signal modelling of electron devices, Electron. Lett. 27 (4) (1991) 357–359.

[89] D.E. Root, S. Fan, J. Meyer, Technology independent large signal non quasistatic FET models by direct construction from automatically characterized device data, in: European Microwave Conference, Stuttgart, Germany, September, 1991, pp. 927–932.

[90] F. Filicori, G. Vannini, V. Monaco, A nonlinear integral model of electron devices for HB circuit analysis, IEEE Trans. Microw. Theory Techn. 40 (7) (1992) 1456–1465.

[91] I. Angelov, N. Rorsman, J. Stenarson, M. Garcia, H. Zirath, An empirical table-based FET model, IEEE Trans. Microw. Theory Techn. 47 (12) (1999) 2350–2357.

[92] E.P. Vandamme, D.M.M.-P. Schreurs, C. van Dinther, G. Badenes, L. Deferm, Development of a RF large signal MOSFET model, based on an equivalent circuit, and comparison with BSIM3v3 compact model, Solid State Electron. 46 (3) (2002) 353–360.

[93] A. Orzati, D.M.M.-P. Schreurs, L. Pergola, H. Benedickter, F. Robin, O.J. Homan, W. Bächtold, A 110-GHz large-signal lookup-table model for InP HEMTs including impact ionization effects, IEEE Trans. Microw. Theory Techn. 51 (2) (2003) 468–474.

[94] F. Filicori, A. Santarelli, P.A. Traverso, A. Raffo, G. Vannini, M. Pagani, Non-linear RF device modelling in the presence of low-frequency dispersive phenomena, Int. J. RF Microwave Comput. Aided Eng. 16 (1) (2006) 81–94.

[95] Y.S. Long, Y.X. Guo, Z. Zhong, A 3-D table-based method for non-quasi-static microwave FET devices modeling, IEEE Trans. Microw. Theory Techn. 60 (10) (2012) 3088–3095.

[96] G. Bosi, G. Crupi, V. Vadalà, A. Raffo, A. Giovannelli, G. Vannini, Nonlinear modeling of LDMOS transistors for high-power FM transmitters, Int. J. Numer. Modell. Electron. Networks Devices Fields 27 (5–6) (2014) 780–791.

[97] G. Avolio, A. Raffo, I. Angelov, V. Vadalà, G. Crupi, A. Caddemi, G. Vannini, D.M. M.-P. Schreurs, Millimetre-wave FET nonlinear modelling based on the dynamic-bias measurement technique, IEEE Trans. Microw. Theory Techn. 62 (11) (2014) 2526–2537.

[98] C. Wang, Y. Xu, X. Yu, C. Ren, Z. Wang, H. Lu, T. Chen, B. Zhang, R. Xu, An electrothermal model for empirical large-signal modeling of AlGaN/GaN HEMTs including self-heating and ambient temperature effects, IEEE Trans. Microw. Theory Techn. 62 (12) (2014) 2878–2887.

[99] TGC4546-SM, upconverter data sheet, TriQuint Semiconductor, 2014.

[100] A. Raffo, A. Santarelli, P.A. Traverso, M. Pagani, F. Palomba, F. Scappaviva, G. Vannini, F. Filicori, Accurate pHEMT nonlinear modeling in the presence of low-frequency dispersive effects, IEEE Trans. Microw. Theory Techn. 53 (11) (2005) 3449–3459.
[101] Nonlinear Devices, Keysight Advanced Design System (ADS), Keysight Technol., Palo Alto, CA, USA, 2011.
[102] I. Angelov, K. Andersson, D. Schreurs, D. Xiao, N. Rorsman, V. Desmaris, M. Sudow, H. Zirath, Large-signal modeling and comparison of AlGaN/GaN HEMTs and SiC MESFETs, in: Asia Pacific Microwave Conference, Yokohama, China, December, 2006, pp. 279–282.
[103] I. Angelov, G. Avolio, D.M.M.-P. Schreurs, Large-signal time-domain waveform based transistor modeling, in: Microwave De-embedding: From Theory to Applications, Academic Press, Oxford, 2013 (Chapter 5).
[104] G. Avolio, D.M.M.-P. Schreurs, A. Raffo, I. Angelov, G. Crupi, G. Vannini, B. Nauwelaers, Waveforms-based large-signal identification of transistor models, in: IEEE MTT-S International Microwave Symposium Digest, Montreal, Canada, June, 2012, pp. 1–3.
[105] A. Raffo, V. Vadalà, D.M.M.-P. Schreurs, G. Crupi, G. Avolio, A. Caddemi, G. Vannini, Nonlinear dispersive modeling of electron devices oriented to GaN power amplifier design, IEEE Trans. Microw. Theory Techn. 58 (4) (2010) 710–718.
[106] A.E. Parker, J.G. Rathmell, Broad-band characterization of FET self-heating, IEEE Trans. Microw. Theory Techn. 53 (7) (2005) 2424–2429.
[107] A. Ferrero, V. Teppati, A. Carullo, Accuracy evaluation of on-wafer load-pull measurements, IEEE Trans. Microw. Theory Techn. 49 (1) (2001) 39–43.
[108] P. Colantonio, F. Giannini, E. Limiti, V. Teppati, An approach to harmonic load-and source-pull measurements for high-efficiency PA design, IEEE Trans. Microw. Theory Techn. 52 (1) (2004) 191–198.
[109] J. Verspecht, Large-signal network analysis, IEEE Microw. Mag. 6 (4) (2005) 82–92.
[110] W. Van Moer, Y. Rolain, A large-signal network analyzer: why is it needed? IEEE Microw. Mag. 7 (6) (2006) 46–62.
[111] D.M.M.-P. Schreurs, Applications of vector non-linear microwave measurements, IET Microwaves Antennas Propag. 4 (4) (2010) 421–425.
[112] G. Paillоncy, G. Avolio, M. Myslinski, Y. Rolain, M. Vanden Bossche, D.M.M.-P. Schreurs, Large-signal network analysis including the baseband, IEEE Microw. Mag. 12 (2) (2011) 77–86.
[113] A. Ferrero, M. Pirola, Harmonic load-pull techniques: an overview of modern systems, IEEE Microw. Mag. 14 (4) (2013) 116–123.

CHAPTER

Radio frequency and microwave linear and nonlinear characterization

2

M. Pirola*, A. Ferrero† and V. Camarchia*

Politecnico di Torino, Torino, Italy Keysight Technologies, Santa Rosa, CA, United States†*

CHAPTER OUTLINE

2.1 INTRODUCTION

The design of complex systems requires careful and accurate characterization of their components to validate behavior and performance of the many parts and to provide the designers with the models required for the synthesis and integration of the several blocks. Moreover, system design and module integration are carried out at different abstraction layers, each with its own distinguished characteristics and semantics, each described with a different figure of merit and asking for specific experimental identification procedures. Wireless systems do not constitute an exception, and this chapter is devoted to provide a concise overview of the characterization strategies suitable for the amplification modules present within wireless transceivers. The characterization at device and amplifier level share a similar background, while system-level characterization instead requires rather different insights and considerations. System level characterizations are in fact aimed to the evaluation

Microwave Wireless Communications. http://dx.doi.org/10.1016/B978-0-12-803894-9.00002-0

of the effects introduced by amplifiers on the base-band information, even if at physical level, they actually operate on high frequency (ie, radio frequency (RF) or microwave) signals [1]. For constant envelope or narrow band modulation schemes, conventional device and amplifier characterizations provide exhaustive information also concerning base-band behavior. In this contest, it has to be considered that, as widely discussed in Chapters 5–8, the varying envelope or the large bandwidth, typically adopted by last-generation communication systems, can dramatically affect system behavior and performance. For these reasons, in the last decade system characterization has become more effective and decisive, not only for the design of wireless communication systems, but also for the development of the digital predistortion schemes which are indispensable to meet the extremely stringent linearization requirements.

This chapter is organized as follows: in Section 2.2, starting from a brief overview of linear circuit properties, the scattering parameters are defined, while Section 2.3 is devoted to the issues related to their measurements. In these sections, we will see how their introduction and comprehension is indispensable for the linear and nonlinear characterization at high frequency. After that, in Section 2.4, the vector network analyzer (VNA) will be introduced as the most effective and important tool for the accurate linear characterization of devices and amplifiers. We will present the most important VNA features and we will see how this instrumentation evolved from the traditional S-parameter only measurement system, towards a key element for the large-signal characterization at high frequency. Sections 2.5 and 2.6 are, respectively, dedicated to downconversion and calibration in VNAs. There we will see how essential these steps are to obtaining accurate and reliable measurements using VNAs. Large-signal characterization is extremely important for device modeling (the reader is referred to Chapter 1), and amplifier base-band assessment. In this context, load-pull, (ie, the evaluation of device behavior and performance under varying termination conditions) is without a doubt the most effective large-signal characterization approach for providing straightforward feedback to device model developers and amplifier designers. The most important aspects concerning load-pull technique will be addressed in Section 2.7. Finally, Section 2.8 will deal with system-level characterization.

2.2 THE SCATTERING PARAMETERS

At low frequencies, measurements are based on voltages and currents, which at high frequencies are difficult to be precisely defined. RF and microwave characterization are handled with a different approach, closer to electromagnetic fields and wave propagations. Moreover, the measurements are block oriented. In fact, they only characterize what happens at the ports of the device under test (DUT), where the electromagnetic interactions with the measurement system take place (Figure 2.1). At high frequency, the description in terms of waves is more suitable due to the presence of distributed components.

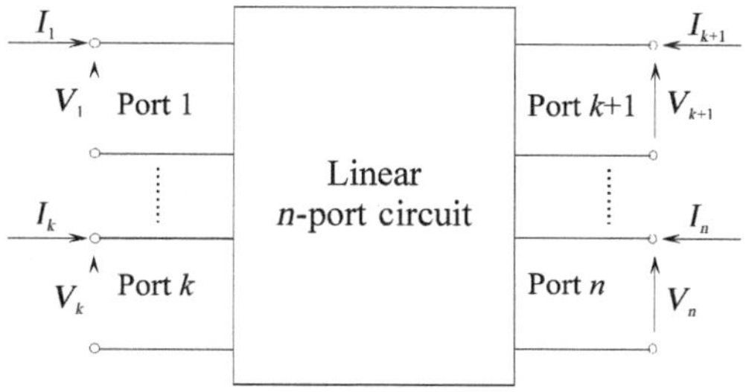

FIGURE 2.1

Linear n-port circuit.

To provide meaningful waves description, two fundamental concepts must be clearly understood and uniquely defined:

- measurement reference planes
- reference impedances

Reference planes are the ideal sections which define the DUT boundaries; in electrical terms, its ports. At these planes, the electromagnetic field is assumed to be monomodal, such that a unique definition of voltages and currents can be achieved.

Reference impedances link voltages and currents with the propagation nature of the typical microwave structures (ie, with the *waves* at the reference plane).

Microwaves have their own characteristic impedance, a concept related to effective propagation, which does not have an equivalent when dealing with lumped electrical circuits. On the other hand, scattering measurement systems as VNA use the characteristic impedance of the calibration devices to precisely set the voltage and current relationships at the measurement ports (ie, the port reference impedances at the reference plane). It is fundamental to have in mind that the port reference impedances are determined by the instrument calibration. In other words, without a proper calibration, the fundamental nature of the measurement process lacks meaning.

To point out this interaction between measurements and physical behavior but also to show the difference between the two environments, Roger Marks and Dylan Williams [2] suggest referring to the new constitutive quantities as *pseudo-waves*, and the corresponding parameters relating these pseudo-waves at the device ports as *pseudo-scattering* parameters. This clarification is regularly ignored and the *pseudo* prefix is almost always omitted. However, this remark should be clearly kept present since the corresponding relationships, shown in Eq. (2.1), are more correct and rigorous than the simpler expressions obtained according to the less sophisticated, and probably more known, approach reported in [3].

$$\begin{aligned} a &= \sqrt{R_r}\frac{V+IZ_r}{2|Z_r|} \\ b &= \sqrt{R_r}\frac{V-IZ_r}{2|Z_r|} \end{aligned} \tag{2.1}$$

In expressions (2.1), a and b are the direct and reflected waves, respectively, while Z_r is the reference impedance, and R_r its real part. Their ratio defines the scattering parameters according to the constitutive equations:

$$b = \Gamma a \tag{2.2}$$

for one-port device, and

$$\underline{\mathbf{b}} = \underline{\underline{\mathbf{S}}}\underline{\mathbf{a}} \tag{2.3}$$

for the more general case of multiport systems, where the scalar representation in Eq. (2.2), is replaced by a multidimensional formalism adopting the port wave vectors $\underline{\mathbf{a}}$, $\underline{\mathbf{b}}$ related through the scattering matrix $\underline{\underline{\mathbf{S}}}$, according to the expression (2.3).

This approach can be extended to the so-called differential scattering parameters to highlight the differential nature of transmission channels and amplifiers, particularly important when dealing with digital communication. Around 20 years ago, the differential S-parameters and the corresponding common and differential mode pseudo-waves were introduced and extended according to concepts in Refs. [4,5] with the expressions:

$$\begin{aligned}
a_d &= \sqrt{R_d}\frac{V_d + I_d Z_{rd}}{2|Z_{rd}|} \\
b_d &= \sqrt{R_d}\frac{V_d - I_d Z_{rd}}{2|Z_{rd}|} \\
a_c &= \sqrt{R_c}\frac{V_c + I_c Z_{rc}}{2|Z_{rc}|} \\
b_c &= \sqrt{R_c}\frac{V_c - I_c Z_{rc}}{2|Z_{rc}|}
\end{aligned} \tag{2.4}$$

where the classical a and b left pace to the common and differential counterparts, respectively a_c, b_c and a_d, b_d. These new quantities depend on the common and differential currents and voltages at the port, V_c, V_d and I_c, I_d, through the corresponding reference impedances Z_{rc} and Z_{rd}.

2.3 SCATTERING PARAMETER MEASUREMENTS

Experimental identification of the elements of n-port linear parameters ideally involves the measurement of quantities that, in case of scattering representation, are the previously introduced *pseudo-waves*, at each port. Considering the transfer function nature of the S-parameters, the two fundamental approaches to their measurement are based on frequency or time domain techniques.

The first is the traditional single-tone frequency swept technique introduced in 1960 in vector voltmeters, which later became VNAs. As shown in Figure 2.2, the basic blocks of these systems are:

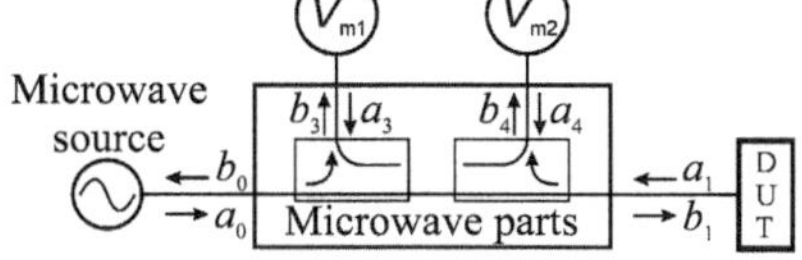

FIGURE 2.2

Principle scheme of a one-port frequency sweep vector network analyzer.

- a sweep frequency source
- a wave separation device
- an homodyne or heterodyne receiver.

The second class of scattering parameter measurement systems relies on time domain solutions, where the main blocks are:

- a narrow pulse source
- a sampling digitizer.

The traditional frequency domain network analyzer represents the standard instrumentation adopted for *S*-parameter measurement, and it will be described in Section 2.3.

The second strategy [6,7] offers a potentially very fast acquisition time because the spectrum components are obtained through mathematical elaboration of time samples which are simultaneously acquired. However, with respect to frequency-based instrumentations, time domain systems exhibit a reduced dynamic range, and need a DC based system, a requirement which is not easy to achieve at high frequency.

In time domain analysis, a narrow pulse is injected on a port of the DUT and the reflected signal is sampled at the other ports.

As in frequency domain, the process is repeated by connecting the source to one port at a time, and closing all the other ports on known terminations.

On the other hand, in the time domain approach, before the scattering parameters could represent a transfer function depending on frequency, time signals have to be properly transformed in their spectrum counterparts. This can be accomplished, in the more simple case, adopting fast Fourier transform (FFT) algorithms, or, in a more actual context, through chirp *Z*-transformation or equivalent.

Again, as in traditional VNAs, the systematic error correction is done in frequency domain by using the same standard devices.

The accuracy of time sample mathematical manipulations, as those required for time to frequency transformation, strongly depends on the causality and passivity of time pulse responses. In fact, their violation, for example, due to finite resolution of the digitalization process, can bring completely nonphysical results on the corresponding spectra, and hence on the scattering parameters. For these reasons, causality and passivity on the measured data have to be carefully evaluated and, if necessary, mathematically enforced.

Finally, the intrinsic ultra-broadband nature of these systems does not allow a high dynamic range.

Although very promising, this technique is still not very common, and for this reason we will focus our attention on the traditional frequency domain VNA architecture.

2.4 TWO-PORT VNAs

Figure 2.2 shows the block scheme of a one-port VNA.

As seen above, the source is swept across the measured frequency band, while the corresponding voltages at the output of a directional coupler or bridge are synchronously acquired.

In this way, the magnitude and phase of V_{m1} and V_{m2} are first obtained. From these readings, with proper calibration procedure, a_1 and b_1 at the DUT reference plane are then computed.

The acquisition process is typically based on traditional heterodyne conversion, where a local oscillator is swept together with the main source to provide an intermediate frequency (IF) which could be easily digitized. Although this basic principle is well known, the extreme broadband characteristic of modern VNAs, along with their huge dynamic range and measurement speed [8], make the process of sweeping and digitizing challenging and far from trivial.

A simplified functional scheme of a two-port VNA with four independent measurement channels is reported in Figure 2.3.

This scheme shows the most common single switched source VNA. Other structures can be adopted with a simplified high frequency section (eg, using three acquisition and downconversion channels instead of four) as outlined in Figure 2.4. Despite this represents a simpler, and above all, cheaper solution (VNA costs strongly depend on the high frequency components, which are by far the most expensive parts), the instrumentation accuracy and flexibility are negatively affected. As an example, for three channel VNAs, the high frequency switches required to change

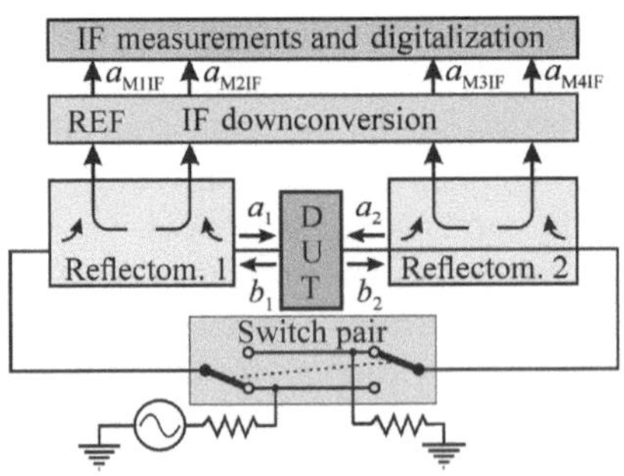

FIGURE 2.3

Two-port vector network analyzer scheme.

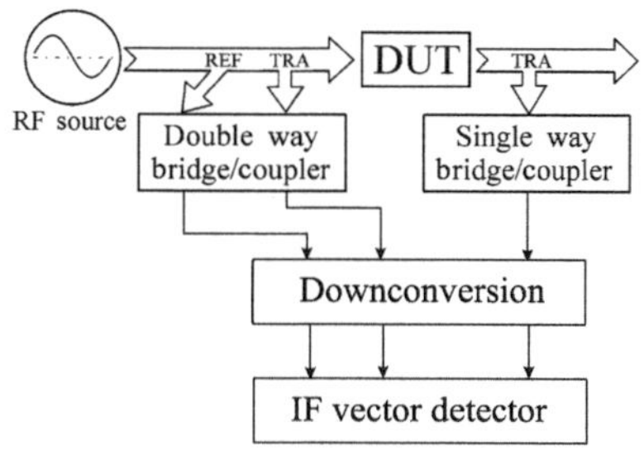

FIGURE 2.4

Block scheme of three sampler vector network analyzers.

the driving conditions from port 1 to port 2 are an integral part of the reflectometer. Only assuming/trusting their repeatability, the errors they introduce can be considered systematic and hence removed with the calibration process [8]. Four channel VNAs instead simultaneously measure the four powers waves at the two DUT ports and the switches can be considered external to the reflectometer pair since the relationships between the measured voltages and the waves at the DUT plans are independent of the reflectometer terminations [8].

2.5 DOWNCONVERSION TECHNIQUES

Downconversion is a nonlinear process where the spectrum of a signal centered on a given frequency is shifted toward lower frequencies, leaving its shape unaltered. There are two practical implementations of this concept. They resort to subsampling or mixing, respectively. VNA manufacturers adopt one or other choice according to their skill, whether on mixer or sampling field. However, it seems that today VNA producers have converged toward mixer-based solutions.

Sampling downconversion relies on time varying switched circuits that operate the spectrum shift at the IF frequency.

The sampling technique basic scheme for four channels is reported in Figure 2.5. A voltage controlled oscillator (VCO) is generating a pulsed wave with width t_W and frequency f_{IF} (in the kHz/MHz range). This signal drives the switch sampling one of the waves at the output of a reflectometer pair (eg, a_{1RF}, which after downconversion represents the system reference channel). If the square wave were an ideal Dirac pulse, the spectrum at the switch output would be given by an infinite series of replicas of the input spectrum, spaced by the sampling rate frequency. In fact, we have

$$\begin{aligned} x(t) &= A\sin\omega_0 t \\ X(\omega) &= jA\pi[\delta(-\omega_0) - \delta(\omega_0)] \\ Y(\omega) &= jA\pi\sum_{n=-\infty}^{\infty}[\delta(\omega+\omega_0-n\cdot f_{IF}) - \delta(\omega-n\cdot f_{IF}-\omega_0)] \end{aligned} \quad (2.5)$$

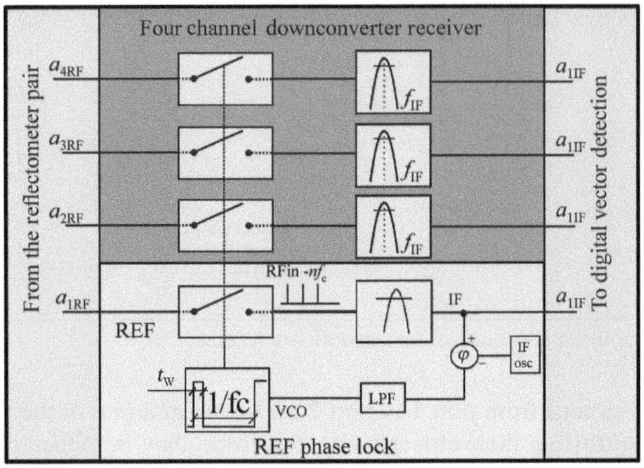

FIGURE 2.5

Subsampling downconversion technique.

where $x(t)$ is the sinusoidal RF input signal, $X(\omega)$ its Fourier transform counterpart, $Y(\omega)$ is the output in frequency domain, and IF is the intermediate frequency. The phase loop locks the VCO frequency to one of the replicas, all with the same amplitude and phase of $X(\omega)$, so to yield after IF band pass filtering an output spectrum exactly at this frequency. As a result, the downconverter output is a perfect replica of $X(\omega)$, but centered around IF. The idea is to downconvert all four waves at the reflectometer pair output using four identical sampling switches driven by the same VCO locked to one of the four waves chosen as the reference (see Figure 2.5). The four signals at IF maintain the same reciprocal phase and modulus relationships of the corresponding waves at high frequency. As a consequence, the phasor ratios, which define the scattering parameters, are unaffected by the downconversion process, and they can hence be measured at IF (see Figure 2.5).

Turning now to the mixer-based downconversion, the key element in this case is the mixer that multiplies the input RF and the local oscillator signals, using a nonlinear device, for example a diode. Mixer basic working principle is sketched in Figure 2.6.

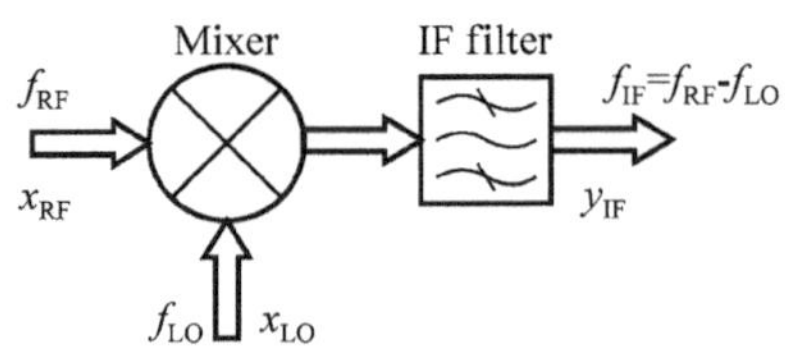

FIGURE 2.6

Basic principle of mixer behavior.

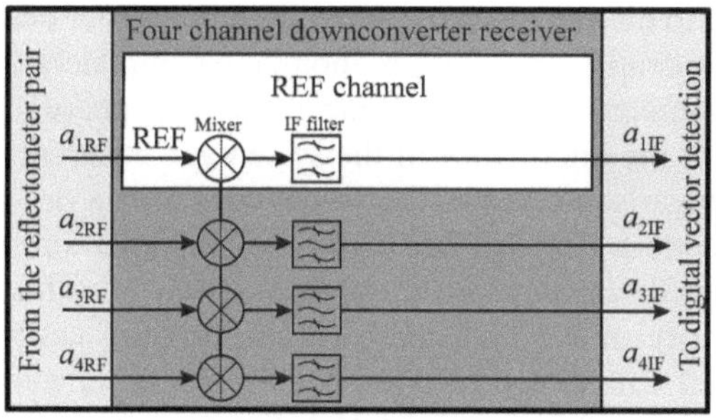

FIGURE 2.7

Mixer downconversion technique.

The spectrum of the mixer output has components at the IF frequency, the difference between the two mixer input frequencies. Other unwanted mixing components are present, but they are rejected with a filter right centered at IF. In formulas, we have

$$\begin{aligned} x_{RF} &= A_{RF} \cos \omega_{RF} t + \psi_{RF} \\ x_{LO} &= A_{LO} \cos \omega_{LO} t + \psi_{LO} \\ y_{IF} &= G_{conv} A_{RF} \cos[(\omega_{RF} - \omega_{LO})t + \psi_{RF} - \psi_{LO}] \end{aligned} \tag{2.6}$$

where A_{LO} and ω_0 are the amplitude and frequency of the local oscillator pump, respectively. G_{conv} is, instead, the conversion gain (ie, the ratio among the amplitude at IF and RF frequency) and depends on the pump power level A_{LO}. Even if the circuit operates in nonlinear conditions, previous relationships display the mixer linear behavior between the input amplitude at RF, and the corresponding output component at IF.

VNAs based on mixer for downconversion adopt four identical mixers driven by a unique local oscillator pump so that all four channels can be coherently downconverted. A block scheme of this solution is shown in Figure 2.7.

The local oscillator is typically a frequency synthesizer with the same time base of the main VNA RF source.

Analogously to what seen for sampling based solutions, adoption of identical mixers yields four IF sinusoidal signals with the same modulus and phase relationship of the waves before downconversion. The assumption of identical mixers seems rather strong, but calibration can take care to minimize the uncertainty introduced by mixer impairment.

2.6 TWO-PORT VNA CALIBRATION

Systematic errors represent the main limitation to the accuracy of microwave VNAs These errors are related to the nonideal behavior of the components present within VNAs.

Reflectometer finite directivity and mismatch, mixer impairment as well as cable phase-shift and losses, alter the magnitude and phase relationships at the receiver section with respect to the actual waves at the DUT ports. For example, the unavoidable presence of few centimeter cables is enough to completely offset measured and DUT scattering parameters, especially concerning their phase.

As an example of the importance of the systematic error correction, Figures 2.8 and 2.9 report a comparison between the so-called raw data (ie, scattering measurements without corrections) and the data with applied calibration. Measurements concern a termination with values close to 50 Ω. As shown in this figure, which also outline the irregularity of the uncalibrated data, original uncorrected data seems completely unrelated with expected ones. The reflection is well below −20 dB in the full frequency range (10 MHz–26.5 GHz). Figure 2.9 evidences the surprisingly capability of the calibration to unearth the phase behavior of the measured sample. In fact, the classical delay behavior introduced by the almost matched load is clearly visible in the phase of the calibrated data.

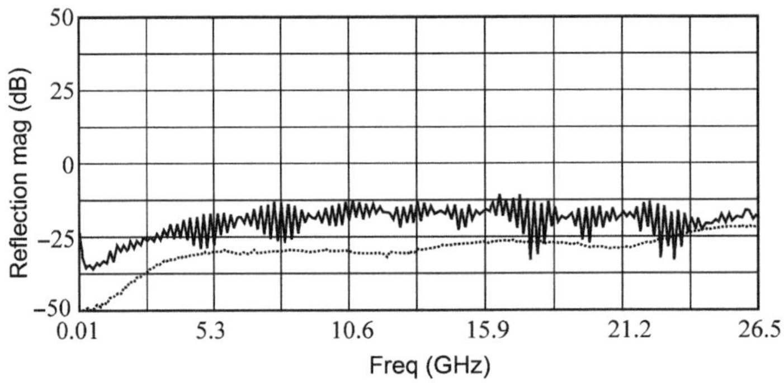

FIGURE 2.8

Modulus in dB of the reflection coefficient of an almost 50 Ω load. Raw (uncalibrated) data, continuous curve, and calibrated data, dotted curve.

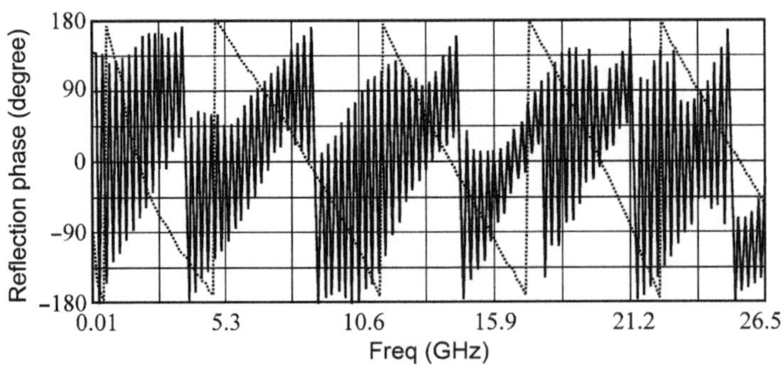

FIGURE 2.9

Phase in degrees of the reflection coefficient of an almost 50 Ω load. Raw (uncalibrated) data, continuous curve, and calibrated data, dotted curve.

As already pointed out, both reference impedance and reference plane are set by the calibration procedure. The VNA systematic error correction process is developed following the traditional metrological path for instrument calibration:

- definition of the error model and related coefficients;
- development of the calibration procedure;
- development of the correction algorithm.

The extremely high linearity response of the VNA receiver along with the passive nature of the front-end VNA components [8] allow developing linear models for the error correction which depend only on the number of ports and how they interact.

Several techniques have been conceived and developed to identify the error model coefficients. They are typically specialized to better suit the different measurement environment (eg, coaxial rather than on wafer), but they are all based on the insertion of devices with fully or partially known frequency behavior, the so-called *calibration standards*.

Each technique uses different combinations of port connections and sets of standards, providing a different level of accuracy. The manual standard insertion and replacement of the standards is giving way to the adoption of electronic calibrators, which greatly improves system user-friendliness, with positive effects on system overall accuracy and measurement repeatability. The VNA calibration topic has been extensively assessed in the last decades as demonstrated by the vast literature in the field.

References from [9–31], sorted chronologically, represent an exhaustive, and diverse set of papers on this subject. Other works dealing with the more sophisticated issue of calibration of leaky VNAs where the direct cross coupling between the reflectometer pairs can no longer be neglected, as in the millimeter frequency range, can be found in Refs. [32–34].

2.7 LOAD- AND SOURCE-PULL CHARACTERIZATION

Small signal characterization, in terms of S-parameters or of any other equivalent representation as a function of frequency, is fully representative of the behavior of passive and active elements working under linear conditions. Linear characterization is basically carried out, exciting the DUT with a single tone which is swept within the considered frequency range, and acquiring the complex phasors of the signal in the points of interest.

Nonlinear characterization requires the evaluation of a device's behavior under large-signal conditions. Resorting to these kinds of measurements is indispensable in assessing the performance of power amplifiers (PAs) or to extract useful information for nonlinear model definition or adjustment. For a detailed discussion of nonlinear transistor modeling, readers are referred to Chapter 1, and to papers [35–38]. Concerning instead the analysis of PAs, useful details can be found in Chapters 4–6, and in papers [39–41]. While linear measurements apply to multiport devices, nonlinear characterization is mostly reserved to two-port systems, with well-defined input and output ports. Moreover, while small signal measurements are carried out in a large range of frequencies, large-signal instrumentations focus on a rather limited set of frequencies. This is due to the amount of characterization time required by any single frequency point, and the huge amount of data to be handled.

Input and output terminations play a crucial role in affecting two-port operations, and they represent the most important design parameters to control and modify the system behavior so as to obtain the required performance. In linear conditions, the role of terminations is accounted for, knowing the small signal parameters, from which all the needed information can be obtained. This is for example the case of the gain maximization of a small signal amplifier.

In large-signal operations, the importance of terminations remains undiminished, but exhaustive characterization is not feasible. Source- and load-pull systems allow for the experimental identification of the input and output terminations, at fundamental frequencies and (some) harmonics which optimize one or more performance. This is accomplished by detecting and monitoring the performance to be optimized under variable user selectable input (source-pull) and output (load-pull) terminations.

Optimum loads can be found with automatically tracked load-pull measurements, but in most cases, considering the high sensitivity on load variations for nonlinear systems, postprocessing of data measurements on a given grid of loads is preferred to more sophisticated approaches.

Nonlinear performance of PAs is described in terms of typical figure of merits: input and output power (at fundamental and harmonic frequencies), power gain, efficiency or power-added efficiency, amplitude modulation (AM)-AM distortion, and AM-phase modulation (PM) conversion, intermodulation distortion, and adjacent channel power measurements. For device characterization, the operative gain is often preferred to transducer one, since it depends only on the output termination.

Different classifications of the source/load-pull systems are possible: scalar (power meter) or vector (VNA) based, nonreal- or real-time, at fundamental only or at fundamental plus harmonics. Another clear distinction concerns how loads are synthesized and controlled, because passive tuner and active load systems are available.

2.7.1 SCALAR SYSTEMS

The simplest source/load-pull system is composed of a single tone signal source, used as a test signal, a pair of tuners to vary the source and load impedance seen by the DUT, and a power meter to measure the output power as shown in Figure 2.10.

Only scalar measurements are possible through power meters in this configuration. It is a low-budget solution, which can be improved by adding a spectrum analyzer at the output to measure harmonics and intermodulation products. The system

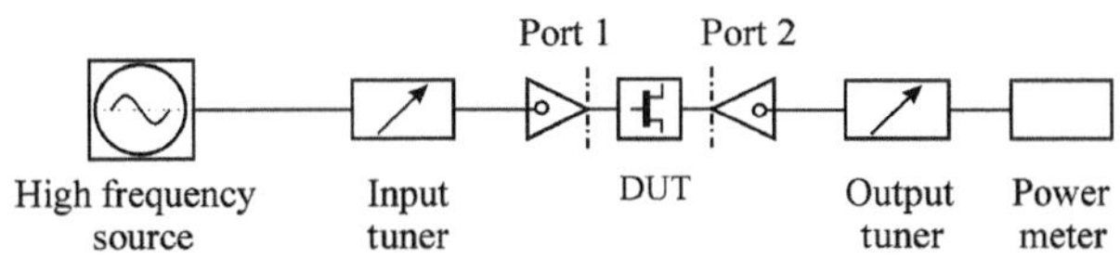

FIGURE 2.10

Scalar source/load-pull system based on power meters.

can also be exploited to test the linearity behavior of amplifiers in the presence of a modulated signal, adding a modulated signal generator together with the corresponding receiver.

The scalar nature of the system does not provide some important measurements like AM-PM conversion, moreover, only the available input power P_{av} is known from the high frequency generator, but not the DUT actual input power P_{in}. Since the source- and load-pull actual loads are not measured, this solution requires time-consuming accurate off-line tuner precalibration. Moreover, the power actually at the DUT output has to be derived using data of the calibration phase and from the power meter reading, according to a de-embedding strategy, details which can be found in [42]. Tuner repeatability in terms of mechanical positions is crucial since any setup modification compromises the accuracy of the measurements. A modified version of this solution, still scalar, also features direct sampling of input and output powers through two directional couplers inserted at the input and output of the DUT, as shown in Figure 2.11.

The complete precharacterization (eg, with VNA S-parameter measurements) of all bench elements (directional coupler, cables, probes) between power meters and DUT reference planes is still required to de-embed the actual powers at the DUT reference planes.

The wideband detection of power meters requires the insertion of high quality filters at the power meter input to filter out the harmonic components, increasing system complexity and cost. The reduced power meter dynamic range reflects on the performance of the system and is one of the main limitations of this setup.

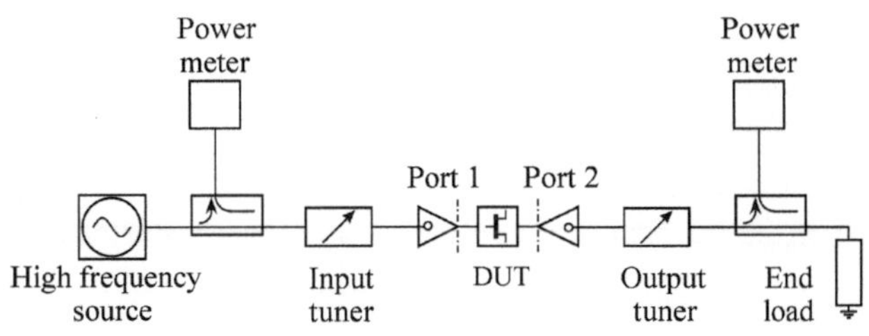

FIGURE 2.11

Scalar source/load-pull system based on power meters. Version with directional couplers.

2.7.2 VECTORIAL SYSTEMS

A different approach consists in measuring the source and load reflection coefficients (normally in real-time) by a VNA. In this case, the scalar power meter and spectrum analyzer are replaced with a VNA acting as a vector receiver. The high linearity, combined with the VNA's huge dynamic range, assisted by vector error correction techniques, provide better accuracy, higher speed, along with waveform capability when a phase reference is used as shown in Figures 2.12 and 2.13.

Broadband systems have been more recently introduced that are capable of processing wideband modern wireless signals. In these systems, the traditional tuners have been substituted with more complex waveform generators behaving as active loads. An example is reported in Figure 2.14. The structure, whose details are reported in Ref. [43], is conceived to be fully modular so that the scheme, customized in figure to control fundamental and second harmonic loads, can be extended, adding more branches, to higher harmonics.

Among the many contributions available in literature reporting on source/load-pull systems evolution, there are references from [43–71], where interested readers can find, sorted chronologically, the most significant papers that pushed forward the state-of-the-art in load-pull characterization.

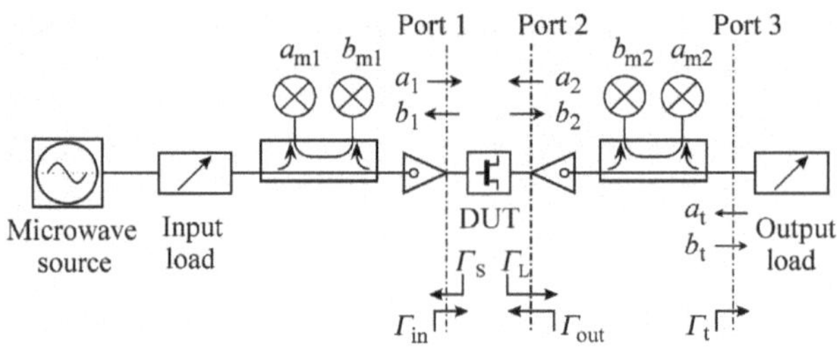

FIGURE 2.12

Vector network analyzer-based vectorial source/load-pull system.

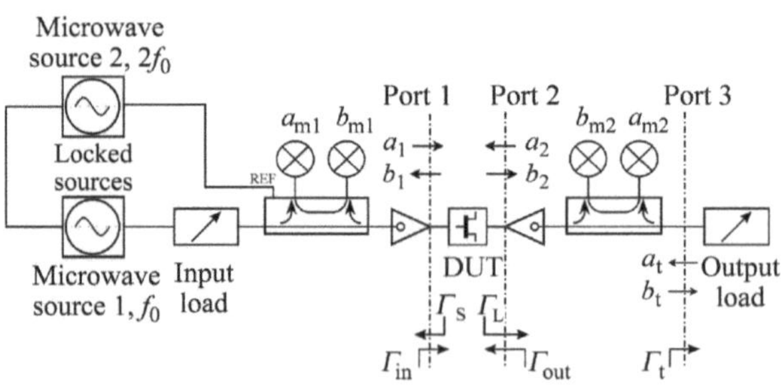

FIGURE 2.13

Vector network analyzer-based vectorial source/load-pull system for harmonic locking.

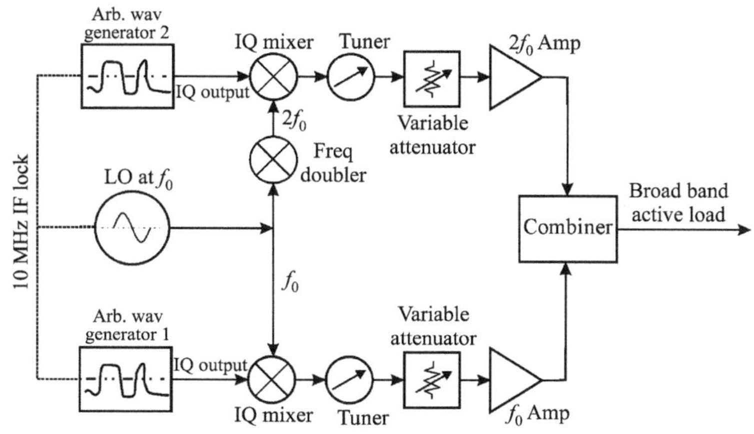

FIGURE 2.14

Scheme of broadband active load-pull system.

2.8 SYSTEM-LEVEL CHARACTERIZATION

System-level characterization of microwave modules is becoming a key step for the correct implementation of transmitter and receiver subsystems both for modeling and verification purposes.

The modern complex modulation schemes ask for specific characterization of the various modules. In fact, the classical continuous wave (CW) techniques described in the previous sections, along with two-tone tests, cannot provide complete and accurate representation of the actual behavior in presence of complex modulated signals instead of periodic and intrinsically narrowband excitations.

System-level measurement setups can be considered at any effect as digital transceivers since the data at the input of the PA have to be digitally modulated, and the base-band constellation has to be reconstructed from the PA output.

This instrumentations yield extremely important results because they are directly reporting the effects at base-band of the distortion introduced by PA nonlinearities. In fact, they can give straightforward insights of PA effects on system-level behavior and performance, such as providing time domain evaluation on the base-band digital in phase (I) and quadrature (Q) symbols.

The simplest approach to these measurements resorts to a basic spectrum analyzer (see Figure 2.15). This characterization provides only information on magnitude, not addressing phase measurement, which is fundamental with today's complex modulations.

In this instrument, the superheterodyne receiver is swept, allowing for power versus frequency signal acquisitions. After the downconversion, the signal is filtered with a high resolution bandwidth filter, and passed through an amplitude detector circuit. By sweeping the tunable filter across the frequency range of interest, the

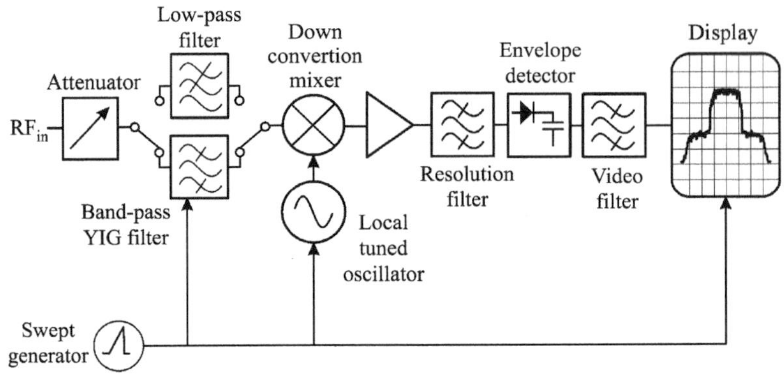

FIGURE 2.15

Spectrum analyzer scheme of principle.

characterization can be carried out on the whole frequency range. Considering the measurement accuracy, it is important to keep the input signals constant and stable during the frequency sweep; a condition which is not always easy to be achieved.

The present generation of spectrum analyzer can embed digital elements such as analog-to-digital conversion, or digital signal processor systems.

In recent years, new ad-hoc setups have been proposed to overcome spectrum analyzer limitations. They are composed essentially of two new instruments: the vector signal generator (VSG) and its counterpart, the vector signal analyzer (VSA). This instrumentation is in continuous evolution, and represents the core of any base-band characterization bench.

The VSG (see Figure 2.16) can generate either CW or arbitrarily modulated waveforms that can be present at the input of the DUT. It is composed of three main blocks:

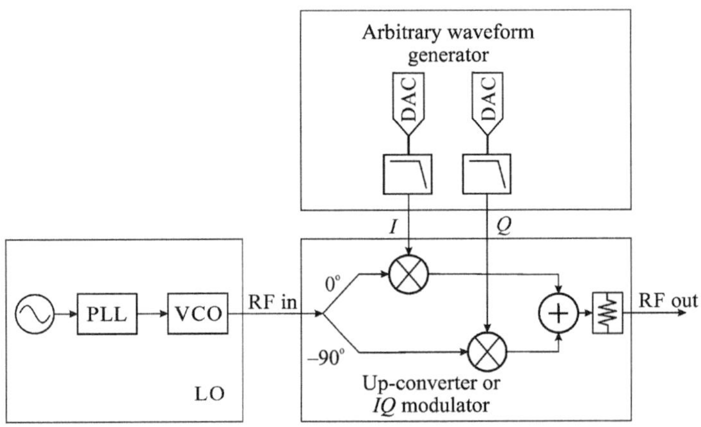

FIGURE 2.16

Vector signal generator scheme of principle.

- arbitrary waveform generator (AWG)
- local oscillator
- upconversion mixer (often a quadrature modulator)

More specifically, it embeds a tunable radio, with the AWG (the modulator) working on the base-band *I* and *Q* components coming from the digital-to-analog converters (ADC) (often differential) operating as a sort of modem. The signal is then filtered, upconverted to a locally generated RF carrier frequency, and eventually amplified to reach the required power level.

The VSA extracts from the high frequency modulated signal the base-band stream of data, typically quadrature demodulated to *I* and *Q* components (see Figure 2.17). It behaves similarly to last-generation spectrum analyzers, at least for the analog high frequency part, while it differs concerning the capabilities of data elaboration operated by the base-band digitizer.

Downconverter, mixer, and digitizer (ADC) in this case represent the three main blocks. According to the scheme of Figure 2.17, the operations of a VSA can be divided in consecutive steps: signal conditioning with frequency translation (attenuation and downconversion), analog-to-digital conversion, quadrature detection, digital filtering and resampling, data storing in memory, and finally, data windowing and FFT analysis.

The main issues of these instruments are those typical of any transmission link: amplitude and phase distortion, *I* and *Q* impairments, and spectral purity of the stimulus. They are normally reduced, and partially compensated for with ad-hoc solutions, both at software and hardware levels [72–76].

2.8.1 MEASUREMENT SYSTEM SYNCHRONIZATION

With proper adjustment and synchronization of VSA and VSG with the rest of the measurement equipment, the setup provides time- and frequency-domain vector measurements of the signals at the output of the DUT (typically an RF amplifier). All instruments are locked with a common 10 MHz oscillator providing the common

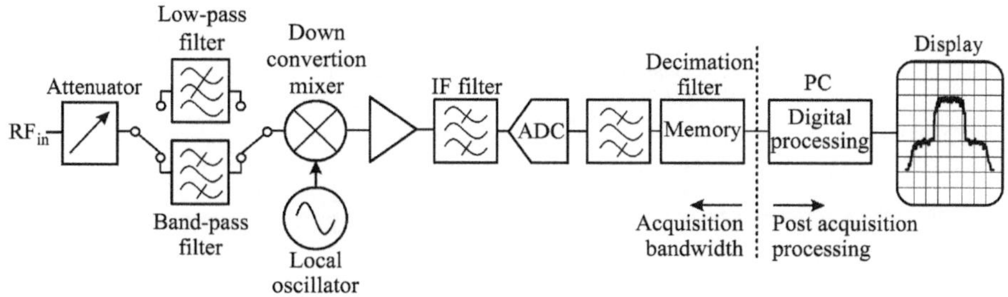

FIGURE 2.17

Vector signal analyzer scheme of principle.

reference signal. Care must be ensured for proper timing and synchronization, to correctly trigger generator and receiver.

The procedure to perform an accurate measurement is not trivial. The sampling frequency must be three/four times higher than the signal bandwidth to detect intermodulation in the adjacent channels. Moreover, the acquisition data blocks need to be large enough to collect the entire waveform generated by the VSG. Another important parameter is the frequency span which has to be chosen according to the *IQ* modulator clock.

In particular, the resolution bandwidth deeply affects sample size (memory allocation) and number of FFT computations (ie, the measurement time). As an example, adopting a modulated signal with 10 MHz sample rate to achieve a resolution bandwidth of 10 Hz, one million points for FFT is needed. The impact on the needed storing memory and on the required measurement time is therefore clear. Choosing the opportune sample rate, antialiasing filters (windowing), and decimation, this issue can be partially overcome when narrow bandwidth are required.

As predicted, a fundamental operation during the VSG-VSA measurements is the proper synchronization of the two instruments. Any misplacement between generation and acquisition streams means loss of data samples, improper sampling, and as a consequence, inaccurate measurements. A common strategy to solve this issue is to insert, at the start of the sequence, some markers (eg, a known sequence of data) to be used during the postprocessing for the correct synchronization. For highly accurate measurements, additional care must be taken to properly correct and transfer the power levels from generator and receiver to the right DUT reference planes. To this purpose, additional calibration steps can be adopted, like in [77], where the losses introduced by the on-waver probe setup of Figure 2.18 are taken into account by adding to the standard calibration a thru measurement. In this case, the VSA is connected to the bench output, applying a single tone input. The power measurements performed by the VSA and by the bench are collected and compared to extract a frequency dependent calibration factor.

Normally, system-level characterization setups are completed by digital predistortion capabilities (either via PC or dedicated hardware, for example based on field programmable gate array (FPGA) integrated circuits).

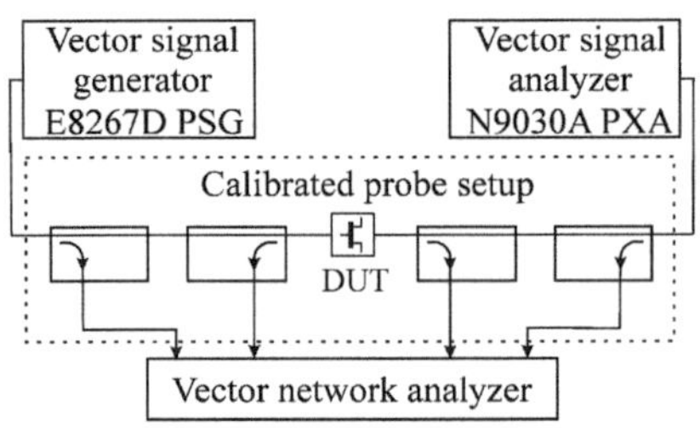

FIGURE 2.18

Example of a modulated signal test-bench scheme for on-wafer characterization.

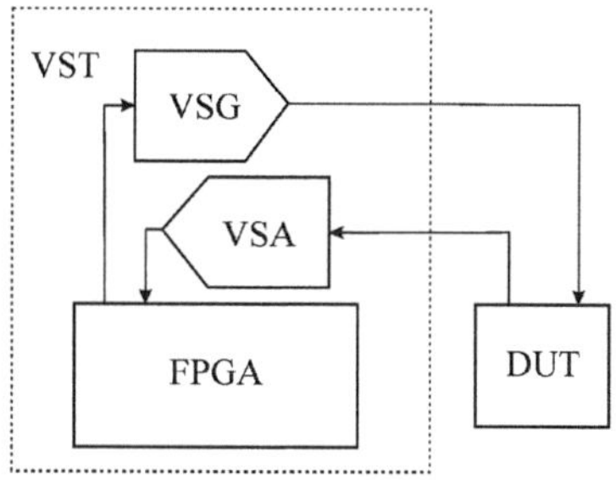

FIGURE 2.19

Example of vector signal transceiver (VST) scheme of principle.

In this context, the vector signal transceiver has emerged for real-time signal processing and control as a new class of instrumentation combining VSG, VSA, and FPGAs. The block scheme is sketched in Figure 2.19. A user-programmable FPGA allows for custom implementation of algorithms at the hardware level of the instrument and, at the same time, can be used to apply specific correction strategies to the system (eg, predistortion).

Commercial off-the-shelf products implementing sophisticated VSG/VSA with often custom software routines to evaluate the various figures of merit of the most adopted wireless standards are available from the main RF and microwave instrument suppliers, see [75] and [78–80] for more details on the available products.

In order to test the quality of the analyzed DUTs, the figures of merit previously introduced are normally rearranged in terms of average quantities.

Regarding linearity, a specific indicator normally employed for complex modulation schemes is the error vector magnitude, which quantifies the differences between the ideal constellation signals and the ones measured at the PA output (see Figure 2.20). The measurement setup represents a real digital transceiver

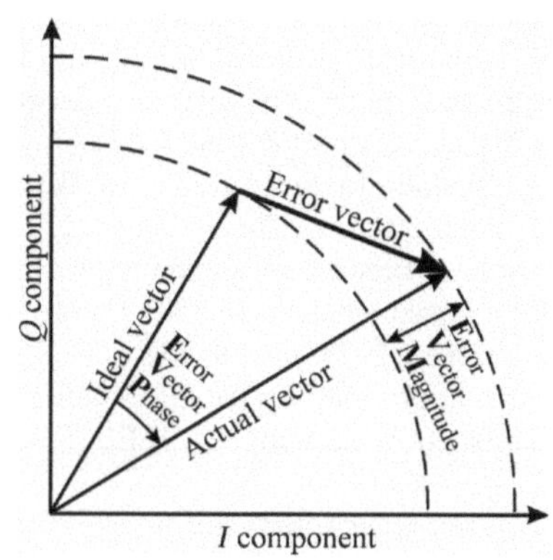

FIGURE 2.20

Error vector magnitude representation.

system. In fact, the data at PA input are digitally modulated, and from the PA output the base-band constellation is reconstructed.

It yields extremely important results since it directly reports the effects at base-band of the distortion introduced, for example, by PA nonlinearities. In fact, it can give straightforward insights into RF effects on system-level behavior and performance, such as providing time domain evaluation of base-band digital *I* and *Q* symbols.

REFERENCES

[1] M. Pirola, V. Teppati, V. Camarchia, Microwave measurements part I: linear measurements, IEEE Instrum. Meas. Mag. 10 (2) (2007) 14–19.

[2] R. Marks, D. Williams, A general waveguide circuit theory, J. Res. Natl. Inst. Stand. Technol. 97 (1992) 533–561.

[3] K. Kurokawa, Power waves and scattering matrix, IEEE Trans. Microw. Theory Techn. 13 (3) (1965) 194–202.

[4] D. Bockelman, M. Eisenstadt, Combined differential and common-mode scattering parameters: theory and simulation, IEEE Trans. Microw. Theory Techn. 43 (7) (1995) 1530–1539.

[5] A. Ferrero, M. Pirola, Generalized mixed-mode S-parameters, IEEE Trans. Microw. Theory Techn. 54 (1) (2006) 458–463.

[6] SPARQ Lecroy Signal Integrity Network Analyzer Datasheet, http://cdn.teledynelecroy.com/files/pdf/sparq_datasheet.pdf.

[7] A. Blankman, SPARQ parameter measurements with impulse response time limiting, Lecroy Technical Note, http://cdn.teledynelecroy.com/files/whitepapers/sparq_impulse_response_limiting_rev2.pdf.

[8] V. Teppati, M. Sayed, A. Ferrero, Modern RF and Microwave Measurement Techniques, Cambridge University Press, Cambridge, 2013.

[9] H.V. Shurmer, New method of calibrating a network analyzer, Electron. Lett. 6 (23) (1970) 733–734.

[10] E.F. Da Silva, M.K. McPhun, Calibration of microwave network analyzer for computer-corrected S parameter measurements, Electron. Lett. 9 (6) (1973) 126–128.

[11] H.V. Shurmer, Calibration procedure for computer-corrected S parameter characterization of devices mounted in microstrip, Electron. Lett. 9 (14) (1973) 323–324.

[12] S. Rehnmark, On the calibration process of automatic network analyzer systems, IEEE Trans. Microw. Theory Techn. 22 (4) (1974) 457–458.

[13] G.F. Engen, Calibration technique for automated network analyzers with application to adapter evaluation, IEEE Trans. Microw. Theory Techn. 22 (12) (1974) 1255–1260.

[14] N.R. Franzen, R.A. Speciale, A new procedure for system calibration and error removal in automated s-parameter measurements, in: 5th European Microwave Conference, Hamburg, Germany, September, 1975, pp. 69–73.

[15] V.G. Gelnovatch, A computer program for the direct calibration of two-port reflectometers for automated microwave measurements, IEEE Trans. Microw. Theory Techn. 24 (1) (1976) 45–47.

[16] G.F. Engen, C.A. Hoer, 'Thru-reflect-line': an improved technique for calibrating the dual six-port automatic network analyzer, IEEE Trans. Microw. Theory Techn. 27 (12) (1979) 987–993.

[17] B. Bianco, A. Corana, L. Gogioso, S. Ridella, M. Parodi, Open-circuited coaxial lines as standard for microwave measurements, Electron. Lett. 16 (10) (1980) 373–374.
[18] M. Roos, V. Sotoudeh, A measurement and calibration technique for accurate measurement of amplifier S parameters, in: IEEE International Microwave Symposium Digest, Las Vegas, USA, June, 1987, pp. 448–451.
[19] H.J. Eul, B. Schieck, Thru-match-reflect: one result of a rigorous theory for de-embedding and network analyzer calibration, in: 18th European Microwave Conference, Stockholm Sweden, September, 1988, pp. 909–914.
[20] R.A. Soares, P. Gouzien, P. Legaud, G. Follot, A unified mathematical approach to two-port calibration techniques and some applications, IEEE Trans. Microw. Theory Techn. 37 (11) (1989) 1669–1674.
[21] K. Silvonen, Calibration of test fixtures using at least two standards, IEEE Trans. Microw. Theory Techn. 39 (4) (1991) 624–630.
[22] H.J. Eul, B. Schieck, A generalized theory and new calibration procedures for network analyzer self-calibration, IEEE Trans. Microw. Theory Techn. 39 (4) (1991) 724–731.
[23] R. Marks, A multiline method of network analyzer calibration, IEEE Trans. Microw. Theory Techn. 39 (7) (1991) 1205–1215.
[24] H.J. Eul, B. Schieck, Reducing the number of calibration standards for network analyzer calibration, IEEE Trans. Instrum. Meas. 40 (4) (1991) 732–735.
[25] A. Ferrero, U. Pisani, QSOLT: a new fast calibration algorithm for two-port S-parameter measurements, in: 38th ARFTG Conference, San Diego, USA, December, 1991, pp. 15–24.
[26] K. Silvonen, A general approach to network analyzer calibration, IEEE Trans. Microw. Theory Techn. 40 (4) (1992) 754–759.
[27] A. Ferrero, U. Pisani, Two-port network analyzer calibration using an unknown 'thru', IEEE Microw. Guided Wave Lett. 2 (12) (1992) 505–507.
[28] K. Silvonen, Calibration of 16-term error model, Electron. Lett. 29 (17) (1993) 1544–1545.
[29] A. Ferrero, U. Pisani, An improved calibration technique for on-wafer large-signal transistor characterization, IEEE Trans. Instrum. Meas. 42 (2) (1993) 360–364.
[30] V. Adamian, A novel procedure for network analyzer calibration and verification, in: 41st ARFTG Conference, Atlanta, USA, June, 1993, pp. 8–17.
[31] D. Bockelman, W. Eisenstadt, Calibration and verification of the pure-mode vector network analyzer, IEEE Trans. Microw. Theory Techn. 46 (7) (1998) 1009–1012.
[32] A. Ferrero, F. Sanpietro, A simplified algorithm for leaky network analyzer calibration, IEEE Microw. Guided Wave Lett. 5 (4) (1995) 119–121.
[33] G. Madonna, A. Ferrero, U. Pisani, A simple NWA calibration algorithm based on a transfer standard, in: IEEE Instrumentation and Measurement Technology Conference, Ottawa, Canada, May, 1997, pp. 32–35.
[34] K. Silvonen, LMR 16 — a self-calibration procedure for a leaky network analyzer, Electron. Lett. 45 (7) (1997) 1041–1049.
[35] J. Horn, D.E. Root, G. Simpson, GaN device modeling with X-parameters, in: IEEE Compound Semiconductor Integrated Circuit Symposium, Monterey, USA, October, 2010, pp. 1–4.
[36] H. Qi, J. Benedikt, P.J. Tasker, Nonlinear data utilization: from direct data lookup to behavioral modeling, IEEE Trans. Microw. Theory Techn. 57 (6) (2009) 1425–1432.

[37] A. Raffo, G. Bosi, V. Vadalà, G. Vannini, Behavioral modeling of GaN FETs: a load-line approach, IEEE Trans. Microw. Theory Techn. 62 (1) (2014) 73–82.

[38] G. Crupi, A. Raffo, D.M.M.-P. Schreurs, G. Avolio, V. Vadalà, S. Di Falco, A. Caddemi, G. Vannini, Accurate GaN HEMT non-quasi-static large-signal model including dispersive effects, Microw. Opt. Tech. Lett. 53 (3) (2011) 692–697.

[39] P. Colantonio, F. Giannini, E. Limiti, V. Teppati, An approach to harmonic load-and-source-pull measurements for high-efficiency design, IEEE Trans. Microw. Theory Techn. 52 (1) (2004) 191–198.

[40] P. Zampardi, A.-V. Pham, Power amplifiers for wireless applications, in: IEEE International Microwave Symposium, Long Beach, USA, June, 2005, pp. 643–645.

[41] W. Chen, S. Zhang, Y. Liu, Y. Liu, F.M. Ghannouchi, A concurrent dual-band uneven Doherty power amplifier with frequency-dependent input power, IEEE Trans. Microw. Theory Techn. 61 (2) (2014) 552–561.

[42] G. Crupi, D. Schreurs, Microwave De-embedding: From Theory to Application, Academic Press, Oxford, UK, 2013.

[43] Y. Takayama, A new load-pull characterization method for microwave power transistor, in: IEEE MTT-S International Microwave Symposium, Cherry Hill, USA, June, 1976, pp. 218–220.

[44] J. Cusack, S. Perlow, B. Perlman, Automatic load contour mapping for microwave power transistors, IEEE Trans. Microw. Theory Techn. 22 (12) (1974) 1146–1152.

[45] G.P. Bava, U. Pisani, V. Pozzolo, Active load technique for load-pull characterization at microwave frequencies, Electron. Lett. 18 (4) (1982) 178–179.

[46] R. Lane, Source and load pull using tuner models, Microw. J. 3 (March) (1983) 149–156.

[47] F. Sechi, R. Paglione, B. Perlman, J. Brown, A computer controlled microwave tuner for automated load pull, RCA Rev. 44 (December) (1983) 566–583.

[48] R. Tucker, P. Bradley, Computer-aided error correction of large-signal load pull measurements, IEEE Trans. Microw. Theory Techn. 32 (3) (1984) 296–300.

[49] I. Hecht, Improved error-correction technique for large-signal load-pull measurements, IEEE Trans. Microw. Theory Techn. 35 (11) (1987) 1060–1062.

[50] F. Ghannouchi, R. Bosisio, Y. Demers, Load-pull characterization method using six-port techniques, in: IEEE Instrumentation and Measurement Technology Conference, Washington, DC, USA, April, 1989, pp. 536–539.

[51] F.M. Ghannouchi, R. Bosisio, Source-pull/load-pull oscillator measurements at microwave/mm wave frequencies, IEEE Trans. Instrum. Meas. 41 (1) (1992) 32–35.

[52] D.D. Poulin, J.R. Mahon, J.-P. Lanteri, A high power on-wafer pulsed active load pull system, IEEE Trans. Microw. Theory Techn. 40 (12) (1992) 2412–2417.

[53] J. Muller, B. Gyselinckx, Comparison of active versus passive on-wafer load-pull characterization of microwave and mm-wave power devices, in: IEEE MTT-S International Microwave Symposium, San Diego, USA, May, 1994, pp. 1077–1080.

[54] D. Le, F. Ghannouchi, Source-pull measurements using reverse six-port reflectomenters with application to MESFET mixer design, IEEE Trans. Microw. Theory Techn. 42 (9) (1994) 1589–1595.

[55] M. Demmler, B. Hughes, A. Cognata, A 0.5–50 GHz on-wafer, inter-modulation, load-pull and power measurement system, in: IEEE MTT-S International Microwave Symposium, Orlando, FL, May, 1995, pp. 1041–1044.

[56] C. Tsironis, System performs active load-pull measurements, Microw. RF (November) (1995) 102–108.

[57] ATN Microwaves, A load-pull system with harmonic tuning, Microw. J. 3 (March) (1996) 128–132.
[58] E. Gebara, J. Laskar, M. Harris, T. Kikel, Development of cryogenic load-pull analysis: power amplifier technology performance trends, in: IEEE MTT-S International Microwave Symposium, Baltimore, USA, June, 1998, pp. 1663–1666.
[59] D. Le, P. Poire, F. Ghannouchi, Six-port-based active source-pull measurement technique, Meas. Sci. Technol. 9 (1998) 1336–1341.
[60] Focus Microwaves, An affordable harmonic load pull setup, Microw. J. (October) (1998) 180–182.
[61] G. Berghoff, E. Bergeault, B. Huyart, L. Jallet, Automated characterization of HF power transistor by source-pull and multiharmonic load-pull measurements based on six-port techniques, IEEE Trans. Microw. Theory Techn. 46 (12) (1998) 2068–2073.
[62] G.L. Madonna, Measurement techniques for microwave device and circuit characterization (Ph.D. thesis), Politecnico di Torino, 1999.
[63] A. Ferrero, G. Madonna, U. Pisani, Recent technological advances for modular active harmonic load-pull measurement systems, in: GAAS99 Gallium Arsenide Applications Symposium, Munich, Germany, October, 1999, pp. 403–406.
[64] R. Doerner, P. Heymann, Harmonic tuning of power transistors by active load-pull measurements, Microw. J. (June) (2000) 22–37.
[65] J. Benedikt, R. Gaddi, P.J. Tasker, M. Goss, High-power time-domain measurement system with active harmonic load-pull for high-efficiency base-station amplifier design, IEEE Trans. Microw. Theory Techn. 48 (12) (2000) 2617–2624.
[66] A. Ferrero, V. Teppati, A. Carullo, Accuracy evaluation of on-wafer load-pull measurements, IEEE Trans. Microw. Theory Techn. 49 (1) (2001) 39–43.
[67] D. Barataud, M. Campovecchio, J.-M. Nebus, Optimum design of very high-efficiency microwave power amplifiers based on time-domain harmonic load-pull measurements, IEEE Trans. Microw. Theory Techn. 49 (6) (2001) 1107–1112.
[68] V. Camarchia, V. Teppati, S. Corbellini, M. Pirola, Microwave measurements — part II non-linear measurements, IEEE Instrum. Meas. Mag. 10 (3) (2007) 34–39.
[69] V. Teppati, A. Ferrero, V. Camarchia, A. Neri, M. Pirola, Microwave measurements — part III: advanced non-linear measurements, IEEE Instrum. Meas. Mag. 11 (6) (2008) 17–22.
[70] M. Marchetti, M. Pelk, K. Buisman, C. Neo, M. Spirito, L. de Vreede, Active harmonic loadpull with realistic wideband communications signals, IEEE Trans. Microw. Theory Techn. 56 (12) (2008) 2979–2988.
[71] A. Ferrero, M. Pirola, Harmonic load-pull techniques: an overview of modern systems, IEEE Microw. Mag. 14 (4) (2013) 116–123.
[72] Agilent vector signal analysis basics, Agilent application note 150–15, Agilent Technologies, Inc., November 21, 2012.
[73] M. Myslinski, K.A. Remley, D. Schreurs, B. Nauwelaers, RF vector measurement testbench for evaluation of behavioral model accuracy under realistic excitation, in: 68th ARFTG Conference, Broomfield, USA, November, 2006, pp. 26–31.
[74] Agilent I/Q modulation considerations for PSG vector signal generators, Agilent application note, Agilent Technologies, Inc., July 24, 2007.
[75] Agilent "X-Series Signal Generators N5171B/72B EXG N5181B/82B MXG Service Guide," Agilent Technologies, Inc., April 2014.
[76] J. Sha, Z. Liu, A virtual implementation of VSG & VSA, in: Internet Seventh International Conference on Computing for Engineering and Science (ICICSE), September, 2013, pp. 77–84.

[77] R. Quaglia, T. Jiang, V. Camarchia, Frequency extension of system level characterization and predistortion setup for on-wafer microwave power amplifiers, in: European Microwave Integrated Circuit Conference, Rome, Italy, October, 2014, pp. 488–491.
[78] RS FSV-K70 Vector Signal Analysis data-sheet, March 2012.
[79] RS CMW500 Wideband Radio Communication Tester Platform overview, December 2009.
[80] Ni pxie-5644r/5645r/5646r, March 2014. http://www.ni.com/white-paper/14027/en/.

CHAPTER

Nonlinear analysis and design of oscillator circuits

3

A. Suarez, S. Sancho and F. Ramirez
University of Cantabria, Santander, Spain

CHAPTER OUTLINE

Microwave Wireless Communications. http://dx.doi.org/10.1016/B978-0-12-803894-9.00003-2

3.1 INTRODUCTION

This chapter will cover all the main aspects of oscillator analysis and design. It is organized as follows. In the first section, essential concepts are presented, such as the mechanism for the oscillation start-up from an unstable *dc* solution [1–4], the coexistence of the oscillatory steady-state solution with the *dc* solution, and the invariance of the periodic steady-state solution under time shifts, which is the origin of the phase-noise problem. Then a harmonic-balance methodology for nonlinear oscillator design is described. Departing from initial values estimated in a small-signal analysis, it is based on the use of an auxiliary (artificial) generator [5,6] to preset relevant characteristics of the steady-state solution, such as oscillation frequency and output power. Stability analysis is fundamental in oscillator circuits, because two different steady-state solutions, *dc* and periodic, necessarily coexist. One must ensure the physical existence (stability) of the steady state and in most cases, guarantee the oscillation start-up from the noise level. When considering the continuous variation of a tuning parameter, the oscillation parameter interval will be delimited by bifurcations or qualitative stability changes [2,3,7,8]. Several examples of these phenomena occurring in the *dc* solution and in the periodic oscillatory regime will be presented throughout the chapter. Another essential characteristic of free-running oscillators is phase noise [9–11], coming from the invariance versus phase shifts, due to the absence of time-varying independent sources. In the chapter, three major methodologies for phase-noise analysis will be described, discussing their advantages and limitations. Finally, a procedure to extract a reduced-order model of the oscillator circuit from harmonic-balance simulations will be presented [12,13], with application examples in coupled-oscillator systems and phase-locked loops (PLLs).

3.2 BASIC CONCEPTS IN OSCILLATOR CIRCUITS

3.2.1 OSCILLATION MECHANISM: START-UP AND STEADY-STATE

A free-running oscillator provides a time-varying solution from *dc* sources only, unlike other circuits, such as amplifiers or mixers, which are excited with time-varying signals. In most cases, the oscillator solution will be periodic at the fundamental frequency ω_o although other types of solutions may be possible in some circuits, which are generally unwanted. Some examples are quasiperiodic regimes, having two or more independent fundamental frequencies, without a rational relationship

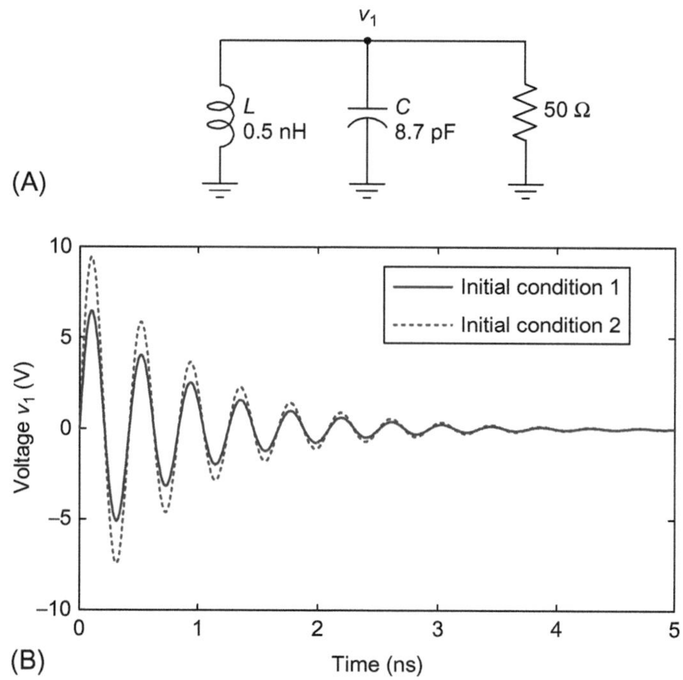

FIGURE 3.1

Passive linear oscillator. (A) Schematic of a linear *R-L-C* resonator. (B) Transient response for two different initial conditions.

[2,3], or chaotic regimes, which are non-periodic and thus exhibit a continuous spectrum in some frequency bands [2,3,7].

Harmonic oscillators, also called LC (inductor-capacitor) oscillators, provide a near-sinusoidal waveform, at a frequency close to a natural circuit resonance. Under any small perturbation, a linear *R-L-C* resonator responds with an oscillatory transient at the resonance frequency. As an example, see the linear circuit in Figure 3.1A and its transient response for two different initial values in Figure 3.1B. Energy oscillates between the capacitor and the inductor, but the waveform amplitude progressively decays as the oscillation energy is transformed into heat by the resistor R. As seen in Figure 3.1B, whatever the initial value, there will be an exponential decay ($e^{\sigma t}$, where $\sigma < 0$) of the oscillation amplitude, due to the energy consumed by the resistor R.

Unlike the above case, in a free-running oscillator circuit, the oscillatory waveform must grow from the noise level and reach a steady state of constant amplitude and frequency. As an example, see the circuit in Figure 3.2A and its transient response in Figure 3.2B. This circuit reacts to perturbations with a transient leading to a steady-state oscillation of constant amplitude V_o and frequency ω_o. These two magnitudes will be the same no matter the particular value of initial condition. The

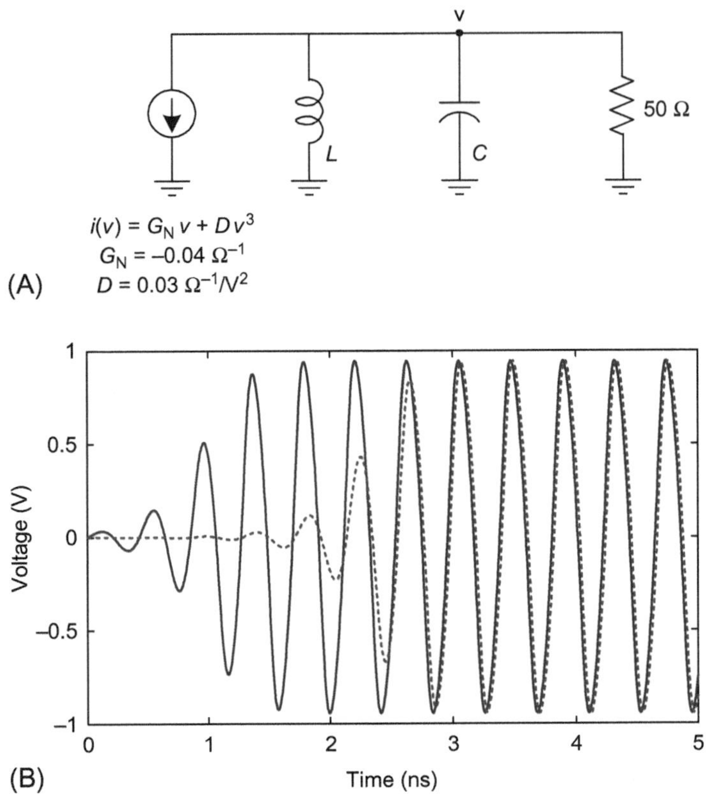

FIGURE 3.2

Free-running oscillator. (A) Schematic of a nonlinear oscillator with a cubic nonlinearity. (B) Transient response leading to a steady-state oscillatory regime for $C=2.9$ pF and $L=1.5$ nH. The oscillation frequency is $f_o=2.4$ GHz.

initial section of the transient has an exponential envelope ($e^{\sigma t}$, where $\sigma > 0$) and the second section corresponds to a transition to the constant amplitude value V_o, which is sometimes modeled with a decreasing positive function $\sigma(V)$, where V is the time-varying amplitude, and $\sigma(V_o)=0$.

To obtain a self-sustained oscillation, such as the one in Figure 3.2B, the circuit must fulfill certain conditions, in addition to the resonance of two reactive elements. The initial values will generally come from noise perturbations, so the circuit must exhibit gain to enable the growth of the oscillating signal from the noise level. In order for the amplitude to stop growing, the circuit gain must be nonlinear and thus sensitive to the oscillation amplitude. In agreement with these two requirements, a classical model for an oscillator circuit consists of a closed loop with a nonlinear gain and a resonant feedback block [14]. The positive feedback enables the signal growth at the resonance frequency until the gain of the nonlinear amplifier saturates and the system reaches the steady state. At circuit level, it is usually more convenient to use a model based on negative resistance [1,4], because it enables a straight forward description of the circuit components in terms of impedance or admittance functions. In fact, a negative-resistance element acts like an energy source. Indeed, positive resistances give rise to energy dissipation and negative resistances deliver energy to the circuit. To understand this, one may note that, by Kirchhoff laws, any independent generator connected to a passive circuit with input impedance $R+jX$ will necessarily exhibit the opposite impedance $-R-jX$. In the oscillator circuit, we do not have any independent generators at the oscillation frequency, but we have *dc* sources and the negative resistance $-R$ is obtained from the energy delivered by these *dc* sources. In fact, no physical device can provide negative resistance without a *dc* power supply.

The circuit in Figure 3.2A is an idealization, as it has no *dc* sources. It contains a voltage-controlled current source with the characteristic $i(v)=G_N v+Dv^3$, such that $G_N<0$ and $D>0$. The coefficient G_N agrees with the small-signal conductance exhibited by the controlled source, which is obtained through the derivative $G_N=\partial i(v)/\partial v$, evaluated at $v=0$. Thus, we have a negative-resistance element. Because there is no time-varying input generator, this circuit can obviously be resolved for a *dc* solution. This property, associated to the absence of input sources, is common to all the free-running oscillator circuits: they can all be resolved for a *dc* solution. The oscillation should start-up from this *dc* solution and should grow in amplitude until reaching the steady-state regime, in which the circuit operates nonlinearly. By linearizing the circuit about the *dc* solution, we can only check for the possible fulfillment of the oscillation start-up conditions.

In small-signal, the negative resistance must exceed the positive one exhibited by the load. This is because the initial value (due to noise perturbations) will be too small and the signal must be amplified. The energy excess (more energy delivered than dissipated by the load) will be employed in the growth in amplitude of the oscillatory signal. Because of the parallel connection of the four elements in the circuit in Figure 3.2A, to have an excess of negative resistance in small-signal, the absolute value $|G_N|$ must be larger than the value of the positive conductance $G_L=1/R_L$. To understand this intuitively, note that the higher the impedance of an element connected in parallel, the smaller the influence of this element on the circuit to which it is connected. In fact, the circuit in Figure 3.2A, with a parallel resonance, is better analyzed in terms of admittance functions, as the voltage variable is common to all the circuit elements. This small-signal analysis is presented in the following.

After introduction of a small perturbation, the circuit is no longer in its *dc* solution. It is not in steady state, so it should be analyzed in terms of the complex frequency s. The small-signal solution will be assumed to have the form Ve^{st}, where the complex coefficient V has a small amplitude value. Through the application of Kirchhoff's laws at the frequency s, one obtains:

$$\left(G_N+G_L+Cs+\frac{1}{Ls}\right)V=0 \tag{3.1}$$

The above system is an homogeneous system, which admits a zero solution, $V=0$, this corresponding to the *dc* solution at which the circuit operates before undergoing any perturbation. To have $V \neq 0$, the homogeneous system must fulfill:

$$s^2 + \frac{(G_L + G_N)}{2C} s + \frac{1}{LC} = 0 \tag{3.2}$$

This is the characteristic polynomial associated to the oscillator in Figure 3.2A. The roots of this polynomial provide the circuit natural frequencies, which are given by:

$$s = -\frac{(G_L + G_N)}{2C} \pm \frac{1}{2}\sqrt{\frac{(G_L + G_N)^2}{4C^2} - \frac{4}{LC}} \tag{3.3}$$

In order to have a near-sinusoidal oscillation of growing amplitude, two conditions must be fulfilled. The radicand must be negative to have an imaginary part $j\omega_o$ and the first term must be positive to have a growing envelope. We must be able to express the two roots as $s = \alpha \pm j\omega_o$, where:

$$\alpha = -\frac{(G_L + G_N)}{2C} > 0 \tag{3.4a}$$

$$\frac{(G_L + G_N)^2}{4C^2} - \frac{4}{LC} < 0 \tag{3.4b}$$

Provided that $\frac{(G_L + G_N)^2}{4C^2} \ll \frac{4}{LC}$, the oscillation frequency will be close to $\frac{1}{\sqrt{LC}}$, which is the resonance frequency of the inductor and capacitor. On the other hand, the circuit must exhibit a dominant negative conductance $G_L + G_N < 0$ for the oscillation amplitude to grow exponentially from the noise level. Note that if the positive and negative conductances cancel each other out $G_L + G_N = 0$, there is no possibility of achieving a self-sustained oscillation. With very low quality factors, such that $C \ll L$, the characteristic polynomial in Eq. (3.2) will exhibit two real roots. For two positive real roots, there will also be an oscillation. However, this oscillation will not be harmonic, but of a relaxation type [7]. In a relaxation oscillator, the system repeatedly alternates between two states with a period $T_o = 1/f_o$, where $f_o = \omega_o/(2\pi)$, depends on the charging of a capacitor. Relaxation oscillators provide non-sinusoidal waveforms and, in general, offer a worse performance at high frequencies, due to higher phase noise (see Section 3.8).

The characteristic polynomial (3.2), under the conditions (3.4a), (3.4b) predicts a solution with an exponentially growing amplitude, of the following form:

$$v(t) = \mathrm{Re}\left[V_i e^{(\alpha + j\omega_o)t}\right] \tag{3.5}$$

where the complex magnitude V_i depends on the initial conditions at $t = t_0$. There is a clear problem in Eq. (3.5). The amplitude in Eq. (3.5) grows unboundedly, and this is different from what we observe in physical oscillators, which exhibit a fixed steady state amplitude and frequency. The problem comes from the linear analysis performed. Indeed, the negative conductance (resistance) of any physical device must eventually decrease with the voltage (current) amplitude across its terminals, due to the finite power consumption from the *dc* sources. Therefore, it must necessarily be nonlinear. In order to predict the oscillation steady state, we must take this nonlinearity into account.

Before performing the full nonlinear analysis, there is another piece of information that can be gathered from Eq. (3.3). This is the speed of the initial transient response, which depends on the real part α of the complex frequency s, given by:

$$\alpha = -\frac{(G_L + G_N)}{2C} \tag{3.6}$$

We will obviously have a quicker initial transient for a higher positive value of α. Assuming a same value of the resonant frequency $\omega_o = 1/\sqrt{LC}$, we will have a quicker transient for a smaller C, that is, for a smaller quality factor of the parallel resonant circuit. This reduction of the quality factor leads to a higher impact of the first term of the square root in Eq. (3.3), so the oscillation frequency can substantially deviate from the resonance frequency $\omega_o = 1/\sqrt{LC}$. The steady-state waveform will also deviate more from a sinusoidal waveform due to a higher harmonic content. This is illustrated in the two simulations of Figure 3.3, comparing the start-up and steady-state waveforms for the two capacitor values $C_1 = 8.5$ pF and $C_2 = 1.3$ pF. In the two cases, the inductance value has been chosen to fulfill $L = 1/(C\omega_o^2)$ at $f_o = 2.4$ GHz. The voltage-controlled current source and linear conductance G_L are the same in the two cases. The transient is quicker for the smaller value C_2 and the waveform deviates more from a sinusoidal one.

The steady-state oscillatory waveform can only be predicted through a nonlinear analysis of the oscillator circuit, taking into account the full instantaneous characteristic of the nonlinear element, providing negative resistance. The cubic nonlinearity considered in Figure 3.2A, having $G_N < 0$ and $D > 0$ enables the prediction of the negative conductance saturation. As an example, a sinusoidal voltage will be introduced into this nonlinear element, $v(t) = V\cos\omega t$, where $V \in R$. Next, we will calculate the ratio between the *first harmonic* component of the current circulating through this element I_1 and the amplitude V. This provides the describing function $G_D(V)$ that models the variation of the nonlinear conductance under assumption of a sinusoidal voltage and current. The result is the following: $G_D(V) = G_N + \frac{3}{4}DV^2$. Due to the positive sign of D, the nonlinear conductance decreases with the excitation amplitude V, as eventually expected in any physical device.

For an indepth study of the nonlinear dynamics of the circuit in Figure 3.2A, this circuit will be formulated in time domain. The number of state variables in the time-domain analysis of a given circuit agrees with the number of reactive elements existing in this circuit [2,15]. The usual choice for these variables corresponds to the

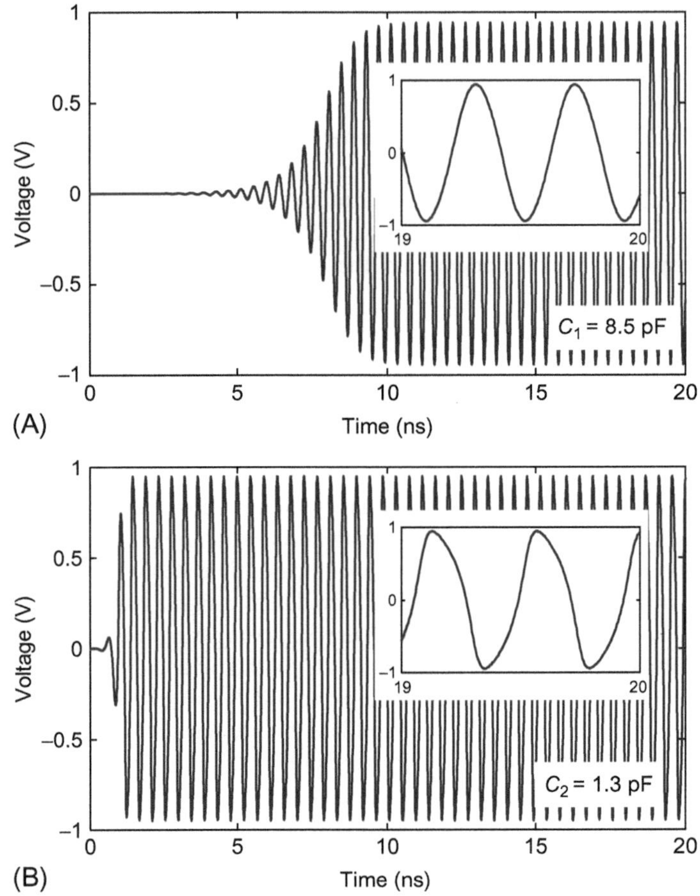

FIGURE 3.3

Transient response for the circuit of Figure 3.2A for $\omega_0 = 2\pi$ (2.4 GHz). (A) Start-up transient obtained for $C_1 = 8.5$ pF, and (B) start-up transient obtained for $C_2 = 1.3$ pF.

currents through the inductors and the voltages across the capacitors. The circuit in Figure 3.2A has two reactive elements, so there will be two state variables: the inductor current $i_L(t)$ and the capacitor voltage $v(t)$. Applying Kirchhoff's laws while taking the nonlinearity into account, one obtains the system:

$$\frac{dv}{dt} = -\frac{G_L}{C}v(t) - \frac{i_L(t)}{C} - \frac{i(v)}{C} \tag{3.7a}$$

$$\frac{di_L}{dt} = \frac{1}{L}v(t) \tag{3.7b}$$

The above system has the general form $\dot{\bar{x}}(t) = \bar{f}(\bar{x})$, where $\bar{x}(t)$ is the state-variable vector, the dot means time differentiation, and $\bar{f}$ is a vector function, whose components agree with the terms on the right-hand side. Note that in the example of Eqs. (3.7a),

(3.7b), the state-variable vector is given by $\bar{x}(t)=[v(t),i_L(t)]^T$, where T indicates transpose. The so-called *state-form* formulation $\dot{\bar{x}}(t)=\bar{f}(\bar{x})$ is not possible for all circuits. A more general formulation is the one derived from the so-called *modified nodal analysis* [15] briefly described in Section 3.6. Before continuing, one should point out a relevant property of free-running oscillators: the absence of explicit time dependence in the functions determining the time derivative $\dot{\bar{x}}(t)$. In a free-running oscillator, the function $\bar{f}(\bar{x})$ does not depend explicitly on time, which is due to the absence of any independent time varying sources, as in the case of Eqs. (3.7a), (3.7b). To illustrate the opposite case, assume that an independent current generator $i_g(t)$ is connected in parallel to the circuit in Figure 3.2A. When deriving the circuit equations, this will require the addition of the term $-i_g(t)/C$ to the right-hand side of Eq. (3.7a), which will gives rise to an explicit time dependence in the state-form equations: $\dot{\bar{x}}(t)=\bar{f}(\bar{x},t)$. Systems without an explicit time dependence $\dot{\bar{x}}(t)=\bar{f}(\bar{x})$, are called *autonomous*. Systems with an explicit time dependence are called *nonautonomous* or *forced*.

To understand the mechanism for the self-sustained oscillator, we will perform a time differentiation on both sides of Eq. (3.7a) and then substitute the derivative di_L/dt, calculated from Eq. (3.7b).

$$\frac{d^2v(t)}{dt^2}+\left(\frac{G_L}{C}+\frac{1}{C}\frac{di}{dv}\right)\frac{dv(t)}{dt}+\frac{1}{LC}v(t)=0 \tag{3.8}$$

Considering the particular nonlinear function $i(v)=G_Nv+Dv^3$ one obtains:

$$\frac{d^2v(t)}{dt^2}+\left(\frac{G_L+G_N}{C}+\frac{3}{C}Dv^2\right)\frac{dv(t)}{dt}+\frac{1}{LC}v(t)=0 \tag{3.9}$$

The term affecting the time derivative is called *damping term*, as it is the term that determines the variations of the oscillation envelope. In a free-running oscillator, this term must be nonlinear, as will be reasoned in the following. Assume the opposite, that is, $D=0$ in Eq. (3.9). One obtains the linear equation:

$$\frac{d^2v(t)}{dt^2}+\left(\frac{G_L+G_N}{C}\right)\frac{dv(t)}{dt}+\frac{1}{LC}v(t)=0 \tag{3.10}$$

which is the time-domain equivalent of Eq. (3.1). Actually replacing V_ie^{st} into Eq. (3.1) and eliminating the common exponentials, one obtains Eq. (3.1). Taking Eq. (3.6) into account, the *damping term* (affecting the time derivative) agrees with 2α, where α is the real part of s. For a positive damping term (excess positive conductance), $\alpha<0$ the oscillation amplitude will decay to zero for any initial condition (any value of V). For a negative damping term, $\alpha>0$, energy is delivered to the system and the oscillation amplitude grows exponentially. The first case does not enable the start-up of the oscillatory solution and the second case is unphysical and does not explain the steady-state oscillation.

The situation is different when the full nonlinear model $i(v)=G_Nv+Dv^3$ is considered, as in the original Eq. (3.9), where the damping term is nonlinear. Assuming the fulfillment of the oscillation start-up conditions (3.4a), (3.4b) and $D>0$, this equation enables a self-sustained periodic oscillation $v_o(t)$, such that $v_o(t)=v_o(t+T_o)$. In sections of the steady-state waveform with small magnitude

$|v_o(t)|$, the damping term is negative and the system delivers energy to the signal. In sections with large $|v_o(t)|$, the damping term becomes positive and energy is dissipated, which prevents the unbounded growth of the oscillation amplitude. This is the mechanism for the self-sustained steady-state oscillation, which is only possible in a nonconservative system, that is, in systems that contain dissipative elements, such as the conductance G_L in the circuit of Figure 3.2A.

The analysis in Eqs. (3.1)–(3.9) is in good with the well-known conditions for oscillation start-up, which require an excess of negative resistance at the desired oscillation frequency ω_o, which should agree with a dominant circuit resonance. Analyzing the circuit from a sensitive observation node, in terms of the total input admittance $Y_T = G_T(V,\omega_o) + jB_T(V,\omega_o)$, these frequency-domain conditions would be written in a general manner as [1]:

$$\begin{aligned} &G_T(V \cong 0, \omega_o) < 0 \\ &B_T(\omega_o) = 0 \\ &\frac{\partial B_T(\omega_o)}{\partial \omega} > 0 \end{aligned} \tag{3.11}$$

where V is the voltage amplitude and $\omega_o = 2\pi f_o$. Note that because we are taking the frequency as a real value (instead of the complex frequency s) and the oscillation steady state is not small-signal, we do not have an equation system in Eq. (3.11) but a set of mathematical conditions. The first condition indicates energy excess and the second and third conditions indicate a physical parallel resonance at ω_o. Particularizing conditions (3.11) to the circuit in Figure 3.2A, these conditions become:

$$\begin{aligned} &G_N(V \cong 0) + G_L < 0 \\ &jC\omega_o - \frac{j}{L\omega_o} = 0 \end{aligned} \tag{3.12}$$

When the voltage amplitude increases (and the circuit is no longer in small-signal regime), the negative conductance in Eq. (3.11) must decrease. If this negative conductance is kept at a constant value, the amplitude of the oscillatory signal would grow unboundedly, as deduced from Eq. (3.5). At a certain voltage amplitude V_o, there will be a perfect balance between power delivered and consumed at the resonance frequency.

Assuming a sinusoidal voltage waveform: $v(t) = \mathrm{Re}\left[V_o e^{j\omega_o t}\right]$ and applying Kirchhoff's laws, the steady-state equations can be approached as:

$$\begin{aligned} &G_T(V_o)\,V_o = (G_N(V_o) + G_L)V_o = 0 \\ &B_T(V_o,\omega_o)\,V_o = \left(jC\omega_o - \frac{j}{L\omega_o}\right)V_o = 0 \end{aligned} \tag{3.13}$$

The above steady-state equation is an approximation, because the nonlinearity of $G_N(V)$ will necessarily give rise to some harmonic content, and the additional frequencies $k\omega_o$, where k is an integer, have not been taken into account in Eq. (3.13). When considering these harmonic frequencies, the oscillation frequency will deviate from the original value ω_o, as will be shown later in this section. But two other essential aspects of the oscillator behavior, reflected by Eq. (3.13) will be discussed first.

System (3.13) is homogeneous and provided Eq. (3.12) is fulfilled, it admits two solutions. First solution is $v(t) = \text{Re}\left[V_o e^{j\omega_o t}\right]$, with $V_o \neq 0$, and the second solution is $V_o = 0$. For this second solution, the circuit does not oscillate. In our simplistic model, it corresponds to a *dc* solution of zero value. In a realistic oscillation model, it would correspond to a *dc* solution depending on the device biasing and resistive loading. This non-oscillatory solution will exist in any free-running oscillator design and is due to the homogeneity of the circuit equations, resulting from the absence any time-varying sources. This *dc* solution is instrumental to achieve the oscillation start-up [4]. Actually, when initially perturbed by noise signals, the circuit, which is only fed by *dc* sources, operates at this *dc* solution. For the oscillation start-up, the *dc* solution must be able to react to perturbations and give rise to an oscillatory transient of growing amplitude. Equivalently, the *dc* solution must be unstable, or non-robust against perturbations, with two complex-conjugate roots of its associated characteristic polynomial (see Eq. (3.2) as an example) having the form $\alpha \pm j\omega_o$, where $\alpha > 0$. Therefore, the unstable *dc* solution is responsible for the oscillation start-up.

As an example, particularizing Eq. (3.13) to the circuit in Figure 3.2A with a fundamental frequency representation of the nonlinear element (describing function), given by $G_D(V) = G_N + \frac{3}{4}DV^2$, one obtains the following steady-state system:

$$\begin{aligned} G_N + \frac{3}{4}DV_o^2 + G_L &= 0 \\ C\omega_o - \frac{1}{L\omega_o} &= 0 \end{aligned} \tag{3.14}$$

The above system provides the following one-harmonic approximation of the oscillation amplitude and frequency:

$$V_o = \sqrt{\frac{-G_L - G_N}{(3D/4)}} \tag{3.15a}$$

$$\omega_o = \frac{1}{\sqrt{LC}} \tag{3.15b}$$

Note that the above relationship will only be valid for a sufficiently large quality factor, such that the harmonic content has little impact on the oscillation frequency and waveform. As gathered from Eq. (3.15a), the oscillation amplitude will be smaller for smaller difference between the magnitudes $|G_N|$ and G_L. No oscillation is possible for $G_L > |G_N|$, because this gives rise to a negative radicand. Therefore, the boundary for the oscillation start-up condition is given by $G_N + G_L = 0$.

Small perturbations of an unstable solution give rise to a transient, such that the system escapes from this solution, so they cannot be physically observed. In a well-behaved free-running oscillator, small perturbations of the *dc* solution give rise to a transient that leads to the steady-state oscillation. This oscillation must be stable, meaning that it must be *robust* against small perturbations, that is, able to recover itself from these perturbations. The *dc* solution and the steady-state oscillation coexist for the same values of the circuit elements, but they have different qualitative stability properties. The stability analysis of the two solutions will be covered later in this chapter.

3.2.2 INVARIANCE VERSUS PHASE SHIFTS

In Eq. (3.13) the solution phase has been assumed to be zero, that is, $V_o e^{j\phi_o} = V_o$. However, one could have assumed any other phase value $\phi_o \neq 0$ and an equally valid solution would have been obtained. This is because the *current-to-voltage* ratio of a nonlinear element at the first harmonic term does not depend on the phase ϕ of the excitation voltage signal [4]. Expressing any possible phase shift in terms of the excitation frequency ω and a delay τ, as $\phi = -\omega\tau$, this is demonstrated as follows:

$$\begin{aligned}Y(V,\omega) &= \frac{\int_0^T i[v(t-\tau)]e^{-j\omega t}\,dt}{Ve^{-j\omega\tau}} = \frac{\int_0^T i[v(t-\tau)]e^{j\omega\tau}e^{-j\omega\tau}e^{-j\omega t}\,dt}{Ve^{-j\omega\tau}} \\ &= \frac{\int_0^T i[v(t-\tau)]e^{-j\omega\tau}e^{-(j\omega t-\tau)}\,dt}{Ve^{-j\omega\tau}} = \frac{e^{-j\omega\tau}\int_0^T i[v(t')]e^{-j\omega t'}\,dt}{Ve^{-j\omega\tau}} \\ &= \frac{\int_0^T i[v(t')]e^{-j\omega t'}\,dt}{V}\end{aligned} \quad (3.16)$$

Equivalently, the *voltage-to-current* ratio does not depend on the phase of the current excitation signal. Taking this into account, Eq. (3.13) would be fulfilled for any phase value ϕ_o, which is due to the circuit autonomy, or absence of an input source establishing the phase origin. To compare, in an amplifier circuit, the solution is defined in terms of the amplitude and phase shift with respect to the input source. The phase of this input source can be anyone but the phase difference with respect to this source is fully defined by the circuit equations. In a free-running oscillator there is an absence of a time-varying source establishing the phase reference. Because of this, the steady-state waveform $v(t)$ can be arbitrarily translated in time, $v(t-\tau)$.

The invariance versus time shifts is the main cause of the phase-noise problem in oscillator circuits [9–11]. In these circuits, any arbitrary phase shift of the steady-state solution provides another valid solution, so, under noise perturbations, there is no restoring mechanism in the phase variable. Perturbations in phase accumulate, according to certain statistical rules, unlike what happens in forced circuit. The phase noise is the dominant contribution to the oscillator noise spectrum, and the phase modulation gives rise to the spreading of the power spectral density (PSD), unlike the distinct spectral lines (at the fundamental frequency and harmonic components $k\omega_o$) obtained in a noiseless oscillator analysis.

3.2.3 IMPACT OF THE HARMONIC CONTENT

As already stated, the oscillator circuit, due to its inherent nonlinear behavior, will necessarily exhibit some harmonic content. Even in a circuit as simple as the one in Figure 3.2A, the consideration of the harmonic content will give rise to differences in the calculated steady-state solution, which will not only affect the voltages and currents, but also the oscillation frequency ω_o. To understand this, one must take into account that in an oscillator circuit, the oscillation frequency behaves essentially like a state variable, as it depends on the values of the circuit elements and parameters. Even in the simple example of Eq. (3.12), the oscillation frequency will vary under any modification of L and C. Taking this property into account, the oscillation frequency ω_o will depend on the number of harmonics considered in the analysis, in a manner similar to the voltages and currents.

To see this more clearly, we will take the example of Figure 3.2A. Let several harmonic components be assumed in the voltage signal, which will be expressed in a Fourier series as $v(t) = \sum_{k=-N}^{N} V_k e^{jk\omega t}$, where V_k are complex numbers. Introducing this signal into the nonlinear current, $i[v(t)]$ and using a Fourier representation, one will have: $i(t) = \sum_{k=-N}^{N} I_k e^{jk\omega t}$. Next, current-to-voltage functions at the various harmonic terms will be defined:

$$Y_k = \frac{I_k(V_{-N}, \ldots, V_{-1}, V_0, V_1, \ldots, V_N)}{V_k}, \quad \text{where } k = -N \text{ to } N \tag{3.17}$$

Due to the nonlinearity of the current, each function Y_k will depend on all the voltage harmonic terms V_l, where $l = -N \ldots N$. Even though the phase of the fundamental harmonic component of the voltage signal V_1 can be arbitrarily set to zero, $\phi_o = 0$, so that $V_1 = |V_1| e^{j0}$, the first harmonic current will be complex, so we will have $Y_1 = Y_1^r + jY_1^i$, where the superindexes stand for real and imaginary parts, and the same will be true for all the rest of admittance functions. Just intuitively, the imaginary part Y_1^i will make the oscillation frequency differ from the linear resonant frequency $1/\sqrt{LC}$ [4].

Taking all the harmonic components into account, the circuit can be formulated:

$$\begin{aligned}
&[Y_{-N}(V_{-N}, \ldots, V_{-1}, V_0, V_1, \ldots, V_N) + Y_L(-N\omega_o)]V_{-N} = 0 \\
&\vdots \\
&[Y_1(V_{-N}, \ldots, V_{-1}, V_0, V_1, \ldots, V_N) + Y_L(\omega_o)]V_1 = 0 \\
&\vdots \\
&[Y_N(V_{-N}, \ldots, V_{-1}, V_0, V_1, \ldots, V_N) + Y_L(N\omega_o)]V_N = 0
\end{aligned} \tag{3.18}$$

As in the case of the single-harmonic analysis in Eq. (3.13), the above higher-dimension system is homogeneous and, therefore, admits a zero solution $V_k = 0, \ \forall k \neq 0$. This is the circuit *dc* solution, which in this case fulfills $V_0 = 0$

due to the absence of bias sources. This nonlinear steady-state solution can only be obtained numerically. In fact, system (3.18) corresponds to the *harmonic-balance* formulation of the oscillator circuit in Figure 3.2A.

The harmonic-balance system is a nonlinear algebraic system [16–18] that is solved through an error-minimization technique, of the Newton-Raphson type. This technique is sensitive to the initial value. In a forced circuit, the input source at a frequency ω_{in} provides an initial value to the circuit variables at this frequency component, and the nonlinear terms naturally give rise to the harmonic content. In an oscillator, there are no sources at the fundamental frequency ω_o, so the error-minimization method will converge, by default, to the non-oscillatory *dc* solution. To understand the problem, one must take into account that the harmonic-balance system results from a particularization of the time-domain equations to a specific type of waveform, corresponding to the Fourier series $v(t) = \sum_{k=-N}^{N} V_k e^{jk\omega t}$. This representation prevents the analysis of transients, so only steady states can be analyzed, without any information on how the particular steady reacts to perturbation. Indeed the perturbation gives rise to a transient waveform that cannot be represented with the Fourier series considered. In order to check for the physical existence of the solution, one must perform a complementary stability analysis.

Obtaining the steady-state oscillatory regimes requires complementary technique, thus the need for "Osc ports" and special boxes in harmonic-balance simulators. To put it as clearly as possible, in system (3.18) the complex steady-state oscillatory solution is in competition with a very simple *dc* solution. Unless a proper initial value, in terms of the oscillation frequency and amplitude, is provided, the error-minimization technique will converge towards the *dc* solution. One must note that unlike the case of forced circuits, the initial value of the free-running oscillator should involve a good estimation of the steady-state oscillation frequency ω_o. In fact, the oscillation frequency is an unknown of the harmonic-balance system, in a manner similar to the one-harmonic analysis in Eq. (3.13), for which this frequency ω_o is also an unknown. In system (3.13) phase does not appear as a variable due to the invariance property that has already been discussed, so we have a mixed system (voltage and frequency variables) that is well balanced, this meaning that we have the same number of equations and unknowns. In an analogous manner, we can set to zero the phase of one of the harmonics in system (3.18), which will allow us to solve for the oscillation frequency ω_o and all the harmonic amplitudes and phases defining the oscillation waveform. Note that we can leave one phase out due to the invariance of the oscillation waveform under time translations.

Table 3.1 present the values of the steady-state oscillation frequency obtained by solving the circuit in Figure 3.2A with the harmonic-balance method and different numbers of harmonic components N. The two cases of a high quality factor C_1 and a lower quality factor C_2 have been considered, for comparison. As expected, the oscillation frequency is more sensitive to the number of harmonic terms for a lower quality factor. From a certain number N, an increase of this number has no effect on the solution, due to the negligible values of the additional harmonic components.

Table 3.1 Variation of the Oscillation Frequency with the Number of Harmonic Terms

	Frequency (GHz)	$N=1$	$N=3$	$N=5$	$N=7$
$C=8.7$ pF $L=0.5$ nH	f_o	2.4131	2.3999798	2.3999798	2.3999798
$C=0.87$ pF $L=5$ nH	f_o	2.4131	2.3979834	2.3979841	2.3979841

3.2.4 PHASE-SPACE REPRESENTATION

The *phase-space* representation [2,3] enables a useful graphical analysis of the circuit solutions, intended for a better comprehension of the behavior of circuits simulated in *time domain*. The phase space is defined by the system state variables, so that each state variable is assigned to one axis. In the case of forced circuits (containing time-varying sources), the set of state variables should include *time* or a time-related value, such as the phase $\theta(t)=\omega_{in}t$, where ω_{in} is the input source frequency, since we have an explicit time dependence $\bar{f}(\bar{x},t)$ in the nonlinear differential equation system $\dot{\bar{x}}=\bar{f}(\bar{x},t)$ [2]. In free-running circuits, time is not a state variable, because we have a function of the form $\bar{f}(\bar{x})$. Note that the phase-space representation will always be limited to a maximum of three variables for obvious reasons. Despite this limitation, the representation usually captures the main properties of the geometry of the system solutions and its dynamics, or evolution in time.

In the case of the two dimensional system in Eqs. (3.7a), (3.7b), we will consider the two state variables $v(t)$, $i_L(t)$. The phase-space representation is obtained by tracing $i_L(t)$ versus $v(t)$, with time as an implicit variable. Results are shown in Figure 3.4A. In this representation, we can distinguish two steady-state solutions: a point, named *equilibrium point* (EP), which corresponds to the *dc* solution and a closed trajectory, named *limit cycle*, which corresponds to the steady-state oscillation. There are also two open trajectories leading from particular initial values to the stable steady-state oscillation.

In a general manner, steady-state solutions give rise to bounded sets in the phase-space representation, termed *limit sets*. Examples of limit sets are the EP (*dc* solution) or *limit cycle* (periodic oscillation). In fact, a periodic solution will give rise to a closed trajectory as the system must return to the same point of the space after one period. A self-sustained oscillation will give rise to a *limit cycle*, which is a closed trajectory that is *isolated* in the phase space, in the sense that there cannot be other closed trajectory at an arbitrarily small distance. This is different from what happens in the case of an LC circuit without any resistances. When changing the initial conditions in such a simple circuit we will obtain different closed trajectories, which may be arbitrarily close to each other. For illustration, Figure 3.4B presents the simulations of an LC linear circuit (without any resistors), having $L=0.5$nH and $C=8.7$pF.

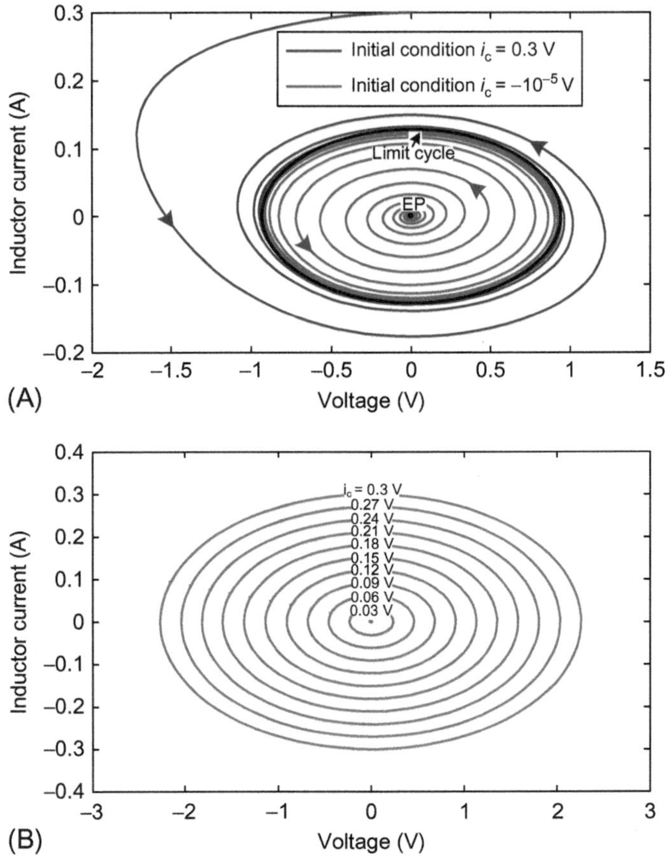

FIGURE 3.4

Phase-space representation of the circuit state variables. (A) Trajectories converging to the limit cycle for two different initial conditions i_C. The two steady-state solutions, ie, the equilibrium point (EP) and limit cycle, are clearly distinguished. (B) Cycles obtained for different initial conditions in a lossless and linear LC circuit.

In the phase-space representation, the oscillation start-up from an unstable *dc* solution will give rise to a spiral trajectory, leading from the *dc* solution to the limit cycle (see Figure 3.4A). A relevant property derives from the fact that any arbitrary time shift of the steady-state oscillation provides a valid solution. All the possible time-shifted solutions give rise to a same limit cycle because the time variable is implicit in the phase-space representation. Remember that time is not a state variable in the case of an autonomous system. The concept of limit cycle will be of fundamental importance for the understanding of phase noise (see Section 3.8).

3.3 NEGATIVE RESISTANCE THROUGH GAIN AND FEEDBACK

The negative resistance required for the oscillation start-up and self-sustained oscillatory regime can be achieved either with a negative-resistance diode (tunnel, Impatt, or Gunn) [19] or with a suitably terminated transistor. The diodes usually exhibit small absolute values of negative resistance, thus requiring application of a strong impedance transformation of the output 50 Ω load to achieve the oscillation start-up conditions. The amount of negative resistance is easier to control in transistor-based oscillators, which also have the advantage of enabling a narrow negative-resistance bandwidth about the desired oscillation frequency, which should avoid additional oscillations. These would lead to unwanted solutions, such as quasiperiodic solutions, in which two independent oscillation frequencies give rise to intermodulation terms [2]. In the following, only transistor-based oscillators will be considered.

Transistor devices have three terminals and two ports, and when properly biased they exhibit transconductance gain. To obtain negative resistance we will combine this gain with a feedback action. The mechanism to obtain negative resistance will be illustrated with the circuit in Figure 3.5, containing a nonlinear transconductance function $i(v)=g_m v+Dv^3$. Other elements include the conductance G_{ds}, the parallel feedback impedance Z_1, the termination impedance Z_2 and the output conductance G_L. In this simplified analysis, transistor capacitances and parasitic elements will be neglected. The aim is to describe the circuit operation with manageable equations, providing insight into the gain and feedback effects. Initially the circuit will be analyzed in small-signal, linearizing the transconductance function about $v=0$, which leads to $i(v)=g_m v$. A simple calculation of the input admittance Y_{in} in Figure 3.5 provides:

$$Y_{in}(\omega)=g_m\frac{Z_2(\omega)}{Z_1(\omega)+Z_2(\omega)}+G_T+\frac{1}{Z_1(\omega)+Z_2(\omega)} \tag{3.19}$$

where $G_T=G_{ds}+G_L$. In general, when designing oscillator circuit, one prefers to limit the intentional use of passive resistances to that of the output load G_L, in order to avoid additional thermal-noise sources and power dissipation. Restricting (3.19) to the case of reactive impedances, $Z_1(\omega)=jX_1(\omega)$, $Z_2(\omega)=jX_2(\omega)$, one obtains:

$$\begin{aligned}Y_{in}(\omega)&=G_N(\omega)+jB_N(\omega)+G_L\\&=g_m\frac{X_2(\omega)}{X_1(\omega)+X_2(\omega)}+G_T-j\frac{1}{X_1(\omega)+X_2(\omega)}\end{aligned} \tag{3.20}$$

The condition to obtain an excess of negative resistance in a certain frequency bandwidth will be:

$$g_m\frac{X_2(\omega)}{X_1(\omega)+X_2(\omega)}+G_T<0 \tag{3.21}$$

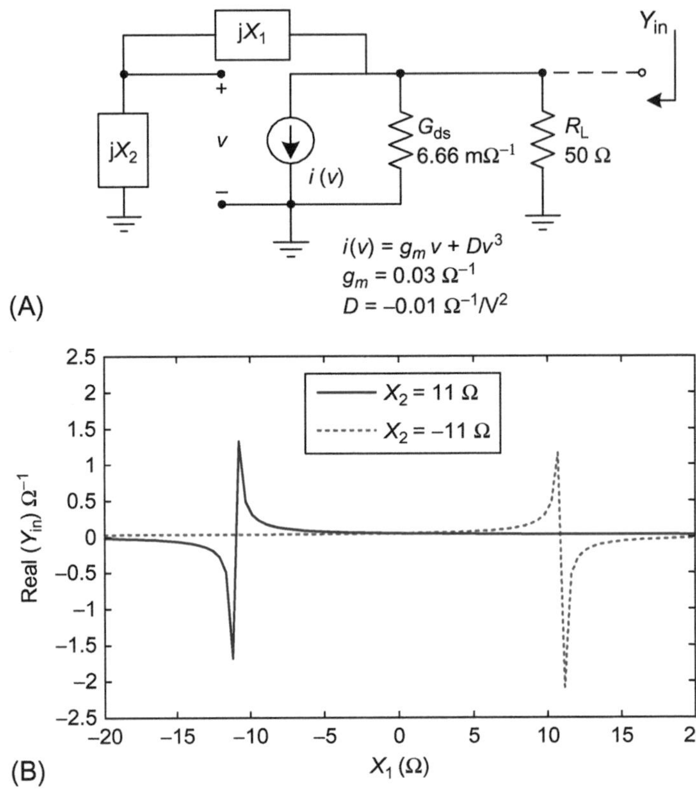

FIGURE 3.5

Negative resistance through gain and feedback. (A) Schematic of a circuit with a circuit with a transconductance function $i(v)$, conductance $G_{ds} = 6.66\ \text{m}\Omega^{-1}$, parallel feedback impedance $Z_1 = jX_1$, terminal impedance $Z_2 = jX_2$, and output load $R_L = 50\,\Omega$. (B) Real part of the total input admittance Re[Y_{in}] versus X_1, for two different values of X_2.

So assuming a positive g_m, one will need:

$$\begin{aligned} &\text{sign}[X_2(\omega)] \neq \text{sign}[X_1(\omega)] \\ &\text{abs}[X_1(\omega)] > \text{abs}[X_2(\omega)] \end{aligned} \tag{3.22}$$

In Figure 3.5, the above conditions are illustrated through simulation. The total conductance looking into port 2 has been traced versus X_2 for two different values of X_1. In total agreement with Eq. (3.22), obtaining negative conductance requires opposite signs in X_1 and X_2, and larger absolute value of X_1. The negative conductance tends to $-\infty$ for equal magnitudes $|X_1| = |X_2|$. Note that in the above analysis the frequency ω is just an implicit variable. Due to the frequency dependence of reactive elements, the relationships (3.22) will have to be fulfilled at the desired oscillation frequency ω_o.

The transistor-based design has several differences with respect to that of the simple LC oscillator in Figure 3.2A. Because all the reactive elements necessarily have frequency-dependent impedance or admittance, the negative conductance will depend on frequency. In the circuit of Figure 3.5A, this negative conductance is achieved by a subnetwork, composed of g_m, X_1, and X_2, which also exhibits an additional imaginary part that also depends on frequency (see Eq. (3.20)). To build an oscillator we must take these dependences into account and synthesize the load network so as to fulfill the oscillation start-up conditions at the desired oscillation frequency ω_o. In fact, the design procedure is similar that of the circuit in Figure 3.2A, though it requires considering the frequency dependence of $G_N(\omega)$ and $B_N(\omega)$.

There are several possibilities to have different signs in $X_1(\omega)$ and $X_2(\omega)$. One simple choice is to use a capacitor and an inductor. Assuming $X_1(\omega) = -1/(C\omega)$ and $X_2(\omega) = L\omega$, the condition to obtain negative resistance will be:

$$G_N(\omega) + G_L = g_m \frac{L\omega}{L\omega - 1/(C\omega)} + G_T < 0 \tag{3.23}$$

The full oscillator design would need additional reactive load $jB_L(\omega)$, such that the oscillation start-up conditions in Eq. (3.11) are fulfilled:

$$\begin{aligned} &G_T(\omega) = G_N(\omega_o) + G_L = g_m \frac{L\omega_o}{L\omega_o - 1/(C\omega_o)} + G_T < 0 \\ &B_T(\omega_o) = \frac{-1}{L\omega_o - 1/(C\omega_o)} + B_L(\omega_o) = 0 \\ &\frac{\partial B_T(\omega_o)}{\partial \omega} > 0 \end{aligned} \tag{3.24}$$

One must note that Eq. (3.11) assimilates the behavior of the transistor-based circuit to a single resonator, as in the case of the circuit in Figure 3.2A, which is only an approximation, as clearly gathered by the frequency dependence of the real and imaginary part. The third condition, on the derivative of the imaginary part, ensures a physical resonance at the frequency ω_o. Parallel (series) resonators exhibit a positive frequency derivative of the susceptance (reactance) at the resonance frequency. Because conditions (3.11) come from an assimilation of the transistor-based circuit to the single resonance oscillator in Figure 3.2A, they are only approximate. A more reliable test for the oscillation start-up should be based on accurate and complete stability-analysis methods (see Section 3.4).

As an example, an oscillation frequency $f_o = 2.4$ GHz has been selected. When choosing this frequency, the inductance L and capacitance C are required to have an excess of negative conductance of $G_T = -5.3\ \text{m}\Omega^{-1}$ are $L = 0.73$ nH and $C = 2.88$pF. The resulting imaginary part is $\text{Im}[Y_{in}] = -[L\omega_o - 1/(C\omega_o)]^{-1} = 83.33\ \text{m}\Omega^{-1}$, so to cancel this imaginary part with $B_L(\omega_o)$ will need a parallel inductive element of value $L_1 = 0.8$ nH. The full oscillator design is shown in Figure 3.6. To verify the fulfillment of the oscillation start-up conditions, the total small-signal admittance observed from node A has been evaluated versus the frequency ω, with the results shown in Figure 3.7A. As can be seen, these conditions are fulfilled at the imposed frequency ω_o.

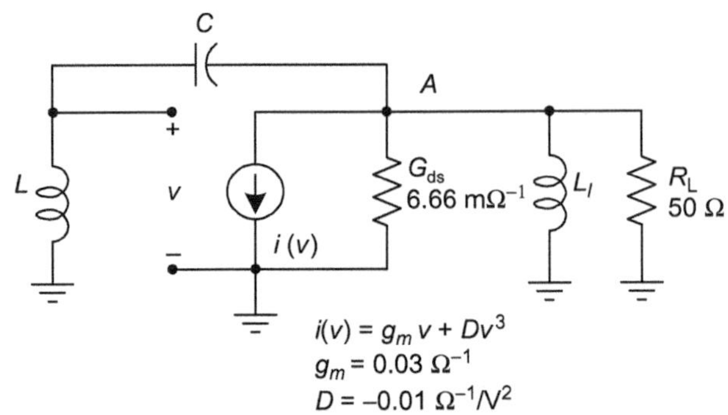

FIGURE 3.6

Oscillator design, after implementation of the reactive impedances in Figure 3.5A and completed with the parallel inductor $L_l = 0.8$ nH.

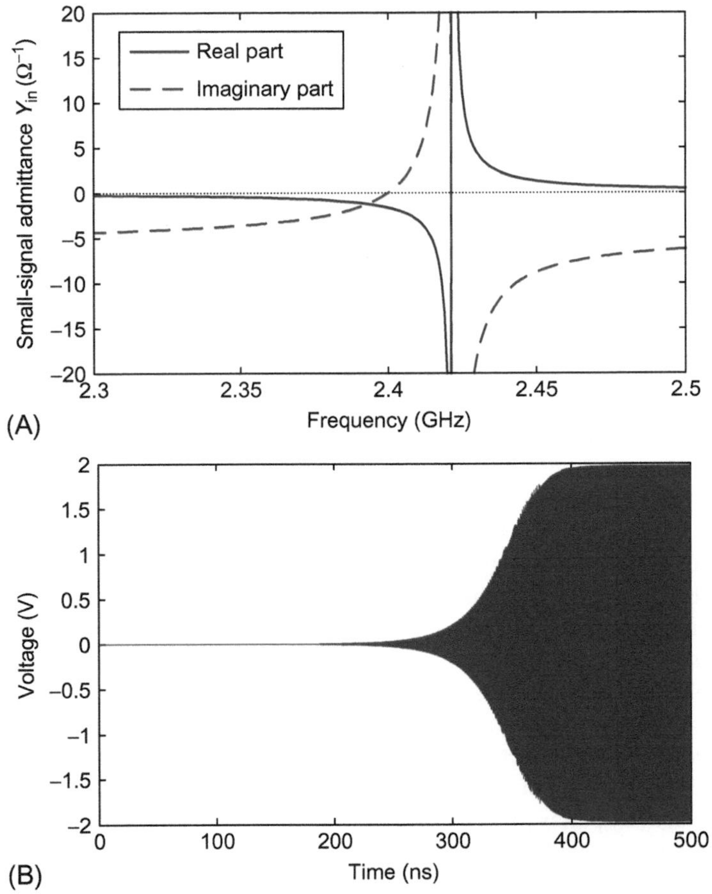

FIGURE 3.7

Oscillation start-up conditions. (A) Total small-signal admittance observed from node *A* in Figure 3.6. The oscillation start-up conditions are fulfilled at the frequency $f_o = 2.4$ GHz. (B) Transient simulation of the start-up and steady-state solution of the oscillator.

To obtain a steady-state oscillation, the full nonlinear model transconductance function $i(v)$ must be considered. Then, the circuit is governed by the following system of nonlinear differential equations:

$$\begin{aligned}
&\frac{dv_c}{dt} = -\frac{i_L}{C} \\
&\frac{di_{L_l}}{dt} = \frac{v}{L_l} - \frac{v_c}{L_l} \\
&\frac{di_L}{dt} = \frac{v}{L} \\
&i_L + i(v) + i_{L_l} + G_T(v - v_c) = 0
\end{aligned} \tag{3.25}$$

The use of the additional variable v avoids the need to introduce a time derivative into the nonlinear function. The time-domain integration of the above system provides the voltage waveform (at node A) shown in Figure 3.7B, where the start-up transient and steady-state waveform are clearly distinguished.

3.4 GENERAL STABILITY ANALYSIS OF OSCILLATOR CIRCUITS

As already indicated, any steady-state solution must be stable (or robust against small perturbation) to be physically observable. Therefore, the steady-state free-running oscillation, desired by the designer, must be stable. This solution always coexists with a *dc* solution, which must be unstable with a pair of complex-conjugate poles $\alpha \pm j\omega_o$ on the right-hand side of the complex plane (RHP) for a regular oscillation start-up from the noise level. The stability of the *dc* solution and the periodic steady-state oscillation are addressed in the two following subsections.

3.4.1 STABILITY OF THE *dc* SOLUTION

In general, an oscillator circuit will have M state variables, instead of the only two variables existing in the circuit of Figure 3.2A. The number M agrees with the number of reactive elements existing in the circuit. Assuming a formulation in state form, the differential equation system governing the circuit behavior will be [2]:

$$\dot{\bar{x}} = \bar{f}(\bar{x}) \tag{3.26}$$

where $\bar{f}$ does not depend explicitly on time, which is due to the absence of any time-varying input sources. The above system admits a *dc* solution that is given by:

$$0 = \bar{f}(\bar{x}_{dc}) \tag{3.27}$$

Under any small perturbation of the *dc* solution, the circuit will get into a transient state, with the circuit variables undergoing a small time-varying increment $\Delta\bar{x}(t)$. As long as $\Delta\overline{x}(t)$ is small, the new solution $\overline{x}_{dc} + \Delta\overline{x}(t)$ of system (3.26) can be calculated linearizing this system about the *dc* point $\overline{x}_{dc}$. This involves expanding the nonlinear function $\overline{f}(\overline{x})$ in first-order Taylor series about $\overline{x}_{dc}$, which provides the linearized system:

$$0 + \Delta\dot{\overline{x}}(t) = \overline{f}(\overline{x}_{dc}) + \frac{\partial\overline{f}(\overline{x}_{dc})}{\partial\overline{x}}\Delta\overline{x}(t) \tag{3.28}$$

which is simplified as:

$$\Delta\dot{\overline{x}} = \frac{\partial\overline{f}(\overline{x}_{dc})}{\partial\overline{x}}\Delta\overline{x} = [Jf(\overline{x}_{dc})]\Delta\overline{x} \tag{3.29}$$

Through the above linearization, the original nonlinear circuit has been approximated with a linear differential equation system with constant coefficients, of the type $\Delta\dot{\overline{x}} = [A]\Delta\overline{x}$, where $[A]$ is a constant matrix. The general solution of this linear system has the form [2,3]:

$$\Delta\overline{x}(t) = c_1\overline{X}_{o1}e^{\lambda_1 t} + c_2\overline{X}_{o2}e^{\lambda_2 t} + \cdots + c_m\overline{X}_{oM}e^{\lambda_M t} \tag{3.30}$$

where the vectors $\overline{X}_{om}$, $m = 1\cdots M$, are the M constant eigenvectors of the matrix $[Jf(\overline{x}_{dc})]$, λ_m are the constant eigenvalues of this matrix, and the coefficients c_m depend on the particular value of the initial perturbation (initial condition). One must note that Eq. (3.30) provides the decomposition of the increment $\Delta\overline{x}(t)$ in the M dimensional space spanned by the vectors $\overline{X}_{om}$. The space dimension agrees with the dimension of the differential equation system that governs the circuit behavior. The vectors are orthogonal and each vector $\overline{X}_{om}$ defines a direction in this space.

The eigenvalues λ_m can be real or complex-conjugate. In the case of a pair of complex-conjugate eigenvalues, their associated eigenvectors and coefficients will be complex-conjugate too, because the increment vector $\Delta\overline{x}(t)$ is composed of real values. In order for the *dc* solution to be stable, the increment $\Delta\overline{x}(t)$ must exponentially decrease in time, that is, it must fulfill: $\lim_{t\to\infty}\Delta\overline{x}(t) = 0$. For a regular oscillation start-up from the noise level, all the eigenvalues must be located on the left-hand side of the complex plane, except a pair of dominant complex-conjugate eigenvalues $\alpha \pm j\omega_o$, located on the right-hand side of this plane. The frequency ω_o must be close to the desired oscillation frequency.

For a frequency-domain analysis, a signal of the form $\Delta\overline{X}e^{st}$, where s is a complex frequency, may be introduced into the linearized system (3.29), which will provide the following characteristic system:

$$\{s[U] - [Jf(\overline{x}_{dc})]\}\Delta\overline{X} = 0 \tag{3.31}$$

The eigenvalues in Eq. (3.31) agree with the roots of the characteristic determinant:

$$\det\{s[U] - [Jf(\overline{x}_{dc})]\} = 0 \tag{3.32}$$

As an example, we will perform a calculation of the eigenvalues associated with the *dc* solution of the circuit in Figure 3.4. Linearizing the nonlinear current $i(v)$ about the *dc* solution and evaluating the system at the complex frequency s, one obtains the following linear system:

$$
\begin{aligned}
s\Delta v_c &= -\frac{\Delta i_L}{C} \\
s\Delta i_{L_l} &= \frac{-1}{L_l(g_m+G_T)}\Delta i_L + \frac{-1}{L_l(g_m+G_T)}\Delta i_{L_l} + \left[\frac{G_T}{L_l(g_m+G_T)} - \frac{1}{L_l}\right]\Delta v_c \\
s\Delta i_L &= \frac{-1}{L(g_m+G_T)}\Delta i_L + \frac{-1}{L(g_m+G_T)}\Delta i_{L_l} + \frac{G_T}{L(g_m+G_T)}\Delta v_c
\end{aligned}
\tag{3.33}
$$

Expressing the above system in the form of Eq. (3.31) and calculating the three eigenvalues, one obtains: $-4.63\times10^{10}\ \mathrm{s}^{-1}$ and $(3.036\times10^{7}\pm \mathrm{j}1.5054\times10^{10}\ \mathrm{s}^{-1})$, which indicates that there is a regular oscillation start-up from the noise level, in agreement with the simulation of Figure 3.7B.

3.4.2 STABILITY OF THE PERIODIC OSCILLATION

For the stability analysis of the periodic oscillation, a small perturbation of the original nonlinear system, operating in this periodic regime $\bar{x}_p(t)$, will be considered. Because the perturbation is small, it will give rise to a small increment of the stable variable $\Delta\bar{x}(t)$, so the circuit equations can be linearized about the periodic solution, in the following manner:

$$\dot{\bar{x}}_p(t) + \Delta\dot{\bar{x}}(t) = \bar{f}(\bar{x}_p(t)) + \frac{\partial\bar{f}(\bar{x}_p(t))}{\partial\bar{x}}\Delta\bar{x}(t) \tag{3.34}$$

which can be simplified as:

$$\Delta\dot{\bar{x}}(t) = \frac{\partial\bar{f}(\bar{x}_p(t))}{\partial\bar{x}}\Delta\bar{x}(t) = \left[Jf(\bar{x}_p(t))\right]\Delta\bar{x} \tag{3.35}$$

The final expression corresponds to a linear differential equation system with periodic coefficients. The general solution of this system has the form [2,3]:

$$\Delta\bar{x}(t) = c_1\bar{x}_{o1}(t)e^{\lambda_1 t} + c_2\bar{x}_{o2}(t)e^{\lambda_2 t} + \cdots + c_M\bar{x}_{oM}(t)e^{\lambda_M t} \tag{3.36}$$

where the constant coefficients c_m depend on the initial value (or value of the initial perturbation). The form is similar to that of Eq. (3.30) with a major difference: the vectors $\bar{x}_{om}(t)$ are periodic with the same period as that of the original periodic solution $\bar{x}_p(t)$. One must note that the number M of independent vectors in Eq. (3.30) is the same as the number of vectors in Eq. (3.30), because this number corresponds to the system dimension, which agrees with the number of reactive elements. Because the vectors $\bar{x}_{om}(t)$ are periodic, they are bounded, so the evolution of the increment $\Delta\bar{x}(t)$ as time tends to infinity depends only on the exponents λ_m. These exponents can be real or complex-conjugate. In the latter case, the constant coefficients and the vectors will also be complex-conjugate, because the increment vector $\Delta\overline{x}(t)$ is composed of real values.

For the periodic solution of a forced circuit (containing one or more time-varying input sources) to be stable, all the exponents must have a negative real part. The case of a free-running oscillation is slightly different, since there is an invariance against time shifts, coming from the fact that the nonlinear function $\overline{f}$ does not depend explicitly on time, that is, it has the form $\overline{f}(\overline{x})$, instead of the form $\overline{f}(\overline{x}, t)$. As a result, one of the exponents is $\lambda_1 = 0$, which is associated with the direction of invariance $\overline{x}_{o1}(t)$. This is the direction of the M dimensional space in which the perturbation neither grows nor decays, and corresponds, at each time value, to the direction tangent to the limit cycle in the phase space, given by $\dot{\overline{x}}_p(t)$. To demonstrate this, we will introduce $\Delta\overline{x}(t) = \dot{\overline{x}}_p(t)$ into Eq. (3.35)

$$\ddot{\overline{x}}_p(t) = \frac{\partial\overline{f}(\overline{x}_p(t))}{\partial\overline{x}_p(t)}\dot{\overline{x}}_p(t) \tag{3.37}$$

Comparing Eq. (3.37) with Eq. (3.36), one will have $\lambda_1 = 0$ in the direction of $\overline{x}_{o1}(t) = \dot{\overline{x}}_p(t)$. Therefore, a free-running oscillation will be stable provided that all the exponents λ_k have negative real parts, except one, which is equal to zero: $\lambda_1 = 0$.

A relevant property of the general expression (3.36) will be discussed next. Expressing the time tending to infinite as an integer number of periods, pT, such that p tends to infinite, any increment of the exponents λ_m in an arbitrary multiple of the oscillation frequency $jn\omega_o$, will not alter the limit value of the perturbation, as gathered from:

$$c_m\overline{x}_{om}(t_o + nT)e^{(\lambda_m + jn\omega_o)pT} = c_m\overline{x}_{om}(t_o)e^{\lambda_m pT} \tag{3.38}$$

In fact, the exponents λ_m are not univocally defined, because the perturbation behavior in any direction $\overline{x}_{om}(t_o)$ is equally determined by any of the exponents $\lambda_m + jn\omega_o$. The so-called *Floquet multipliers* [8] avoid the problem of non-univocity and enable the characterization of the stability of a periodic solution by as many numbers as the system dimension M. The Floquet multipliers are defined as:

$$F_m = e^{\lambda_m T} \tag{3.39}$$

where T is the period of the steady-state solution. Clearly, the whole set of exponents $\lambda_m + jn\omega_o$ provide a same Floquet multiplier F_m, associated to a particular vector $\overline{x}_{om}(t)$. In the case of a free-running oscillator, one of these multipliers is necessarily $F_1 = 1$, due to the invariance under time shifts. To be stable, the remaining $M-1$ Floquet multipliers must have magnitude smaller than one.

The stability of a periodic solution can also be analyzed in the frequency domain, where it will be essential to keep in mind the non-univocal relationship between the Floquet multipliers and the exponents λ_m For the frequency-domain analysis, we will consider a small perturbation $\Delta\overline{X}e^{st}$, where s is a complex frequency. Because the circuit is operating in a periodic regime at the frequencies $k\omega_o$, introduction of the complex frequency s will give rise to the mixing frequencies $jk\omega_o + s$. Assuming N harmonic components in the periodic solution, the frequency-domain expression of Eq. (3.34) is the following:

$$\left\{\begin{bmatrix} [-jN\omega_o+s] & \cdots & 0 & \cdots & 0 \\ \vdots & \vdots & \vdots & \vdots & \vdots \\ 0 & \cdots & [s] & \cdots & 0 \\ \vdots & \cdots & \vdots & \cdots & \vdots \\ 0 & \cdots & 0 & \cdots & [jN\omega_o+s] \end{bmatrix} - \begin{bmatrix} [JF]_0 & \cdots & [JF]_{-N} & \cdots & [JF]_{-2N} \\ \vdots & \vdots & \vdots & \vdots & \vdots \\ [JF]_N & \cdots & [JF]_0 & \cdots & [JF]_{-N} \\ \vdots & \cdots & \vdots & \cdots & \vdots \\ [JF]_{2N} & \cdots & [JF]_N & \cdots & [JF]_0 \end{bmatrix}\right\} \times \begin{bmatrix} \Delta\overline{X}(-jN\omega_o+s) \\ \vdots \\ \Delta\overline{X}(s) \\ \vdots \\ \Delta\overline{X}(jN\omega_o+s) \end{bmatrix} = 0 \tag{3.40}$$

where the $\Delta\overline{X}(jk\omega_o+s)$ is the perturbation component at the mixing frequency $jk\omega_o+s$ and $[JF]_k$ is the kth harmonic component of $\left[Jf\left(\overline{x}_p(t)\right)\right]$. Defining the vector $\Delta\overline{X}(\omega_o,s)=\left(\Delta\overline{X}_{-N}\ldots\Delta\overline{X}_0\ldots\Delta\overline{X}_N\right)^t$, one can rewrite the above system in a compact manner as:

$$\{[jk\omega_o+s]-[JF(\omega_o)]\}\Delta\overline{X}(\omega_o,s)=0 \tag{3.41}$$

where $[JF(\omega_o)]$ represents the constant matrix in Eq. (3.40). The above system is homogeneous, so in order to have $\Delta\overline{X}(\omega_o,s)\neq 0$, the matrix affecting this vector must be singular. This leads to the following characteristic equation:

$$\det\{[jk\omega_o+s]-[JF(\omega_o)]\}=0 \tag{3.42}$$

Unlike Eq. (3.32), the dimension of the above system is not defined by that of the original time-domain system. It depends on the number N of harmonic components considered in the frequency-domain analysis, and it is given by $(2N+1)\times M$. For $N\to\infty$, the roots of Eq. (3.42) agree with the exponents in Eq. (3.36) [20,21]. For a finite harmonic number N, there will be clusters of roots of Eq. (3.42), each associated with one exponent λ_k in Eq. (3.36), having the approximate values $\lambda_k+jn\omega_o$, where n is an integer. In the case of a free-running oscillator, $\lambda_1=0$ is a solution of Eq. (3.34), so the value $s=0$ must be a solution of Eq. (3.42). Because of this root of zero value, the Jacobian matrix associated to the free-running periodic oscillation is singular, which agrees with Eq. (3.37). This singularity was already noticed in the formulations (3.13), (3.18).

3.4.3 APPROXIMATE STABILITY ANALYSIS OF THE PERIODIC SOLUTION

As has been shown, stability of a *dc* solution of a system with M dimensions is characterized by M eigenvalues, whereas stability of a periodic solution is characterized by M Floquet multipliers. As an example, the single resonator oscillator, with only one inductor and one capacitor, is a system of two dimensions. The stability of the *dc* solution of this oscillator is determined by two eigenvalues. Because the system has only two dimensions, the stability of the free-running oscillation of this circuit will be characterized by two Floquet multipliers. One will be $m_1 = 1$, associated to the invariance under phase shifts and the other, m_2, must be necessarily real too, $m_2 \in R$ though it may fulfill $0 < m_2 < 1$ for a stable steady-state oscillation or $m_2 > 1$ for an unstable steady-state oscillation. A Floquet multiplier $m_1 = 1$ is associated with a real pole $\gamma_1 = 0$ or, equivalently, with the set of poles $jn\omega_o$. On the other hand, a real Floquet multiplier $0 < m_2 < 1$ ($m_2 > 1$) is associated with a real pole $\gamma_2 < 0$ ($\gamma_2 > 0$), and, in each case, with the full set of poles $\gamma_2 + jn\omega_o$.

In a practical oscillator, either due to intentionally introduced additional reactive elements (eg, those used as transistor feedback elements to achieve negative resistance) or due to linear or nonlinear parasitic elements, the system order will be $M > 2$, unlike the case of the ideal oscillator in Figure 3.2A. Therefore, we will have $M > 2$ Floquet multipliers characterizing the stability of the periodic steady-state solution. Even with $M > 2$, besides the pole $\gamma_1 = 0$ (associated with the circuit autonomy), we will also have a dominant real pole γ_2, at least in certain intervals of the circuit element values or parameters. This is explained in the following.

It is easy to figure out that an oscillator circuit will only operate as an oscillator within certain parameter intervals. Usual parameters are the transistor or diode bias voltage or an element value. As an example, Figure 3.8A and B show the variation of the oscillation amplitude in the circuit of Figure 3.2A with the conductance G_L, acting as an analysis parameter. Two different nonlinear functions $i(v)$ have been considered, one corresponding to the cubic polynomial $i(v) = G_N v + Dv^3$ and the other, to the fifth order polynomial $i(v) = G_N v + Ev^3 + Fv^5$. In the latter case, the constant coefficients must fulfill $G_N < 0$, $F > 0$ to enable the oscillation start-up and the establishment of the steady-state oscillatory regime. With the two nonlinear functions, the boundary between stable and unstable *dc* solutions corresponds to the same G_L value. This is obtained by solving Eq. (3.13) for $V_o = 0$, which agrees with the condition $G_N + G_L = 0$, depending only on the small-signal approximation of the nonlinear function.

A similar situation would be obtained in any general oscillator circuit, under variations of a sensitive parameter that here will be denoted η. Assume that the circuit does not oscillate for $\eta < \eta_o$, but it does oscillate for $\eta > \eta_o$. The qualitative change in the physical solution is due to a qualitative variation of the stability of the *dc* solution. For $\eta < \eta_o$, the *dc* solution is stable, with all of its eigenvalues on the left-hand side of the complex plane. However, when continuously increasing the parameter η towards η_o, from certain η value, a pair of complex-conjugate eigenvalues $\sigma \pm j\omega_o$ must approach the imaginary axis and cross this axis at η_o.

The crossing of a pair of complex-conjugate poles $\sigma \pm j\omega_o$ through the imaginary axis will give rise to a qualitative change in the stability of the *dc* solution. For $\eta > \eta_o$, the pair of dominant complex-conjugate poles $\sigma \pm j\omega_o$, where $\sigma > 0$, will be responsible for the start-up of an oscillation from the noise level. This start-up should lead to a steady-state oscillation of certain amplitude. The value η_o at which the pair of complex-conjugate poles is on the imaginary axis corresponds to the boundary between stable and unstable *dc* solutions.

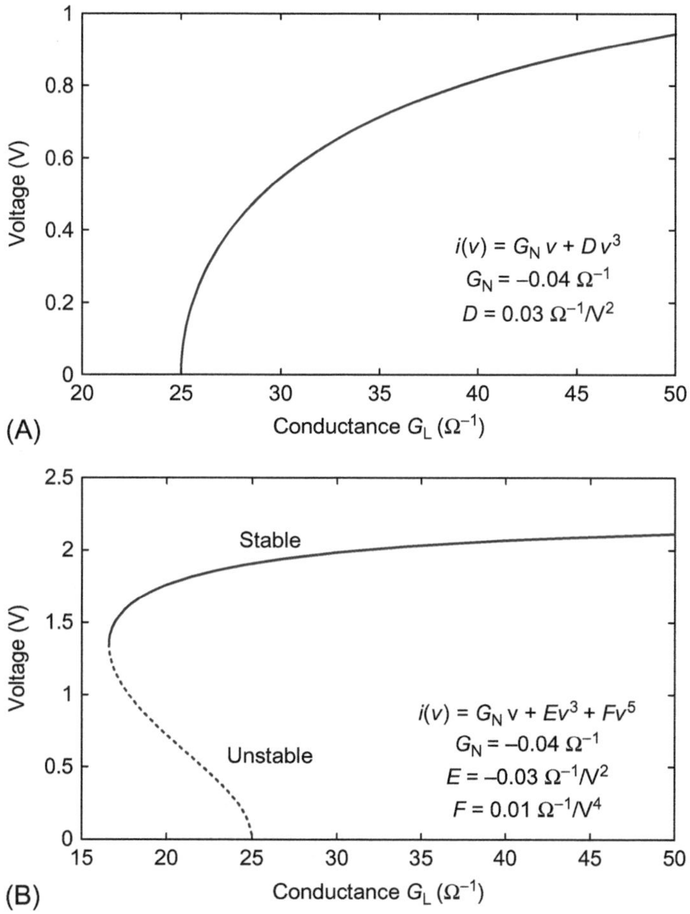

FIGURE 3.8

Variation of the oscillation voltage amplitude versus the conductance $G_L = 1/R_L$ for the circuit in Figure 3.2A. In both cases, the oscillation amplitude tends to zero at the G_L boundary (25 Ω^{-1}) between stable and unstable *dc* solutions. (A) Cubic polynomial $i(v) = G_N v + Dv^3$, and (B) fifth order polynomial: $i(v) = G_N v + Ev^3 + Fv^5$.

Due to the continuity of the differential equation system governing the circuit behavior, the amplitude of this steady-state oscillation must tend to zero at the limit value η_o (see Figure 3.8), this corresponding to a degenerate oscillation [4]. This is because at η_o the linearized system is fulfilled at the real frequency $\pm j\omega_o$, because the critical eigenvalues are exactly on the imaginary axis ($\sigma = 0$). These two critical eigenvalues of the *dc* solution involve two dimensions of the differential equation system. In consistency with this, the critical subspace of the degenerate oscillation must also have two dimensions, associated to two Floquet multipliers. Due to the phase invariance of the free-running oscillatory regime one of them will be $m_1 = 1$, so the second critical multiplier m_2 must be real too, just like in the single resonator oscillator.

The multiplier $m_1 = 1$ will be associated to a real pole $\gamma_1 = 0$. The other one will also be $m_2 = 1$ (associated to $\gamma_2 = 0$) at the boundary parameter value η_o as both arise from a same pair of eigenvalues of the *dc* solution, given by $\pm j\omega_o$. Considering the variation of the steady-state oscillation amplitude versus the analysis parameter η (see Figure 3.8, as an example), the eigenvalue $\gamma_2 = 0$ gives rise to an infinite slope of the oscillation curve at η_o. In this manner, we will have the necessary transition from the linear circuit operation at the oscillation boundary η_o (insensitive to the oscillation amplitude) to the nonlinear operation (sensitive to the amplitude), which is an essential condition to obtain a steady-state oscillation [4]. When slightly varying the parameter from η_o towards the oscillation region, the second real pole γ_2 will no longer be zero and may become either negative, for a stable oscillation, or positive, for an unstable one.

One can conclude that in an oscillation region sufficiently close to η_o, the dominant pole of any oscillator circuit will be real, regardless of the dimension of the differential equation system governing the circuit behavior. At a larger distance from the boundary value η_o, other poles may dominate the circuit stability properties.

In the assumption of only one dominant real pole γ_2, associated with a main dominant circuit resonance, it is possible to obtain an approximate value for this pole from a simple perturbation analysis of steady-state oscillation condition at the fundamental frequency. This condition is written in terms of the total admittance observed from a particular circuit node, in the following manner.

$$Y_T(V_o, \omega_o)V_o = [G_T(V_o, \omega_o) + B_T(V_o, \omega_o)]V_o = 0 \tag{3.43}$$

Next, a small perturbation of complex frequency s will be considered. This perturbation will give rise to a small amplitude increment $\Delta V(t)$ with respect to the steady-state variable. On the other hand, because we aim at detecting a dominant real pole, we can also assume that the complex frequency increment s has a small value. This assumption enables a Taylor series expansion of the perturbed equations about the steady-state solution (V_o, ω_o), which is carried out as follows [4]:

$$\begin{aligned} Y_T(V_o + \Delta V, j\omega_o + s)(V_o + \Delta V(t))e^{j\phi(t)} &= 0\left[Y_T(V_o, \omega_o) + \frac{\partial Y_T}{\partial V_o}\Delta V(t) + \frac{\partial Y_T}{\partial j\omega_o}s\right](V_o + \Delta V(t)) \\ \times e^{j\phi(t)} &= \left[\frac{\partial Y_T}{\partial V_o}\Delta V(t) + \frac{\partial Y_T}{\partial j\omega_o}s\right](V_o + \Delta V(t))e^{j\phi(t)} \\ &= 0 \end{aligned} \tag{3.44}$$

where the steady-state value $Y_T(V_o, \omega_o)$ has been taken into account and the product by s indicates a time-differentiation operation. Performing this differentiation, and neglecting higher order terms, one obtains:

$$\begin{aligned} \frac{\partial Y_T}{\partial V_o}\Delta V(t)V_o e^{j\phi(t)} + \frac{\partial Y_T}{\partial j\omega_o}s\left[V_o e^{j\phi(t)} + \Delta V(t)e^{j\phi(t)}\right] &= \frac{\partial Y_T}{\partial V_o}\Delta V(t)V_o e^{j\phi(t)} \\ &+ \frac{\partial Y_T}{\partial j\omega_o}\left[V_o j\dot{\phi}(t)e^{j\phi(t)} + \Delta\dot{V}(t)e^{j\phi(t)}\right] \\ &= 0 \end{aligned} \tag{3.45}$$

Dividing by the common term $V_o e^{j\phi(t)}$, Eq. (3.45) is simplified as:

$$\frac{\partial Y_T}{\partial V_o}\Delta V(t) + \frac{\partial Y_T}{\partial \omega_o}\left[\dot{\phi}(t) - j\frac{\Delta \dot{V}(t)}{V_o}\right] = 0 \tag{3.46}$$

Splitting this complex equation into real and imaginary parts one obtains:

$$\begin{aligned} &\frac{\partial Y_T^r}{\partial V_o}\Delta V(t) + \frac{\partial Y_T^i}{\partial \omega_o}\frac{1}{V_o}\Delta \dot{V}(t) + \frac{\partial Y_T^r}{\partial \omega_o}\dot{\phi}(t) = 0 \\ &\frac{\partial Y_T^i}{\partial V_o}\Delta V(t) - \frac{\partial Y_T^r}{\partial \omega_o}\frac{1}{V_o}\Delta \dot{V}(t) + \frac{\partial Y_T^i}{\partial \omega_o}\dot{\phi}(t) = 0 \end{aligned} \tag{3.47}$$

As derived below, the above system can be written in the matrix form: $\Delta\dot{\overline{X}}(t) = [A]\Delta\overline{X}(t)$, where $\Delta\overline{X}(t) = [\Delta V(t)\Delta\phi(t)]^T$, where T indicates "transpose." The resulting matrix system is the following:

$$\begin{bmatrix}\Delta \dot{V}(t) \\ \dot{\phi}(t)\end{bmatrix} = V_o^2 \frac{\begin{bmatrix} -\dfrac{\partial Y_T^r}{\partial V_o}\dfrac{\partial Y_T^i}{\partial \omega_o}\dfrac{1}{V_o} + \dfrac{\partial Y_T^r}{\partial \omega_o}\dfrac{\partial Y_T^i}{\partial V_o}\dfrac{1}{V_o} & 0 \\ -\dfrac{\partial Y_T^r}{\partial \omega_o}\dfrac{\partial Y_T^r}{\partial V_o} - \dfrac{\partial Y_T^i}{\partial \omega_o}\dfrac{\partial Y_T^i}{\partial V_o} & 0 \end{bmatrix}}{\left(\dfrac{\partial Y_T^r}{\partial \omega_o}\right)^2 + \left(\dfrac{\partial Y_T^i}{\partial \omega_o}\right)^2} \begin{bmatrix}\Delta V(t) \\ \phi(t)\end{bmatrix} = [A]\begin{bmatrix}\Delta V(t) \\ \phi(t)\end{bmatrix} \tag{3.48}$$

The two eigenvalues of the matrix $[A]$ are $\lambda_1 = 0$, associated with the system autonomy, and the following real eigenvalue, λ_2:

$$\lambda_2 = -V_o \frac{\dfrac{\partial Y_T^r}{\partial V_o}\dfrac{\partial Y_T^i}{\partial \omega_o} - \dfrac{\partial Y_T^r}{\partial \omega_o}\dfrac{\partial Y_T^i}{\partial V_o}}{\left(\dfrac{\partial Y_T^r}{\partial \omega_o}\right)^2 + \left(\dfrac{\partial Y_T^i}{\partial \omega_o}\right)^2} = -V_o \frac{\chi_o}{\left(\dfrac{\partial Y_T^r}{\partial \omega_o}\right)^2 + \left(\dfrac{\partial Y_T^i}{\partial \omega_o}\right)^2} \tag{3.49}$$

where the coefficient χ_o has been introduced. In many cases, the above real eigenvalue will dominate the stability properties of the steady-state free-running oscillation. For stability, this eigenvalue must fulfill $\lambda_1 < 0$, which is equivalent to the well-known condition [1]:

$$\chi_o = \frac{\partial Y_T^r}{\partial V_o}\frac{\partial Y_T^i}{\partial \omega_o} - \frac{\partial Y_T^r}{\partial \omega_o}\frac{\partial Y_T^i}{\partial V_o} > 0 \tag{3.50}$$

From inspection of the above condition, a high value of the frequency derivative of the susceptance, $\dfrac{\partial Y_T^i}{\partial \omega_o} > 0$, facilitates the stability of the oscillation. This is in good agreement with the condition (3.11) for the oscillation start-up. As an example, condition (3.50) will be tested in the two oscillators in Figure 3.8A and B, respectively, based on nonlinear functions $i(v) = G_N v + Dv^3$ and $i(v) = G_N v + Ev^3 + Fv^5$.

The describing function associated to $i(v) = G_N v + Dv^3$ is $G_N(V) = G_N + 3DV^2/4$ and the total function $Y_T(V, \omega)$ is:

$$Y_T(V, \omega) = G_N + G_L + \frac{3}{4}DV^2 + j\left(C\omega - \frac{1}{L\omega}\right) \tag{3.51}$$

Calculating the function in Eq. (3.50), one obtains:

$$\chi = \frac{\partial Y_T^r}{\partial V}\frac{\partial Y_T^i}{\partial \omega} - \frac{\partial Y_T^r}{\partial \omega}\frac{\partial Y_T^i}{\partial V} = 3bVC > 0$$

Because the magnitude V is necessarily positive, the real pole will be negative and the oscillation will be stable for any G_L.

The describing function associated to $i(v) = G_N v + Ev^3 + Fv^5$ is $G_N(V) = G_N + 3EV^2/4 + 5FV^4/8$ and the total function $Y_T(V, \omega)$ is:

$$Y_T(V, \omega) = G_N + G_L + 3EV^2/4 + 5FV^4/8 + j\left(C\omega - \frac{1}{L\omega}\right) \quad (3.52)$$

where $G_N + G_L < 0$ and $F > 0$. Calculating the function in Eq. (3.50) one obtains:

$$\chi = \frac{\partial Y_T^r}{\partial V}\frac{\partial Y_T^i}{\partial \omega} - \frac{\partial Y_T^r}{\partial \omega}\frac{\partial Y_T^i}{\partial V} = 3EVC + 5FV^3C \quad (3.53)$$

So the boundary between negative and positive values of χ is given by:

$$V_b = \sqrt{-\frac{3E}{5F}} \quad (3.54)$$

For $V < V_b$, the coefficient χ is negative and the real pole predicted with Eq. (3.49) fulfills $\lambda_2 > 0$, so the steady-state oscillation is unstable. On the other hand, for $V > V_b$, the coefficient χ is positive and the real pole predicted with Eq. (3.49) fulfills $\lambda_2 < 0$, so the steady-state oscillation is stable.

In the following, a whole design procedure to obtain a free-running oscillator at a specified frequency ω_o will be described. After an initial choice of the transistor device, the procedure will be composed of three stages: initial linear design to fulfill the oscillation start-up condition, correction of the initial design through a large-signal analysis of the steady-state oscillation, using harmonic balance, and final verification of the oscillation start-up condition and stable behavior of the steady-state oscillation.

The synthesis technique will be illustrated by means of its application to the design of a single-ended oscillator circuit, based on the transistor NE3210S01. The desired oscillation frequency is $f_o = 5$ GHz.

3.5 INITIAL LINEAR DESIGN TO FULFILL THE OSCILLATION START-UP CONDITIONS

As has been shown, the oscillation start-up from the noise level requires an excess of negative resistance at a resonance frequency, sufficiently close to the desired oscillation frequency. The oscillation starts-up from small-signal amplitude, so the initial oscillator design can be carried out by means of the scattering parameters. In transistor-based designs, one can use a combination of gain and feedback to obtain the negative resistance, as shown in the simple example of Figure 3.4. Here a detailed description of the transistor device, including parasitics, will be considered, in terms of the scattering matrix $[S]$.

Even if the transistor is unconditionally stable for the chosen bias point at the desired oscillation frequency (fulfilling the condition $\mu > 1$ [22]), it will be possible to obtain negative resistance if we introduce a feedback network. Redefining the scattering matrix so as to include this additional feedback network (see Figure 3.9A), the new scattering matrix $[S']$ should exhibit $\mu < 1$ at the desired oscillation frequency ω_o. Condition $\mu < 1$ implies the possibility to select passive loads, connected either at port 1 or port 2, such that the two-port network exhibits negative resistance when looking into the other port (port 2 or port 1, respectively).

Because the oscillator circuit has no time-varying input generator, we can perform the initial design using a single observation port (see Figure 3.9B). We will synthesize the feedback element(s) and the termination at the other transistor port (*terminal impedance*) so as to obtain negative resistance when looking into the single observation port. The procedure is similar to that carried out for the design of the simple oscillator circuit, based on a transconductance element, in Figure 3.5A. However, in this case we will consider the whole model of the active device, including capacitive effects and parasitics. A comprehensive treatment of the transistor modeling and characterization [23–28] is provided in Chapters 1 and 2. We will finalize the design by obtaining the load that should be connected to the single observation port (see Figure 3.9C), in order to fulfill the oscillation start-up conditions.

As already stated, both the feedback and terminal impedances are generally restricted to reactive values, to avoid unnecessary dissipation. Note, however, than in some cases, it may be necessary to consider a real part in the terminal impedance, especially when the aim is to synchronize the oscillator at a later stage, because the forcing generator, connected at port 1, will necessarily introduce a resistive component.

Unlike the case of the circuit in Figure 3.5A, employing parallel feedback, a series reactive feedback element X_{fb}, at the transistor source, will be considered here (see Figure 3.9A), which will generally enable a simpler implementation in practical transistor-based designs. Redefining the two-port network so as to include this feedback element X_{fb}, one should be able to obtain $\mu < 1$, for some values of this element, at the desired oscillation frequency ω_o. Port 1 will be used to connect the reactive terminal impedance X_T. Through a convenient choice of the values of these two reactive elements, negative resistance should be obtained when looking into port 2, defined between the transistor drain terminal and ground (see Figure 3.9B). This will be our single observation port. Both the feedback element and the terminal impedance are reactive, so they can be represented in terms of their reflection coefficients as: $\Gamma_T = 1e^{j\theta_T}$ and $\Gamma_{fb} = 1e^{j\theta_{fb}}$. The corresponding impedances will be inductive for positive phase values and capacitive for negative phase values. This distinction will be useful when a feedback or terminal element of a particular type is desired, as in the case of voltage-controlled oscillators (VCOs), in which a tunable capacitance is to be included as either terminal or feedback element.

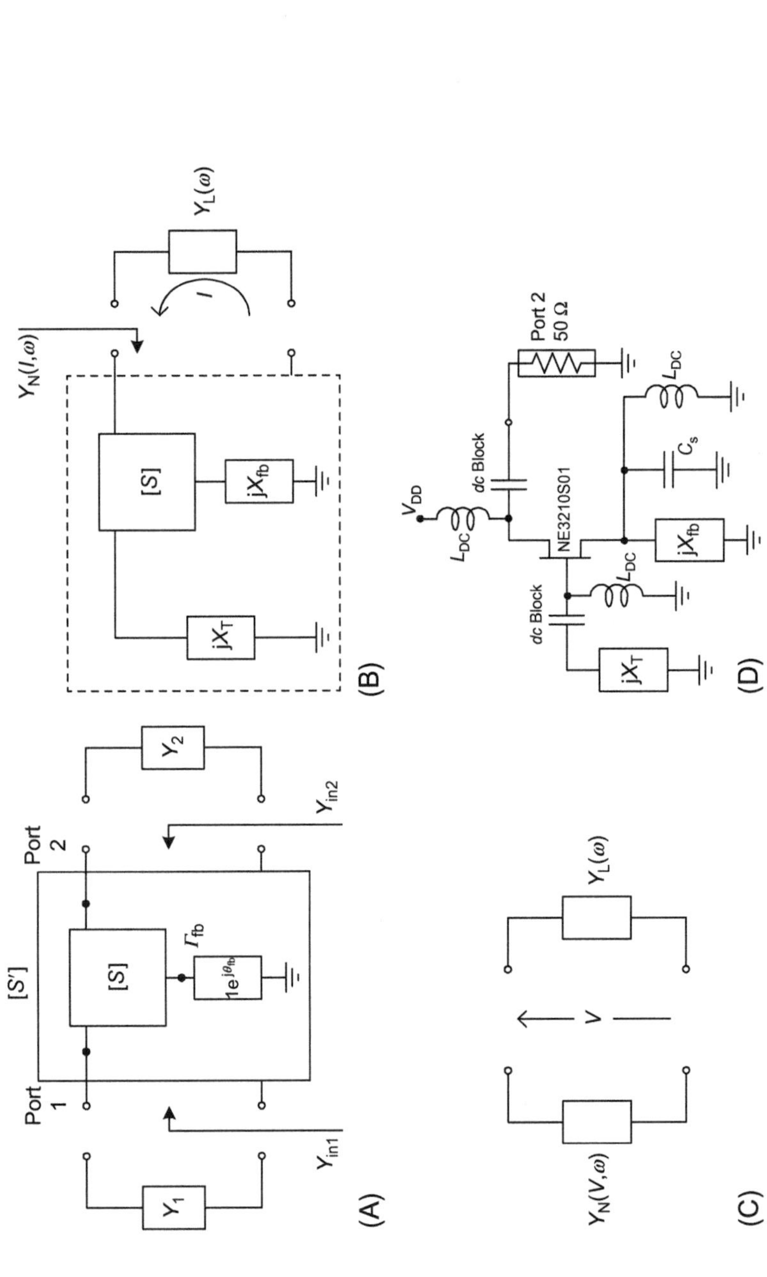

FIGURE 3.9

Stages of the single-ended oscillator design. (A) Schematic of the circuit containing the NE3210S01 transistor, a terminal load with reflection coefficient Γ_T and feedback load with reflection coefficient Γ_{fb}. (B) Analysis as a one-port network. Negative conductance should be obtained when looking into this port. (C) Addition of the linear load Y_L to complete the oscillator design. (D) Final oscillator schematic. The capacitor C_s will be later replaced with a varactor diode for oscillator tunability.

In general, the objective will be to maximize the negative resistance, so a double sweep will be carried out in θ_T and θ_{fb}, analyzing the resulting impedance or admittance when looking into port 2. If the oscillation design is to be completed by means of a parallel (series) resonance, this analysis is best carried out here calculating the input admittance (impedance) when looking into port 2, at the constant frequency ω_o. Note that to complete the small-signal oscillator design, we can formulate the oscillation start-up conditions either in terms of the total impedance or total admittance. A positive slope of the total reactance should be obtained in the case of a series resonance, and a positive slope of the total susceptance should be obtained in the case of a parallel resonance. Without loss of generality, we will consider a parallel resonance and thus evaluate the input admittance when looking into port 2, which will be called $Y_N = G_N + jB_N$. At the constant analysis frequency ω_o, we can write: $Y_N(\theta_T, \theta_{fb})$.

As an example, the analysis described has been applied to the NE3210S01 transistor biased at $V_{DD} = 4$ V and $V_{GG} = 0$ V (see Figure 3.9D). Because we have a double dependence on the angles θ_T and θ_{fb} we have carried out a representation through contour plots, joining all the pairs (θ_T, θ_{fb}) that provide the same conductance value $G_N(\theta_T, \theta_{fb})$. This leads to the diagram shown in Figure 3.10. We choose the minimum value corresponding to $\theta_{To} = 187$ degree and $\theta_{fbo} = 53$ degree. At the oscillation frequency ω_o, the reflection coefficients $\Gamma_{To} = 1e^{j\theta_{To}}$ and $\Gamma_{fbo} = 1e^{j\theta_{fbo}}$ correspond to the reactive elements: $L_{To} = 3.2$ nH and $C_{fb} = 10.6$ pF. For practical reasons, a microstrip implementation of these elements (using RO4003C substrate) will be used here. The capacitive feedback C_{fb} will include the contributions of a varactor diode SMV1232,

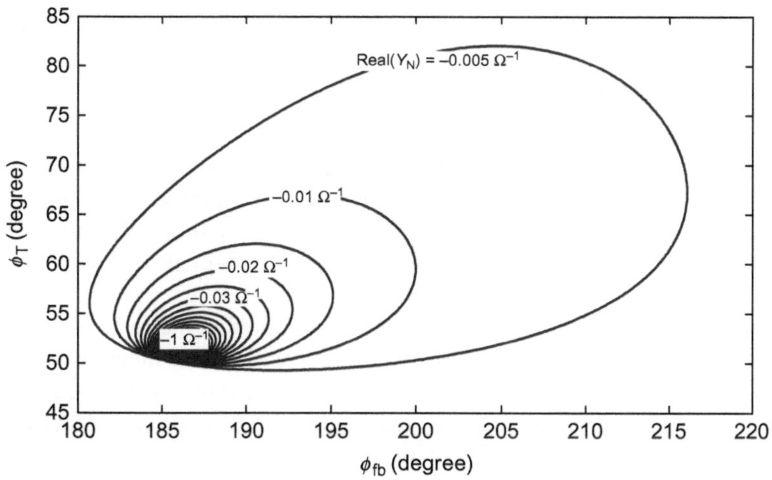

FIGURE 3.10

Input admittance observed at port 2, obtained versus variations in ϕ_{fb} and ϕ_T. Contour plots of constant $G_N(\theta_T, \theta_{fb})$ have been traced in the plane defined by ϕ_{fb} and ϕ_T.

used for the tenability of the oscillation frequency. Thus, the capacitive feedback will decompose as $C_{fb} = C_{var} + C_{add}$, where C_{var} is the varactor capacitance and C_{add} is the additional capacitance implemented in microstrip. This results in transmission lines with the following dimensions: $w_{To} = 0.55$ mm, $l_{To} = 6$ mm, $w_{fb} = 0.55$ mm, and $l_{fb} = 18.5$ mm.

The subnetwork including the implemented feedback and terminal elements will exhibit negative conductance at ω_o when looking into port 2. Now the design must be completed with additional linear elements, so as to fulfill the oscillation start-up conditions at the desired oscillation frequency ω_o. As already indicated, a parallel resonance is chosen, so the load admittance $Y_L = G_L(\omega) + jB_L(\omega)$ must fulfill:

$$G_T(V \cong 0, \omega_o) \equiv G_N(V \cong 0, \omega_o) + G_L(\omega_o) < 0 \tag{3.55a}$$

$$B_T(V \cong 0, \omega_o) \equiv B_N(V \cong 0, \omega_o) + B_L(\omega_o) = 0 \tag{3.55b}$$

$$\frac{\partial B_T(V \cong 0, \omega_o)}{\partial \omega} = \frac{\partial B_N(V \cong 0, \omega_o)}{\partial \omega} + \frac{\partial B_L(V \cong 0, \omega_o)}{\partial \omega} > 0 \tag{3.55c}$$

where $V \cong 0$ indicates small-signal voltage. From inspection of Eq. (3.55b), one can ensure a high frequency derivative of the total susceptance B_T by choosing a load impedance with a high frequency selectivity. The resulting high value of the oscillator quality factor will enable low phase noise of the steady-state oscillation. Dielectric resonators can be used to obtain this high quality factor. They are made of a ceramic material, being stable in temperature, with low loss and high dielectric constant. Here a simple planar design will be carried out instead.

In the case of the circuit in Figure 3.9, for the loads selected X_T and X_{fb} one obtains, when looking into the output load, the admittance $Y_N(V \cong 0, \omega_o) = -0.016 + j0.013\ \Omega^{-1}$. In order to fulfill conditions (3.55a), (3.55b), (3.55c), an inductive element $L = 2.529$ nH can be introduced in parallel, together with the output conductance $G_L = 50\ \Omega$. The microstrip implementation of this inductive element results in a transmission line with dimensions $w_d = 0.55$ mm and $l_d = 14$ mm. Evaluation of the total admittance $Y_T(V \cong 0, \omega) = Y_N(V \cong 0, \omega) + Y_L(\omega)$ versus frequency provides the result in Figure 3.11. Thus, the oscillation start-up conditions (3.55a), (3.55b), (3.55c) are fulfilled. Note that these conditions are only approximate in a circuit containing multiple reactive elements, so the oscillation start-up will also be verified with an accurate stability-analysis method.

With a small-signal analysis it is not possible to predict the steady-state oscillatory solution, since, as already demonstrated, this oscillation is inherently nonlinear. The only way to predict the steady-state output power and oscillation frequency is to perform a large-signal simulation of the oscillator circuit. At microwave frequencies, this simulation is usually carried out in the frequency domain, by means of the harmonic-balance method [15–17]. This method, described briefly in the following section, can be used for both oscillator analysis and synthesis.

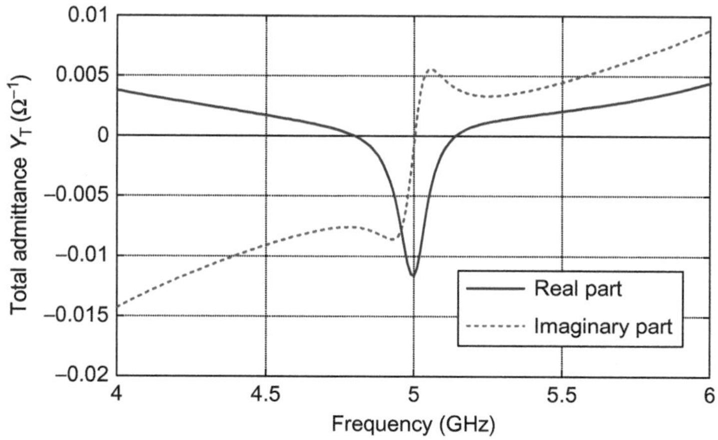

FIGURE 3.11

Total admittance of the implemented oscillator traced versus frequency. The real part and imaginary part of this admittance have been represented, which shows the fulfillment of the approximate conditions for the oscillation start-up at the frequency 5 GHz.

3.6 OSCILLATOR DESIGN WITH HARMONIC-BALANCE SIMULATIONS

The harmonic-balance method enables an accurate oscillator design considering the effect of all the nonlinearities contained in the oscillator model, such as nonlinear currents and capacitances, and the impact of the harmonic content. As discussed in Section 3.2.3, neglecting relevant harmonic components can give rise to a wrong prediction of both the oscillation power and oscillation frequency. Reliability of the results obtained with harmonic balance strongly depends on the accuracy of the transistor models. In fact, these models, together with those of the passive circuitry, determine the nonlinear algebraic system to be resolved, so under inaccurate models, the simulated solution will deviate from the measured one. In this section, the harmonic-balance method is initially described, pointing out some fundamental characteristics of the oscillator analysis. Then, the way how this method can be used for oscillator synthesis is presented.

3.6.1 HARMONIC BALANCE

In harmonic balance, the circuit variables are expressed in a Fourier series as $x_i(t) = \sum_k X_{i,k} e^{j\omega_k t}$, which due to the purely imaginary exponents restricts the analysis to steady-state solutions. The nodal harmonic balance formulation is obtained by introducing these Fourier expressions for the circuit variables in the so-called *modified nodal approach* [15], resulting from the application of Kirchhoff's laws to the circuit nodes and loops. Just as a reminder, the modified nodal approach resolves a set of nonlinear integro-differential equations in an n-dimension vector $\bar{x}(t)$ of state variables [4,5,15]. Using the modified nodal analysis [15], the equations are the following:

$$\bar{i}\,[\bar{x}(t)] + \frac{d}{dt}\bar{q}[\bar{x}(t)] + \int_{-\infty}^{t} c\left(t-t'\right)\bar{x}(t')dt' + \bar{i}_s = \bar{0} \tag{3.56}$$

where $\bar{x}(t)$ is composed by the node voltages and inductor currents, $\bar{i}[\bar{x}(t)]$ contains the sums of resistive currents (that enter each node) and loop voltages, $\bar{q}[\bar{x}(t)]$ the linear and nonlinear charges and fluxes of the circuit, $\bar{i}_s$ the bias sources and $c(t)$ the impulse-response matrix of the distributed elements. At the periodic steady state, each state variable in $\bar{x}(t)$ can be expanded in a Fourier series as:

$$x_i(t) = \sum_{k=-N}^{N} X_{ik}^0 e^{jk\omega_o t}, \quad \omega_o = \frac{2\pi}{T}, \qquad i = 1,\ldots,M \tag{3.57}$$

where T is the period of the oscillation and $2N+1$ is the number of harmonic components X_{ik}^0 employed to approximate the periodic solution.

Introducing the expressions (3.57) into Eq. (3.56), equating harmonic components of the same order and eliminating the exponentials, one obtains the harmonic-balance system:

$$\bar{H}\left[\bar{X},\bar{k}\omega_o\right] \equiv \bar{I}\left[\bar{X}\right] + \left[j\bar{k}\omega_o\right]\bar{Q}\left[\bar{X}\right] + C\bar{X} + \bar{I}_s = \bar{0}, \; \bar{k} = [-N\cdots N]^t \tag{3.58}$$

where the upper-case vector/matrix variables contain the harmonic components of the corresponding lower-case time-varying vector/matrix signals, and the matrix $\left[j\bar{k}\omega_o\right]$ multiplies each kth harmonic component by $jk\omega_o$. The vector function $\bar{H}\left[\bar{X},\bar{k}\omega_o\right]$ allows expressing the harmonic-balance system in a compact manner.

The nonlinear circuit will contain a number of voltage-controlled currents $i_1(v_1)\ldots i_p(v_p)$ and voltage-dependent charges $q_1(v_1)\ldots q_s(v_s)$. Note that some currents may depend on two voltages, such as the transconductance current of a FET transistor, which depends on v_{gs} and v_{ds}, and some of the currents and charges may depend on the same voltages. These nonlinear relationships $[i_p(v_p)$ or $q_s(v_s)]$ are instantaneous and defined in time domain. However, the harmonic-balance system is written in the frequency domain. To cope with this situation, we will have to make use of the inverse and forward Fourier transforms [4,15–17]. On the other hand, the equations composing the system (3.58) are not independent but coupled. This is because each harmonic of each nonlinear function $I_{p,k}$ and $Q_{s,k}$ depends, in general, on all the harmonic components of its control voltage or voltages, with k going from $-N$ to N. Therefore, system (3.58) is a nonlinear algebraic system that can only be resolved through error minimization, usually with the Newton-Raphson algorithm.

In an oscillator analysis, the fundamental frequency of the Fourier series expressions of the circuit variables will be the oscillation frequency ω_o, which will have to be determined during the analysis procedure. Because of the invariance with respect to time shifts (or, equivalently, with respect to phase shifts), the phase of one harmonic of one state variable can be set arbitrarily to zero. In this way, we will replace one of the original unknowns (real or imaginary part of one control voltage) with the oscillation frequency ω_o. By this procedure, we will have a well-conditioned system of $(2N+1)\times M$ equations in $(2N+1)\times M$ unknowns. Because the set of unknowns mixes voltages and currents with a frequency variable, this harmonic-balance analysis is known as *mixed harmonic balance*.

A serious problem in the harmonic-balance analysis of free-running oscillations is the coexistence of the steady-state oscillation with the circuit *dc* solution. Due to the absence of time-varying sources, the vector $\bar{I}_s$ will only contain *dc* components, which will be unable to initialize the components of the state-variable vector $\bar{X}$ at the oscillation frequency ω_o. The Newton-Raphson algorithm, used to resolve the harmonic-balance system, is very sensitive to the initial value. Thus, unless an accurate estimation of the oscillation frequency and amplitude are provided, it will converge towards the *dc* solution [4–6]. To understand the problem, one should take into account that the two solutions are equally valid from a mathematical point of view, but the *dc* solution is much simpler. The situation is identical to that described for the case of the simplified harmonic-balance system of Eq. (3.18). A useful manner to cope with this problem will be described in the next subsection.

3.6.2 USE OF AN AUXILIARY GENERATOR FOR OSCILLATOR ANALYSIS AND SYNTHESIS

Auxiliary generators are artificial generators fulfilling certain conditions that are used for simulation purposes only [4–6]. In the case of troublesome simulation of oscillator circuits, an auxiliary generator operating at the unknown oscillation frequency ω_o (a variable to be calculated in the analysis procedure) can be introduced into the circuit to avoid the undesired harmonic balance convergence towards the *dc* solution. The auxiliary generator can be of the voltage or current type. It operates at the oscillation frequency, so we will have $\omega_{AG}=\omega_o$. A voltage auxiliary generator of amplitude A_{AG} should be connected in parallel at a sensitive circuit node (see Figure 3.12). To prevent short-circuiting frequencies different from $\omega_{AG}=\omega_o$, an ideal band-pass filter must be connected in series with the voltage source. The filter should exhibit zero impedance at $\omega=\omega_{AG}$ and an infinite impedance at $\omega\neq\omega_{AG}$. Due to the invariance with respect to phase shift that characterizes the free-running oscillation, the phase of the auxiliary generator may be arbitrarily set to zero $\phi_{AG}=0$.

The use of an auxiliary generator will allow prefixing the desired steady-state oscillation frequency ω_o and even the first harmonic output power P_{out_o}. With the load circuit in Figure 3.11, the oscillator output power at the first harmonic component can be expressed as $P_{out\_o}=1/2G_LV_o^2$. Therefore, we can have different choices of G_L and V_o providing the same output power P_{out_o}. On the other hand, the steady-state solution must fulfill:

$$\begin{aligned} G_N(V_o,\omega_o)+G_L&=0 \\ B_N(V_o,\omega_o)+B_L&=0 \end{aligned} \tag{3.59}$$

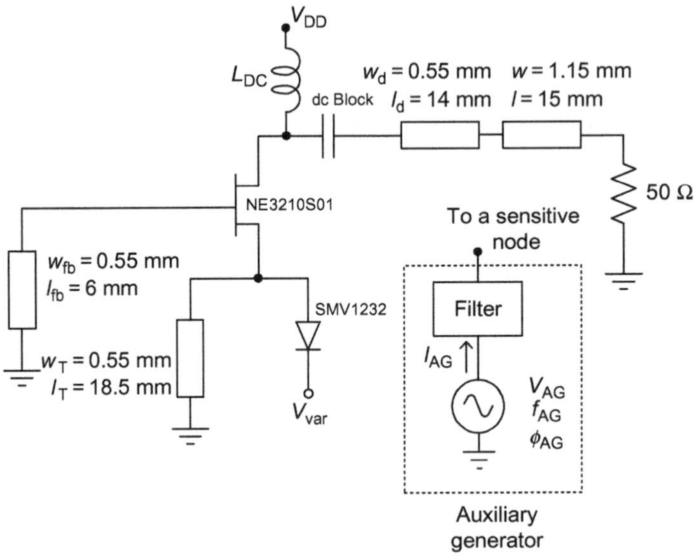

FIGURE 3.12

Schematic of the implemented oscillator including the auxiliary generator.

Taking the output power requirement, as well as the steady-state equations (3.59) into account, we may estimate the load conductance G_L required to obtain the output power P_{out_o} by tracing two different curves. One will correspond to the function $G_N(V_o,\omega_o)$, which will be calculated with a harmonic-balance simulation and represented versus V_o. This will be done by connecting a voltage auxiliary generator in parallel at the circuit port 2. For this analysis, we will load the subcircuit comprised between the terminal impedance and this observation port with a very high resistor R, instead of the original oscillator load $G_L + jB_L$. Note that the purpose now is to obtain a design providing the specified output power P_{out_o}, instead of just fulfilling the oscillation start-up conditions. To obtain the curve $G_N(V_o,\omega_o)$, amplitude of the auxiliary generator will be $A_{AG} = V_o$ and the frequency will be $\omega_{AG} = \omega_o$. Next, A_{AG} will be swept from a near zero to sufficiently large value, which will provide the function $G_N(V_o,\omega_o)$. Because we are performing only an approximate analysis, a simple one-harmonic simulation may be carried out. To obtain the possible V_o values providing the specified output power $P_{out\_o} = 10\,\text{dBm}$, a second curve must be obtained, given by $-G_L(V_o) = -2P_{out\_o}/V_o^2$. The possible values of V_o will correspond to the intersections of the curves $G_N(V_o,\omega_o)$ and $-G_L(V_o) = -2P_{out\_o}/V_o^2$. We must take into account that the conductance G_L must also enable the fulfillment of the oscillation start-up condition, $G_L < |G_N(V \cong 0,\omega_o)|$. The intersection points will be called V_{ox}, G_{Lox}. Application of this method to the circuit in Figure 3.12 provides the two curves in Figure 3.13. There are two intersections, corresponding to two different designs with the same output power $P_{out\_o} = 10\,\text{dBm}$. By choosing a load conductance

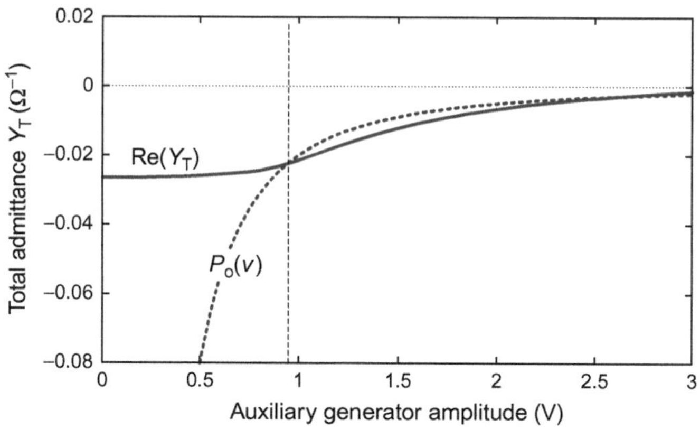

FIGURE 3.13

Approximate method to obtain a steady oscillation providing specific output power. The approximate solutions correspond to the intersections of the two curves $G_N(V_o, \omega_o)$ and $-G_L(V_o) = -2P_{out\_o}/V_o^2$, where $P_{out\_o} = 10\,mW$. The value chosen is $V_{ox} = 0.955\,V$.

$G_L = 0.022\,\Omega^{-1}$, the estimated oscillation point will correspond to the first intersection, indicated with dashed line. Note that harmonic balance may exhibit convergence problems when limiting the number of harmonic terms under large-signal operation. If this is the case, the number of harmonic components may be increased, even though the use of the second curve $-G_L(V_o) = -2P_{out\_o}/V_o^2$ is only fully consistent for an analysis limited to one-harmonic term. In any case, it should be emphasized that the aim of this analysis is only to get insight into the circuit requirements for oscillation at a certain output power. The resulting approximate solution will be used as initial guess for a full harmonic-balance analysis, using an auxiliary generator.

In a second stage, we will perform a full harmonic balance analysis of the circuit, with a sufficiently high number N of harmonic terms. The estimated output load $Y_L(\omega) = G_{Lox} + jB_L(\omega)$ will be connected to the circuit. The initial amplitude and frequency values of the auxiliary generator will be $A_{AG} = V_{ox}$ and $\omega_{AG} = \omega_o$. These values will next be optimized in order to fulfill the oscillation condition $Y_T = Y_N + Y_L = 0$.

The total admittance function at the fundamental frequency Y_T agrees with the opposite of the ratio between the current flowing through the auxiliary generator and the auxiliary-generator voltage $Y_T = -I_{AG}/V_{AG} = -Y_{AG} = 0$. Note that the auxiliary generator only operates at the fundamental frequency, so the rest of harmonic terms are naturally self-sustained by the active device and load of the nonlinear circuit. When the oscillation condition $Y_{AG} = 0$ is fulfilled, the auxiliary generator has no influence over the steady-state solution, since due to zero value of the current at a voltage different from zero, the impedance exhibited by the auxiliary generator tends to infinite.

To obtain the steady-state oscillation, an optimization of the auxiliary-generator amplitude A_{AG} and frequency ω_{AG} will be carried out, in order to fulfill $Y_{AG}(A_{AG}, \omega_{AG}) = 0$ with $N = 5$ harmonic terms, one obtains $f_{AG} = f_o = 5.002$ GHz and the first harmonic amplitude $V_{AG} = V_o = 1.044$ V. To illustrate the accuracy, a validation analysis using the auxiliary-generator amplitude has been carried out. The auxiliary-generator amplitude has been set constant to the optimized value, $A_{AG} = V_o$, performing a sweep in the auxiliary-generator frequency ω_{AG} about the optimized value $\omega_{AG} = \omega_o$. As can be seen, both the real and imaginary part of the admittance function take zero value at $\omega_{AG} = \omega_o$, so the steady-state oscillation condition is fulfilled, as shown in Figure 3.14. Compare with the start-up conditions in Figure 3.11. That figure, corresponding to a small-signal analysis, shows an excess negative conductance at the resonance frequency, as required for the oscillation start-up. In the analysis of Figure 3.14, performed for an amplitude equal to that of the steady-state oscillation $V_{AG} = V_o = 1.044$ V, both the total conductance and total susceptance are zero at the oscillation frequency $f_{AG} = f_o = 5.002$ GHz.

As gathered from the above analysis, after optimization in order to fulfill the steady-state oscillation condition $Y_{AG}(A_{AG}, \omega) = 0$, the amplitude and frequency of the auxiliary generator will be slightly different from the initial ones: $V_{ox} = 0.95$ V and $f_o = 5$ GHz, which is due to the impact of the harmonic terms, which were not taken into account in the estimation of the required amplitude value V_{ox}. With the values calculated through a multiharmonic simulation, the output power is $P_{out} = 10.4$ dBm.

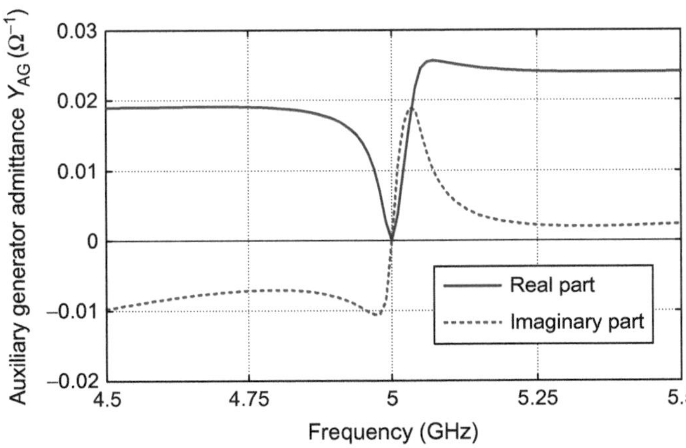

FIGURE 3.14

Validation of the results of the harmonic-balance simulation using an auxiliary generator. The auxiliary-generator amplitude has been set constant to the optimized value, $A_{AG} = V_o$, performing a sweep in the auxiliary-generator frequency ω_{AG} about the optimized value $\omega_{AG} = \omega_o$.

The fundamental aspect of the analysis method based on the use of an auxiliary generator derives from the fact that using this generator the oscillator circuit can be simulated as if it were a *forced system*. An additional advantage is that we can make use of this "forcing" capability to impose the desired steady-state oscillation frequency and amplitude. We can set the auxiliary-generator amplitude and frequency to the desired values $A_{AG}=V_o$ and $\omega_{AG}=\omega_o$ and, instead of using this amplitude and frequency as optimization variables, optimize other circuit elements or parameters. These may belong to the terminal load, the feedback network or the output load. Considering the parameters $\eta_1 \ldots \eta_M$ the condition to impose is $Y_{AG}(\eta_1 \ldots \eta_M)=0$. For an efficient optimization procedure, the number of elements M cannot be too large. On the other hand, we must take into account that an arbitrary change of the circuit element values or parameters may lead to admittance functions for which the oscillation start-up conditions are not fulfilled. In fact, in an oscillator design, we must always cope with two solutions: the *dc* solution and the periodic steady-state solution.

The described optimization procedure has been applied to the oscillator in Figure 3.12, simulated with $N=0.5$ harmonic contents. The imposed oscillation frequency and fundamental output voltage are $\omega_o=2\pi 5\times 10^9$ and $V_o=1\,\text{V}$, corresponding to our desired output power value $P_{out}=10\,\text{dBm}$. The optimization variables are $\eta_1=w_T$, $\eta_2=w_{fb}$, and $\eta_3=w_d$. The resulting output spectrum is shown in Figure 3.15. As can be seen, the specified values of oscillation frequency and output power have been achieved with an accurate multiharmonic analysis.

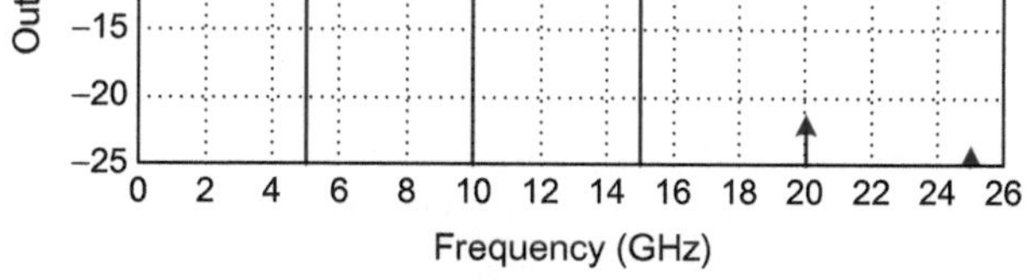

FIGURE 3.15

Oscillator spectrum after use of an auxiliary generator to obtain the specified values of oscillation frequency and output power. The analysis has been carried out with $N=5$ harmonic components.

3.7 STABILITY ANALYSIS

This section describes a *practical* methodology for stability analysis of oscillator circuits, compatible with harmonic-balance simulators [29,30]. Two different types of analysis will be considered: the stability analysis of a particular steady-state solution (either *dc* or periodic), known as *local stability* analysis, and the stability analysis of the set of solutions obtained when varying continuously a circuit parameter, such as a *dc* bias voltage or an element value. The latter analysis aims at detecting qualitative changes of the stability properties or *bifurcations*.

3.7.1 LOCAL STABILITY ANALYSIS

In a free-running oscillator one must always keep in mind the coexistence of two steady-state solutions, the *dc* solution and the oscillatory solution. For the oscillation start-up from the noise level the *dc* solution must be unstable, with a pair of complex-conjugate poles, at a frequency close to the desired oscillation frequency ω_o, located on the right-hand side of the complex plane. The steady-state oscillatory solution must be stable. In the two cases, the stability analysis will rely on the introduction of a small perturbation at the complex frequency s into the system.

We will initially consider the case of the *dc* solution. Prior to the introduction of the perturbation in the nodal harmonic balance formulation, we will only have a system of *dc* equations:

$$\overline{H}\left[\overline{X}_{dc}, dc\right] \equiv \overline{I}\left[\overline{X}_{dc}\right] + C\overline{X}_{dc} + \overline{I}_{s,dc} = \overline{0} \tag{3.60}$$

When introducing the small perturbation, the reactive terms will get into action at the complex frequency s and the variables will undergo a small increment $\Delta\overline{X}$. The vector $\Delta\overline{X}$ refers to the increment undergone by the harmonic components of the state variables in the nodal harmonic-balance formulation, corresponding to the node voltages and inductor currents Because the perturbation is small, we can linearize the circuit nonlinear functions about the *dc* solution $\overline{X}_{dc}$. The third consideration is that there are no sources at the perturbation frequency. These three considerations lead us to the following homogeneous system:

$$\overline{H}\left[\Delta\overline{X}, s\right] \equiv \left[\frac{\overline{I}\left(\overline{X}_{dc}\right)}{\partial\overline{X}} + [s]\frac{\partial\overline{Q}\left(\overline{X}_{dc}\right)}{\partial\overline{X}} + C(s)\right]\Delta\overline{X} = \overline{0} \tag{3.61}$$

In order to have $\Delta\overline{X} \neq 0$, the matrix affecting this vector must be singular, so its associated determinant, named *characteristic determinant*, must be zero. The system generalized eigenvalues are given by the roots of this determinant:

$$\det\left[\frac{\overline{I}\left(\overline{X}_{dc}\right)}{\partial\overline{X}} + [s]\frac{\partial\overline{Q}\left(\overline{X}_{dc}\right)}{\partial\overline{X}} + C(s)\right] = \overline{0} \tag{3.62}$$

In a practical harmonic balance analysis, we can estimate these roots by introducing a small-signal current generator I_p at a sensitive circuit node [29,30], operating at a frequency ω. In the presence of this small-signal generator, the circuit nonlinear functions can also be linearized, which provides:

$$\bar{H}[\Delta\bar{X},\omega] \equiv \left[\frac{\bar{I}(\bar{X}_{dc})}{\partial\bar{X}} + [j\omega]\frac{\partial\bar{Q}(\bar{X}_{dc})}{\partial\bar{X}} + C(j\omega)\right]\Delta\bar{X} = \bar{I}_p \tag{3.63}$$

where $\bar{I}_p$ is a vector containing only the small-signal generator I_p at the position required by the system ranging. Calculation of whatever the output signal will involve the inversion of the matrix on the left-hand side. Therefore, whatever the transfer function we may calculate, the denominator will be the same and will formally agree with the characteristic determinant, evaluated at $j\omega$ instead of s. This transfer function can be calculated as the ratio between the voltage at the node where the current source I_p has been introduced and the corresponding current value. In terms of $j\omega$, the resulting transfer function can be expressed as:

$$Z(j\omega) = \frac{V_p(j\omega)}{I_p(j\omega)} = A\frac{(j\omega - z_1)(j\omega - z_2)\ldots(j\omega - z_q)}{(j\omega - p_1)(j\omega - p_2)\ldots(j\omega - p_r)} \tag{3.64}$$

where the transfer function has been approached as a quotient of polynomials. Note that the above expression would be formally identical in terms of s, since the zeroes and poles are the constant values $z_1, z_2 \ldots z_q$ and $p_1, p_2 \ldots p_r$, which would equally appear in $Z(s)$. The stability analysis based on pole-zero identification consists of fitting the function $Z(j\omega)$ with a quotient of polynomials, as shown in Eq. (3.64), which can be done through a minimum square algorithm.

Following the above description, the stability of the *dc* solution can be analyzed introducing a small-signal current generator at a sensitive location, such as a device terminal, performing a sweep in ω to obtain the ratio $Z(j\omega) = V_p(j\omega)/I_p(j\omega)$ and applying pole-zero identification to this function. This analysis has been applied to the *dc* solution of the circuit in Figure 3.12, after the final design improvement that led to the spectrum

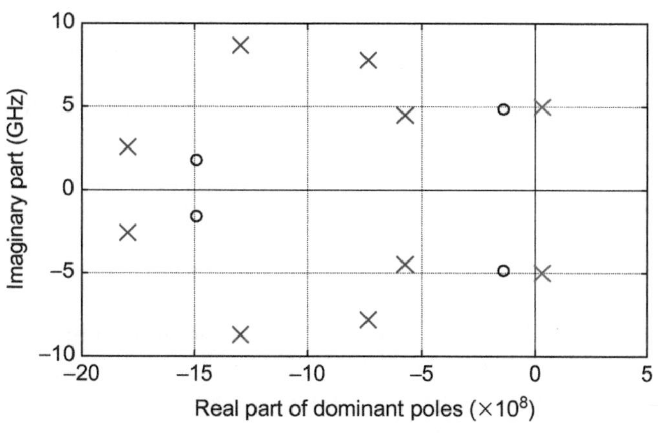

FIGURE 3.16

Pole-zero locus of the *dc* solution of the circuit in Figure 3.12.

in Figure 3.15. The pole-zero locus is shown in Figure 3.16. As can be seen, there are several poles, in agreement with the relatively high order of the differential equation system governing the circuit behavior. The pair of complex-conjugate poles on the right-hand side of the complex plane at 5 GHz is responsible for the oscillation start-up.

The stability analysis of the free-running steady-state oscillation is a bit more involved. The steady-state solution contains the harmonic components $k\omega_o$, where $-N \leq k \leq N$, so after the application of the small perturbation at the complex frequency s, we will have the mixing terms: $k\omega_o + s$ [29,30]. In a manner similar to the case of the *dc* solution, because the perturbation is small, we can linearize the circuit about the steady-state solution, which, in this case, corresponds to the periodic regime at $k\omega_o$. This provides the following homogeneous system:

$$\overline{H}\left[\Delta\overline{X}, jk\omega_o + s\right] \equiv \left[\frac{\overline{I}(\overline{X}_s)}{\partial\overline{X}} + [jk\omega_o + s]\frac{\partial\overline{Q}(\overline{X}_s)}{\partial\overline{X}} + C(jk\omega_o + s)\right]\Delta\overline{X} = \overline{0} \tag{3.65}$$

The derivative matrixes are composed by the harmonic components of the derivatives of the currents and charges with respect to the harmonic components of their control variables, which can be easily calculated from the derivatives of the instantaneous nonlinear functions with respect to their control voltages, for instance, $\partial i_p/\partial v_p$ or $\partial q_s/\partial v_s$.

In order to have $\Delta\overline{X} \neq 0$ in Eq. (3.65), the matrix affecting this vector must be singular, so its associated determinant, named *characteristic determinant* must be zero. Obviously, if we introduce a small-signal current generator I_p at any of the mixing frequencies $jk\omega_o + \omega$, and calculate any possible transfer function, as the ratio of whatever output signal $\Delta X(jl\omega_o + \omega)$, where $-N \leq l \leq N$, the denominator will be the same and will formally agree with the characteristic determinant associated with the system (3.65).

Therefore, following similar reasoning to that of the stability analysis of the *dc* solution, we can introduce a small-signal current source at a frequency ω that is nonharmonically related to the fundamental frequency of the steady-state solution of the free-running oscillator ω_o. Next, the frequency ω is swept, calculating a convenient transfer function (a function enabling good numerical accuracy), such as the ratio $Z(\omega)$ between the voltage $V_p(\omega)$ at the node where the current source is connected and the current value I_p. Selection of the baseband component $V_p(\omega)$, corresponding to $k = 0$, usually provides good results, although the denominator of the transfer function should be the same, regardless of the selected output variable and sideband component. The poles associated to the periodic solution are calculated applying pole-zero identification to the function $Z(\omega)$, in a manner similar to Eq. (3.64).

The analysis described has been applied to the steady-state solution of the free-running oscillator in Figure 3.12, after the final design improvement that led to the spectrum in Figure 3.15. The pole locus obtained is shown in Figure 3.17. Note that the poles on the imaginary axis are due to the autonomy of the steady-state oscillation. All the rest of the poles are located on the left-hand side of the complex plane, so this oscillation is stable.

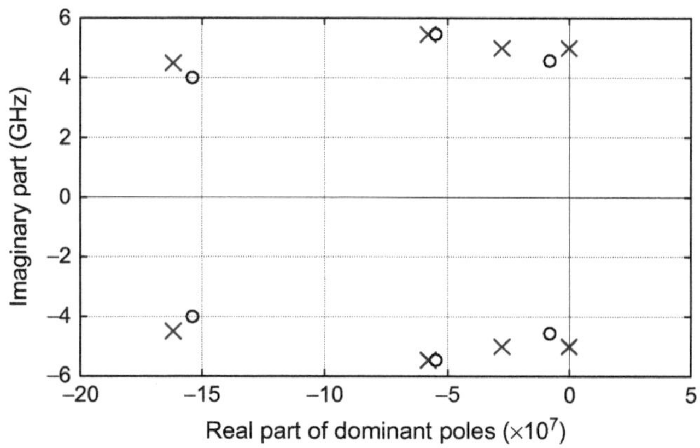

FIGURE 3.17

Pole locus of the periodic steady-state oscillation of the circuit in Figure 3.12. The pair of complex-conjugate poles on the imaginary axis is due to the autonomy of the steady-state solution.

3.7.2 BIFURCATIONS

A bifurcation [2,3,7] is a qualitative change in the stability properties of a steady-state solution or in the number of steady-state solutions, obtained when continuously varying a circuit parameter, such as a bias voltage [4–6]. In this section, we will focus on the bifurcations that can be expected in free-running oscillators, considering both the *dc* solution and the free-running oscillatory solution.

3.7.2.1 Bifurcation from a dc *solution*

Because the oscillation start-up requires the fulfillment of certain mathematical conditions, we can easily figure out that these conditions will not be fulfilled for all the possible values of the circuit parameters, as already shown in Figure 3.8, where the oscillation amplitude of the circuit in Figure 3.2A has been traced versus the linear conductance G_L in the two cases of a cubic nonlinearity and a nonlinearity of fifth order. As another example, in the case of a transistor-based oscillator, the oscillation will obviously be suppressed when biasing this transistor device below conduction threshold, due to the associated gain reduction.

If we vary a sensitive parameter η continuously, the *dc* solution will be different for each η value and so will be its stability properties, determined by the linearization equation (3.61). We will have a *bifurcation* at the particular parameter value(s) at which either a real pole or a pair of complex-conjugate poles cross the imaginary axis. The most common type of bifurcation from *dc* regimes is the Hopf bifurcation, at which a pair of complex-conjugate poles $\sigma \pm j\omega$ cross the imaginary axis.

Assume that increasing a parameter η, a pair of complex-conjugate poles $\sigma \pm j\omega$ crosses to the RHP at the particular value η_{o1}. The *dc* regime will be unstable from the parameter value η_{o1}, because under perturbation it will exhibit a pair of

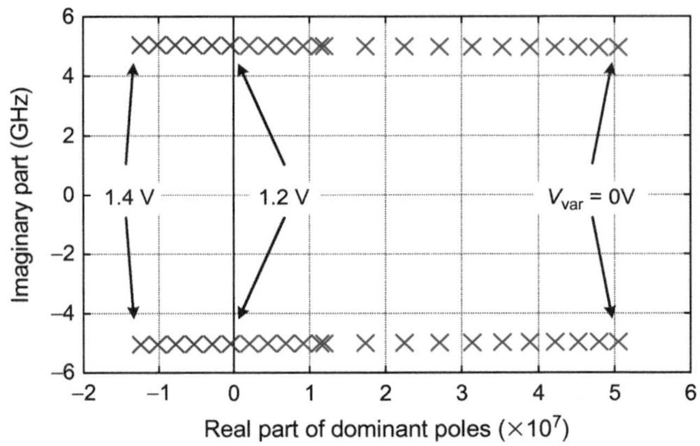

FIGURE 3.18

Evolution of the dominant complex-conjugate poles versus the varactor bias voltage V_{var}. A bifurcation is obtained at $V^b_{var} = 1.2\,V$.

complex-conjugate poles on the RHP. The *dc* regime may eventually become stable again if when further varying the parameter, the pair of complex-conjugate poles cross again to the left-hand side of the complex plane (LHP) at a different parameter value η_{o2}. As an example, Figure 3.18 presents an analysis of the dominant poles of the *dc* solution of the circuit in Figure 3.12 performed versus the bias voltage of the varactor diode. For the original value $V_{var} = 0\,V$, the *dc* solution is unstable and we will have the start-up of an oscillation at about the frequency of the pair of complex-conjugate poles on the RHP. When increasing V_{var}, this pair of complex-conjugate poles approach the imaginary axis and cross this axis at the bifurcation value $V^b_{var} = 1.2\,V$. Because the complex-conjugate poles cross from the RHP to the LHP, we have an *inverse* Hopf bifurcation. For $V_{var} > V^b_{var}$ all the poles associated to the *dc* solution are on the LHP, so this solution is stable. There will be no oscillation start-up from the noise level.

At the bifurcation parameter, the perturbation system is fulfilled for $s = \pm j\omega$, or equivalently, for a real frequency ω. Therefore, at the Hopf bifurcation points, the steady-state oscillation condition is fulfilled for amplitude tending to zero. Indeed, if we trace the variation of the steady-state oscillation amplitude versus the parameter, we will observe that this amplitude decays to zero at the Hopf bifurcation points. The oscillation curve has slope tending to infinite at the Hopf bifurcation [4], in agreement with the two poles at zero value of the periodic regime at the point where the oscillation is generated. As already shown, it is very convenient to use an auxiliary generator to obtain the steady-state oscillatory solution. To cope with the problem of tracing the high-slope sections of the oscillation curve, we can sweep the amplitude of the auxiliary generator A_{AG} and solve the non-perturbation condition for the parameter and the oscillation frequency, which due to the autonomous behavior of the circuit will always be an unknown of the problem. These high-slope sections will be obtained by sweeping A_{AG} and solving $Y_{AG}(\eta, \omega_{AG}) = 0$.

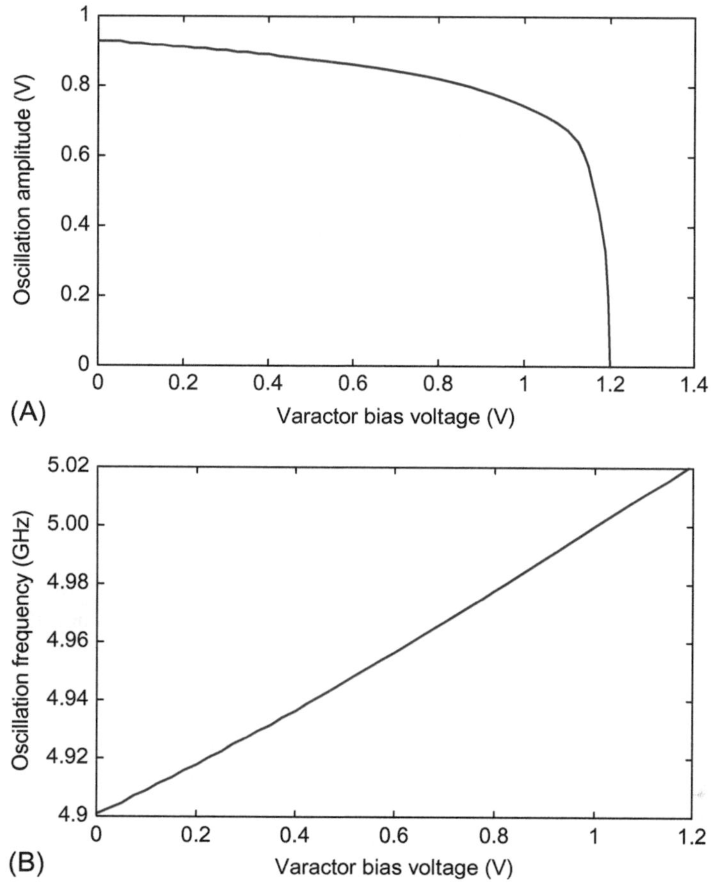

FIGURE 3.19

Variation of steady-state oscillatory solution of the circuit in Figure 3.12 versus the varactor bias voltage. (A) Oscillation amplitude. (B) Oscillation frequency.

Figure 3.19 presents the variation of the steady-state oscillation amplitude (A) and frequency (B) of the circuit in Figure 3.12, traced versus the varactor bias voltage. As can be seen, the oscillation amplitude tends to zero at the bifurcation point (see Figure 3.19A), obtained at the varactor voltage $V_{var}^{b} = 1.2\,\text{V}$.

3.7.2.2 Bifurcations from a periodic solution

The periodic solution may undergo three main types of bifurcation depending on the frequency value of the poles crossing the imaginary axis and its relationship with the frequency of the periodic regime. These three main types of bifurcation are turning points, Hopf bifurcations and flip bifurcations [4–6].

Turning point

Assume that a circuit parameter η is varied continuously. We will obtain a turning-point bifurcation if a real pole pair γ crosses the imaginary axis at a particular parameter value η_o. Due to the non-univocal relationship between poles and multipliers, when the real pole γ crosses the axis, the whole family of poles $\gamma + jk\omega_o$, will also cross the axis, where k is an integer and ω_o is the steady-state oscillation frequency. All of them are associated with a same positive real multiplier. When the real pole γ takes a zero value, the perturbed system degenerates into the Jacobian matrix of the harmonic-balance system. In fact, the characteristic determinant evaluated at $s=0$ agrees with the determinant of the Jacobian matrix of the error vector of the harmonic-balance system, given by $\overline{H}\left[\overline{X}, \overline{k}\omega_o\right]$ in Eq. (3.58). Therefore, the harmonic-balance system becomes singular when $\gamma = 0$, which gives rise to an infinite slope of the solution curve at this bifurcation point. An example of turning-point bifurcation can be seen in the oscillation curve of the single resonator circuit with a polynomial function of fifth order, presented in Figure 3.6B.

Hopf bifurcation

Assume that a circuit parameter η is varied continuously. We will obtain a Hopf bifurcation if a pair of complex-conjugate poles $\sigma \pm j\omega_o'$ crosses the imaginary axis at a particular value η_o, under the condition that the frequency ω_o' is nonharmonically related with the fundamental frequency ω_o of the periodic regime. The pair of complex-conjugate poles $\sigma \pm j\omega_o'$ is associated with a pair of complex-conjugate Floquet multipliers.

Once the periodic solution at the frequency ω_o has the pair of complex-conjugate poles $\sigma \pm j\omega_o'$ on the right-hand side of the complex plane, this solution (which will continue to exist as a mathematical solution) will no longer be physically observable. The pair of poles $\sigma \pm j\omega_o'$ on the right-hand side of the complex plane will give rise to the onset of a second oscillation, leading to a steady-state quasiperiodic regime at two independent fundamental frequencies ω_o and ω_o'.

Flip bifurcation

Assume that a circuit parameter η is varied continuously. We will obtain a flip bifurcation if a pair of complex-conjugate poles $\sigma \pm j\omega_o/2$ crosses the imaginary axis at the particular value η_o. The pair of complex-conjugate poles $\sigma \pm j\omega_o/2$ is associated with a negative real Floquet multiplier.

Once the periodic solution at the frequency ω_o has the pair of complex-conjugate poles $\sigma \pm j\omega_o/2$ on the right-hand side of the complex plane, this solution (which will continue to exist as a mathematical solution) will no longer be physically observable. The pair of poles $\sigma \pm j\omega_o/2$ on the right-hand side of the complex plane will give rise to a frequency division by 2.

Hopf and flip bifurcations are less usually obtained in free-running oscillators than in forced nonlinear circuits such power amplifiers or frequency multipliers. However, it is not unlikely to obtain the anomalous generation of undesired additional spectral lines when varying a circuit parameter. The frequency of spectral lines will be subharmonic, in the case of a flip bifurcation, or nonrationally related with that of the previously existing oscillation, in the case of a Hopf bifurcation.

3.8 PHASE NOISE

In oscillator circuits the phase noise introduces an undesired phase modulation in the oscillating signal. This signal is often used in the frequency-conversion stages of communication systems, to up- or down-convert the carrier frequency of a modulated signal. In the mixing procedure the phase noise of the local oscillator (LO) corrupts the modulation signal, which can give rise to demodulation errors. Another consequence of the phase noise is that it broadens the spectrum of the oscillating signal. As a consequence, adjacent spectra from different down-converted signals containing modulations may overlap and become corrupted [31]. This phenomenon may degrade the performance of a system of wireless communications, as it is illustrated in the following example. Consider a generic receiver which down-converts an input modulated signal with carrier frequency f_1 to the baseband at $f_1 - f_o$. This procedure can be carried out by mixing the input signal with a LO with frequency $f_o = f_1 - f_{IF}$ and filtering the mixer output. In practice, the input signal may be accompanied by a higher power interferer in an adjacent channel with carrier frequency f_2. If the LO output contains phase noise, both down-converted signals result corrupted. As a consequence, the down-converted band consists of two overlapping spectra, with the wanted signal at f_{IF} suffering from significant noise due to tail of the interferer which is centered at $f_2 - f_o$. This effect is called "reciprocal mixing" [31]. The oscillator phase noise depends on the kind of active device on which the design is based, as transistors usually contain the noise sources with the highest spectral densities. The flicker noise, or $1/f$ noise, which is a low frequency noise, is mostly due to random trapping and de-trapping of charge carriers, and exhibits lower values in hetero-junction bipolar transistors (HBT) than in field-effect transistors [32]. Due to the nonlinear behavior of the oscillator, this noise will be up-converted to the carrier frequency and will be the dominant noise contribution at low offset frequency. On the other hand, the shot noise, with a white spectrum, is due to the fluctuation in the number of carriers traversing a potential barrier, and will be higher in HBT devices. Regarding the circuit design, as a general rule, high frequency selectivity will be needed because of the direct relationship between phase and frequency fluctuations. Thus, the phase noise will decrease with the oscillator quality factor, as predicted by the semiempirical Leeson's model [33]. Attention must be paid to basic circuit elements, such as inductors, and their quality factors, not to degrade the overall performance. The values of the device noise sources depend on the bias point and there is also an interaction with the periodic large-signal voltages or currents, leading to a cyclostationary behavior [32]. The device operation point must also be suitably chosen to avoid both high values of the noise sources and a strong up-conversion of this noise.

In the following, a formal derivation of the phase-noise mathematical modeling in the frequency domain will be carried out. The resulting formulation will provide useful equations that can be applied to simulate the oscillator phase noise spectrum. The analysis will be illustrated by means of its application to the FET based oscillator used in Section 3.5.

3.8.1 FREQUENCY-DOMAIN TECHNIQUES

Low phase noise is a key requirement in modern radiofrequency and microwave communication systems in order to ensure the adequate recovery of the transmitted signals. Low phase-noise oscillators are needed and so are reliable simulation tools enabling an efficient design. The frequency-domain methods, such as harmonic balance, are well suited for the noise analysis [1], because the noise sources are usually described by their power spectral densities. For the frequency-domain analysis of phase noise, two different phenomena are considered: the modulation of the carrier frequency by the noise sources and the frequency conversion, as in a mixer. When using harmonic balance, the oscillator phase noise is calculated through two different linearizations of the harmonic-balance system, versus the noise perturbations. They respectively constitute the so-called matrix-conversion and carrier-modulation approaches.

3.8.2 PHASE-NOISE DYNAMICS

In order to obtain a formulation describing the phase-noise dynamics, it is useful to apply the modified nodal approach (3.56), modeling the microwave oscillator with a set of nonlinear integro-differential equations in an n-dimension vector $\overline{x}(t)$ of state variables [6,10].

The trajectory traced by the periodic steady-state solution $\overline{x}_s(t)$ as time evolves can be represented in a coordinate system provided by the circuit state variables. At the steady state, this representation provides a one-dimensional closed curve called *limit cycle*. Assuming that the oscillating solution is stable, its response to noise perturbations can be illustrated by considering a noise source that perturbs the state variables in $t=0$, pushing the system trajectory away from the limit cycle during an infinitesimal time interval (see Figure 3.20). When removing the noise source, the trajectory is attracted again towards the limit cycle, eventually reaching a time-shifted solution $\overline{x}_s(t+\tau)$ which, due to the autonomy of the oscillator system, is also a solution to system (3.56). According to expansion (3.57), each kth harmonic component of $\overline{x}_s(t+\tau)$ is phase shifted by an amount $\Delta\phi_k = k\omega_o\tau$ with respect to those of the unperturbed solution $\overline{x}_s(t)$, whereas the amplitudes of the harmonic components in both signals remain equal.

In practice, when a typical noise source is introduced in the oscillator, the steady-state solution is constantly perturbed and, due to the oscillator autonomy, the phase perturbation at each harmonic is accumulated. In the frequency domain, the phase

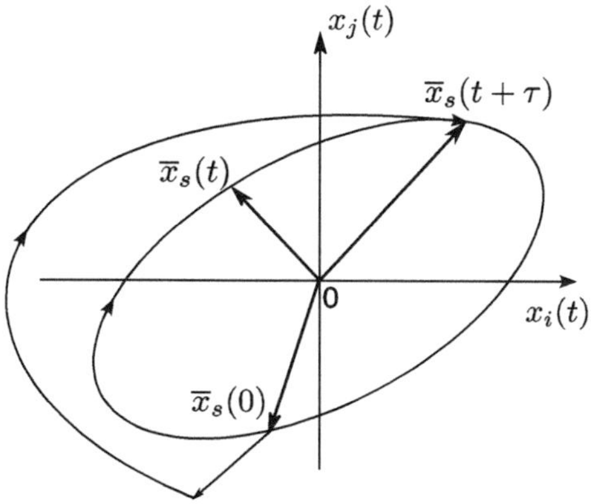

FIGURE 3.20

Schematic of the limit cycle in the space provided by two state variables $x_i(t)$ and $x_j(t)$, together with the evolution of the perturbed trajectory.

noise is the dynamical phase perturbation of each harmonic component. As it will be shown in the following analyses, the phase perturbations $\Delta\phi_k(t)$ can grow unboundedly. The amplitude noise is related to the perturbations in the amplitude of the harmonics. Because these perturbations are attenuated by the oscillator dynamics, their contribution to the oscillator noisy spectrum can be neglected when compared to the phase noise.

The harmonic-balance system (3.58) corresponding to oscillator equation (3.56) is of great utility for the phase-noise study. For a given oscillating steady-state solution $\overline{x}_s(t)$ to system (3.56), the oscillator autonomy states that $\overline{x}_s(t+\tau)$ is also a solution for any constant time shift value τ. In the frequency domain, this statement provides the following property of the function $\overline{H}\left[\overline{X},\overline{k}\omega_0\right]$:

$$\overline{H}\left[\overline{X}_s(\tau),\overline{k}\omega_o\right]=\overline{0},\ \ \forall\tau\rightarrow\left.\frac{\partial\overline{H}\left[\overline{X}_s(\tau),\overline{k}\omega_o\right]}{\partial\tau}\right|_{\tau=0}=H_X|_o\overline{U}_1=\overline{0},$$
$$H_X\left[\overline{X},\overline{k}\omega\right]\equiv\frac{\partial\overline{H}\left[\overline{X},\overline{k}\omega\right]}{\partial\overline{X}},\ \ H_X|_o\equiv H_X\left[\overline{X}_o,\overline{k}\omega_o\right],\ \ \overline{U}_1\equiv\left.\frac{\partial\overline{X}_s(\tau)}{\partial\tau}\right|_{\tau=0} \tag{3.66}$$

where $\overline{X}_s(\tau)$ is the vector containing the harmonic components of the vector signal $\overline{x}_s(t+\tau)$. Then, due to the oscillator autonomy, the Jacobian matrix evaluated at the steady-state solution is singular with at least one zero eigenvalue, which will be denoted by $\lambda_1=0$, and with $\overline{U}_1$ being the eigenvector associated to λ_1. In the presence of noise sources, Eq. (3.56) becomes:

$$\overline{i}[\overline{x}(t)]+\frac{d}{dt}\overline{q}[\overline{x}(t)]+\int_{-\infty}^{t}c(t-\tau)\overline{x}(\tau)d\tau+\overline{i}_s=g_w[\overline{x}(t)]\overline{n}_w(t)+g_\gamma[\overline{x}(t)]\overline{n}_\gamma(t) \tag{3.67}$$

where the matrices $g_w[\overline{x}(t)]$, $g_\gamma[\overline{x}(t)]$ represent the first-order system response to the vectors of white $\overline{n}_w(t)$ and colored $\overline{n}_\gamma(t)$ noise sources. For compactness of the formulation, the terms of Eq. (3.67) involving the noise sources will be included in a unique term:

$$g[\overline{x}(t)]\overline{n}(t) \equiv g_w[\overline{x}(t)]\overline{n}_w(t) + g_\gamma[\overline{x}(t)]\overline{n}_\gamma(t) \tag{3.68}$$

where the vector $\overline{n}(t)$ contains all the white and colored noise sources. These noise sources are modeled as stochastic processes of small amplitude and narrow band PSD [11] about each harmonic component $k\omega_0$. Thus, the vector $\overline{n}(t)$ can be expressed in the following manner:

$$\overline{n}(t) = \sum_{k=-N}^{N} \Delta\overline{N}_k(t)e^{jk\omega_0 t}, \quad \Delta\overline{N}_k(t) = \int_{-BW/2}^{BW/2} \Delta\overline{N}_k(\Omega)e^{j\Omega t}d\Omega \tag{3.69}$$

where the harmonic components $\Delta\overline{N}_k(t)$ are the vectors of narrow-band stochastic processes and BW is the considered noise bandwidth. The effect of the noise sources is to perturb the harmonic components of the oscillating steady-state solution, whose state variables can be expressed as:

$$x_i(t) = \sum_{k=-N}^{N} X_{ik}(t)e^{jk\omega_o t}, \quad X_{ik}(t) = \int_{-BW/2}^{BW/2} X_{ik}(\Omega)e^{j\Omega t}d\Omega \tag{3.70}$$

Each of the perturbed harmonic components $X_{ik}(t)$ is a complex time-dependent function that can be decomposed by means of a pair of real numbers $(\Delta A_{ik}, \Delta\phi_{ik})$ representing the phase and amplitude perturbations. In the following, two phase-noise techniques that use different harmonic decompositions will be presented.

3.8.3 CONVERSION MATRIX APPROACH

This approach uses the harmonic decomposition shown in Figure 3.21, where the phase and amplitude perturbations $(\Delta\phi_{ik}, \Delta A_{ik})$ are responsible for the harmonic

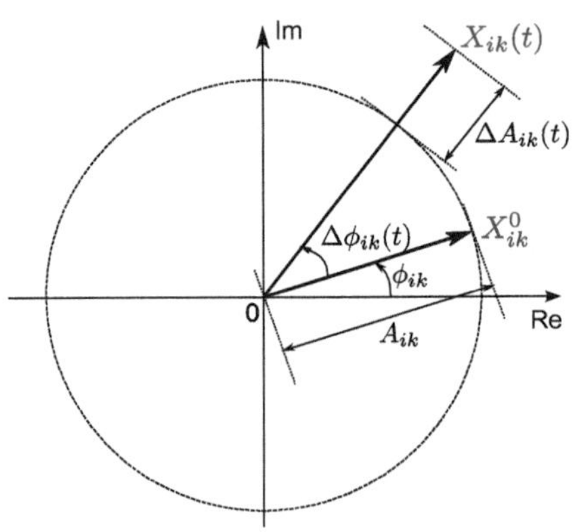

FIGURE 3.21

Conversion matrix approach. Harmonic decomposition in the complex plane.

rotation and elongation, respectively. Each harmonic component is then expressed as [34,35]:

$$X_{ik}(t) = [A_{ik} + \Delta A_{ik}(t)]e^{j\Delta\phi_{ik}(t)}e^{j\phi_{ik}} \tag{3.71}$$

where the unperturbed steady-state harmonics are given by $X_{ik}^{o} = A_{ik}e^{j\phi_{ik}}$. Assuming small amplitude and phase perturbations, expression (3.71) can be approached by:

$$X_{ik}(t) \approx [A_{ik} + \Delta A_{ik}(t)][1 + j\Delta\phi_{ik}(t)]e^{j\phi_{ik}} \approx A_{ik}e^{j\phi_{ik}} + [\Delta A_{ik}(t) + jA_{ik}\Delta\phi_{ik}(t)]e^{j\phi_{ik}} \tag{3.72}$$

Due to the small amplitude of the noise sources, the superposition principle can be applied, analyzing separately the effect of each noise harmonic component $\Delta\overline{N}_k(\Omega)e^{j\Omega t}$. The amplitude and phase perturbation response to each of these components are:

$$\Delta A_{ik}(t) = \Delta A_{ik}(\Omega)e^{j\Omega t}, \quad \Delta\phi_{ik}(t) = \Delta\phi_{ik}(\Omega)e^{j\Omega t} \tag{3.73}$$

Introducing expressions (3.73) in Eq. (3.72) the perturbed solution (3.70) becomes:

$$x_i(t) = \sum_{k=-N}^{N} A_{ik}e^{j(k\omega_o t + \phi_{ik})} + \Delta X_{ik}(\Omega)e^{jk(\omega_o + \Omega)t}, \quad \Delta X_{ik}(\Omega) = [\Delta A_{ik}(\Omega) + jA_{ik}\Delta\phi_{ik}(\Omega)]e^{j\phi_{ik}} \tag{3.74}$$

Eq. (3.74) implies that each noise component $\Delta\overline{N}_k(\Omega)e^{j\Omega t}$ produces in the perturbed solution spectrum a single sideband $\Delta X_{ik}(\Omega)$ at a frequency offset Ω from each carrier at $k\omega_o$. These sidebands can be calculated by introducing expression (3.74) in Eq. (3.67), obtaining the following system in the frequency domain [34–36]:

$$H_X\left[\overline{X}_o, \overline{k}\omega_o + \Omega\right]\Delta\overline{X}(\Omega) = T_G\Delta\overline{N}(\Omega) \tag{3.75}$$

where $\overline{X}_o$, $\Delta\overline{X}(\Omega)$ and $\Delta\overline{N}(\Omega)$ contain, respectively, the harmonics terms X_{ik}^{o}, $\Delta X_{ik}(\Omega)$ and $\Delta\overline{N}_k(\Omega)$, and T_G is the Toeplitz matrix corresponding to the time-domain periodic matrix $g[\overline{x}_s(t)]$. The vector of sidebands is obtained from Eq. (3.75) as:

$$\Delta\overline{X}(\Omega) = H_X\left[\overline{X}_o, \overline{k}\omega_o + \Omega\right]^{-1}T_G\Delta\overline{N}(\Omega) \tag{3.76}$$

The phase and amplitude perturbations can be extracted from the sideband components by the following procedure:

$$\Delta\phi_{ik}(\Omega) = \frac{\Delta X_{ik}(\Omega)e^{-j\phi_{ik}} - \Delta X_{ik}(-\Omega)^{*}e^{j\phi_{ik}}}{2jA_{ik}}, \quad \Delta A_{ik}(\Omega) = \frac{\Delta X_{ik}(\Omega)e^{-j\phi_{ik}} + \Delta X_{ik}(-\Omega)^{*}e^{j\phi_{ik}}}{2} \tag{3.77}$$

where expression (3.74) together with the relations $\Delta A_{ik}(-\Omega) = \Delta A_{ik}(\Omega)^{*}$ and $\Delta\phi_{ik}(-\Omega) = \Delta\phi_{ik}(\Omega)^{*}$ have been applied.

The oscillator state variables become stochastic processes in the presence of noise sources. Then, the oscillator output spectrum is provided by PSD of the state variable $x_o(t)$ corresponding to the oscillator output signal. Expressing the harmonics of $x_o(t)$

in the form (3.71) and assuming that $x_o(t)$ is a stationary process, its power spectrum is given by the Fourier transform of the correlation function:

$$S_o(\Omega)=F\{\langle x_o(t)x_o(t+\tau)\rangle\}\approx\sum_{k=-N}^{N}\left|X_{ok}^{o}\right|^2 S_{k\phi}(\Omega-k\omega_o),\quad S_{k\phi}(\Omega)\equiv F\left\{\left\langle e^{j[\Delta\phi_{ok}(t+\tau)-\Delta\phi_{ok}(t)]}\right\rangle\right\} \tag{3.78}$$

where F is the Fourier transform operator and the amplitude perturbation components $\Delta A_{ok}(t)$ have been neglected. The spectral components $S_{k\phi}(\Omega)$ are narrow-band frequency-dependent functions. Then, expression (3.78) states that the oscillator output spectrum due to phase perturbations is composed by each of the functions $S_{k\phi}(\Omega)$ placed about the corresponding kth harmonic component, with Ω representing the frequency offset from each carrier at $k\omega_o$. According to the analysis [37], for Ω high enough these functions can be approached by:

$$S_{k\phi}(\Omega)\approx|\Delta\phi_{ok}(\Omega)|^2 \tag{3.79}$$

3.8.4 CARRIER-MODULATION APPROACH

In this case, the perturbed state variables are expressed as:

$$x_i(t)=\sum_{k=-N}^{N}\left[X_{ik}^{o}+\Delta X_{ik}(t)\right]e^{jk[\omega_o t+\Delta\phi(t)]}=\sum_{k=-N}^{N}X_{ik}(t)e^{jk\omega_o t} \tag{3.80}$$

where $\Delta X_{ik}(t)$ is the complex amplitude perturbation and $\Delta\phi(t)$ is a common phase perturbation. This approach uses the harmonic decomposition shown in Figure 3.22. As a difference from the conversion matrix approach, in this case a common phase perturbation $\Delta\phi(t)$ equally affecting all the harmonic components is considered [36].

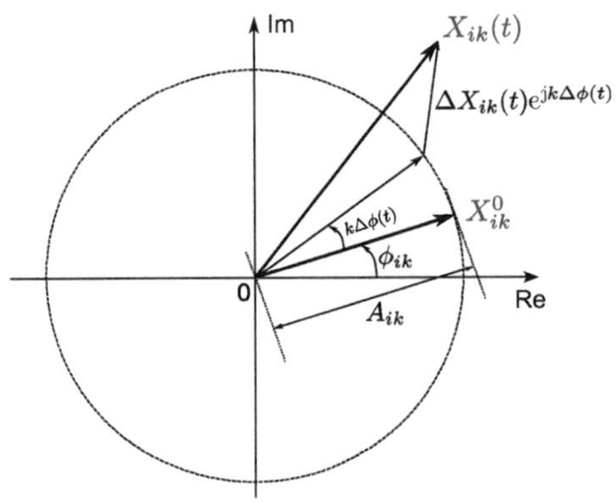

FIGURE 3.22

Carrier-modulation approach. Harmonic decomposition in the complex plane.

Introducing expression (3.80) in Eq. (3.67), the following time-frequency domain system is obtained:

$$\overline{H}\left[\overline{X}_o + \Delta\overline{X}(t), \overline{k}(\omega_o + \Delta\omega(t))\right] = T_G \Delta\overline{N}(t), \quad \Delta\omega(t) \equiv \Delta\dot{\phi}(t) \tag{3.81}$$

where $\Delta\overline{X}(t)$, $\Delta\overline{N}(t)$ contain, respectively, the time-varying harmonic terms $\Delta X_{ik}(t)$ and $\Delta\overline{N}_k(t)$, and $\Delta\omega(t)$ is the instantaneous frequency perturbation. Since both $\Delta\overline{X}(t)$ and $\Delta\omega(t)$ may be considered of small amplitude, the vector function in Eq. (3.81) can be approached by a first-order Taylor series obtaining:

$$H_X|_o \Delta\overline{X}(t) + \overline{H}_\omega\left[\overline{X}_o, \overline{k}\omega_o\right]\Delta\omega(t) = T_G \Delta\overline{N}(t) \tag{3.82}$$

where matrix $H_X|_o$ was defined in Eq. (3.66), the steady-state condition $H\left[\overline{X}_o, \overline{k}\omega_o\right] = \overline{0}$ has been applied and time derivative terms $\Delta\dot{\overline{X}}(t)$ have been neglected. The frequency perturbation $\Delta\omega(t)$ can be calculated from system (3.82) by making use of the singularity of matrix $H_X|_o$:

$$\Delta\omega(t) = \Delta\dot{\phi}(t) = \frac{\overline{V}_1^+ T_G}{\overline{V}_1^+ \overline{H}_\omega\left[\overline{X}_o, \overline{k}\omega_o\right]}\Delta\overline{N}(t) = \overline{B}_w^+ \Delta\overline{N}_w(t) + \overline{B}_\gamma^+ \Delta\overline{N}_\gamma(t) \tag{3.83}$$

where the left eigenvector $\overline{V}_1^+ H_X|_o = \overline{0}^+$ associated to the eigenvalue $\lambda_1 = 0$ has been used. In Eq. (3.83) the harmonic components of the noise sources have been separated into white and colored noise sources. Then, the sensitivity vectors $\overline{B}_w^+ = \overline{V}_1^+ T_{G_w}$ and $\overline{B}_\gamma^+ = \overline{V}_1^+ T_{G_\gamma}$ have been introduced, with T_{G_w} and T_{G_γ} being the Toeplitz matrixes corresponding to the time-domain periodic matrixes $g_w[\overline{x}_s(t)]$ and $g_\gamma[\overline{x}_s(t)]$.

The noisy spectrum can be calculated following a procedure similar to the one used in the conversion matrix analysis. Neglecting the complex amplitude terms in expression (3.80), the PSD of the oscillator output signal is:

$$S_o(\Omega) = F\{\langle x_o(t)x_o(t+\tau)\rangle\} \approx \sum_{k=-N}^{N} \left|X_{ok}^o\right|^2 S_{k\phi}(\Omega - k\omega_o), \quad S_{k\phi}(\Omega) \equiv F\left\{\left\langle e^{jk[\Delta\phi(t+\tau)-\Delta\phi(t)]}\right\rangle\right\} \tag{3.84}$$

As in the previous analysis, the spectral components $S_{k\phi}(\Omega)$ are narrow-band frequency-dependent functions and $S_o(\Omega)$ is composed by each of these functions placed about the corresponding kth harmonic component. For high enough frequency offset Ω, the phase-noise spectrum about the kth harmonic component can be approached by [37]:

$$S_{k\phi}(\Omega) \approx \frac{k^2}{\Omega^2} F\{\langle\Delta\omega(t)\Delta\omega(t+\tau)\rangle\} = \frac{k^2}{\Omega^2}\overline{B}_w^+ \Gamma \overline{B}_w + \frac{k^2}{\Omega^2}\sum_{i=1}^{m}\left|B_{\gamma 0}^i\right|^2 S_{\gamma i}(\Omega) \tag{3.85}$$

where expression (3.83) has been applied, Γ is the frequency-domain correlation matrix of the white noise sources, $B_{\gamma 0}^i$ are the *dc* components of the sensitivity vector $\overline{B}_\gamma^+$ and each $S_{\gamma i}(\Omega)$ is the PSD of the ith colored noise source. In this analysis, baseband uncorrelated colored noise sources have been considered.

3.8.5 NEAR CARRIER SPECTRUM DUE TO PHASE NOISE

The phase-noise spectrum near to each kth carrier is given by each function $S_{k\phi}(\Omega)$ calculated for small values of the frequency offset Ω. Assuming that the noise sources are constituted by Gaussian processes, these functions have the form [9,10,38,39]:

$$S_{k\phi}(\Omega) = F\{R_k(\tau)\}, \quad R_k(\tau) = \exp\left\{-\frac{k^2\sigma_\phi^2(|\tau|)}{2}\right\}, \quad \sigma_\phi^2(\tau) = \left\langle[\Delta\phi(t+\tau) - \Delta\phi(t)]^2\right\rangle \tag{3.86}$$

In the case of a common phase perturbation, the variance term $\sigma_\phi^2(\tau)$ can be calculated from Eq. (3.83), obtaining [9]:

$$\sigma_\phi^2(\tau) = \frac{1}{\pi}\int_{-\infty}^{\infty}\int S_{\Delta\omega}(\Omega)\frac{1-e^{j\Omega|\tau|}}{\Omega^2}d\Omega \tag{3.87}$$

where $S_{\Delta\omega}(\Omega)$ is the PSD of the frequency perturbation process $\Delta\omega(t)$ that, from Eq. (3.83), depends on the noise sources present in the circuit.

The procedure to obtain the near carrier spectrum is the following. Once the noise sources are identified, the function $S_{\Delta\omega}(\Omega)$ can be analytically obtained. Then, this function is introduced in Eq. (3.87) to numerically calculate the variance term $\sigma_\phi^2(\tau)$. There are several numerical methods [39] to approach expression (3.87) by means of sine and cosine integral functions, instead of performing the numerical resolution of the integral. Once $\sigma_\phi^2(\tau)$ is calculated for a long enough interval in τ, the correlation function $R_k(\tau)$ is calculated and the Fourier transform is applied to obtain the corresponding noise spectrum component $S_{k\phi}(\Omega)$.

3.8.6 APPLICATION EXAMPLE

The presented techniques have been applied to FET based oscillator used in Section 3.5. A voltage-noise source has been introduced in series with the intrinsic gate terminal, modeling the flicker noise source. Thermal-noise generators have also been added to all of the resistive elements. The simulated phase noise spectra are presented in Figure 3.23. When using the conversion matrix technique, once the phase components $\Delta\phi_{ok}(\Omega)$ have been calculated, the phase-noise spectrum has been approached by expression (3.79). Note that, as the frequency offset from the carrier is reduced towards zero, the predicted spectrum grows unboundedly. This is because, due to the oscillator autonomy, the frequency-dependent Jacobian matrix inverted in Eq. (3.76) becomes singular when $\Omega \to 0$, as demonstrated in Eq. (3.66). Indeed, for small frequency offset values approach (3.79) fails and can provide inaccurate results.

In this oscillator, the result of the carrier-modulation technique agrees with that of the conversion matrix. In general, due to the different definition of the phase perturbation considered in both techniques, both phase-noise spectra may present discrepancies, especially in low stability margin configurations. As in the previous analysis, the phase-noise spectrum obtained from approach (Eq. (3.85)) grows

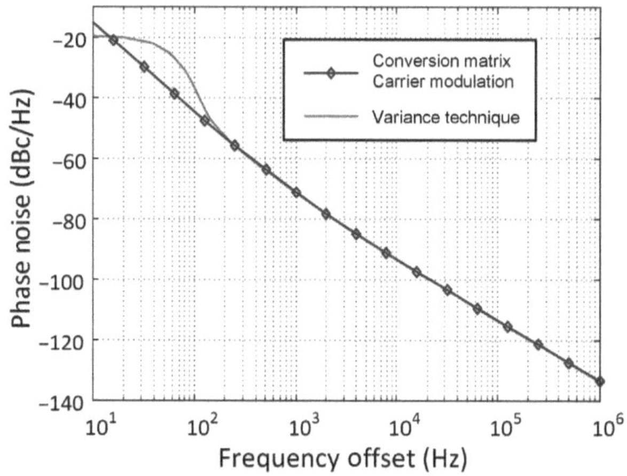

FIGURE 3.23

Phase-noise spectra calculated by the three techniques proposed.

unboundedly as $\Omega \to 0$. Note that this approach fails for small frequency offset values. As stated in Ref. [9], the total power of the oscillating signal cannot be altered by the influence of the common phase perturbation. For this reason, the functions $S_{k\phi}(\Omega)$ must flatten as $\Omega \to 0$. This behavior is observed when obtaining the complete spectral components through the variance term $\sigma_\phi^2(\tau)$ applying the procedure (Eqs. (3.86), (3.87)).

3.9 REDUCED-ORDER MODELS FOR OSCILLATOR CIRCUITS

The oscillator circuit can be part of a larger system, as in the case of PLLs, having a VCO inserted into a feedback loop, usually with the aim to reduce its phase noise through phase comparison with a low noise reference oscillator. Other examples of systems containing oscillator elements are coupled oscillators, applied for power combination and beam steering of phased-array antennas. Provided that once inserted into a given system, the oscillator frequency and amplitude only undergo small increments with respect to the original free-running values, it will be possible to describe the oscillator circuit by means of a reduced-order model, extracted through harmonic-balance simulations.

For illustration of the use of these reduced-order models we will consider the case of a system of three coupled oscillators. The oscillators are coupled by means of a simple network, consisting of a transmission line, bounded by high values resistances R. As has been demonstrated in Refs. [40,41], one can change the constant phase shift between the oscillator elements by tuning the outermost oscillators only through variations of their tuning voltages η_1 and η_3. This procedure substantially reduces the number of tuning parameters, in comparison with the conventional architectures, based

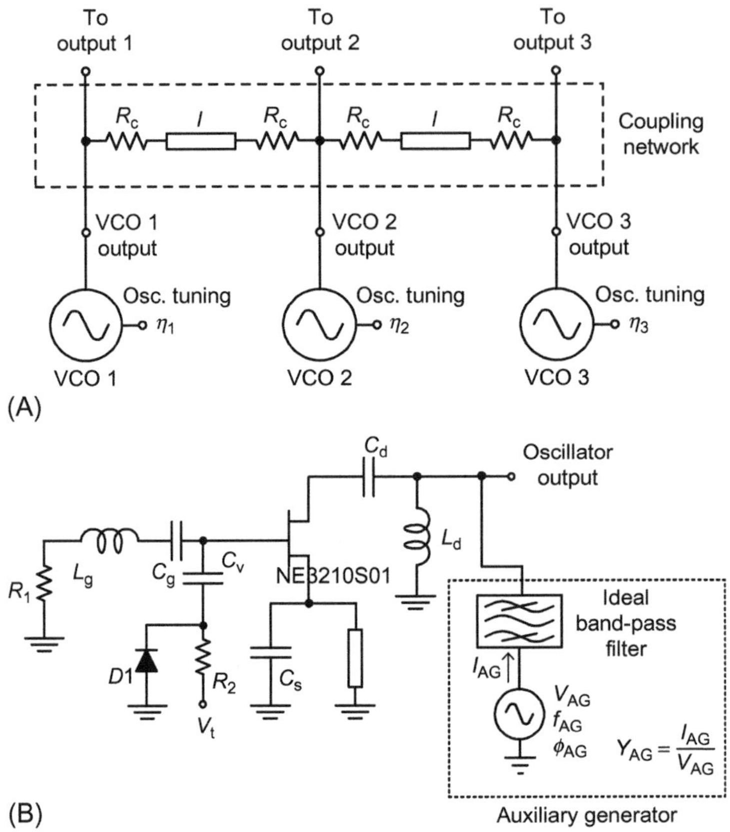

FIGURE 3.24

Coupled-oscillator system with three oscillator elements. (A) Coupled configuration. (B) Elementary oscillator, including the auxiliary generator used to obtain the reduced-order model.

on the use of phase shifters. The coupled system considered, based on the oscillator in Figure 3.12, is shown in Figure 3.24.

Assuming weak coupling conditions (a high R value in the coupling networks), the coupled-oscillator system is analyzed with a two-level procedure, described in the following.

3.9.1 INNER LEVEL

When isolated from the coupled system, the total admittance function of the oscillator circuit (current-to-voltage ratio) at any analysis node is equal to zero. When introduced into the coupled system, this admittance function will be different from zero since there will be current entering through the coupling networks (see Figure 3.23A). However, provided that this current has small amplitude (as expected under weak coupling conditions) it will only undergo a small variation. Therefore, it can be modeled with a first-order Taylor series about the free-running solution. Thus, each oscillator element can be described with the partial derivatives of its admittance function Y (current-to-voltage ratio) with respect to the amplitude V, frequency ω, and tuning parameter η, respectively, given by Y_V, Y_ω, Y_η. Note that the admittance function is zero ($Y=0$) when the oscillators are isolated from the system.

Assuming equal oscillator elements, the derivatives Y_V, Y_ω, Y_η are calculated by introducing a voltage auxiliary in the free-running oscillator [12,13], that is, in the elementary oscillator isolated from the system (see Figure 3.24B). This auxiliary generator must be connected to the same node to which the coupling network will be connected at a later stage (see Figure 3.24A). At the quiescent point, the frequency of the auxiliary generator agrees with the free-running oscillation frequency, $\omega_{AG}=\omega_o$, and its amplitude agrees with the node-voltage amplitude in free-running regime, $A_{AG}=V_o$. The admittance-function derivatives are calculated through finite differences. The calculation procedure is described in the following.

To calculate the derivative Y_V, the auxiliary-generator frequency and the oscillator tuning voltage are fixed at their steady-state variables, that is, $\omega_{AG}=\omega_o$ and $\eta=\eta_o$. Then, a sweep must be carried out in the auxiliary-generator amplitude in a small interval A_{AG1} to A_{AG2}, about the free-running value, $A_{AG}=V_o$ [12] (see Figure 3.24B), ensuring that the admittance function exhibits linear behavior with respect to A_{AG} in this interval. For each A_{AG} a full harmonic-balance analysis is carried out, so that the pure harmonic balance system acts as an inner tier. Then, the amplitude derivative is calculated through finite differences by doing: $Y_V=(Y_{AG2}-Y_{AG1})/(A_{AG2}-A_{AG1})$.

To calculate the frequency derivative Y_ω, the auxiliary-generator amplitude and the oscillator tuning voltage are fixed at their steady-state variables, that is, $A_{AG}=V_o$ and $\eta=\eta_o$. Then, a sweep is carried out in the auxiliary-generator frequency in a small interval ω_{AG1} to ω_{AG2} about the free-running value, $\omega_{AG}=\omega_o$, ensuring that the admittance function exhibits linear behavior with respect to ω_{AG} in this interval. Then, frequency derivative is calculated through finite differences as: $Y_\omega=(Y_{AG2}-Y_{AG1})/(\omega_{AG2}-\omega_{AG1})$.

Finally, to calculate the derivative with respect to the tuning parameter, the auxiliary-generator amplitude and frequency are fixed at their steady-state variables, that is, $A_{AG}=V_o$ and $\omega_{AG}=\omega_o$. Then, a sweep must be carried out in the small tuning parameter interval η_1 to η_2 about the original value η_o, ensuring that the admittance function exhibits linear behavior with respect to η in this interval. Then, the derivative with respect to the tuning parameter is calculated through finite differences as: $Y_\eta=(Y_{AG2}-Y_{AG1})/(\eta_2-\eta_1)$.

This technique has been applied to the individual oscillator of Figure 3.24 at 5 GHz, which is based on the FET NE3210S01. It provides the derivative values: $Y_V=0.044\angle{-5.75}$ degree $\Omega^{-1}\,V^{-1}$, $Y_\omega=2\times10^{-11}\angle 47.2$ degree $\Omega^{-1}\,rad^{-1}$, $Y_\eta=0.0068\angle{-140}$ degree $\Omega^{-1}\,V^{-1}$.

3.9.2 OUTER LEVEL

The oscillators are coupled by means of equal subnetworks [40,41], which generally consist of a transmission-line section bounded with series resistors (see Figure 3.1). Each subnetwork is described [12,13] by a symmetrical 2×2 admittance matrix $[Y_{c2}]$:

$$[Y_{c2}] = \begin{bmatrix} Y_e & Y_{nb} \\ Y_{nb} & Y_e \end{bmatrix} \tag{3.88}$$

Applying Kirchhoff's laws and assuming a constant interstage phase shift ϕ, one obtains, for three oscillator elements, the following system in the increment $\Delta\omega$ with respect to the free-running frequency, and the increments ΔV_i, $\Delta\eta_i$ (where $i=1$–3) with respect to the free-running amplitude and tuning parameter:

$$\begin{aligned} &Y_V V_o \Delta V_1 + Y_\omega V_o \Delta\omega + Y_\eta V_o \Delta\eta_1 = -Y_e V_o - Y_{nb} V_o e^{j(\phi_2-\phi_1)} \\ &Y_V V_o \Delta V_2 + Y_\omega V_o \Delta\omega + Y_{nb} V_o e^{j(\phi_1-\phi_2)} = -2Y_e V_o - Y_{nb} V_o e^{j(\phi_3-\phi_2)} \\ &Y_V V_o \Delta V_3 + Y_\omega V_o \Delta\omega + Y_\eta V_o \Delta\eta_2 = -Y_{nb} V_o e^{j(\phi_2-\phi_3)} - Y_e V_o \end{aligned} \tag{3.89}$$

where all the second order terms: ΔV_i^2, $\Delta V_i \Delta\omega$, $\Delta V_i \Delta\eta$ have been neglected. The system still has a nonconstant amplitude distribution, due to the terms ΔV_i. We are interested in the solution with a constant interstage phase shift, so we will impose $\phi_{i+1} - \phi_i = \phi$. Dividing also both sides of the equation by the constant amplitude V_o, we will obtain the following system:

$$\begin{aligned} &Y_V \Delta V_1 + Y_\omega \Delta\omega + Y_\eta \Delta\eta_1 = -Y_e - Y_{nb} e^{j\phi} \\ &Y_V \Delta V_2 + Y_\omega \Delta\omega = -Y_{nb} e^{-j\phi} - 2Y_e - Y_{nb} e^{j\phi} \\ &Y_V \Delta V_3 + Y_\omega \Delta\omega + Y_\eta \Delta\eta_3 = -Y_{nb} e^{-j\phi} - Y_e \end{aligned} \tag{3.90}$$

For each precise value of the constant interstage phase shift ϕ we need particular values of $\Delta\eta_1$ and $\Delta\eta_3$, determined by the above equations. On the other hand, the coupled-oscillator system, without any external time-varying source, behaves as an autonomous system, so for each pair of parameter values $\Delta\eta_1$, $\Delta\eta_3$ we will have a different steady-state oscillation frequency ω, expressed as: $\omega = \omega_o + \Delta\omega$.

When using the derivatives obtained numerically through harmonic-balance simulations, Eq. (3.90) constitute a "semi-analytical formulation" of the coupled-oscillator system, that is, the coupled system is formulated analytically, but the oscillator elements are described with numerical models. The semi-analytical formulation has the advantage of providing insight into the variation of the oscillation-frequency deviation $\Delta\omega$ with the imposed interstage phase shift ϕ or the parameters required $\Delta\eta_1$, $\Delta\eta_3$ for each ϕ value.

Splitting system (3.90) into real and imaginary parts, we can solve for $\Delta\omega$ [13], as a function of the constant phase shift ϕ, which provides:

$$\Delta\omega = \Delta\omega_o - \frac{2|Y_{nb}|\sin\alpha_{v,nb}}{|Y_\omega|\sin\alpha_{v,\omega}}\cos\phi; \quad \Delta\omega_o = -2\frac{|Y_e|\sin\alpha_{v,e}}{|Y_\omega|\sin\alpha_{v,\omega}} \tag{3.91}$$

where $\alpha_{a,b} = \text{phase}(Y_b) - \text{phase}(Y_a)$, with Y_a, Y_b any of the magnitudes Y_V, Y_ω, Y_η, Y_{nb}, and Y_e. Note that according to Ref. [1], for a stable free-running oscillation, we should have $\sin\alpha_{v,\omega} > 0$. The semi-analytical expression (3.91) demonstrates two different effects on the oscillation frequency when connecting the oscillator elements through the coupling network: a change of this oscillation frequency due to loading effects through the parameter Y_e and a change due to the influence of the nearest neighbors through the parameter Y_{nb}. The latter depends on the oscillator state, more precisely, on the phase shift between these oscillators, which is varied with the tuning parameters $\Delta\eta_1$, $\Delta\eta_3$.

We can also solve for the parameter values $\Delta\eta_1$, $\Delta\eta_3$ required to obtain each value ϕ of the constant interstage phase shift [13]. When solving Eq. (3.90) for $\Delta\eta_1$ one obtains:

$$\Delta\eta_1 = \frac{|Y_e|\sin\alpha_{v,e}}{|Y_\eta|\sin\alpha_{v,\eta}} - \frac{|Y_{nb}|\sin\left(-\phi + \alpha_{v,nb}\right)}{|Y_\eta|\sin\alpha_{v,\eta}} \tag{3.92}$$

Solving for $\Delta\eta_2$, one obtains that the two parameter varies symmetrically about $\phi = 0°$.

The semi-analytical formulation has been applied to the coupled system in Figure 3.24. Figure 3.25 compares the results obtained with the semi-analytical model and with a full harmonic-balance simulation of the coupled system, at circuit level, with a high computational cost. Figure 3.24A presents the oscillation-frequency deviations versus the imposed phase shift ϕ, Figure 3.24B presents the tuning voltages in the two outermost oscillators required for each value of the constant interstage phase shift ϕ. Finally, Figure 3.24C presents the excursions in the three individual oscillation amplitudes traced versus the imposed phase shift ϕ. In all cases, there is very good agreement with the costly circuit-level harmonic-balance simulations.

This example demonstrates the usefulness of combining analytical descriptions of oscillator based systems with numerical models of the oscillator elements. The following example presents the formulation of a PLL based on a numerical model of the VCO, extracted through harmonic-balance simulations.

3.10 PHASE-LOCKED LOOPS

In order to fulfill the strict phase-noise specifications of current communication systems, the phase dynamics of the oscillator circuit output signal are usually controlled by means of a PLL. In this loop, the output signal of the VCO, after being frequency-divided, is compared with a reference signal $v_{in}(t)$ of low phase noise by means of a phase detector (PD). The resulting low pass filtered voltage signal controls the VCO, stabilizing the output frequency [42] (see Figure 3.26). The nonlinear nature of the PLL elements, together with the system feedback, often gives rise to very complex nonlinear behavior [43].

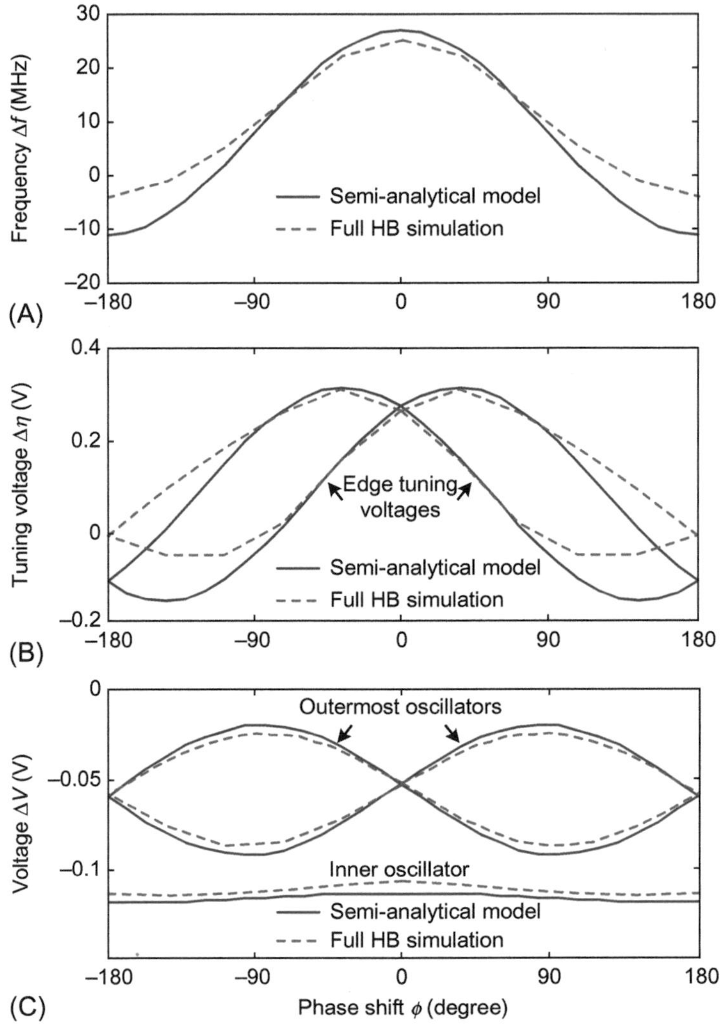

FIGURE 3.25

Application of the semi-analytical formulation to the coupled system in Figure 3.24. Results are compared with those obtained with a full harmonic-balance simulation of the coupled system, at circuit level. (A) Oscillation-frequency deviations versus the imposed phase shift ϕ. (B) Tuning voltages in the two outermost oscillators required for each value of the constant interstage phase shift ϕ. (C) Excursions in the three individual oscillation amplitudes traced versus the imposed phase shift ϕ.

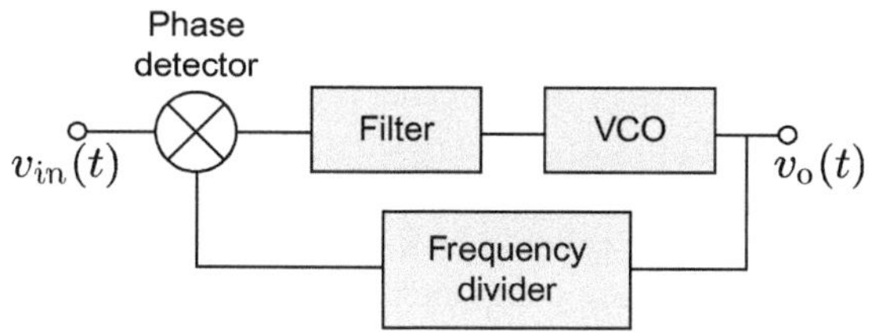

FIGURE 3.26

Schematic of the phase-locked loop.

3.10.1 VCO FORMULATION

When isolated from the loop, the VCO fulfills the following complex steady-state equation at any observation node:

$$Y\left(V_T^0, V_1, j\omega_o\right) = 0 \tag{3.93}$$

where Y is the total admittance function at the selected observation node n, V_T^0 is the *dc* value of the control voltage, V_1 is the first harmonic amplitude and $\omega_o = 2\pi f_o$ is the free-running frequency corresponding to the tuning value $v_T = V_T^0$. It is convenient to obtain the outer tier equation (3.93) in a node n corresponding to one of the nonlinear device terminals, where there is higher sensitivity to the voltage amplitude.

It is assumed that the VCO amplitude, frequency and control voltage only undergo small variations from the free-running values V_T^0, V_1, $j\omega_o$. The control signal will be time-varying $v_T = V_T^0 + \Delta V_T(t)$ and the oscillator signal, at the observation node n, can be expressed as: $[V_1 + \Delta V(t)]e^{j\phi_n(t)}$, where $(\phi_n, \Delta V)$ are the first harmonic phase and amplitude increments. In the dynamic regime, the VCO state variables $(\phi_n, \Delta V)$ are time-varying, and the operation frequency will be expressed as $j\omega_o + s$, with s being the time differentiator [4,44]. Then, Eq. (3.93) becomes:

$$Y\left[V_T^0 + \Delta V_T(t), V_1 + \Delta V(t), j\omega_o + s\right][V_1 + \Delta V(t)]e^{j\phi_n(t)} = 0 \tag{3.94}$$

Since the increments in Eq. (3.94) are small, the admittance function can be approached by a first-order Taylor series about the free-running values V_T^0, V_1, $j\omega_o$, which provides the following differential equation [45]:

$$f_{VCO}^c\left(\Delta V, \Delta\dot{V}, \dot{\theta}_n\right) \equiv Y_{V_T}\Delta V_T(t) + Y_V \Delta V(t) + Y_\omega\left(\dot{\theta}_n - \omega_0 - j\frac{\Delta\dot{V}}{V_1}\right) = 0 \tag{3.95}$$

where Y_{V_T}, Y_V and Y_ω are, respectively, the admittance-function derivatives with respect to V_T, V_1 and ω at the quiescent point. This point belongs to the linear region of the frequency characteristic versus the bias voltage. The derivatives in Eq. (3.95) are calculated in circuit-level HB through finite differences as explained in Section 3.9.1. In Eq. (3.95) higher order terms have been neglected and $\theta_n(t) = \phi_n(t) + \omega_o t$ is the instantaneous phase at the node n. System (3.95) will be called the *complex* model of the VCO, since it is constituted by a complex equation. In contrast, the standard VCO model $\dot{\theta}_n = \omega_o + K_v \Delta V_T$, with K_v being the VCO tuning sensitivity, will be called the *real* model.

3.10.2 PLL FORMULATION

The entire analysis presented here is applicable regardless of the type of PD (analogue or digital). However, for simplicity, a PD of mixer type will be considered. For other PDs, such as tri-stage phase-frequency comparator, convenient behavioral models such as those in Refs. [42,46,47] should be introduced in the formulation. In the case of a mixer PD, the low frequency component of the output signal has the form:

$$u(t) = K_{\mathrm{d}} \sin\left(\omega_i t - \frac{\theta_{\mathrm{o}}(t)}{N}\right) = K_{\mathrm{d}} \sin\phi, \quad \phi = \omega_i t - \frac{\theta_{\mathrm{o}}(t)}{N} \tag{3.96}$$

where $\omega_i = 2\pi f_i$ is the reference frequency, $\theta_{\mathrm{o}}(t)$ is the VCO instantaneous output phase, N is the division order, and $\phi(t)$ is the phase-error variable. The influence of the high frequency component of the PD output has been neglected due to the presence of the loop filter. Note that the observation node considered in the admittance model (3.95) may be different from the oscillator output node. In that case, a constant phase shift $\Delta\varphi$ will exist between the two nodes, such that:

$$\theta_{\mathrm{o}}(t) = \theta_n(t) + \Delta\varphi \tag{3.97}$$

In order to simplify the formulation, without loss of generality, it will be assumed that an offset voltage V_{T}^0 is added to the filter output to bias the VCO. Then, according to Eq. (3.94), the filter output voltage agrees with the perturbation variable $\Delta V_{\mathrm{T}}(t)$. Here, the following first-order filter will be considered:

$$F(s) = \frac{\tau_1 s + 1}{\tau_2 s + 1} = \frac{\Delta V_{\mathrm{T}}(s)}{u(s)} \tag{3.98}$$

Combining the Eqs. (3.95)–(3.98), the system of differential equations governing the PLL dynamics is:

$$\begin{aligned} &\tau_1 K_{\mathrm{d}} \dot{\phi} \cos\phi + K_{\mathrm{d}} \sin\phi = \tau_2 \Delta \dot{V}_{\mathrm{T}} + \Delta V_{\mathrm{T}}, \\ &Y_{V_{\mathrm{T}}} \Delta V_{\mathrm{T}} + Y_V \Delta V + Y_\omega \left(N\omega_i - \omega_{\mathrm{o}} - N\dot{\phi} - \mathrm{j}\frac{\Delta \dot{V}}{V_1}\right) = 0 \end{aligned} \tag{3.99}$$

System (3.99) is composed by three real nonlinear differential equations in the state variables $(\phi, \Delta V, \Delta V_{\mathrm{T}})$, because the admittance equation is complex. It can be written in a matrix form as:

$$M_{\dot{X}} \dot{\overline{x}}(t) + M_X \overline{x}(t) + \overline{G}[\overline{x}(t)] = \overline{0}, \quad \overline{x} = (\phi, \Delta V, \Delta V_{\mathrm{T}})^t \tag{3.100}$$

with:

$$M_{\dot{X}} = \begin{pmatrix} \tau_1 K_{\mathrm{d}} \cos\phi & 0 & -\tau_2 \\ -N Y_\omega^{\mathrm{r}} & Y_\omega^{\mathrm{i}}/V_1 & 0 \\ -N Y_\omega^{\mathrm{i}} & Y_\omega^{\mathrm{r}}/V & 0 \end{pmatrix}, \quad M_X = \begin{pmatrix} 0 & 0 & -1 \\ 0 & Y_V^{\mathrm{r}} & Y_{V_{\mathrm{T}}}^{\mathrm{r}} \\ 0 & Y_V^{\mathrm{i}} & Y_{V_{\mathrm{T}}}^{\mathrm{i}} \end{pmatrix}, \quad \overline{G} = \begin{pmatrix} K_{\mathrm{d}} \sin\phi \\ Y_\omega^{\mathrm{r}}(N\omega_{\mathrm{i}} - \omega_{\mathrm{o}}) \\ Y_\omega^{\mathrm{i}}(N\omega_{\mathrm{i}} - \omega_{\mathrm{o}}) \end{pmatrix} \tag{3.101}$$

where the superindexes r and i mean real and imaginary parts. Then, Eq. (3.99) can be expressed in the following compact form:

$$\dot{\overline{x}}(t) = -M_{\dot{X}}^{-1}\{M_X \overline{x}(t) + \overline{G}[\overline{x}(t)]\} = \overline{f}[\overline{x}(t)] \tag{3.102}$$

System (3.102) can be used to simulate the evolution of the PLL variables during the transient to phase-locked state. The phase-locked solution is an EP of system (3.102), which is given by:

$$\dot{\overline{x}}(t) = \overline{f}[\overline{x}(t)] = \overline{0} \rightarrow \overline{x}(t) = \overline{x}_0 \tag{3.103}$$

Note that $\dot{\overline{x}}(t) = \overline{0}$ implies that the phase-locked solution $\overline{x}_0$ has constant frequency and amplitude. System (3.103) allows the prediction of the variation of the phase-locked solution amplitude with the reference frequency or any other parameter.

3.10.3 APPLICATION EXAMPLE

The previous formulation has been applied to a PLL at the frequency $f_i = 5.275$ GHz. The VCO used is based on the transistor CFY30 at the bias point $V_{GS} = -0.5$ V and $V_{DS} = 2.5$ V. Figure 3.27 shows the evolution of the phase-error value at the phase-locked solution versus the operation frequency. For each frequency value, two phase-locked solutions coexist, stable and unstable, respectively. The range of existence of these two solutions is called the hold-in range, and it is constrained between the two turning points T_1 and T_2. Beyond the turning points, the system is no longer phase locked to the reference oscillator [43,48]. The results obtained with Eq. (3.103) have been compared with measurements.

Figure 3.28 shows the transient to phase-locked state obtained for the bias voltages $V_{DD} = 3.5$ V and $V_{GG} = -1.1$ V. As can be seen, the measured transient exhibits oscillatory behavior in agreement with the prediction of the *complex* model. The *real* model cannot predict this behavior. This difference can be explained by analyzing the VCO model (3.95). This is a linear system whose dynamics is determined by

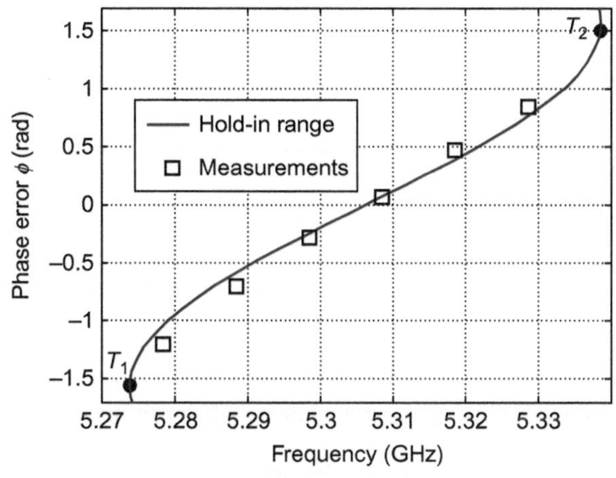

FIGURE 3.27

Steady-state analysis in phase-locked conditions. Phase error and voltage amplitude in the gate node versus the operation frequency.

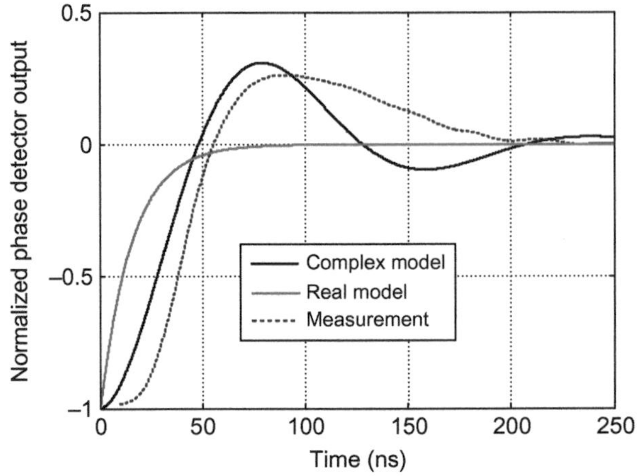

FIGURE 3.28

Transients to phase-locked state for $V_{DD}=3.5$ V, $V_{GG}=-1.1$ V.

a pair of real poles $\lambda_1=0$ and $\lambda_2<0$. The transient behavior can be approached by analyzing the linear system obtained through linearization of system (3.102) about the steady-state solution $\bar{x}_0$. On the one hand, when using the *complex* model the dynamics of this linearized system in the neighborhood of $\bar{x}_0$ is determined by $p=3$ poles whose values are influenced by λ_1 and λ_2.

In the analysis of Figure 3.26, it is clear that the linearized system has one real pole and two complex-conjugate poles whose imaginary part determines the transient oscillation frequency. On the other hand, the *real* model does not take into account the influence of the pole λ_2 on the PLL dynamics. This can prevent the detection of the pair of complex-conjugate poles producing the oscillatory transient, as it is the case in Figure 3.26. Note that, in most cases, both models will agree, but for a small value of the real pole λ_2 of the VCO, there can be discrepancies [45].

REFERENCES

[1] K. Kurokawa, Some basic characteristics of broadband negative resistance oscillators, Bell Syst. Tech. J. 48 (6) (1969) 1937–1955.

[2] T.S. Parker, L.O. Chua, Practical Numerical Algorithms for Chaotic Systems, Springer-Verlag, New York, NY, 1989.

[3] J. Guckenheimer, P. Holmes, Nonlinear Oscillations, Dynamic Systems, and Bifurcations of Vector Fields, Springer-Verlag, New York, NY, 1983.

[4] A. Suárez, Analysis and Design of Autonomous Microwave Circuits, Wiley-IEEE Press, Hoboken, NJ, 2009.

[5] A. Suarez, J. Morales, R. Quéré, Synchronization analysis of autonomous microwave circuits using new global-stability analysis tools, IEEE Trans. Microw. Theory Techn. 46 (5) (1998) 494–504.

[6] A. Suárez, R. Quéré, Stability Analysis of Nonlinear Microwave Circuits, Artech House, Norwood, MA, 2003.

[7] J.M.T. Thompson, H.B. Steward, Nonlinear Dynamics and Chaos, Wiley, Chichester, UK, 1986.

[8] G. Ioos, D. Joseph, Elementary Stability and Bifurcation Theory, Springer, New York, NY, 1980.

[9] A. Demir, Phase noise in oscillators: DAEs and colored noise sources, in: IEEE/ACM International Conference on Computer-Aided Design, San Jose, CA, USA, 1998, pp. 170–177.

[10] F.X. Kaertner, Analysis of white and $f^{-\alpha}$ noise in oscillators, Int. J. Circ. Theor. App. 18 (5) (1990) 485–519.

[11] A. Suárez, S. Sancho, S.V. Hoeye, J. Portilla, Analytical comparison between time and frequency-domain techniques for phase-noise analysis, IEEE Trans. Microw. Theory Techn. 50 (10) (2002) 2353–2361.

[12] F. Ramírez, E. de Cos, A. Suárez, Nonlinear analysis tools for the optimized design of harmonic-injection frequency dividers, IEEE Trans. Microw. Theory Techn. 51 (6) (2003) 1752–1762.

[13] A. Suárez, F. Ramírez, S. Sancho, Stability and noise analysis of coupled-oscillator systems, IEEE Trans. Microw. Theory Techn. 59 (4) (2011) 1032–1046.

[14] K. Ogata, Modern Control Engineering, Prentice-Hall, Englewood Cliffs, NJ, 1980.

[15] K.S. Kundert, Introduction to RF simulation and its application, IEEE J. Solid-State Circ. 34 (9) (1999) 1298–1319.

[16] V. Rizzoli, A. Neri, State of the art and present trends in nonlinear microwave CAD techniques, IEEE Trans. Microw. Theory Techn. 36 (2) (1988) 343–356.

[17] R. Quéré, E. Ngoya, M. Camiade, A. Suarez, M. Hessane, J. Obregon, Large signal design of broadband monolithic microwave frequency dividers and phase-locked oscillators, IEEE Trans. Microw. Theory Techn. 41 (11) (1993) 1928–1938.

[18] F. Filicori, V.A. Monaco, C. Naldi, Simulation and design of microwave class-C amplifiers through harmonic analysis, IEEE Trans. Microw. Theory Techn. 27 (12) (1979) 1043–1051.

[19] S.M. Sze, K.K. Ng, Physics of Semiconductor Devices, John Wiley & Sons, New York, NY, 2006.

[20] J.M. Collantes, I. Lizarraga, A. Anakabe, J. Jugo, Stability verification of microwave circuits through Floquet multiplier analysis, in: IEEE APCCAS, Taiwan, 2004, pp. 997–1000.

[21] F. Bonani, M. Gilli, Analysis of stability and bifurcations of limit cycles in Chua's circuit through the harmonic-balance approach, IEEE Trans. Circuits Syst.-I 46 (8) (1999) 881–890.

[22] D. Woods, Reappraisal of the unconditional stability criteria for active 2-port networks in terms of S parameters, IEEE Trans. Circuits Syst. 23 (2) (1976) 73–81.

[23] J. Verspecht, Large signal network analysis, IEEE Microw. Mag. 6 (4) (2005) 82–92.

[24] M. Marchetti, M.J. Pelk, K. Buisman, W. Neo, M. Spirito, L.C.N. de Vreede, Active harmonic load-pull with realistic wideband communications signals, IEEE Trans. Microw. Theory Techn. 56 (12) (2008) 2979–2988.

[25] A. Ferrero, V. Teppati, A. Carullo, Accuracy evaluation of on-wafer load-pull measurements, IEEE Trans. Microw. Theory Techn. 49 (1) (2001) 39–43.

[26] V. Rizzoli, A. Costanzo, An accurate FET model suitable for general nonlinear and power applications, Int. J. RF Microw. C. E. 10 (1) (2000) 43–62.

[27] G. Avolio, A. Raffo, I. Angelov, V. Vadalà, G. Crupi, A. Caddemi, G. Vannini, D.M.M.-P. Schreurs, Millimetre-wave FET nonlinear modelling based on the dynamic-bias measurement technique, IEEE Trans. Microw. Theory Techn. 62 (11) (2014) 2526–2537.
[28] G. Crupi, D.M.M.-P. Schreurs, A. Raffo, A. Caddemi, G. Vannini, A new millimeter wave small-signal modeling approach for pHEMTs accounting for the output conductance time delay, IEEE Trans. Microw. Theory Techn. 56 (2008) 741–746.
[29] J. Jugo, J. Portilla, A. Anakabe, A. Suárez, J.M. Collantes, Closed-loop stability analysis of microwave amplifiers, Electron. Lett. 37 (4) (2001) 226–228.
[30] A. Anakabe, J.M. Collantes, J. Portilla, et al., Analysis and elimination of parametric oscillations in monolithic power amplifiers, in: IEEE MTT-S International Microwave Symposium Digest, Seattle, WA, USA, 2002, pp. 2181–2184.
[31] B. Razavi, A study of phase noise in CMOS oscillators, IEEE J. Solid-State Circuits 31 (3) (1996) 331–343.
[32] C. Florian, P.A. Traverso, M. Feudale, F. Filicori, AC-band GaAs-pHEMT mmIC low phase noise VCO for space applications using a new cyclostationary nonlinear noise model, in: IEEE International Microwave Symposium Digest, Anaheim, CA, USA, 2010.
[33] D.B. Leeson, A simple model of feedback oscillator noise spectrum, Proc. IEEE 54 (1966) 329–330.
[34] J.M. Paillot, J.C. Nallatamby, M. Hessane, R. Quere, M. Prigent, J. Rousset, A general program for steady state, stability, and FM noise analysis of microwave oscillators, in: IEEE MTT-S International Microwave Symposium Digest, Dallas, TX, USA, 1990, pp. 1287–1290.
[35] J.C. Nallatamby, M. Prigent, J.C. Sarkissian, R. Quere, J. Obregon, A new approach to nonlinear analysis of noise behavior of synchronized oscillators and analog-frequency dividers, IEEE Trans. Microw. Theory Techn. 46 (8) (1998) 1168–1171.
[36] V. Rizzoli, F. Mastri, D. Masotti, General noise analysis of nonlinear microwave circuits by the piecewise harmonic-balance technique, IEEE Trans. Microw. Theory Techn. 42 (5) (1994) 807–819.
[37] J.A. Mullen, D. Middleton, Limiting forms of FM noise spectra, Proc. IRE 45 (6) (1957) 874–877.
[38] S. Sancho, A. Suárez, F. Ramirez, General phase-noise analysis from the variance of the phase deviation, IEEE Trans. Microw. Theory Techn. 61 (12) (2012) 472–481.
[39] S. Sancho, A. Suárez, J. Dominguez, F. Ramirez, Analysis of near-carrier phase-noise spectrum in free-running oscillators in the presence of white and colored noise sources, IEEE Trans. Microw. Theory Techn. 58 (3) (2010) 587–601.
[40] R.A. York, Nonlinear analysis of phase relationships in quasi-optical oscillator arrays, IEEE Trans. Microw. Theory Techn. 41 (10) (1993) 1799–1809.
[41] R.A. York, P. Liao, J.J. Lynch, Oscillator array dynamics with broadband N-port coupling networks, IEEE Trans. Microw. Theory Techn. 42 (11) (1994) 2040–2042.
[42] U.L. Rohde, Microwave and Wireless Synthesizers, John Wiley & Sons, New York, NY, 1997.
[43] S. Sancho, A. Suárez, T. Fernandez, Nonlinear dynamics of microwave synthesizers — stability and noise, IEEE Trans. Microw. Theory Techn. 49 (10) (2001) 1792–1803.
[44] E. Ngoya, R. Larcheveque, Envelope transient analysis: a new method for the transient and steady state analysis of microwave communication circuits and systems, in: IEEE MTT-S International Microwave Symposium Digest, San Francisco, CA, USA, 1996, pp. 1365–1368.

[45] S. Sancho, M. Pontón, A. Suárez, F. Ramirez, Analysis of injection pulling in phase-locked loops with a new modeling technique, IEEE Trans. Microw. Theory Techn. 61 (3) (2013) 1200–1214.
[46] S. Sancho, A. Suárez, J. Chuan, General envelope-transient formulation of phase-locked loops using three time scales, IEEE Trans. Microw. Theory Techn. 52 (4) (2004) 1310–1320.
[47] B. De Smedt, G. Gielen, Nonlinear behavioral modeling and phase-noise evaluation in phase-locked loops, in: IEEE Custom Integrated Conference, Santa Clara, CA, USA, 1998, pp. 53–56.
[48] E. Bradley, Using chaos to broaden the capture range of a phase-locked loop, IEEE Trans. Circuits Syst.-I 40 (11) (1993) 808–817.

CHAPTER

Microwave power amplifiers: Design and technology

4

P. Colantonio, E. Cipriani and F. Giannini
University of Rome Tor Vergata, Rome, Italy

CHAPTER OUTLINE

4.1 INTRODUCTION

The power amplifier, PA, is an essential electronic component in nearly all microwave and millimeter-wave applications, and in particular in any transmitting system, with the aim to increase the power level of the signal at its input up to a predefined level at its output.

The PA application space includes a broad range of systems [1], spanning from military and space applications [2–9] (radar, electronic warfare, satellite imaging, and communication) to scientific and medical applications [10–22] (heating, microwave

Microwave Wireless Communications. http://dx.doi.org/10.1016/B978-0-12-803894-9.00004-4

imaging, and implantable devices), and civil uses [23–27] (base station transceivers, digital television broadcasts, and mobile handsets).

Apart from transmission of information, PAs would be of fundamental use in future systems aiming at the transmission of microwave energy for the remote supply of unmanned vehicles or for the development of space solar power systems [28,29].

This diverse and evolving landscape dictates the strategy for the design, fabrication, and optimization of RF power amplifiers, RFPAs. Moreover, as contrasted to low-level amplifiers, generally specified in terms of their gain, the absolute output power is the key feature of a PA. Hence, the active devices used in PA design are selected on the basis of their output power capabilities, which are fully exploited by operating the devices under large-signal conditions, simultaneously maximizing current and voltage swings.

A PA is therefore a nonlinear system component, whose large-signal operating conditions often lead to detrimental effects on the output signal, which is no more a linear replica of the input, but presents some distortion effects that should be taken into account during the design. Hence, the linear approximation underlying small-signal amplifier design techniques (eg, S-parameter based) is no more strictly valid. Dedicated design methodologies have to be adopted, although first-guess and preliminary approaches exist and often lead to a good estimation of a PA's final performance [30]. Hence, the design of a PA is usually the result of a tradeoff in order to mitigate conflicting requirements such as linearity versus efficiency or high output power level versus low distortion. Moreover, designing approaches depend on operating frequency and bandwidth, available device technology, application (fixed, mobile or satellite communications, modulated continuous wave, CW, or pulsed signal, etc.) and many other factors [31,32].

This chapter aims to give an organic, although not exhaustive, view of the major factors involved in the design process of a PA. Starting from the definition of power match condition in Section 4.2 and after defining the main figures of merit of a PA in Section 4.3, some design strategies to achieve high efficiency will be analyzed in Section 4.4. Then, a technological overview about PA fabrication will be addressed, taking into account the capabilities of the active device (according to the semiconductor), and discussing pros and cons of standard hybrid (hybrid microwave integrated circuits, HMIC) and monolithic (monolithic microwave integrated circuit, MMIC) fabrication in Section 4.5. A system-level view of the PA is provided, introducing a behavioral description from the point of view of linearity in Section 4.6 and an example of PA for communication system (the Doherty amplifier) is reported in Section 4.7. Finally, the last section is dedicated to summarizing some concepts involved in the PA design using CAD tools in Section 4.8.

4.2 DEVICE CHARACTERISTICS AND POWER MATCH CONDITION

The maximum power obtainable from a transistor, regardless the semiconductor material used, depends on its maximum output current and voltage swing capabilities.

Hence, the power performance of a given device type may be readily demonstrated to exhibit a well-known behavior if considered in their limit frequency region (ie, following the Pf^2 law [33]):

$$P \cdot f^2 = \text{constant} \tag{4.1}$$

In RF and microwave applications, solid-state devices typically operate either at high power and low frequency, or at low power and high frequency, being extremely challenging, from the reliability point of view, to realize devices able to operate simultaneously at both high power and high frequency.

This technology concern can be easily understood by analyzing the power gain G of an active device, defined as the ratio between the output power P_{out} that the active device can provide and the input power P_{in} required to drive it:

$$G = \frac{P_{\text{out}}}{P_{\text{in}}} = \left(\frac{I_{\text{out}}}{I_{\text{in}}}\right)^2 \cdot \frac{\mathcal{R}\{Z_{\text{L}}\}}{\mathcal{R}\{Z_{\text{in}}\}} \tag{4.2}$$

where I_{out} is the output current provided from the device to an external load Z_{L}, while I_{in} is the input current flowing into the device, characterized by the equivalent input impedance Z_{in}.

Defining the ratio between the output and input currents as the device current gain A_I:

$$A_I = \frac{I_{\text{out}}}{I_{\text{in}}} \tag{4.3}$$

which typically shows a low-pass behavior in frequency, the device power gain can be simply represented by [31,32]:

$$G(f) \approx K \cdot \left(\frac{f_T}{f}\right)^2 \cdot \frac{\mathcal{R}\{Z_{\text{L}}(f)\}}{\mathcal{R}\{Z_{\text{in}}(f)\}} \tag{4.4}$$

where K is a constant and f_T is the cutoff frequency of the device, defined as the frequency for which the current gain A_I falls down to unity.

Accordingly, to increase device power performance it is mandatory to increase its capabilities to provide higher output currents, which is typically achieved by increasing the number (ie, arraying) of emitter/gate fingers. However, the ability of a device to deliver higher currents diminishes as the frequency increases due to de-phasing of the input signal along gate/bias fingers and across multigate manifolds. Similarly, the use of long fingers implies an increase of parasitic effects, thus reducing the cutoff frequency f_T. To avoid such issues, higher doping profiles could be adopted trying to preserve the output current I_{out} level, but this would result in smaller breakdown voltage (ie, limited voltage swing) and consequent reduced power capabilities, if the device geometry and technology are not properly optimized [34].

Moreover, if the device output power increases, the power dissipated in the device itself increases proportionally too, resulting in nonuniform heat dissipation across the device fingers, leading to hotspots eventually causing device failure [35].

The output power P_{out} delivered from the active device to the external load Z_{L} (usually 50 Ω) at a specified frequency f is given by:

$$P_{\text{out}}(f) = \frac{1}{2} \cdot \mathcal{R}\{V_{\text{out}}(f) \cdot I^*_{\text{out}}(f)\} \tag{4.5}$$

being V_{out} and I_{out} the frequency components of the output voltage and current, respectively, and "*" representing the complex conjugate operation.

In order to deliver the output power at frequency f, the PA has to be driven with an input power P_{in} at the same frequency f, given by:

$$P_{in}(f) = \frac{1}{2} \cdot \mathcal{R}\{V_{in}(f) \cdot I_{in}^{*}(f)\} \tag{4.6}$$

being V_{in} and I_{in} the corresponding frequency components of the input voltage and current, respectively.

The measure of the power amplification capability is given by the power gain G, defined as:

$$G(f) = \frac{P_{out}(f)}{P_{in}(f)} \tag{4.7}$$

The power gain G, due to the nonlinear behavior of the devices used in the PA, is clearly dependent on the input signal level also. For very small drive levels, the amplifier behaves almost linearly, and it is usually possible to refer to this linear gain G_L, defined as:

$$G_L = \lim_{P_{in} \to 0} G(f) \tag{4.8}$$

When the input drive is increased, output current and voltage swings allowed by the active device tend to be limited by its nonlinearities. Thus output power tends to saturate also to the so-called saturation value:

$$P_{sat} = \lim_{P_{in} \to \infty} P_{out}(f) \tag{4.9}$$

while the corresponding power gain asymptotically tends to zero.

The power limiting mechanisms in active devices reside in their inherent physical constraints, which limit the voltage and current swings. In the case of field effect transistor (FET) devices, the output IV characteristics of which are shown as an example in Figure 4.1, the output current swing is limited by the input junction forward conduction (a) and the device channel pinch-off (b), while the output voltage swing is limited by the ohmic behavior (c) and the breakdown (d). Similarly, for bipolar-junction transistor (BJT) devices, the output IV characteristics of which are shown as an example in Figure 4.2, the output current swing is limited by the breakdown (a) and cutoff (b) limitation of the base-emitter junction, while the voltage swing is limited by the saturation (c) and breakdown (d). It is remarkable that in both cases the maximum dissipation power capability, in static or dynamic operating conditions, generically represented by the curve (e), further limits the I-V swings.

Regardless of the device technology used, an ideal and simplified model, shown in Figure 4.3, can be used to clarify the operating principle of a PA and focus on its mode of operation [32], based on the device intrinsic I-V output characteristics depicted in Figure 4.4.

There are several remarks to be made in order to avoid misleading and to clarify discrepancies between measured results and theoretical performance predictions.

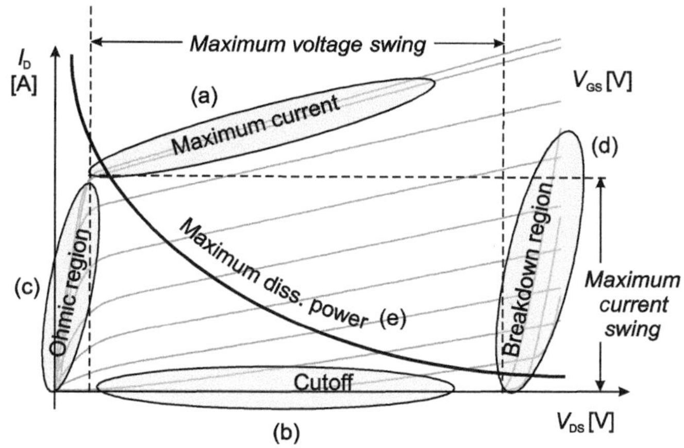

FIGURE 4.1

Example of FET device output IV characteristics.

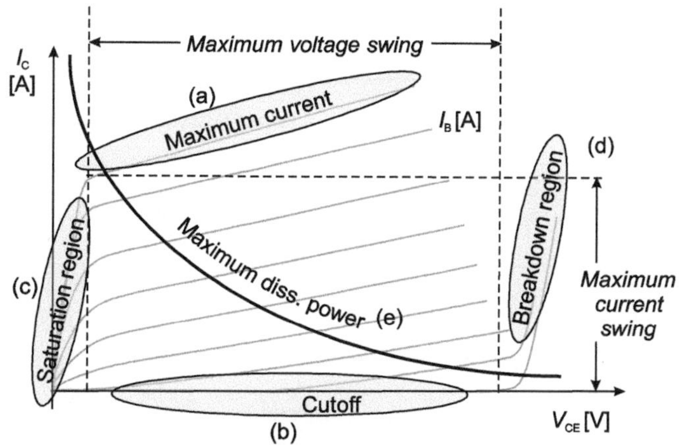

FIGURE 4.2

Example of BJT device output IV characteristics.

First of all, a complete nonlinear model of the active device accounts for all its capacitive, inductive, and resistive parasitic effects, as well as any possible packaging parasitic elements. As discussed in Chapter 1, the extrinsic elements can be straightforwardly extracted by using the well-known "cold" approach [36–40]. However, it is always possible to recognize an intrinsic part, whose output reference plane is at the device intrinsic current source $I_{out,int}$, and an extrinsic part, including the resistive and reactive parasitic elements [41,42]. Of course, the use of a complete nonlinear model is mandatory to effectively simulate the device behavior, but the simplified model is useful to understand some basic concepts on PA design.

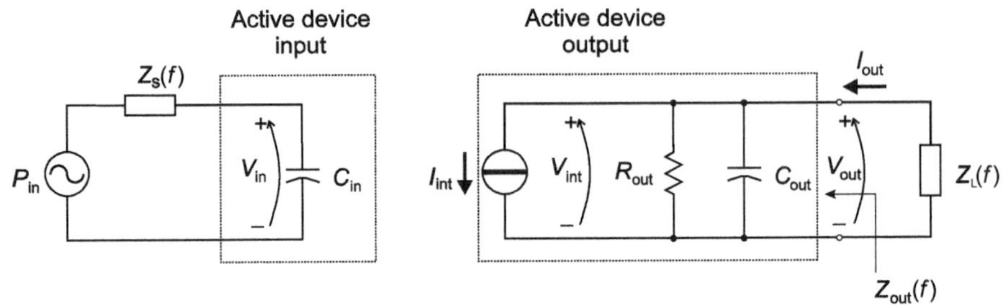

FIGURE 4.3

Simplified model of active device output, connected to an external load Z_L.

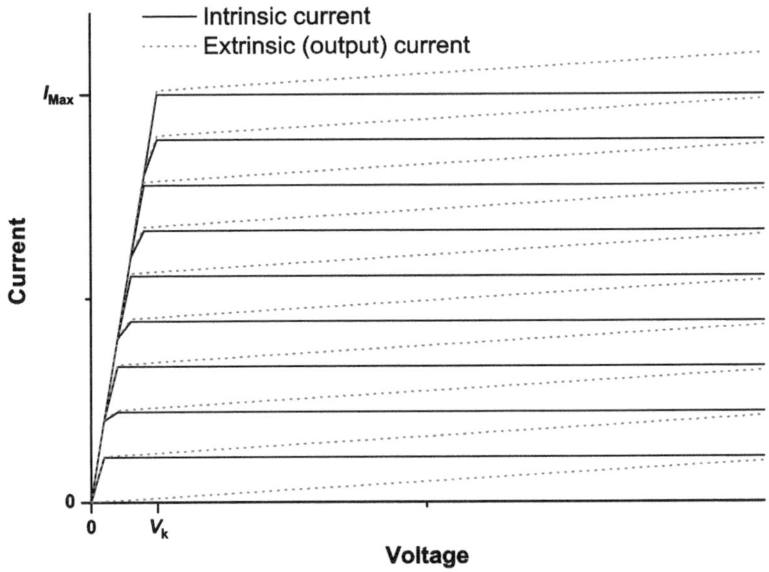

FIGURE 4.4

Ideal device output I-V characteristics.

A second assumption is that in the simplified model, the feedback elements are temporary neglected to further simplify the analysis. This assumption, accounting for the small value of the reactive (capacitive) feedback element in an active device is roughly valid for operating frequency well below the device cutoff frequency, while dramatically affects the theoretical results at operating frequencies close to f_T.

All the remaining parasitic elements of the device are roughly embedded in the input, Z_S, and output, Z_L, impedances, except for the device output resistive, R_{out}, and capacitive, C_{out}, components, which are useful at this stage to clarify some aspects. It is useful to note that the intrinsic output IV characteristics (continuous lines) reported in

Figure 4.4 represent the $I_{out,int}$ behavior with respect to the output voltage V_{out}. In the same Figure 4.4 are also represented the extrinsic output characteristics $I_{out,ext}$ (dotted gray lines), which slope accounts for the device output resistance R_{out}.

Finally, the intrinsic IV characteristics are based on some further ideality assumption, as summarized in the following:

- The active device behaves as a voltage (FET) or current (BJT) controlled current source in the active region, ie, above the turn-on and the knee region ($V_{out} > V_k$ in Figure 4.4).
- A linear relationship is assumed between the output current and the input controlling signal in the active region (identified as quasilinear region).
- The input controlling signal is assumed sinusoidal at the intrinsic port, represented by V_{in} in Figure 4.3.
- The current abrupt goes to zero (cutoff) when the controlling signal drops below a threshold value.
- The maximum device current is limited to I_{Max} (it saturates in FET type device or is limited due to reliability reasons in BJT).
- A quasi-static behavior is considered (ie, the same IV characteristics regardless of signal frequency sweep considered).

By using this simplified model, it is possible to highlight the basic difference in designing a PA with respect to a linear one (ie, a gain amplifier). In fact, from the small-signal analysis it can be inferred that the maximum power transfer from the device output to the external load is achieved fulfilling the conjugate match condition (compatibly with device stability) [43], that is, the output load Z_L has to fulfill the following condition:

$$Z_L = Z_{out}^* = \left(\frac{1}{R_{out}} - j \cdot \omega \cdot C_{out}\right)^{-1} \tag{4.10}$$

This condition implies the compensation of the active device output reactive part, C_{out}, and a resistive load matching its output resistance R_{out}. Because the latter corresponds to the slope of the output IV characteristics, the resulting dynamic load line, DLL, (ie, the curve describing the relationship between output current and voltage) superimposed on the device output characteristics (eg, FET example) becomes the curve A reported in Figure 4.5.

As can be noted, the conjugate match condition maximizes the device voltage gain, while it is obvious that a reduced current swing outcomes when the device is driven into large-signal regime, resulting in a sudden compression of the device output power due to the attainment of the voltage-limited operation.

Similarly, if the external load were selected to fully exploit the maximum current swing, as for example shown in Figure 4.5 with the curve B, then a reduced voltage swing would result when the current-limited operation is reached, still driving the active device output power into compression.

The optimum condition for power purposes is clearly achieved simultaneously maximizing both current and voltage swings, as shown with curve C in

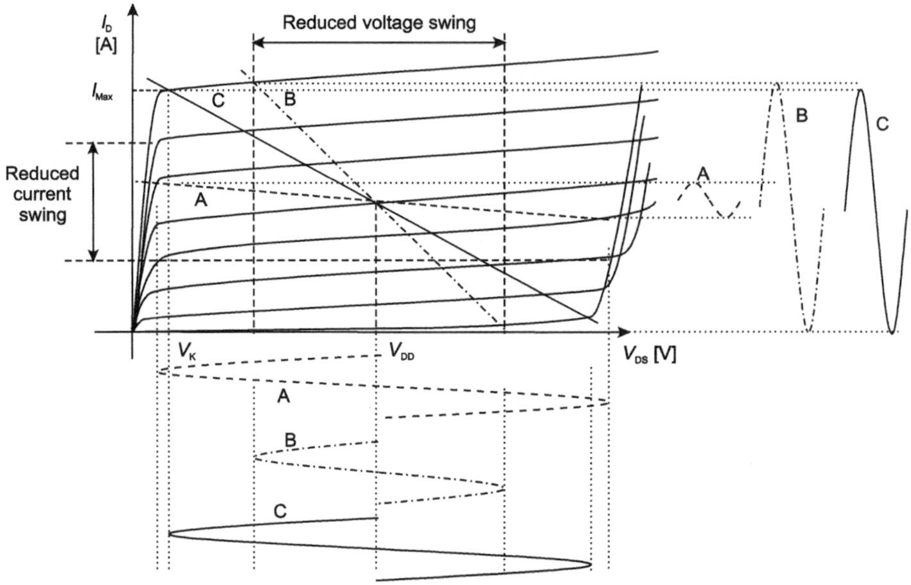

FIGURE 4.5

Active device output characteristics with superimposed conjugate match load line (A), current-limited (B), and optimum loading (C).

Figure 4.5. This loading condition is usually referred by PA designer as the *load-line matching condition* or simply the *power matching condition* [32].

The device output power behavior expected for the three loading conditions described in Figure 4.5, as a function of the input driving power P_{in} in a logarithmic scale (dBm), is reported in Figure 4.6. Clearly, the conjugate match condition results in a higher power gain at small-signal (curve A) but in a lower saturation power P_{sat} compared to the power match condition case (curve C).

The use of a simplified model serves to highlight an important aspect in PA design, referred to as the intrinsic versus extrinsic behavior. In fact, referring to the intrinsic device characteristics, and assuming a bias condition at the center of the characteristics (ie, a DC current $I_D = I_{Max}/2$ and a DC voltage $V = V_{DC}$), the intrinsic current i_{int} can be represented as a sinusoidal waveform:

$$i_{int} = I_D + I_P \cdot \cos(\omega t) \tag{4.11}$$

while the intrinsic voltage v_{int} (corresponding to the extrinsic voltage also in this simplified model) can be represented as:

$$v_{int} = V_{DC} - I_P \cdot |Z_{int}| \cdot \cos(\omega t + \phi_{int}) \tag{4.12}$$

where

$$Z_{int} = |Z_{int}|^{j\phi_{int}} \tag{4.13}$$

is the impedance seen at the intrinsic reference plane.

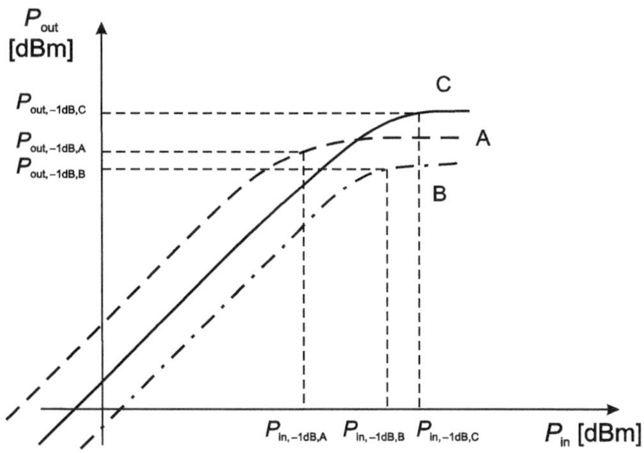

FIGURE 4.6

Output power for the three loading conditions of Figure 4.5: voltage limited (A), current-limited (B), and optimum loading (C).

Consequently, the power delivered from the device at the intrinsic reference plane is given by:

$$P_{out,int} = \frac{1}{2} \cdot I_p^2 \cdot |Z_{int}| \cdot \cos(\phi_{int}) \tag{4.14}$$

which is obviously maximized when $\phi_{int} = 0$ (ie, if the termination loading the controlled current source is purely resistive).

The resulting load curve at the intrinsic reference plane in this case results in a strict line, as depicted in Figure 4.7. Conversely, when referring to the extrinsic (eg, external) load line, the contribution of the reactive part of the load modifies the shape of the load curve, transforming it into an ellipse, as also shown in Figure 4.7.

The power match condition highlights other important aspects in PA design, which typically exhibits poor output return loss and Voltage Standing Wave Ratio, VSWR, being the resistive part of the loading condition, R_L, usually very different, and typically smaller, from the device output resistance, R_{out}. The output return loss of a PA could be easily improved by adopting an output isolator or transformers (which clearly decrease output power by their losses) or resorting to balanced configurations [31,32]. Also in this case, however, the combining structure losses could affect the overall power performance.

4.3 POWER AMPLIFIER FIGURE OF MERITS

From an energetic point of view, and regardless of the specific application considered, a PA acts as a DC-to-RF converter: it increases (amplifies) the radio-frequency RF power of the signal at its input by converting supplied DC power P_{dc} into the RF or microwave output power P_{out}, as graphically represented in Figure 4.8.

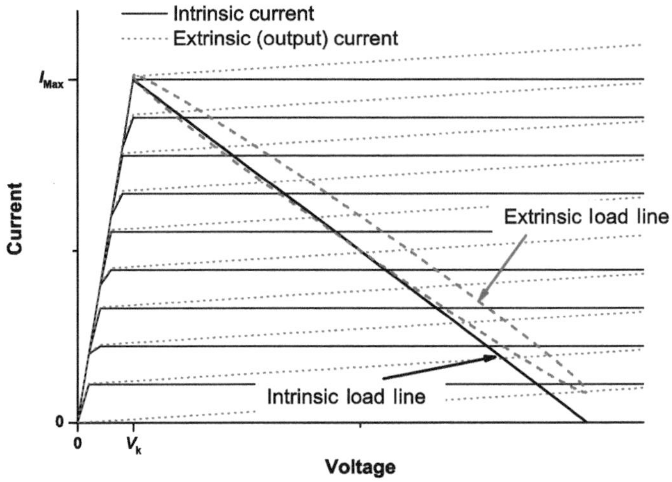

FIGURE 4.7

Load curve at the intrinsic (*continuous line*) and extrinsic (*dotted line*) reference planes, superimposed on the corresponding output IV characteristics.

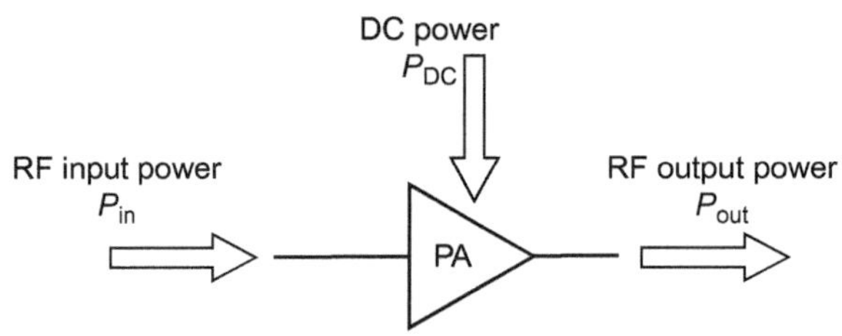

FIGURE 4.8

Graphical representation of a PA operation from the energetic perspective.

The supplied DC power can be computed as:

$$P_{dc} = V_{bias} \cdot \frac{1}{T} \cdot \int_0^T i(t) \cdot dt \tag{4.15}$$

being V_{bias} and $i(t)$ the supplied DC voltage and current, respectively, with the latter that can vary with the applied input power level, integrated over a period, T, of the RF applied input signal.

The effectiveness of the conversion process in a PA is usually described by means of the amplifier's drain (FET) or collector (BJT) efficiency, usually simply referred as efficiency, η, defined as the ratio between output RF and supplied DC power:

$$\eta = \frac{P_{out}}{P_{dc}} \tag{4.16}$$

and typically expressed as percentage.

The efficiency is one of the key parameters in specifying overall system performance, since for a given output power requested to the PA, it actually determines the DC power budget, and hence, the supply power dimensioning. It is significant in mobile devices that are typically battery-operated, whose operating time strictly depends on the transmitting section power requests.

Moreover, a high-efficiency performance implies, in turn, a low power dissipated in the active device. In fact, the conversion from DC to RF power implies that a fraction of the supplied power is lost and actually dissipated in the active power device. The major part of such loss is located at the active device output, and is given by:

$$P_{\mathrm{diss}} = \frac{1}{T} \cdot \int_0^T v_{\mathrm{out}}(t) \cdot i_{\mathrm{out}}(t) \cdot dt \tag{4.17}$$

Thus, a high-efficiency performance implies a low-power dissipation in the active device. therefore reducing actual size and weight of the heat sinks eventually required, and improving device reliability.

It is to note that in the linear region (ie, small-signal), where G and P_{dc} remain practically constant and thus independent from P_{in}, the efficiency increases exponentially while increasing the input drive:

$$\eta = \frac{P_{\mathrm{out}}}{P_{\mathrm{dc}}} = \frac{G \cdot P_{\mathrm{in}}}{P_{\mathrm{dc}}} = \frac{G}{P_{\mathrm{dc}}} \cdot \frac{10^{\frac{P_{\mathrm{in,dBm}}}{10}}}{1000} \tag{4.18}$$

If P_{in} is further increased, due to the gain compression phenomenon related to the nonlinear active device behavior, both gain G and DC power P_{dc} start depending on P_{in}, and efficiency usually tends to saturate to a maximum value η_{sat} (at large-signal condition, G tends to decrease while P_{dc} to increase).

In a cascade of amplifiers, as shown in Figure 4.9, the overall system efficiency is directly affected by the respective values of the last stages, being:

$$\eta_{\mathrm{tot}} = \frac{\eta_N}{1 + \sum_{i=1}^{N-1} \frac{\eta_N}{\left[\prod_{j=i+1}^{N} G_j\right] \cdot \eta_i}} \tag{4.19}$$

where η_i and G_i indicate the efficiency and power gain of the stage i ($i = 1,\ldots,N$), respectively.

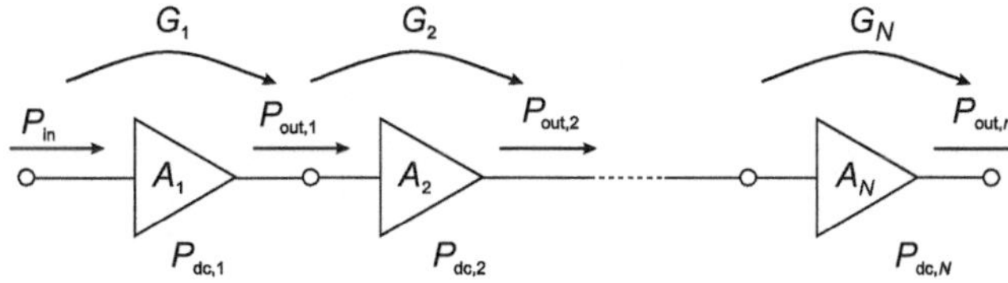

FIGURE 4.9

Cascade connection of amplifiers.

As the operating frequency increases, the device power gain decreases (roll-off behavior) and the contribution to the output power from the input drive cannot be neglected yet. Hence, a more meaningful parameter, namely power-added efficiency, PAE or η_{add}, is defined as the ratio between the RF added power and the supplied DC power:

$$\eta_{add} = \frac{P_{out} - P_{in}}{P_{dc}} = \eta \cdot \left(1 - \frac{1}{G}\right) \tag{4.20}$$

A typical behavior of the previous parameters versus input power levels is shown in Figure 4.10, where the linear (small-signal) and saturation limits (large-signal) are also indicated.

The power gain compression behavior relies on the type of active device adopted and its bias point, and an eventual gain expansion from G_L can also be observed (in particular for bias condition close to the threshold voltage) before a gain compression is experienced.

A widely used parameter to simply represent the compression behavior is the 1 dB compression point, 1 dBcp, which is defined as the output power level $P_{out,1dBcp}$ corresponding to a 1 dB deviation from the ideal linear behavior (ie, from the small signal gain value). Nevertheless, it is always possible to define and consequently determine the amplifier's features corresponding to any gain compression level, as required by the particular application (eg, in radar systems the PA are usually operated in deep saturation, resulting in 3–5 dB of gain compression).

As can be noted from Figure 4.10, the maximum efficiency is attained when the amplifier is driven into compression (typically PAE maximum is accomplished between 1 and 3 dB of gain compression in GaAs devices, 6–8 dB in GaN devices),

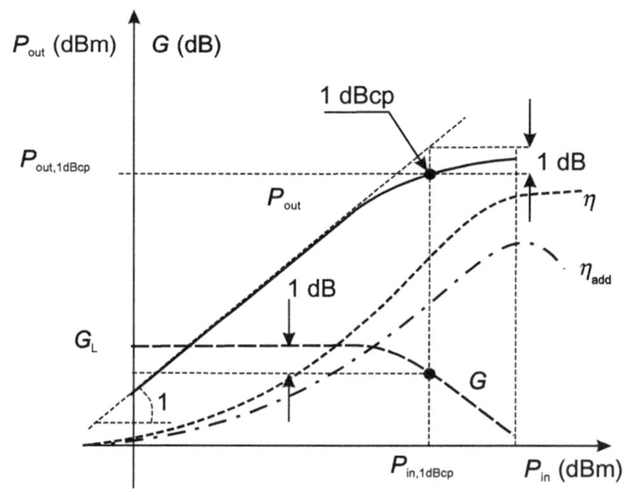

FIGURE 4.10

Typical performance in a PA as a function of input drive.

thus leaving its almost linear region and approaching the nonlinear active device physical limitations.

Unfortunately, the gain compression underlies a nonlinear behavior of the active device, resulting in a not purely current (and voltage) generation from the intrinsic current source, thus generating harmonic components at multiple frequencies of the carrier frequency f.

In general, for small amplitudes of the input level P_{in}, however, sufficient to generate harmonic distortion, HD, the output power delivered at the generic harmonic frequency nf can be demonstrated to be proportional to the nth power of P_{in} [32], as shown in Figure 4.11, that is,

$$P_{out}(nf) \propto P_{in}(f)^n \qquad P_{out}(nf)_{dBm} \propto n \cdot P_{in}(f)_{dBm} \tag{4.21}$$

In order to quantify the harmonic generation, the corresponding harmonic distortion HD_{nf} parameter is defined as:

$$HD_{nf} = \frac{P_{out}(nf)}{P_{out}(f)} \tag{4.22}$$

while the total harmonic distortion, THD, is defined summing up all HD components in the output signal, that is,

$$THD = \frac{\sum_{n>1} P_{out}(nf)}{P_{out}(f)} \tag{4.23}$$

The above quantities are typically expressed in logarithmic units (decibel) over the carrier power (ie, in dBc or in percentage).

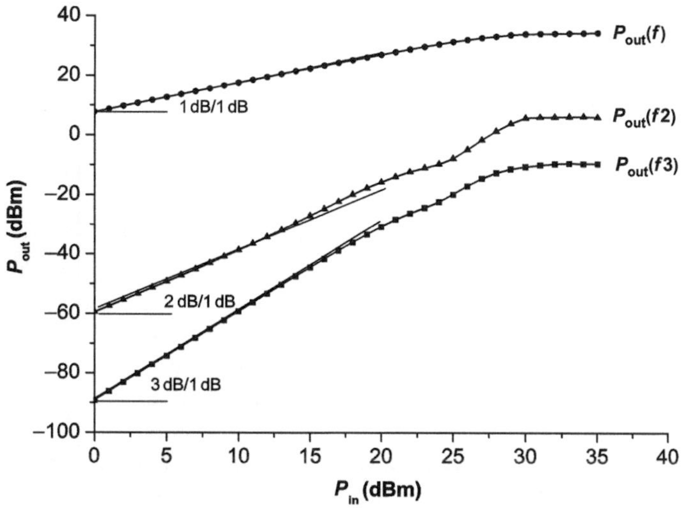

FIGURE 4.11

Example of measured output power in a single-tone test, at fundamental frequency (*circle*), second (*triangle*), and third (*squared*) harmonic components.

The previous parameters quantify the nonlinear distortion in terms of amplitude only. However, actual amplifiers are dynamic systems characterized also by memory effects, whose nonlinear behavior affects not only the amplitude but also the phase of the output signal.

A simplified representation of such nonlinear memory effects can be adopted by using the amplifier's behavior described by the amplitude-to-amplitude compression, AM/AM, and the phase-to-amplitude conversion, AM/PM [44]. It is based on the assumption that if the input signal is expressed in the form:

$$s_{\text{in}}(t) = A(t) \cdot \cos(\omega t + \phi) \tag{4.24}$$

the in-band (fundamental frequency) output signal can be represented as:

$$s_{\text{out}}(t) = G[A(t)] \cdot \cos\{\omega t + \phi + \Psi[A(t)]\} \tag{4.25}$$

where its amplitude $G(\cdot)$ is a nonlinear function related to the amplitude $A(t)$ of the input signal, giving rise to the AM/AM compression, while also its phase change $\Psi(\cdot)$ is depending on the input signal drive level, resulting in an AM/PM conversion effect too.

A typical example of AM/AM compression and AM/PM conversion plot in the case of a CW signal applied is shown in Figure 4.12. The latter can be represented by plotting the absolute values of the output signal phase or, as reported in the figure, normalizing its phase to the small-signal value, thus representing its variation.

As extensively discussed in Chapter 8, the information gathered via a single tone test (CW) is usually not sufficient to fully characterize the nonlinear behavior of the PA, since typically the input signals are modulated rather than single tone ones, thus

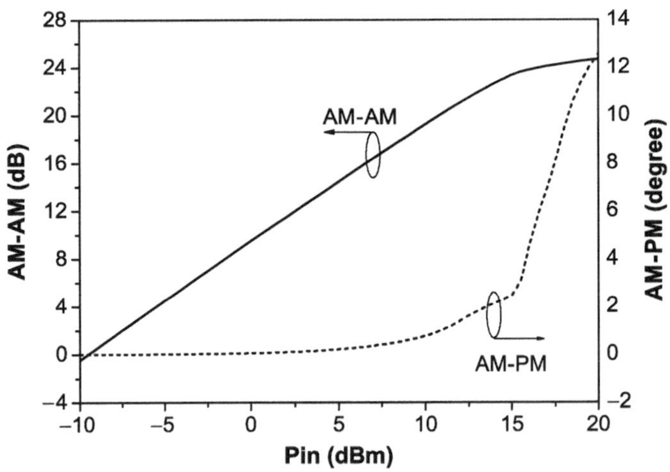

FIGURE 4.12

Typical AM/AM compression and AM/PM conversion curves for a PA.

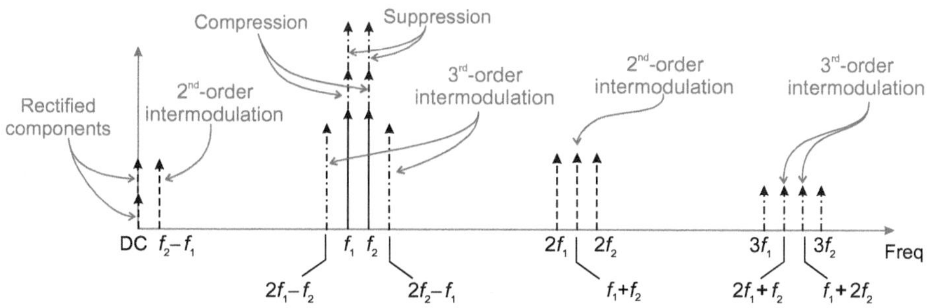

FIGURE 4.13

Frequency allocation of the output components originated in a two-tone test.

resulting in a populated spectrum over a frequency band. Other figures are therefore inferred to characterize the nonlinear PA behavior.

Starting from a simple two-tone test, used to simulate the simultaneous treatment of two different signals, and therefore their mutual interaction in the nonlinear PA, the third-order intercept point, *IP*3, and the carrier over intermodulation, *C*/*I*, parameters are introduced [44]. Obviously, this is a simplified approximation of a modulated signal, being two tones much more representative than a simple sinusoid, and sufficiently easy to be experimentally characterized.

In this case the output spectrum is shown in Figure 4.13, where the intermodulation, *IM*, signals are generated by the interaction of each frequency component, resulting in additional signals at frequencies that are not just at harmonic frequencies (integer multiples) of either frequency, but at any linear combination of the original frequencies and their multiples, that is, at frequencies given in general by:

$$f_{IM} = k_1 \cdot f_1 + k_2 \cdot f_2 \qquad k_1, k_2 = \pm 1, \pm 2, \ldots \tag{4.26}$$

where f_1 and f_2 are the frequencies of the input signal and $O = |k_1| + |k_2|$ represents the order of intermodulation.

Obviously, the intermodulation is unattractive in communication systems representing spurious emissions in the form of sidebands. If most of *IM* products can be eliminated by filtering, nonetheless the contribution of the third-order intermodulation related to the terms at frequencies $2f_2 - f_1$ and $2f_1 - f_2$ (commonly referred to as third-order intermodulation components, *IM*3) are relatively close to the input signal frequencies f_1 and f_2 (in-band distortion) and usually cannot be removed by filtering.

In particular, considering the two input frequencies f_1 and f_2, the third-order intercept point, *IP*3, is defined as the output, *OIP*3, or input, *IIP*3, power level at which the third-order *IM* component level (ie, the output signal at frequency $2f_1 - f_2$ or $2f_2 - f_1$) is equal to the ideal linear output power of the PA at f_1 or f_2. Clearly this parameter cannot be directly measured and it is inferred ideally extrapolating the behavior of the output power at one carrier and at *IM*3 frequency, ideally rising by 1 and 3 dB per dB increase in input power, respectively, as graphically shown in Figure 4.14.

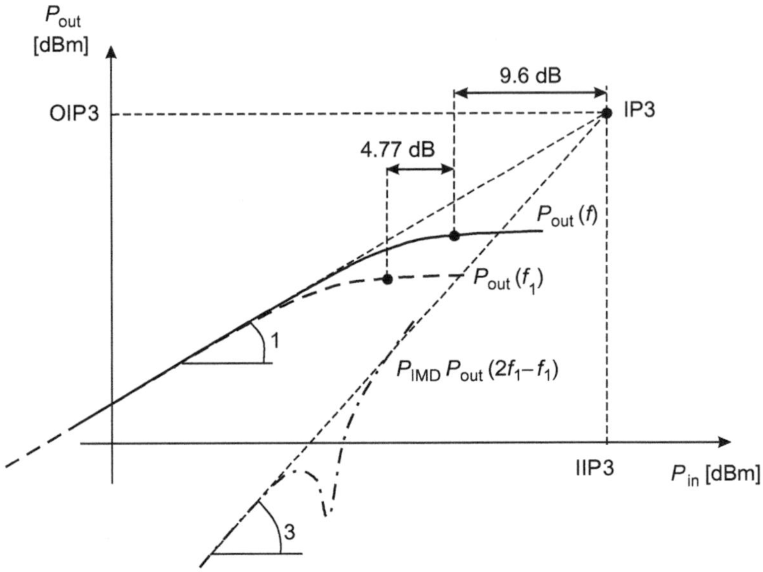

FIGURE 4.14

Third-order intercept point definition.

It is of note that the intermodulation products at lower $2f_1 - f_2$ or upper $2f_2 - f_1$ side bands could present an asymmetric amplitude behavior with the increasing level of input signal [45–48]. Consequently, in the definition of *IP*3 the side band considered should be clearly stated.

Another used parameter to quantify the *IM* distortion is the Carrier-to-Intermodulation Ratio, *C/I*, defined as:

$$C/I = \frac{P_{out}(f_i)}{P_{out}(2f_{2i} - f_j)} \qquad i = 1,2 \;\; j = 1,2; \;\; j \neq i \tag{4.27}$$

Also this parameter is usually measured by using logarithmic units, thus expressed as decibel relative to the carrier (dBc). An example of a typical *C/I* behavior in a PA is shown in Figure 4.15.

The figures of merit above defined for linearity evaluation are related to single or two-tone tests, trying to resemble in this way the PA behavior in response to narrow-band or multi-carrier input. However, actual input signals to a PA may deviate substantially from the single- or two-tone approximation, since modulation formats and bandwidth occupation may differ significantly, especially in communication systems. For this purpose, more complex figures of merits are introduced, as the adjacent channel power ratio, ACPR, or similar, which will be discussed in Chapter 6.

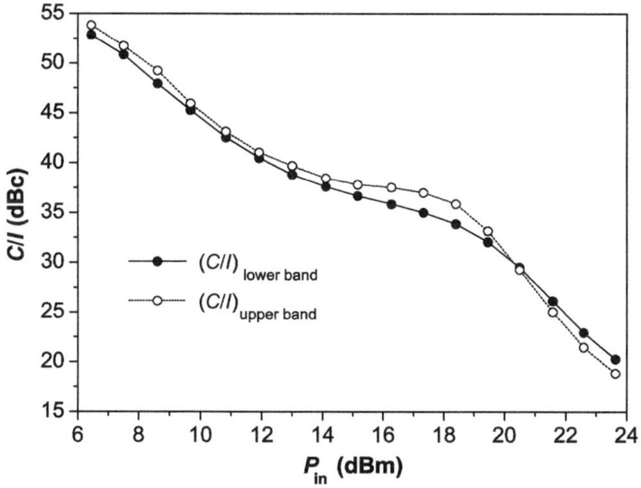

FIGURE 4.15

Example of measured *C/I* for a PA.

4.4 DESIGN STRATEGIES FOR HIGH-EFFICIENCY PAs

In the design of a PA, conflicting requirements usually have to be simultaneously fulfilled, as high power, gain, and efficiency with high linearity. A high gain (and output power) level is typically required to reduce the number of stages. However, as previously discussed, a high gain, resulting in the conjugate matching condition, is in disagreement with a high output power, which conversely results in the power match condition. Similarly, a high efficiency implies to operate the device in large-signal conditions (at 1 dB of compression or over), which clearly is contrasting with a high linearity. Moreover, as will be discussed in this section, highest efficiency levels can be attained for low bias conditions, further sacrificing some amount of power gain.

Consequently, proper design solutions and suitable compromises are demanded from the designers. The active device nonlinear behavior (intrinsically related to the large-signal operating conditions) has to be properly taken into account, but unfortunately this feature does not allow obtaining closed form relationship or design equations as happen in the linear case, where the S-parameters representation and related linear methods are adopted via closed-form expressions for a small-signal linear amplifier.

However, simplified device equivalent models have been developed and refined at different levels to account large-signal phenomena and to infer guidelines for PA design [49–52].

The most simple model useful to infer the theoretical maximum output power and efficiency for different bias conditions and the corresponding optimum loads for harmonics also is based on the equivalent circuit shown in Figure 4.3, where only the drain-source (FET case) or collector-emitter (BJT case) controlled current source is assumed to be nonlinear, neglecting, at a first approximation, all parasitics and feedback elements.

The nonlinear behavior, highlighted in the output IV characteristics also, is simplified assuming a piecewise linear approximation, as shown in Figure 4.16, further simplifying the model by assuming a constant transconductance, g_m. It is to note that the IV characteristics shown in this figure are representing the behavior of the intrinsic current source, i_{int}, thus differing from the extrinsic IV characteristics, which accounts for the device output resistance R_{out} (see Figure 4.3).

In order to account for the g_m variation in actual devices, which is responsible of the distortion and gain expansion observable for Class B bias condition, a linear transconductance could be considered, as made in [50–52]. Nevertheless, for our purpose the most simple model is sufficient to highlight the main guidelines in PA design without extra complication in computation (accounting for that in any case the results have to be considered as a rough approximation of an actual full nonlinear system).

The main parameter that has to be taken into account are the maximum device output current I_{Max}, the knee voltage V_k, the breakdown maximum voltage V_{BR}, eventually the maximum input voltage V_{bi} (for which the current is I_{Max}), and the input pinch-off voltage V_p (for which the output current becomes zero).

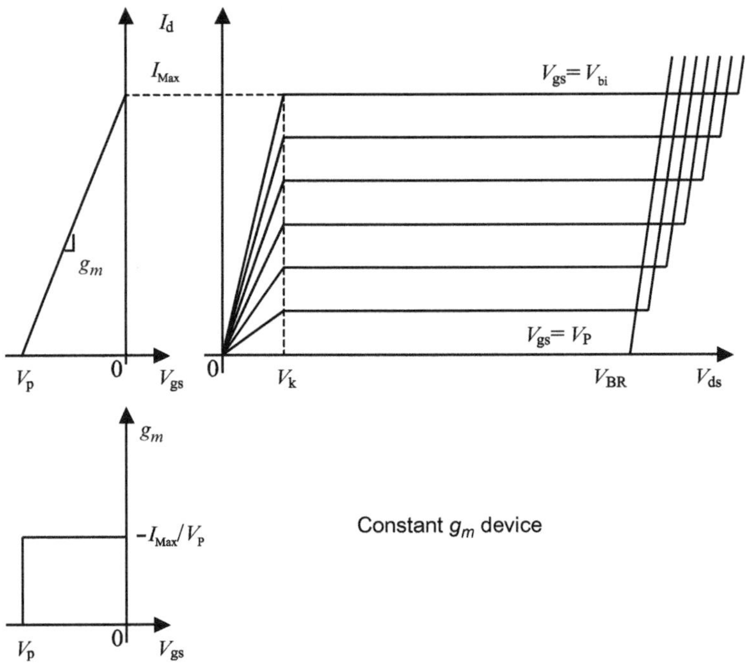

FIGURE 4.16

Piecewise linear approximation of the device output characteristics for the case of constant transconductance.

It is convenient, at this point, to define the class of operation of a PA. If the device is treated as a controlled current source, delivering a certain amount of current in correspondence with an input signal, a possible classification can be made on the basis of the current conducting angle, CCA, assuming that the device is driven by a sinusoidal signal, independently of the output loading condition. Hence, the Class A condition will correspond to a 2π conduction angle, meaning that the device is conducting current during the whole period of input signal; the Class B condition to a π conduction angle, meaning that the device is conducting current only for half period; the Class AB condition to a conduction angle between π and 2π; and finally the Class C to a conduction angle smaller than π. Being this definition related to the amplitude of the input signal (as a Class AB condition could correspond actually to a Class A condition for low levels of input signal), it would be more convenient to determine the same bias classes considering the value of the quiescent current I_{DC} (eg, in absence of RF signal applied), with respect to the maximum value, I_{Max}. In this definition, Class A will correspond to have $I_{DC} = I_{Max}/2$, Class B to $I_{DC} = 0$, Class AB to an intermediate value ($0 < I_{DC} < I_{Max}/2$) and Class C to a condition in which the theoretical quiescent current should assume negative values (obviously, this is a mathematical artifact).

Then, assuming a Class A bias condition and a DC voltage $V_{ds,DC}$, driving the device up to its maximum linear excursion (ie, without overdriving it into nonlinear regimes) [49], the maximum output power $P_{out,A}$ is accomplished when the intrinsic current source achieves the maximum current excursion, which is purely sinusoidal with amplitude $I_1 = I_{Max}/2$, and simultaneously the voltage across it is maximized also, that is, $V_1 = V_{ds,DC} - V_k$:

$$P_{out,A} = \frac{1}{2} \cdot I_1 \cdot V_1 = \frac{I_{Max}}{4} \cdot \left(V_{ds,DC} - V_k\right) \tag{4.28}$$

This condition is obtained if the load seen by the intrinsic current source Z_{int} is given by a purely resistive impedance R_A given by:

$$R_A = \frac{V_1}{I_1} = 2 \cdot \frac{\left(V_{ds,DC} - V_k\right)}{I_{Max}} \tag{4.29}$$

resulting in a dynamic load line, DLL, as shown in Figure 4.17.

If a complex impedance Z_{int} is considered, that is, $Z_{int} = |Z| \cdot e^{j\phi}$, then even if the voltage and current have the maximum excursion, they are phase shifted and the power degrades to:

$$P_{out} = \frac{I_{Max}}{4} \cdot \left(V_{ds,DC} - V_k\right) \cdot \cos(\phi) \tag{4.30}$$

while the load line becomes an elliptic curve, as shown in Figure 4.18 (traveled clockwise or counterclockwise depending if the reactive part of Z_{int} is capacitive or inductive, respectively).

The DC supplied power is constant and given by:

$$P_{dc,A} = V_{ds,DC} \cdot \frac{I_{Max}}{2} \tag{4.31}$$

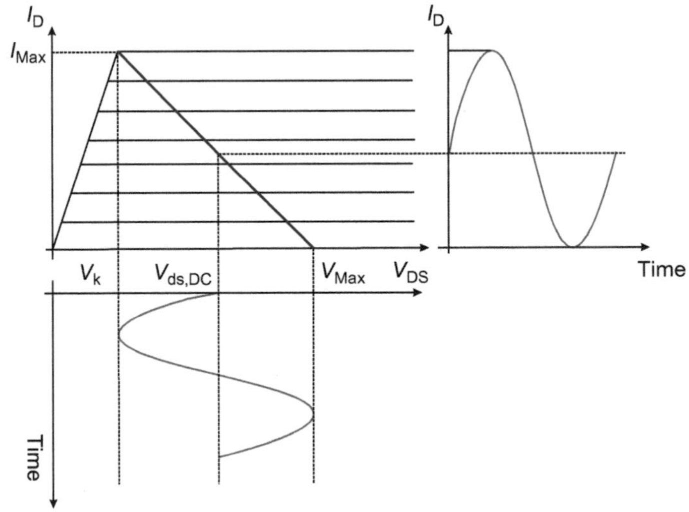

FIGURE 4.17

Load curve for a Class A bias condition under an optimum (and purely resistive) load condition across the intrinsic current source.

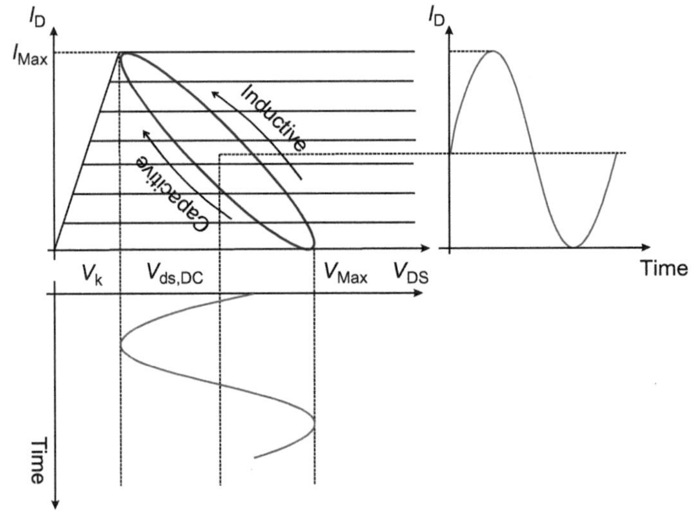

FIGURE 4.18

Load curve for a Class A bias condition under a complex load condition across the intrinsic current source.

and finally, the theoretical efficiency can be computed as:

$$\eta_A = \frac{I_{Max}}{4} \cdot \frac{V_{ds,DC} - V_k}{V_{ds,DC} \cdot I_{Max}/2} = \frac{1}{2} \cdot \left(1 - \frac{V_k}{V_{ds,DC}}\right) \tag{4.32}$$

resulting in the well-known 50% "theoretical magic number" if V_k is neglected. This equation highlights also the critical effects of the knee voltage if compared to the bias point in the effective efficiency, as for instance happens in low-voltage PA design.

However, such quasilinear approach and the Class A bias point does not represent the best compromise, since potential beneficial effects of the harmonic load terminations are not exploited at all, while in actual cases they play a fundamental role to improve both efficiency and output power levels [53–56].

Moving away from Class A bias, the current starts to be truncated at the pinch-off. The output current is imposed by a suitable input signal, usually a sinusoidal shape as external source, which eventually can be re-shaped across the controlling element C_{gs} in Figure 4.3. Assuming a sinusoidal driving signal (eg, in the simplified model assuming a sinusoidal voltage V_{gs} in Figure 4.3), the output current, when its maximum swing is reached, can be assumed as a truncated sinusoid:

$$i_D(t) = \begin{cases} \dfrac{I_{Max}}{1 - \cos(\Phi/2)} \cdot [\cos(\omega t - \Phi) - \cos(\Phi/2)] & \text{if } |\omega t| \leq \Phi/2 \\ 0 & \text{otherwise} \end{cases} \tag{4.33}$$

as shown in Figure 4.19, where Φ represents the output current conduction angle (CCA); that is, $\Phi = 2\pi$ Class A, $\Phi = \pi$ Class B, $\pi < \Phi < 2\pi$ Class AB, $\Phi < \pi$ Class C.

The CCA can be related to the ratio between the quiescent DC bias current I_{DC} and the maximum achievable value I_{Max}, that is

$$\cos\left(\frac{\Phi}{2}\right) = \frac{I_{DC}}{I_{DC} - I_{Max}} \tag{4.34}$$

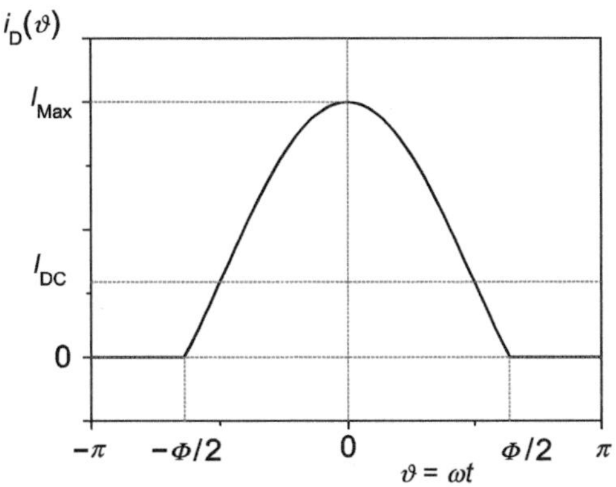

FIGURE 4.19

Output current waveform assuming a constant transconductance g_m.

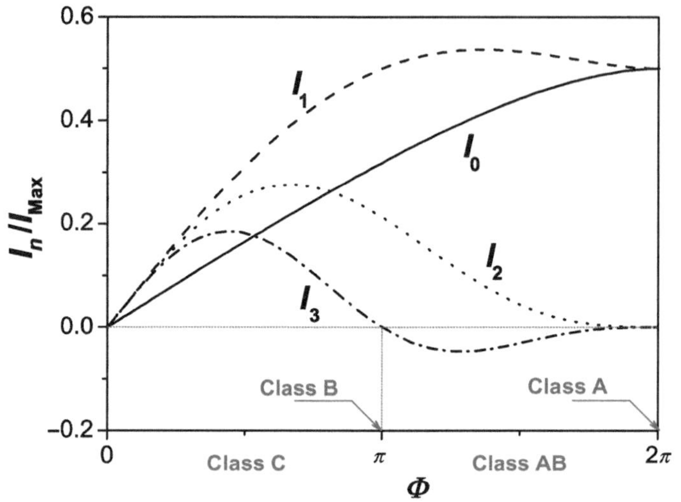

FIGURE 4.20

Fourier components I_0, I_2, I_3 normalized to the device maximum current I_{Max} as functions of the drain CCA Φ.

Since the output current for a generic CCA contains harmonics, as shown in Figure 4.20, in order to compute the device performance is required to define how these harmonics are terminated to generate the output voltage.

In fact, the output voltage is related to the impedances presented across the current source and can be generically represented as in the following:

$$\begin{aligned} v_{DS} &= V_{DD} - Z_1 \cdot I_1 \cdot \cos(\omega t + \phi_1) - Z_2 \cdot I_2 \cdot \cos(2\omega t + \phi_2) + \\ &\quad - Z_3 \cdot I_3 \cdot \cos(3\omega t + \phi_3) + \cdots \\ &= V_{DD} - V_1 \cdot [\cos(\omega t + \phi_1) + k_2 \cdot \cos(2\omega t + \phi_2) + \\ &\quad + k_3 \cdot \cos(3\omega t + \phi_3)] + \cdots \end{aligned} \tag{4.35}$$

being the intrinsic impedance in frequency $Z_{int}(f)$ given by:

$$Z_{int}(nf) = Z_n \cdot e^{j\phi_n} \tag{4.36}$$

and defined

$$V_1 = Z_1 \cdot I_1 \tag{4.37}$$

$$k_n = \frac{Z_n \cdot I_n}{Z_1 \cdot I_1} \tag{4.38}$$

Simple relationships can be computed if a purely resistive load is assumed for all the harmonics (ie, assuming that the intrinsic current source sees a constant resistance R at all harmonics):

$$Z_{int,Res}(nf) = R \ \forall \ n \tag{4.39}$$

or a short-circuit condition (still at the intrinsic terminal) is assumed for all the harmonics, known as tuned load (TL) condition [32]:

$$Z_{\text{int,TL}}(nf) = \begin{cases} R_{\text{TL}} & n = 1 \\ 0 & n > 1 \end{cases} \tag{4.40}$$

The obtained results are summarized in Tables 4.1 and 4.2 for the resistive or tuned load conditions, respectively.

An example of load curves for the resistive case and the tuned load, assuming a Class AB bias condition, for different input driving signals, are shown in Figures 4.21 and 4.22, respectively.

For a generic CCA Φ the resulting expression for the output power, efficiency, and optimum load condition are reported in the following:

$$P_{\text{out,TL}} = \frac{P_{\text{out,A}}}{\pi} \cdot \frac{\Phi - \sin(\Phi)}{1 - \cos\left(\frac{\Phi}{2}\right)} \tag{4.41}$$

Table 4.1 Single-Device PA Performance With Resistive Loading for Class A and B Bias and Constant Transconductance

	Class A	Class B
Optimum intrinsic impedance	R_A	$\frac{\pi}{\pi-1} \cdot \frac{R_A}{2}$
Output power	$\frac{(V_{\text{ds,DC}} - V_K) \cdot I_{\text{Max}}}{4}$	$\frac{\pi(V_{\text{ds,DC}} - V_K) \cdot I_{\text{Max}}}{8 \cdot (\pi - 1)}$
Dissipated power	$\frac{(V_{\text{ds,DC}} - V_K) \cdot I_{\text{Max}}}{4}$	$\frac{(0.4V_{\text{ds,DC}} + 0.87V_K) \cdot I_{\text{Max}}}{4}$
Efficiency	$\frac{1}{2} \cdot \left(1 - \frac{V_K}{V_{\text{ds,DC}}}\right)$	$0.58 \cdot \left(1 - \frac{V_K}{V_{\text{ds,DC}}}\right)$

Table 4.2 Single-Device PA Performance With Tuned Load Condition for Class A and B Bias and Constant Transconductance

	Class A	Class B
Optimum intrinsic impedance	$Z_{\text{int}} = \begin{cases} R_A & \text{at } f \\ 0 & \text{at } nf, n > 1. \end{cases}$	$Z_{\text{int}} = \begin{cases} R_A & \text{at } f \\ 0 & \text{at } nf, n > 1. \end{cases}$
Output power	$\frac{(V_{\text{ds,DC}} - V_K) \cdot I_{\text{Max}}}{4}$	$\frac{(V_{\text{ds,DC}} - V_K) \cdot I_{\text{Max}}}{4}$
Dissipated power	$\frac{(V_{\text{ds,DC}} + V_K) \cdot I_{\text{Max}}}{4}$	$\frac{\left[V_{\text{DD}}\left(\frac{4}{\pi} - 1\right) + V_K\right] \cdot I_{\text{Max}}}{4}$
Efficiency	$\frac{1}{2} \cdot \left(1 - \frac{V_K}{V_{\text{ds,DC}}}\right)$	$\frac{\pi}{4} \cdot \left(1 - \frac{V_K}{V_{\text{ds,DC}}}\right)$

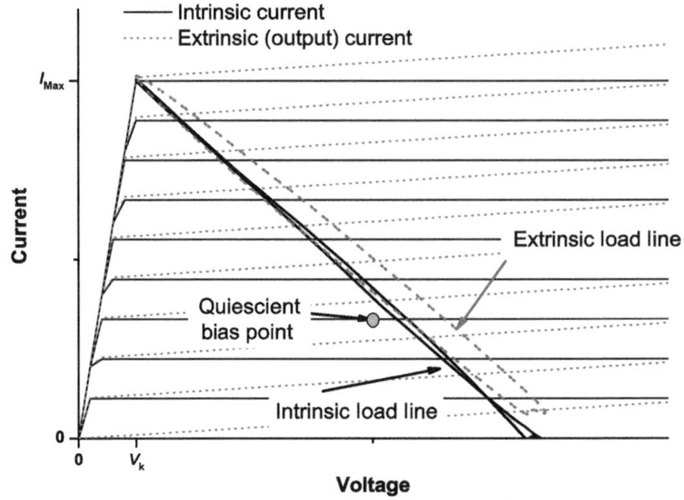

FIGURE 4.21

Load line for the resistive case: intrinsic (*solid line*) and extrinsic (*dotted line*), for increased level of the input driving signal.

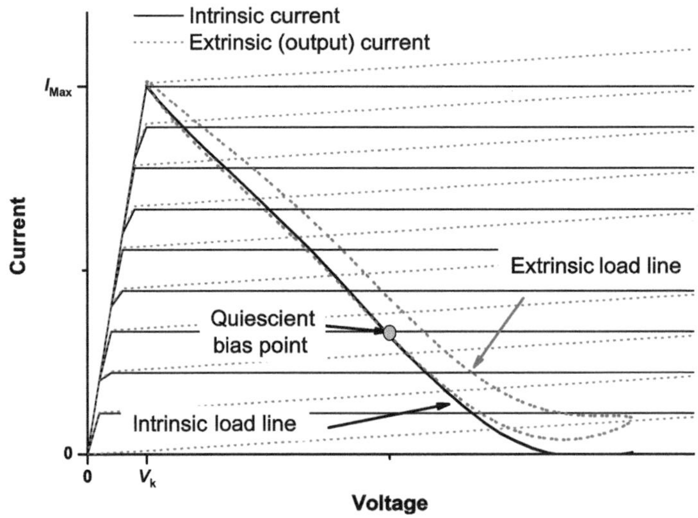

FIGURE 4.22

Load line for the tuned load case: intrinsic (*solid line*) and extrinsic (*dotted line*), for increased level of the input driving signal.

$$\eta_{TL} = \eta_A \cdot \frac{\Phi - \sin(\Phi)}{2 \cdot \sin\left(\frac{\Phi}{2}\right) - \Phi \cdot \cos\left(\frac{\Phi}{2}\right)} \tag{4.42}$$

$$R_{TL} = R_A \cdot \pi \cdot \frac{1 - \cos\left(\frac{\Phi}{2}\right)}{\Phi - \sin(\Phi)} \tag{4.43}$$

A comparison of the theoretical performances obtained with resistive or tuned load cases, varying the bias point, is shown in Figure 4.23.

As can be noted, moving the bias point from Class A to Class B, roughly the same output power levels are achieved for the tuned load, while for the resistive case a decrease is expected. Moving from Class A to B the efficiency increases in both cases.

From a simple comparison, it is clear that the tuned load approach shows better results since the output power does not degrade while the efficiency increases largely if compared to the resistive case. Moreover, from a practical point of view, the design of a matching network to fulfill the tuned load condition (ie, short circuit at all harmonics across the intrinsic current source) is much more simple to perform if compared to the resistive case, where it is required to resonate the output device impedance (C_{ds}) at all harmonics, thus increasing dramatically the complexity of the matching network.

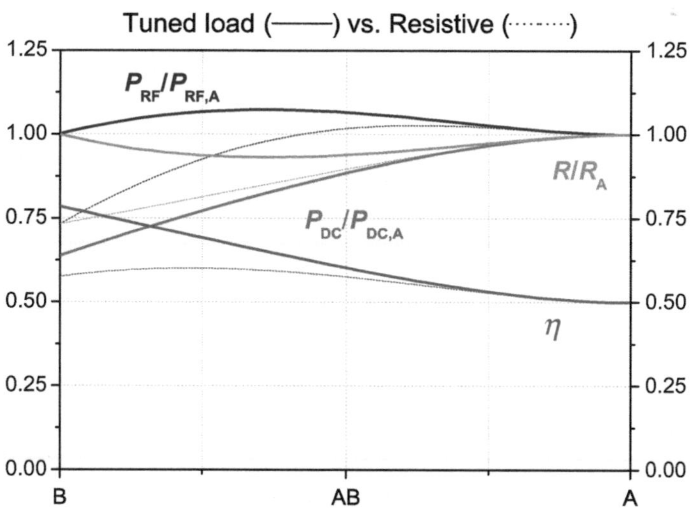

FIGURE 4.23

Comparison between resistive or tuned load cases, or different bias points.

Thus, the first conclusion that could be drawn from this simple analysis results in the possibility to increase the device performances, changing the bias point (from Class A) and use suitable harmonic terminations (short circuit). However, there is another parameter that was not still taken into account that unfortunately decreases when the bias point is moved toward Class B: the gain. In fact, it is possible to demonstrate that the gain theoretically decreases of 6 dB moving from Class A to Class B [32], thus sometimes preventing the adoption of Class B (or Class C) conditions due to the resulting lower gain. Moreover, since the PAE is related to the gain value also (see Eq. (4.19)), it follows that the maximum PAE is achieved with a Class AB bias point intrinsically depending on the device gain.

A further consideration can be made analyzing the behavior of the output power and efficiency as functions of the input drive signal, as theoretically reported in Figure 4.24.

Moving from Class A ($I_{DC} = 0.5 \cdot I_{Max}$, $\Phi = 2\pi$) to Class B ($I_{DC} = 0$, $\Phi = \pi$), the input power back-off required to retrieve a linear behavior increases, while simultaneously a larger reduction in drain efficiency is experienced.

Up to now only the output harmonic terminations have been considered, while in actual cases also the input harmonic terminations should be accounted for, due to the intrinsic nonlinear behavior of the active device at the input port. Unfortunately, no simplified models are actually available to easily represent the input nonlinearities, thus no simplified relationships can be derived. However, some obvious considerations can

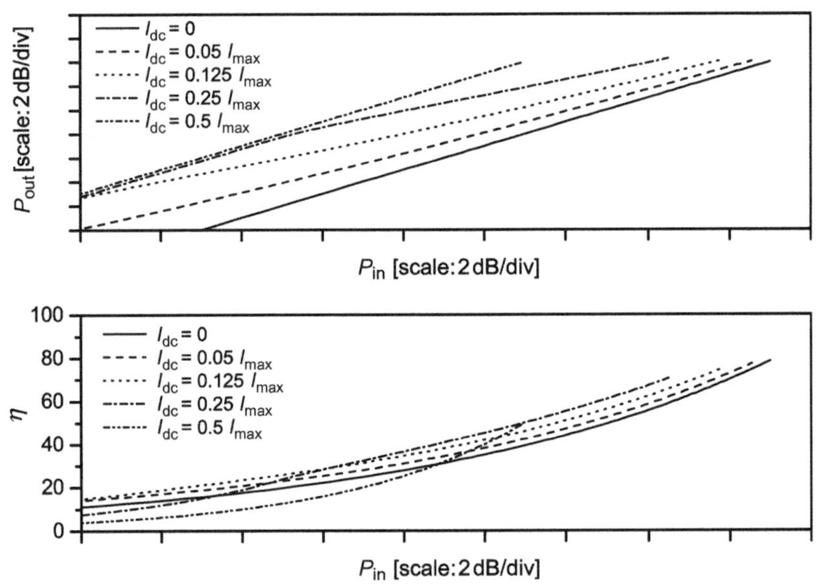

FIGURE 4.24

Tuned load output power (*top*) and efficiency (*bottom*) as function of the input power, for different bias conditions.

be inferred. For example, in order to assure the maximum power transfer from the external source to the active device, a conjugate matching condition for the fundamental frequency shall be adopted. For the harmonics, the previous theoretical results are achieved if the intrinsic driving signal across C_{gs} (ref. to Figure 4.3) is purely sinusoidal. Thus a suitable choice is to adopt a tuned load configuration also for the input, assuring a short-circuit condition for all the harmonics across the intrinsic C_{gs} element [57,58].

The TL approach represents a first and preliminary attempt to exploit the benefits of harmonic tuning (HT) mechanisms. Contributions in this field range from experimental observation of performance improvements [59–61], the introduction of novel topologies as in the case of Harmonic Reaction amplifiers [62,63], to systematic investigations of experimental performance [55] and design methods deserved for special HT configurations [64].

In order to clarify the HT approaches and how they can be helpful to improve the device performances, and in particular the efficiency capabilities, a simple power balance analysis can be performed, considering the physical mechanisms involved in the amplifying process. From this simple consideration, the efficiency can be represented as [32]:

$$\eta = \frac{P_{\text{out},f}}{P_{\text{DC}}} = \frac{P_{\text{out},f}}{P_{\text{diss}} + P_{\text{out},f} + \sum_{n=2}^{\infty} P_{\text{out},nf}} \tag{4.44}$$

where $P_{\text{out},nf}$ is the output power at the generic harmonic frequency nf, given by:

$$P_{\text{out},nf} = \frac{1}{2} V_n \cdot I_n \cdot \cos(\phi_n) \tag{4.45}$$

P_{DC} is the supplied DC power given by:

$$P_{\text{DC}} = V_{\text{DD}} \cdot \frac{1}{T} \int_0^T i_{\text{D}}\, dt = V_{\text{DD}} \cdot I_0 \tag{4.46}$$

and P_{diss} is the power dissipated in the active device, which can be expressed as:

$$P_{\text{diss}} = \frac{1}{T} \int_0^T v_{\text{DS}} \cdot i_{\text{D}} dt = V_{\text{DD}} \cdot I_0 - \sum_{n=1}^{\infty} V_n \cdot I_n \cdot \cos(\phi_n) \tag{4.47}$$

From Eq. (4.44) it follows that the theoretical maximum efficiency of 100% is achieved if the two following conditions are simultaneously fulfilled:

- the power dissipated in the active device is nulled

$$P_{\text{diss}} = \frac{1}{T} \int_0^T v_{\text{DS}} \cdot i_{\text{D}}\, dt = 0 \tag{4.48}$$

- the power delivered at harmonic frequencies is nulled

$$P_{\text{out},nf} = \frac{1}{2} \cdot V_n \cdot I_n \cdot \cos(\phi_n) = 0 \quad n > 1 \tag{4.49}$$

Eq. (4.48) is known as nonoverlapping condition between the drain (collector) voltage and current waveforms. It can be accomplished by properly shaping the two waveforms, forcing the current on the device to vanish for a nonzero voltage and vice-versa. This condition alone is not sufficient to ensure a 100% of efficiency [32]: also Eq. (4.49) has to be fulfilled simultaneously with the previous one.

The latter condition highlights two possible design strategies, that are at the basis of two classes of PAs:

- null the current I_n or voltage V_n harmonic components, by selecting open- or short-circuit terminations, respectively, originating the class of *current-mode amplifiers*; and
- null the cosine term in Eq. (4.49), that is, by using purely reactive (capacitive or inductive) harmonic terminations, originating the class of *switched-mode amplifiers*.

4.4.1 TUNED LOAD

In this case $V_n = 0$ for $n > 1$ due to the short-circuiting conditions previously discussed. However, an overlap between output voltage and current waveforms occurs and the theoretical maximum efficiency (100%) can be achieved only for a deep Class C condition, namely for a current conduction angle Φ dropping to zero, which in turn implies no power transferred to the external load, thus not representing a practical solution. The amplifier schematic is shown in Figure 4.25.

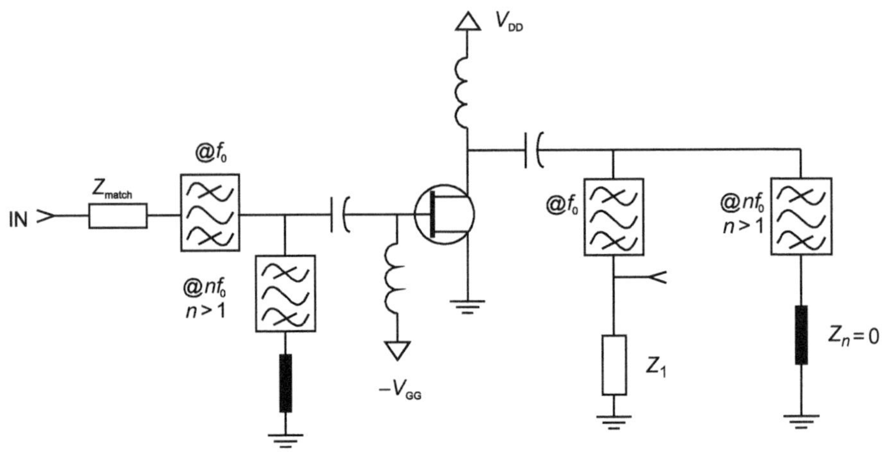

FIGURE 4.25

Schematic of the ideal tuned load power amplifier.

4.4.2 IDEAL CLASS F OR INVERSE CLASS F (CLASS F^{-1})

In order to have a fundamental output power different from zero, while achieving 100% of drain efficiency, a proper waveform shaping is mandatory. From a mathematical point of view, if the current and voltage waveforms assume the shape reported in Figure 4.26, no overlapping occurs and, while considering the harmonic content, it follows that the odd harmonics of the waveform $a(t)$ are zero as well as the even harmonic of the waveform $b(t)$, thus fulfilling the condition (4.49).

If the intrinsic output current $i(t)$ is represented by the waveform $a(t)$, then the proper set of harmonic terminations becomes:

$$i(t) = a(t) \Rightarrow \begin{cases} Z_{2n} = 0 \\ Z_{2n+1} = \infty \end{cases} \tag{4.50}$$

which corresponds to the Class F [56,65–68] strategy, whose a schematic example is shown in Figure 4.27 (actual implementations can be found in [69–73]).

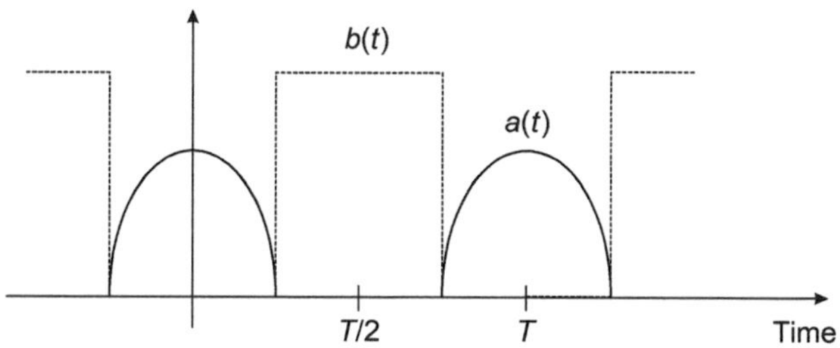

FIGURE 4.26

Ideal voltage and/or current waveforms assumed at the intrinsic device output.

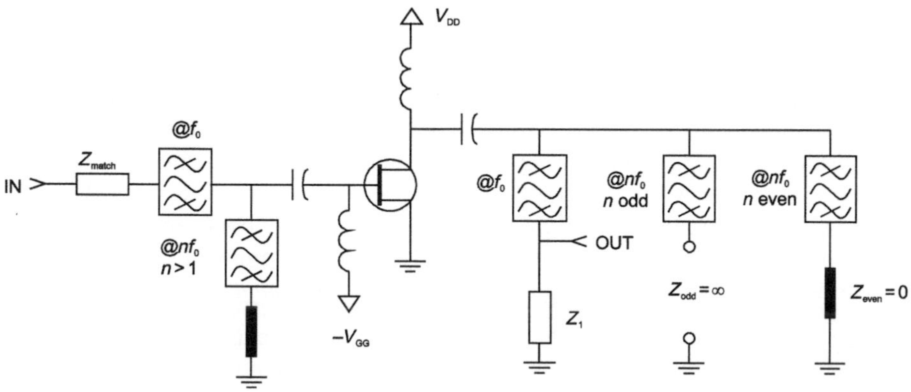

FIGURE 4.27

Schematic of the ideal Class F power amplifier.

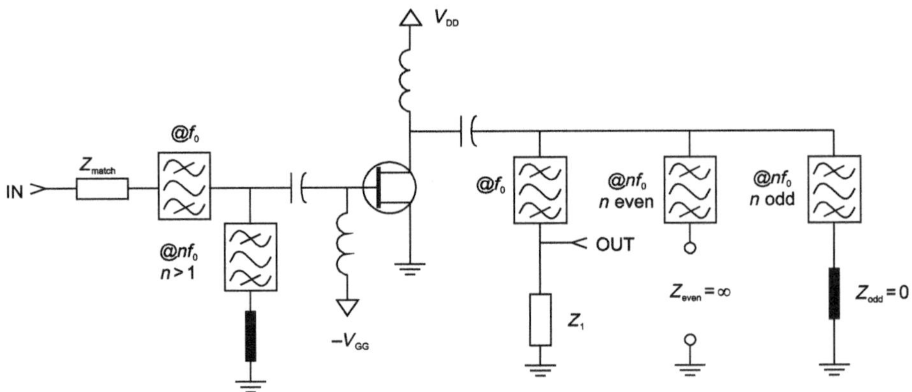

FIGURE 4.28

Schematic of the ideal inverse Class F power amplifier.

Conversely, if the current waveform $i(t)$ corresponds to $b(t)$, then the harmonic terminations are given by:

$$i(t) = b(t) \Rightarrow \begin{cases} Z_{2n} = \infty \\ Z_{2n+1} = 0 \end{cases} \tag{4.51}$$

which corresponds to the inverse Class F (or Class F^{-1}) strategy [74], whose a schematic example is shown in Figure 4.28.

4.4.3 IDEAL CLASS E

A different possibility to achieve a theoretical 100% efficiency is to resort reactive harmonic terminations (for $n > 1$), while ensuring a proper phase shift between voltage and current harmonics, that is,

$$\Phi_n = \pm\frac{\pi}{2} \tag{4.52}$$

Obviously, the fulfillment of these conditions is not sufficient to ensure maximum efficiency, being also necessary to select the proper "stimulus" (current waveform) from which the proper out-coming signal (voltage waveform) can be generated, avoiding any overlap, through the proper choice of harmonic loads (including DC and fundamental frequency). Such a condition is typically obtained operating the active device as a switch, and adopting purely reactive terminating impedances at all harmonic frequencies, leading to the *switching-mode amplifiers*.

The fulfillment of condition $\Phi_n = \frac{\pi}{2}$ corresponds to the classical Class E configuration proposed by Sokal [75], whose schematic is shown in Figure 4.29, while the condition $\Phi_n = -\frac{\pi}{2}$ corresponds to the alternative Class E configuration proposed by Grebennikov [76], whose schematic is shown in Figure 4.30.

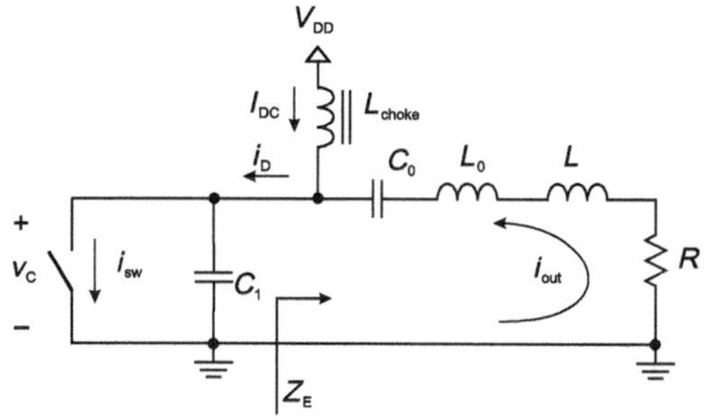

FIGURE 4.29

Schematic of the ideal Class E power amplifier.

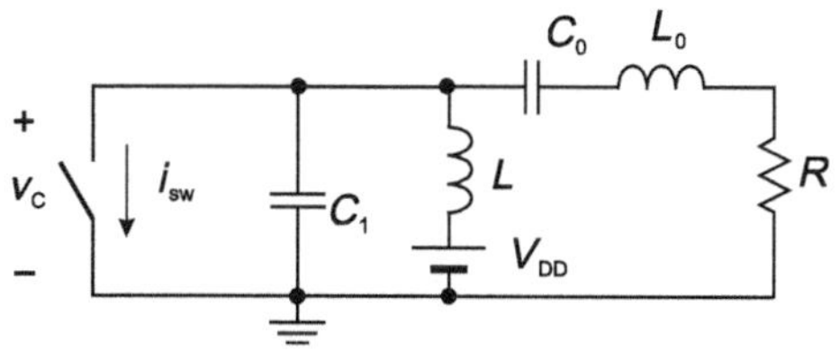

FIGURE 4.30

Schematic of the ideal Class E power amplifier with a shunt inductor.

4.4.4 HIGH-FREQUENCY HT APPROACHES

In low-frequency applications, a large number of harmonic terminations (therefore well approximating the assumption of an infinite number of harmonics) can be effectively controlled. It is so possible to achieve significant power efficiency levels, closer to the ideal figures, especially if the selected active device exhibits very low knee voltage V_k and/or high breakdown voltage V_{BR}. The picture considerably modifies if the operating frequency range is moved toward the microwave region and beyond. In this case, in fact, the application of both design approaches (ideal Class E and ideal Class F) results in degraded performance as compared to the ideally expected ones, due to the limited number of harmonics that can be effectively controlled.

As an example, actual Class F amplifiers are usually designed using two or three harmonic traps to control output harmonic impedances at $2f_0$ (in the following referred as second) and $3f_0$ (in the following referred as third) only. In fact, the benefits attainable controlling higher-order harmonic terminations are usually negligible, if compared to the growing circuit complexity and to the resulting higher losses [62,63]. Moreover, as frequency increases, the control of higher-order output terminations become unfeasible, because active device output capacitive behavior practically short-circuits higher-frequency components, not allowing the desired wave shaping.

Consequently, the aforementioned issues suggest, for a PA stage, to control the fundamental and the first two harmonic terminations only. The higher-order harmonics are assumed to be terminated through the shunting effect of the output device capacitive behavior, thus assuming (as a reasonable approximation) a short-circuit condition for all the frequency components higher than the third one.

As a first consequence, the lack of the contribution of such higher-order ($n > 3$) harmonics, clearly results in a nonnegligible overlap between current and voltage waveforms. The power dissipated in the active device P_{diss} increases, consequently lowering maximum achievable efficiency, even if the power delivered to the second and third harmonic is still nulled.

To infer the generic HT design criteria, by referring to the simplified model shown in Figure 4.3, we could assume that the output current is at a first approximation imposed by the input signal, while the output voltage can be properly shaped by using suitable loading conditions. Moreover, considering a limited number of harmonics to be controlled, this voltage can be written as already reported in Eq. (4.35):

$$v_{ds}(t) = V_{DD} - V_1 \cdot \cos(\omega t + \phi_1) - V_2 \cdot \cos(2\omega t + \phi_2) - V_3 \cdot \cos(3\omega t + \phi_3) \qquad (4.53)$$

where the voltage harmonic components V_n and their phases ϕ_n ($n = 1, 2, 3$) are related to the corresponding harmonic currents through the complex impedances $Z_{int,n}$ synthesized across the intrinsic current sources:

$$V_n = |Z_{int,n}| \cdot I_n$$
$$\phi_n = \text{phase}(Z_{int,n}) \qquad (4.54)$$

The aim of the HT is to select proper terminations $Z_{int,n}$ at fundamental and harmonic frequencies, in order to maximize the fundamental component (V_1) of the v_{ds} voltage swing, while fulfill the intrinsic device limiting condition, that is:

$$V_k \le v_{ds}(t) \le V_{BR} \qquad (4.55)$$

In the case of tuned load, where $Z_{int,2} = Z_{int,3} = 0$ and $Z_{int,1} = R_{opt}$, the maximum value for V_1 is:

$$V_1 = V_{1,TL} = V_{DD} - V_k \qquad (4.56)$$

The goal of the HT is to increase the fundamental voltage amplitude (V_1) over such value $V_{1,TL}$ by means of appropriately selected harmonic components.

Forgetting for the moment the feasibility of the optimum impedance values and the complexity for their implementation in practical cases, ideally voltage waveforms which optimize V_1 component, depending on the number of harmonic controlled (the uncontrolled are assumed short circuited), are graphically shown in Figure 4.31.

The theoretically optimum design parameters and performance's improvement are summarized in Table 4.3 [32].

From the results shown in Table 4.3, several remarks are mandatory to avoid misunderstanding interpretation.

First of all, when the required impedance $Z_{n,HT}$ (for $n = 1, 2, 3$) is higher than the device output impedance R_{ds} (see Figure 4.3), which represent the upper limit to the impedance values that can be effectively presented to the active device internal

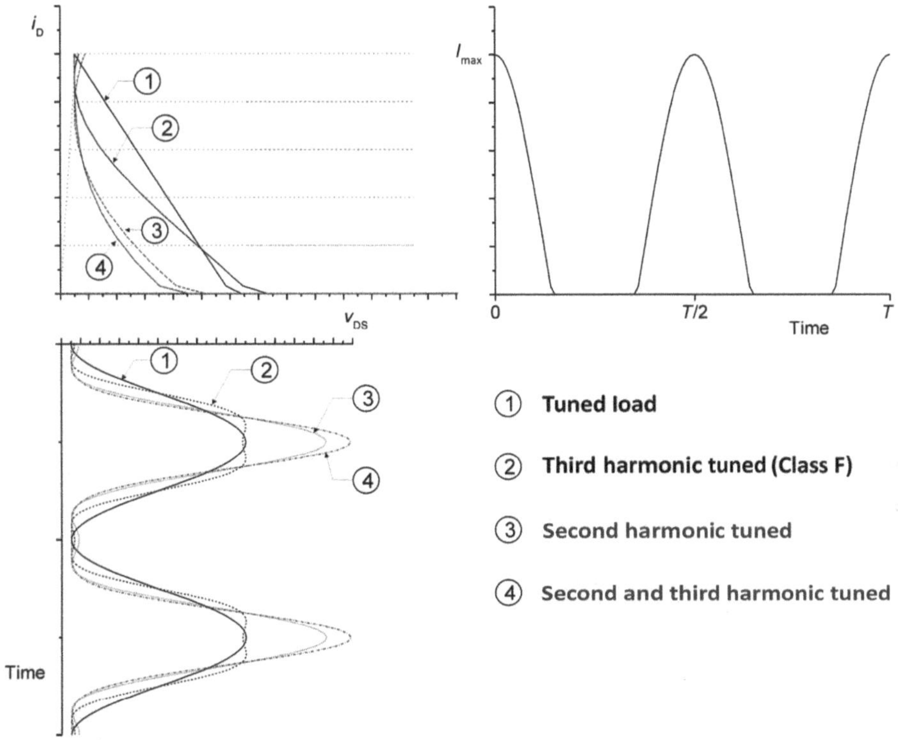

FIGURE 4.31

Ideal DLL for different HT approaches.

Table 4.3 Design Parameter and Performance's Improvement

	Tuned Load	Third HT	Second HT	Second and Third HT
$Z_{1,HT}$	R_{TL}	$1.155 \cdot R_{TL}$	$1.414 \cdot R_{TL}$	$1.618 \cdot R_{TL}$
$Z_{2,HT}$	0	0	$-0.5 \cdot R_{TL} I_1/I_2$	$-0.894 \cdot R_{TL} \frac{I_1}{I_2}$
$Z_{3,HT}$	0	$-0.192 \cdot R_{TL} \frac{I_1}{I_3}$	0	$0.276 \cdot R_{TL} \frac{I_1}{I_3}$
$\frac{P_{out,HT}}{P_{out,TL}}$	1	1.155	1.414	1.618
$\frac{\eta_{out,HT}}{\eta_{out,TL}}$	1	1.155	1.414	1.618

current source, then the practical impedance to be synthesized at the external device output port is the inductive load required to resonate C_{ds} (thus the intrinsic current source sees only its R_{ds} impedance).

For the third HT strategy, also known as high-frequency Class F, this leads to the erroneously simplified concept of open the third harmonic impedance (erroneously since in practical it is required an imaginary part of the load to resonate C_{ds}, if the latter was not negligible).

For the second and second & third HT, accounting for the values of I_n for the different biasing classes, graphically reported in Figure 4.20, the results in Table 4.3 imply intrinsic impedances which should be negative resistance, thus no feasible in actual cases. This lead many designers in the past to the erroneous conclusion that the control of a second harmonic impedance, different from the short-circuit condition, will degrade the PA performance, due to the wrong resulting voltage waveform shaping. Clearly this is true if the simplified truncated sinusoidal output current model is assumed. Nevertheless, in actual cases, different current waveforms could be considered, for example changing the shape of the input driving signal, or injecting harmonics [77,78], or more practically exploiting the device input nonlinearity to properly modify the current harmonic contents. Unfortunately, up to now there are no simplified models to describe the effects of the input nonlinearities and to account for them. However, several experimental results and design are available in the literature [46,56], demonstrating the effectiveness and the proper approach to design harmonic tuned PAs.

A further remark, highlighted in Table 4.3, is that the optimum fundamental frequency termination, when playing with the harmonic impedances also, must be properly increased to fully exploit the device capabilities [46].

Finally, it should be stressed that the optimum impedances should be synthesized in the entire PA bandwidth. Consequently, since resonating structures are usually adopted in the design of harmonic tuned PA to properly set the harmonic impedance values, it is obvious that narrowband results are expected. Practical results on HT PAs shown maximum relative bandwidths of 10–15%. Thus the HT approaches are usually not considered in broadband PA design and different optimization procedure (based on the resistive load concept) are resurrected [79,80] or distributed design concepts are adopted [81,82].

Different sub-optimum HT strategies can also be adopted (ie, by introducing phase shifts ϕ_n different from 0 degrees or 180 degrees), as for instance made in the Class J amplifier [83–86], thus working only on the design of the output harmonic matching impedances, while neglecting the input nonlinearities. This approaches could lead to broadband frequency capabilities, accounting for the inherent phase shifting introduced by C_{ds}, even if the expected performances in this case are reduced to the ones achievable with the tuned load.

4.5 TECHNOLOGIES FOR PAs REALIZATION

Due to the extremely diversified fields of application, the characteristics and the specification of solid-state PAs greatly differ on operating, technological and design point of view, depending on the particular case.

While solid-state PAs are claiming to substitute vacuum electron devices in a variety of applications (transmitting/receiving T/R modules for synthetic aperture radar SAR, air traffic control systems, and warfare applications [4,5,87]), at the same time their use in wireless communication systems deserves particular notice, since their development increased in the last decades at the same rate as communication and information technology impacts on life.

This gave rise to a variegated ensemble of PA in different technologies, evolving year by year.

Hence, a generic but comprehensive classification of PAs from a technological point of view can be made, according to two different criteria:

- The specific *semiconductor* technology, that is, the particular type of active device.
- The *physical realization* of the circuit, in a monolithic (MMIC) or hybrid (HMIC) form.

These criteria cannot be considered as independent from each other, but are often correlated, in the sense that a particular semiconductor technology is more frequently used in a particular form or, conversely, a monolithic realization can be available in a specific semiconductor technology.

Giving some examples, laterally diffused metal oxide semiconductor (LDMOS) based hybrid PAs are widely employed in digital TV broadcasting [88,89] and cellular base stations [90–93], while military and space-borne systems often recourse to employ monolithic PAs on gallium arsenide (GaAs) or gallium nitride (GaN) substrate, although these solutions are having brought to civil systems too [94–96]. Approaching the topic from the users' point of view, the need of miniaturization in mobile — and even wearable — devices push researchers and industry to the full exploitation of silicon (Si) capabilities, leading to the realization of high performance integrated CMOS PAs manufactured in high volumes [97–99].

A more detailed description of semiconductor technologies for PAs and their realization is provided in the following.

4.5.1 SEMICONDUCTOR TECHNOLOGIES FOR PAs

The appropriate choice of semiconductor technology plays a key factor in the design of a PA, since its final performances are strictly dependent, if not totally influenced, by the particular active device adopted. Hence, a careful evaluation of semiconductor technologies would be decisive to properly balance the different features in order to meet the design requirements of the PA, which are not only specified in terms of output power, but which often include more precise conditions on minimum efficiency, linearity at a given compression level, minimum bandwidth, back-off operation, and so on.

Large-scale RF and microwave power device production for commercial applications, including internally matched transistors, is mainly based on Si and GaAs compounds. Nonetheless, a great research interest has been devoted, in the last years, to the development of high power density devices using wide-bandgap materials such as SiC (silicon carbide) and GaN [32,100–102].

The latter semiconductors still have prohibitive costs for a wide diffusion, since the technological process is not mature and reliable enough to allow a large-scale production with an acceptable yield. On the other hand, if GaAs is widely employed at high-frequency range (typically starting from the C-band) and is generally qualified for space applications, silicon is still the lead material in civil communication applications as digital TV broadcasting and cellular base stations, while the miniaturization demanded in commercial electronics is pushing silicon CMOS technology beyond communications, to high-frequency PA developed for automotive radar applications.

In Table 4.4 relevant parameters of each semiconductor are summarized: these quantities directly affect the active device physical properties and, finally, the performance of the PA. Semiconductors having a higher band-gap, as GaN and SiC, can withstand higher breakdown fields with an improved robustness and reliability. Moreover, considering comparable current level, this feature allows large drain voltages to be used, leading to high output impedance per watt of RF power, resulting in easier matching during the circuit synthesis. Similarly, the dielectric constant is an indication of the capacitive loading of a device, thus a low value is generally desired. Higher values of charge density can improve the power handling capabilities of the transistor, leading to more compact realizations. However, high charge density is often associated with a lower thermal conductivity, making the heat dissipation a critical issue for such semiconductors: particular care is required in the dimensioning of cooling system to prevent the power dissipated in the active device from increasing the junction temperature and from degrading the device performance and reliability. High-frequency operation can take advantage from high saturation velocity and high charge mobility: they determine the electrically ON resistance and knee voltage of a power device. A low mobility results in increased parasitic resistance, increased losses, and reduced gain, thus clearly limiting the frequency of operation.

Table 4.4 Significant Parameters of Semiconductors for Power Amplification

Parameter	Unit	Si	GaAs	SiC	GaN
Bandgap, E_g	eV	1.1	1.42	3.25	3.49
Breakdown field, E_{br}	MV/cm	0.3	0.4	3.0	3.3
Thermal conductivity, Θ	W/(cm · K)	1.5	0.5	4.5	1.5
Charge density, n_i	10^{13}/cm^2	0.3	0.3	0.4	1.0
Saturation velocity, v_{sat}	10^7 cm/s	1.0	1.3	2.0	2.5
Electron mobility at 300 K, μ_n	cm^2/(V · s)	1500	8500	700	2000
Rel. dielectric constant, ϵ_r		11.8	12.8	10.0	9.0
Johnson's FoM, $E_{br} \cdot v_{sat}/2\pi$		1.0	2.7	20.0	27.5

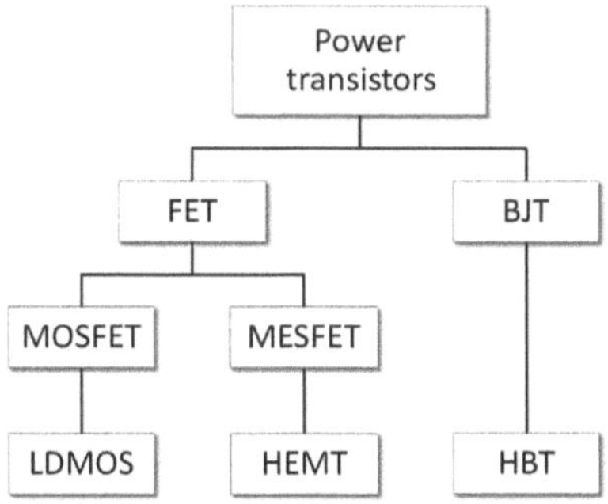

FIGURE 4.32

RF and microwave power device family tree.

Solid-state active devices for microwave power amplification can be grouped into two main classes, as shown in Figure 4.32: BJTs and FETs. As their development is an open research field in continuous evolution, every further detailed classification would be incomplete. By the way, the main structures descending from the two major classes are well-defined transistor families optimized for microwave operation: heterojunction bipolar transistors (HBTs), LDMOS FETs, and high electron mobility transistors (HEMTs) [32].

An extensive literature is available on each different technology: not pretending to be exhaustive, Figure 4.33 is only an example of power and frequency performance reached by state-of-the-art PAs different semiconductor technologies. From the graph is possible to see that the use of LDMOS devices is mostly concentrated in the lower microwave frequency band (L- and S-bands), from moderate to high output power levels (40–50 dBm). On the contrary, HBT transistors are mainly employed up to the X-band for limited power levels (20–30 dBm). At present, HEMTs are probably the most promising transistors for microwave applications, being used in a wide range of frequency bands (from L- to X-band and above) and demonstrating, especially for the GaN technology, the highest power levels (up to 53 dBm).

4.5.2 HYBRID MICROWAVE PAs

As the word suggests, hybrid microwave PAs are fabricated recurring to different forms and technologies for the active and the passive parts: in particular, a hybrid PA adopts discrete active devices while recurring to external boards for input and output matching networks.

A hybrid PA is therefore composed by different elements, as Figure 4.34 shows:

- a passive substrate, usually in the form of a printed circuit board, on which matching networks are realized in a distributed form, using quasi-TEM (transverse electromagnetic) waveguide structures as microstrip transmission lines, striplines, or coplanar waveguides;

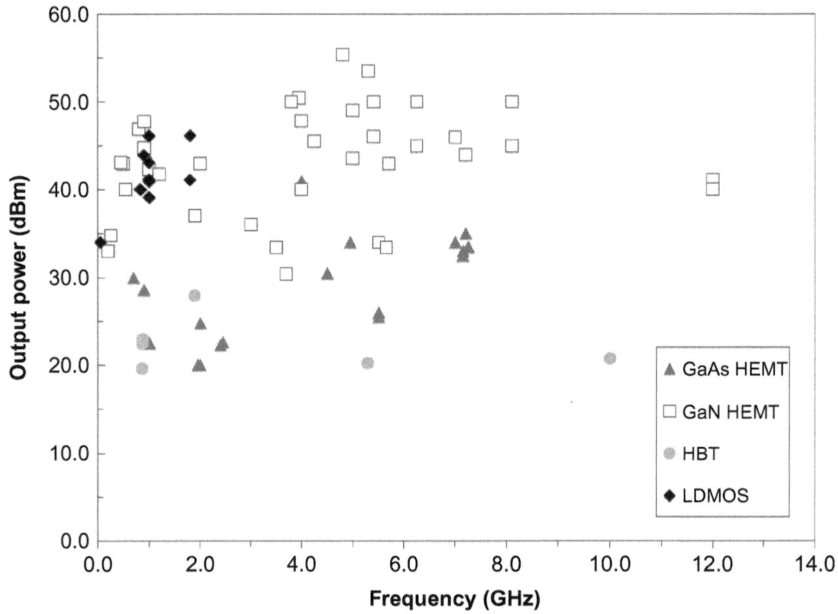

FIGURE 4.33

Achievable power level versus frequency for different active devices' technologies.

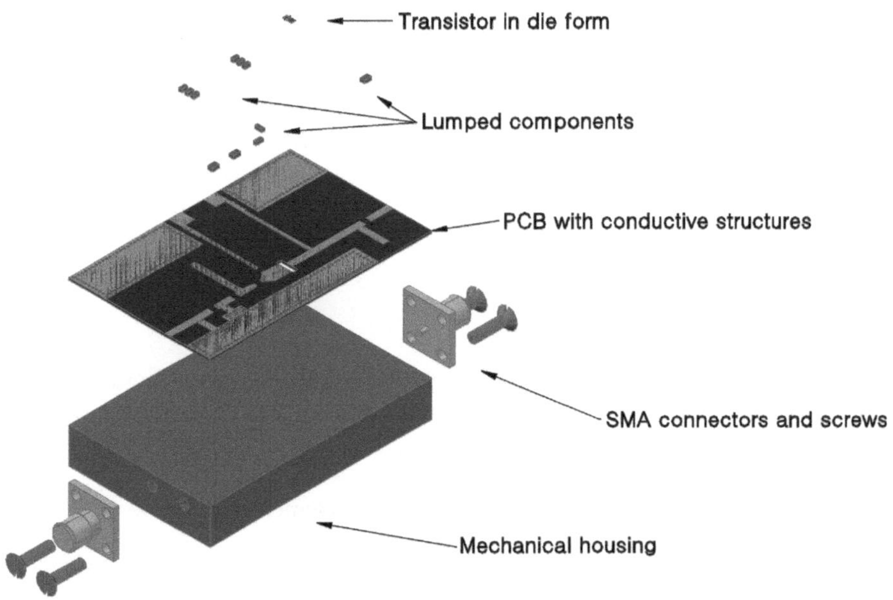

FIGURE 4.34

Explosed view of a hybrid power amplifier prototype.

- some lumped components, if necessary (especially in the case of low-frequency applications) as decoupling capacitors, stabilization resistors, and bias inductors;
- an active device, in a die or packaged form, connected to the matching network through bond wires or by soldering;
- a mechanical metallic structure supporting all the above elements, providing RF connections and often sealing and heat sinking.

A photograph of a hybrid PA prototype is shown in Figure 4.35.

Thanks to their composite structures, hybrid PAs can be fabricated quite easily with moderated costs if compared to their integrated counterparts. Moreover, the hybrid technology provides higher flexibility in the choice of lumped elements and active devices, as well as manual fine tuning of conductive structures, allowing a fully custom optimization of PA performance also during assembly and testing procedures.

On the other side, the poor repeatability due to several boundary conditions (eg, differences in bond wire length, inaccuracies in the soldering, tolerances of lumped components, defects in the drilling) prevents high volume production of hybrid PAs. As a consequence, the market of hybrid PAs is far from being a niche market,

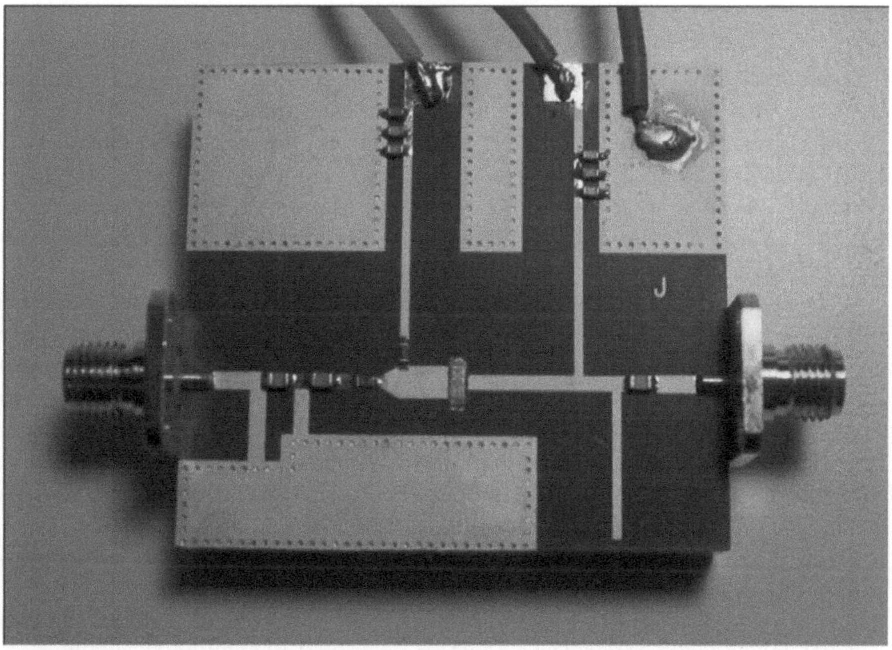

FIGURE 4.35

Photograph of a hybrid power amplifier prototype.

since these devices are employed in several applications, as high-energy physics particle accelerators [103], mobile phone base stations, satellite radar and communication systems (to replace vacuum tubes amplifier), and in the field of defense electronic [35].

Today, most of hybrid PAs' realizations are concentrated in the low RF and microwave frequency bands (L- to S-band), reserved to most medium to very high power applications [104–110].

Some examples are available at intermediate frequency bands (C- to K-bands), concerning PA prototyping for research scopes [111–113], as well as some dedicated space-born applications, while at higher frequency bands the number of hybrid PAs decreases dramatically for different reasons, as:

- the behavior of packaged active devices, which experiences a steep degradation as the frequency increases;
- nonnegligible uncertainties in the fabrication of distributed structures, which become critical as the frequency increases;
- the significant effect introduced by interconnections, as soldering and bonding; and
- the frequency-limited behavior of lumped elements.

For these reasons, monolithic solutions are generally preferred above the K-band, provided the suitable technology.

4.5.3 MICROWAVE MONOLITHIC PAs

In monolithic PAs — and in general in MMIC — all the circuit elements are realized in an integrated form on the same semi-conductive substrate.

GaAs was historically the first semiconductor employed in the fabrication of MMIC in the late 1970s [35]: since then, it has been the leading material in military and space systems up to W-band [26,114], until its supremacy was threatened at the beginning of new century with the advent of GaN-based devices and the improvements related to this new disruptive technology [115–118].

Aside from compound semiconductors, Si-based device technologies as CMOS [119–121], BiCMOS and SiGe HBT [122–124] are promising and cost-effective solution for commercial applications, as wireless and automotive radar [125, 126]. They can fully profit from the advantages of silicon integration (smaller circuit size, reliability of well-established processes, larger and cheaper wafer dimensions); on the other hand, their development counts several design and fabrication steps and, in term of performance, output power is not comparable with that obtained from compound semiconductors.

Figure 4.36 is a conceptual view of a MMIC, reporting all the main components present in such kind of circuit:

- active components, typically FETs as HFET or pHEMT, although bipolar transistors (HBT) are being also used;
- conductive structures, generally realized as microstrip lines or coplanar waveguides;

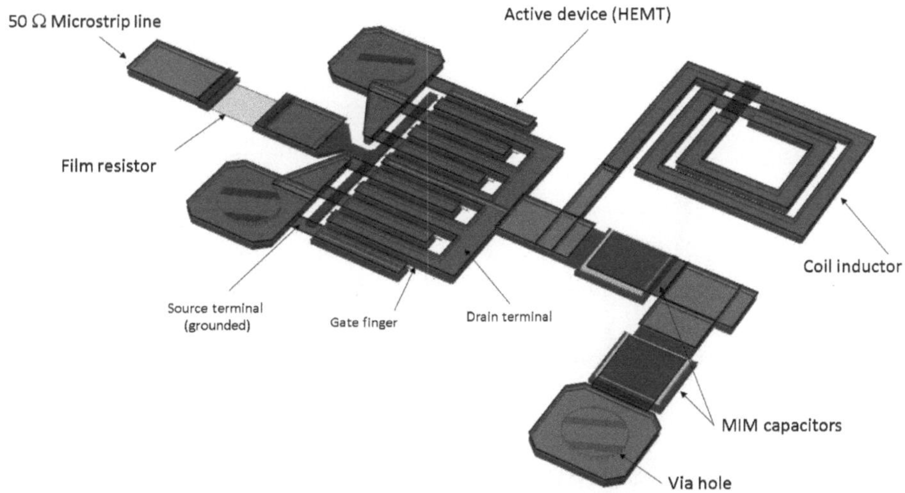

FIGURE 4.36

Three-dimensional structure of an MMIC circuit, showing the key elements.

- capacitors, in the form of MIM (Metal-Insulator-Metal) capacitors or interdigitated capacitors;
- planar coil inductors; and
- metal filled via holes, providing low loss connection with the ground plane.

Employed from low microwave frequency range (L-band) to millimeter-waves, monolithic realization is a versatile solution for the realization of PAs, presenting several advantages with respect to the hybrid technology, being preferred for high volume applications once the design has been well established. Hence, MMIC realization implies minimum unit cost, small size and light weight, design flexibility, high reproducibility, and reliability.

A typical MMIC PA is depicted in Figure 4.37 [127]. Depending on design requirements, a multi-device realization is the common choice in MMIC PA, in order to fulfill both the constraints on output power (proportional to the number of the active devices in the last stage) and gain (related to the number of stages, generally two or three). Hence, active devices, their combination, and the number of stages are thoroughly chosen by the designer in order to exploit the technology at its best.

Matching and inter-stage networks are generally realized using a lumped approach, but distributed or semi-distributed solutions are not uncommon at very high frequency or when a high current is required, as when combining several active devices in the last stage of a PA. As a consequence, thanks to miniature dimensions, MMIC realization allows for a higher gain in a smaller area and consequently for a higher overall PAE in multistage realizations.

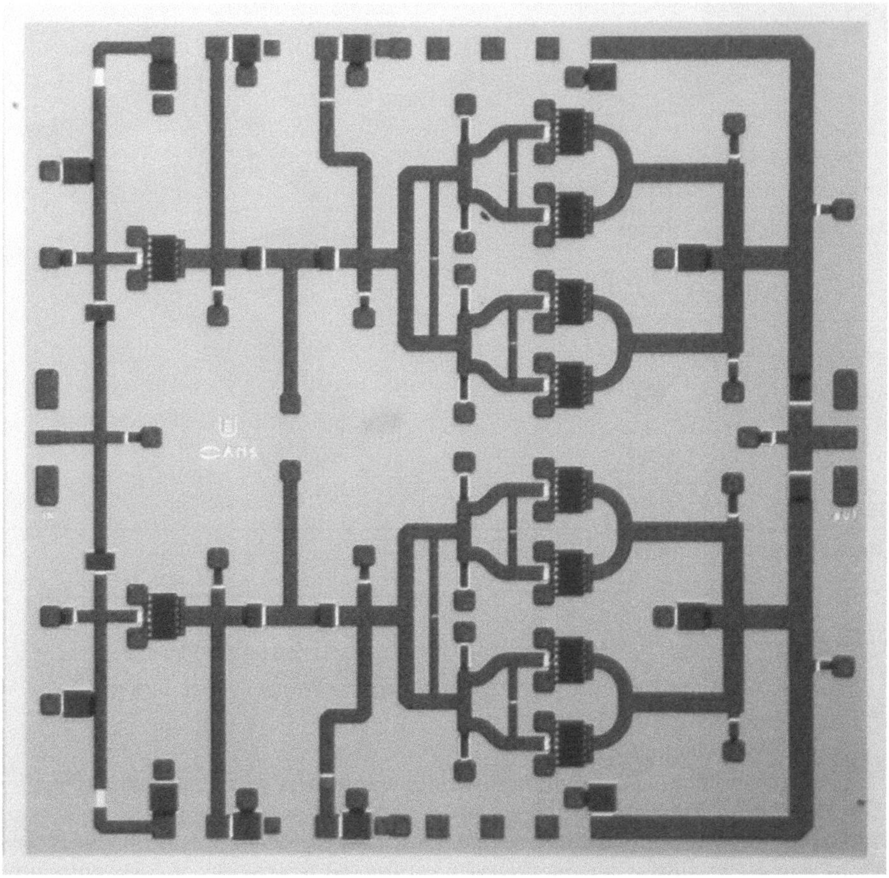

FIGURE 4.37

Photograph of an MMIC GaAs PA at X-band. Dimensions are 4.7 × 4.7 mm^2 [127].

MMIC PAs require suitable packaging to be managed and to be integrated in their destination system. Being not a trivial aspect of the design, the package can have an important effect on final performance, cost, and reliability of the product [35,128–130]. Basically, it should offer a mechanical shield to the circuit, an easy RF and DC access, and should guarantee an adequate thermal management of the chip [131]. Plastic low-cost packages are the most diffused in commercial applications [132], whereas for high performance applications, as military and spacecraft, custom ceramic packages are often realized [133], as depicted in Figure 4.38.

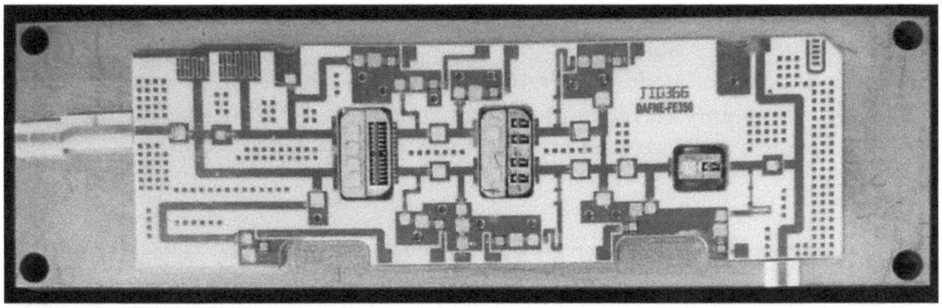

FIGURE 4.38

Example of different MMIC integrated in a custom package for spacecraft applications.

4.6 LINEARITY ISSUES

As previously asserted, the aim of a PA is to increase, in a given frequency range, the power level of a signal applied to its input up to a predefined output level. Obviously in a transmitting system it is required to perform such an operation without affecting the information content (ie, without introducing unwanted or uncontrolled distortion phenomena which could drastically modify the quality of the signal to be transmitted). Unfortunately, the high linearity requirement is inherently in contrast with the requirement of high efficiency, being the latter maximized in compression (see Figure 4.10) which intrinsically implies a distortion issue [134].

However, a clear distinction has to be made between the design of a stand-alone PA or a power amplifying system, which is directly transferred to the design strategies adopted at circuit or system level, respectively. In the first case, the designers efforts are mainly focused to maximize the active device performance (typically power and efficiency) at some compression level. In this case, the linearity optimization is a secondary aspect, if not specifically declared. It is to stress that the linearity optimization is critically affected by the prediction capability of the nonlinear model, as well as to the complex and time-consuming simulation at circuit level for modulated signals (as will be briefly discussed at the end on this chapter). In this case the design goals are accomplished by suitable nonlinear design approaches based, for instance, on a suitable selection of biasing conditions and matching networks at the device input and the output ports.

Conversely, from a system perspective, the designers' challenge becomes the realization of an amplifying system, usually based on an already existing PA, with the aim to achieve the highest linearity performance to maintain the quality of the traveling signals over a fixed minimum level. In this case, the system architecture includes both a PA (or more than one PA) and other circuits (devices) with the aim to improve the linearity of the resulting transmitter unit. For example, the adoption of linearization approaches, such as predistortion (at baseband or at RF), or the proper combination of PA and circuital solutions able to cancel the distortion products, like the feed-forward or feed-back architectures, could be useful to alleviate the linearity issues.

Accounting for the wide topic referred as highly linear amplifier stage, in this chapter we will mainly focus on the design strategies at circuit level, while linearization techniques are discussed in Chapters 6–8 [135].

4.6.1 SYSTEMS CLASSIFICATION (MEMORYLESS VS. MEMORY PA)

A PA can be considered as a nonlinear system where its output signal $y(t)$ is related to its input signal $x(t)$ through a nonlinear transfer function f:

$$y(t) = f\left\{t, x(t), \frac{d[x(t)]}{dt}, \frac{d^2[x(t)]}{dt^2}, \ldots\right\} \tag{4.57}$$

Depending on the transfer function f, the PA (or in general a system) can be referred to as a *memoryless system* (or zero memory system) or to as a system with memory (or dynamic system) [136]. In the former case, the output signal $y(t)$ is a function of the input $x(t)$ in an instantaneous (or static) way (ie, not depending on the time behavior of $x(t)$), as for example in a resistor:

$$y(t) = f\{t, x(t)\} \tag{4.58}$$

Conversely, in a memory system, the output response is dependent upon the input signal $x(t)$ and its history (ie, past behavior represented in Eq. (4.57) by the derivative terms), as for example in a capacitor.

In electronic circuits, the memory phenomena are associated with electronic charge or magnetic flux storage, or in delay effects (as for instance in the transmission line cases). Thus, any circuit containing capacitors, inductors, time delay or in general distributed elements, is inherently a system with memory as well.

However, a further classification of memory systems is possible referring to their memory property in a time scale. In particular, it is possible to identify two main families, referred to as short-term or long-term memory systems. Considering for example an input signal composed by an RF component (at angular frequency ω_{RF}) modulated by a slowly carrying envelope signal $M(t)$ (at angular frequency $\omega_{env} \ll \omega_{RF}$) as in the following:

$$x(t) = M(t) \cdot \cos(\omega_{RF} t) \tag{4.59}$$

in a memory system the output signal $y(t)$ depends on the actual value of the input signal $x(t)$ and on its past values. In the short-term case the dependence on past samples is limited in a time scale comparable to the period of the RF component of the input signal $x(t)$, typically in the order of nanoseconds. In this case, the impulse response has a short-time duration, as for example shown in Figure 4.39.

Conversely, in a long-term memory system, sometimes referred to as envelope memory system, the output response depends on the past samples at a time scale comparable with the envelope components, typically in the order of milli/microseconds, and the impulse response exhibits a longer time tail behavior, as shown in Figure 4.40.

FIGURE 4.39

Example of impulse response of a short-term system.

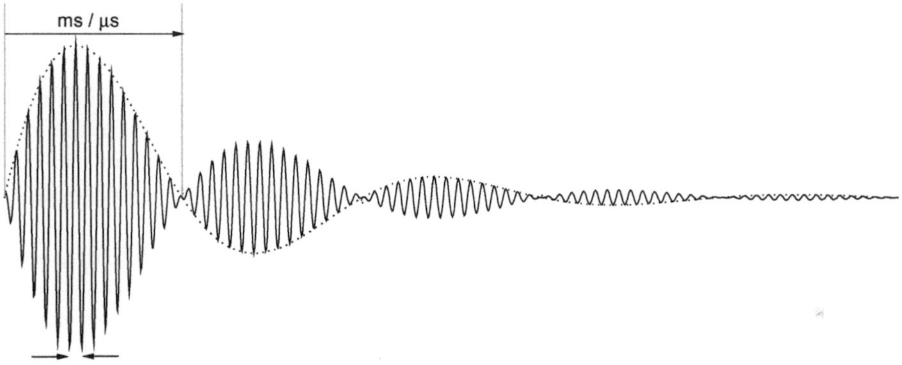

FIGURE 4.40

Example of impulse response of a long-term system.

In a PA, the memory effects originate from different sources, both depending on the active device physics and the circuit components and topologies used in the PA realization. For instance, sources of short-term (ie, high-frequency) phenomena are typically related to:

- charge storage in semiconductor devices (ie, intrinsic capacitances whose values are in the order of few pF or fF); and
- transit time in semiconductor devices.

Similarly, long-term (ie, low-frequency) phenomena are ascribed to:

- self-heating effects in semiconductor devices (thermal effects [137,138]);
- dynamic trapping phenomena in semiconductors [139]; and
- biasing and matching circuits [140–143].

The nonlinearity characterization of a PA is clearly depending on the properties of the driving signal and on the behavior of the PA itself. For example, the AM/AM compression and AM/PM conversion plots can be helpful for quasi-memoryless (or narrow-band) systems, while HD or more in general intermodulation distortion figures are useful for memory systems. Therefore, depending on the type of nonlinear behavior that would be investigated, a corresponding characterization figure and approach is adopted, as outlined in Chapter 2.

A simple and useful way to highlight the long-term memory effects in a PA is based on the simple two-tone characterization. Injecting two tones at angular frequencies ω_1 and $\omega_2 > \omega_1$, with $\omega_1 \approx \omega_2$:

$$x(t) = A \cdot \cos[(\omega_1 t) + \cos(\omega_2 t)] = \frac{A}{2}\cos(\omega_{RF} t) \cdot \cos(\omega_{env} t) \tag{4.60}$$

where

$$\omega_{RF} = \frac{\omega_1 + \omega_2}{3} \qquad \omega_{env} = \omega_1 - \omega_1 = \Delta\omega \tag{4.61}$$

it is possible to emphasize the RF and envelope components.

As discussed in Section 4.3, the third-order intermodulation distortion (IM3) is generated at the two angular frequencies $2\omega_1 - \omega_2$ and $2\omega_2 - \omega_1$, spaced apart by $2\Delta\omega$. Thus, the different behavior of these two products as a function of the tone spacing ($2\Delta\omega$), as shown in Figure 4.41, can be related to long-term memory effects [44]. However, it is to note that the same IMD products behavior does not necessarily imply the system to be without long-term memory.

The accurate estimation of the distortion behavior in the active device by a nonlinear model is a very time-consuming task [144–149]. Moreover, the analysis (simulation) of distortion behavior in a PA driven with complex signal request a long computation time, becoming practically unfeasible during the design process and

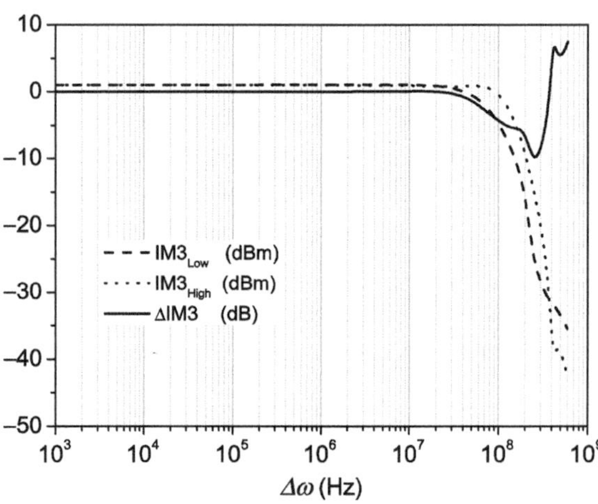

FIGURE 4.41

Example of IMD variations with tone spacing.

usually postponed at the final step of the design procedure, providing quite negligible or straightforward feedback to the design itself.

It is clear that some simplified design criteria, as used for the high-efficiency guidelines, could be highly appreciated to reduce design efforts, while identifying an initial estimate for the PA design parameters. Unfortunately, up to now there is no available simplified model able to predict in the same simple way the main linearity features of the amplifier, as for instance the intermodulation behavior.

Approaches suggested in literature have been derived from a weakly nonlinear assumption to models the active device, and are based on the adoption of the Volterra analysis trying to infer useful design guidelines [44,150]. Note that the Volterra series becomes a local approximation and it cannot be expected to give results with the same accuracy attainable using a full nonlinear active device model and analysis method. However, Volterra approaches were used to understand the generating mechanisms of intermodulation products in active devices, even if limited in a mildly nonlinear operating condition (ie, far enough from the 1 dB compression point [151–154]). Even if more complex representation were adopted for large-signal conditions [154,155], unfortunately they result inefficient for easily inferring design guidelines at circuit level.

Assuming for the device output nonlinear current a generic dependence on both the input and output voltages, that is:

$$i_{\text{out}} = \sum_{m=1}^{3}\sum_{n=1}^{3} g_{m,n} \cdot v_{\text{in}}^{m} \cdot v_{\text{out}}^{n} \tag{4.62}$$

where

$$g_{m,n} = \frac{\partial^{m,n} i_{\text{out}}}{\partial v_{\text{in}}^{m} \partial v_{\text{out}}^{n}} \qquad m,n = 0,1,2,3 \ (m+n \geq 0) \tag{4.63}$$

the main conclusions arising from the research activities can be summarized in the following statements.

4.6.2 INFLUENCE OF BIAS POINT

To understand the influence of bias point selection on the linearity performance, the transfer function can be further simplified, representing the output current i_{out} of an active device through a power series of the input signal v_{in} only, so neglecting to a first approximation the dependence of i_{out} on the output voltage v_{out} [156–158].

Clearly the $g_{m,0}$ coefficients are strongly dependent on the active device selected bias point, as shown in Figure 4.42 for a typical case.

Assuming a two equal tones input signal with amplitude A

$$v_{\text{in}} = A \cdot (\cos(\omega_1 t) + \cos(\omega_2 t)) \tag{4.64}$$

the third-order intermodulation component is given by:

$$i_{\text{out},3} = \frac{3}{4} \cdot A_1^3 \cdot g_{3,0} \cdot \cos[(2\omega_2 - \omega_1)t] \tag{4.65}$$

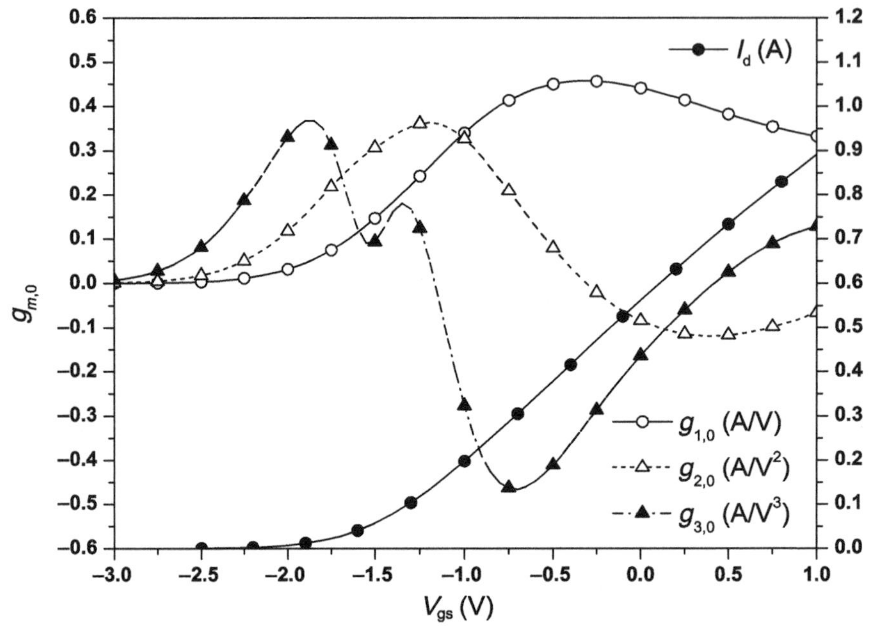

FIGURE 4.42

Example of $g_{m,0}$ behavior as a function of active device bias point.

Thus, from this simple relationship it follows that the third-order intermodulation product, which is proportional to $i^2_{out,3}$, is strictly related to the values assumed by the coefficient $g_{3,0}$. Since such coefficient takes values which can be positive or negative as a function of the bias (see Figure 4.42), it follows that could exist a particular bias conditions for which $g_{3,0} = 0$, thus nulling the IM3 products, resulting in the condition known as sweet spot (ie, zero of IM3) [154].

To validate this concept, as an example in Figure 4.43 is shown the IM3 behavior on a FET-based PA as a function of the input power P_{in} for different bias conditions.

One more time, it is important to stress that the adoption of a simplified model is useful only to highlight the possible existence of a sweet spot related to the bias point selection, while the real dependency of the sweet spot condition on the bias, and on the input drive, cannot be accurately predicted through such a simplified approach. Moreover, the cubic model assumed does not take into account that IM3 output signals can be generated also by nonlinearities with higher (odd) orders. Thus the simulation results on sweet spot position are not effectively considered as a design target.

4.6.3 INFLUENCE OF HARMONIC LOADINGS

Increasing the model complexity and resurrecting Eq. (4.62), it has been demonstrated that the third-order IM product generation depends on the second-order output current nonlinearity (ie, on the values of the coefficients $g_{2,0}$, $g_{0,2}$ and the cross

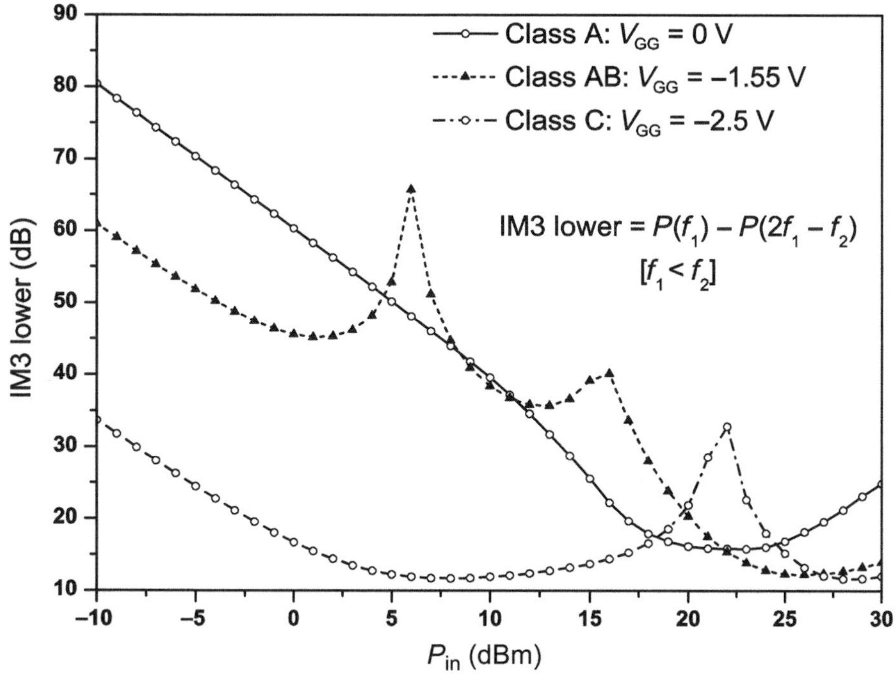

FIGURE 4.43

Example of IM3 behavior as a function of injected signal P_{in} for Class A, Class AB, and Class C bias.

term $g_{1,1}$), as well on the loading impedances presented at both input Z_{in} and output Z_L ports of the active device.

Usually, the main contributions to IM are related to the transfer coefficients $g_{m,0}$, while the output $g_{0,n}$ and transfer $g_{m,n}$ coefficients produce quite negligible effects [156–158].

The exploitation of device's input nonlinearities, and therefore the effect of proper terminations for different harmonics at the active device input port, has been the subject of a series of investigations, leading to experimental and theoretical studies, with often contrasting results [53,54,56,159–162]. In fact, the generation of an input second harmonic component, while reducing the IM3 through $g_{2,0}$, could affect in a detrimental way the PA efficiency, if not properly accounted for at the output side. Conversely, the exploitation of input nonlinearity phenomena, thus the proper harmonic loading condition at the input, could be useful also to increase the efficiency performance [156,163].

There are several research studies analyzing the effects of the harmonic terminations on the IM distortion generation and in particular on its asymmetric behavior [45,47,48,140–143, 156,157,163]. The main conclusions arising from these studies can be summarized in the following statements.

- The dependence of the third-order IM product with baseband impedance is strictly dependent on the second-order drain current nonlinearity (ie, on the existence and on the values of the coefficients $g_{2,0}$, $g_{0,2}$ and the cross term $g_{1,1}$). As a consequence, in terms of device nonlinear characterization, it becomes mandatory to carefully approximate the second-order transconductance $g_{2,0}$, the second-order output conductance $g_{0,2}$ and the cross term $g_{1,1}$.
- Being the input nonlinearities usually negligible, the impedance of the output termination $Y_L(\Delta\omega)$ exhibits the prevalent effect, whereas the baseband impedance of the input network $Y_{in}(\Delta\omega)$ does not show a significant impact [141].
- The IM3 level will change with the tone spacing $\Delta\omega$ if the baseband impedance changes with $\Delta\omega$. It clearly affects the intercept point identification, since it can vary widely depending on the tone spacing.
- The asymmetrical behavior of IM3, and thus the long-term memory effects, is strongly dependent, in a weakly nonlinear regime, on the values of the baseband impedances (at $\Delta\omega$) loading the active device (ie, on the selected biasing circuitry). Moreover, the asymmetry of IM3 increases with the imaginary part of the baseband impedance, while intermodulation levels will be lower if the output baseband impedance is not zero [45,140,141,164]. Finally, the memory effect can be reduced if the baseband output impedance does not change over the desired operating bandwidth.
- A different condition to null intermodulation asymmetry is obtained by nulling the susceptance of the second and third harmonics output loads $B_L(\omega_c) = B_L(2\omega_c) = 0$, that is, providing a purely resistive value for $Y_L(\omega_c)$ and $Y_L(2\omega_c)$ [48]. Such condition suggests an interesting alternative solution, by exploiting the selection of proper harmonic terminations to optimize at the same time efficiency, output power and gain performances, as discussed in previous paragraphs.

4.7 PA SOLUTIONS FOR COMMUNICATION SYSTEMS: THE DOHERTY EXAMPLE

Up to now we have considered the PAs and their related design strategies for CW operating conditions. However, in modern wireless communication systems, the increasing request of highest data rates and the optimization of channel efficiency pushed toward the development of complex modulations techniques (WCDMA, OFDM, etc.). From the RF transmitter site, and in particular for the PA, the new signal coding implies the amplification of signals characterized by larger time-varying envelope behaviors, with typical peak-to-average power ratios (PAPRs) between 6 and 12 dB [165].

Such different operating conditions cause a lower average efficiency in classical power stages (Class AB, B, F, etc.), which becomes the major issue in modern RF applications [166], with critical implications on thermal reliability [167]. Consequently, solutions to improve the average efficiency are becoming mandatory in modern PA design. The proposed solutions, which will be further discussed in Chapter 7, can be grouped as complex architectural approaches, like envelope elimination and restoration (EER [168]), envelope tracking (ET [169]), linear amplification with nonlinear components (LINC [170]), which also involve a baseband processing of the incoming signal, or in different PA architectures, fully developed in the RF domain, like in the Doherty power amplifier (DPA [171]).

In this context, we will focus on the latter approach and in particular on the DPA solution. In fact, the original idea proposed by W.H. Doherty in 1936 [171] actually seems to be a promising and feasible solution to improve the average efficiency in transmitters [31,32,172].

The DPA is based on the basic idea to modulate the load presented to an active device, namely Main (or Carrier), exploiting the active load concept, by using a second active device, namely Auxiliary (or Peaking). To understand the Doherty concept, we could refer to the behavior of an active device, biased for example in Class B and in a tuned load condition. When the maximum voltage swing is achieved, regardless the maximum current attained, as for example shown in Figure 4.44, the resulting efficiency is maximized and given by:

$$\eta = \frac{\pi}{4} \cdot \frac{V_{\text{out}}}{V_{\text{DD}}} \tag{4.66}$$

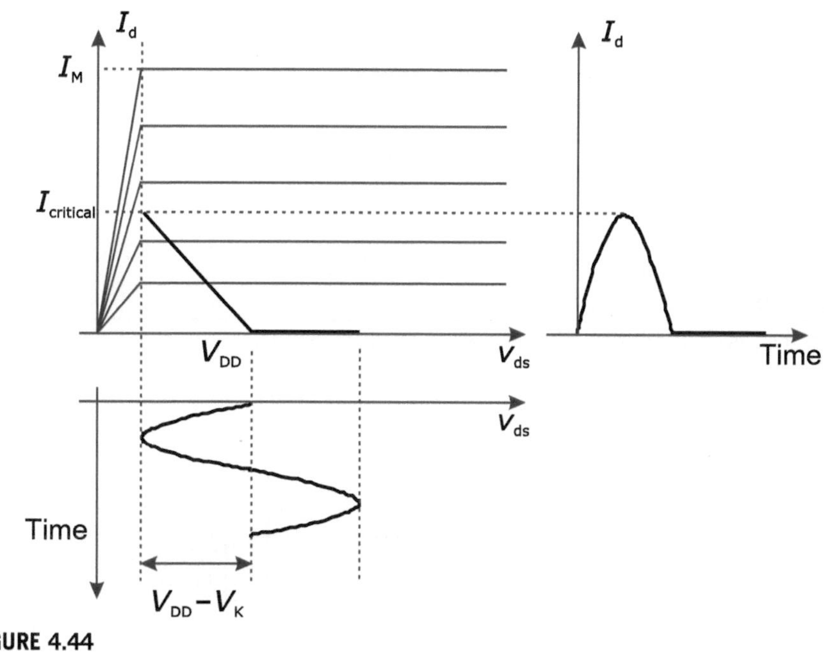

FIGURE 4.44

DLL for the device maximum efficiency when $I_{critical}$ is achieved.

If the peak current value is $I_{critical}$ (see Figure 4.44), the load condition to obtain this situation is given by:

$$R = 2 \cdot \frac{V_{DD} - V_k}{I_{critical}} \tag{4.67}$$

If now, increasing the input power, it were possible to accordingly reduce the load value, in order to maintain fixed the voltage swing, as for example shown in Figure 4.45, the device would operate with a constant (and maximum) efficiency given by Eq. (4.66).

In order to modulate the load with respect to the input power, we could refer to the behavior of the circuit shown in Figure 4.46.

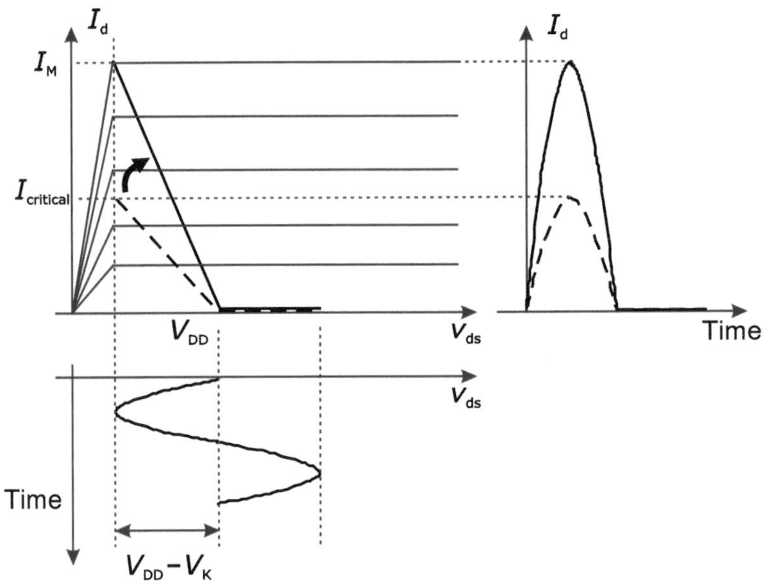

FIGURE 4.45

DLL for the device maximum efficiency when I_{Max} is achieved.

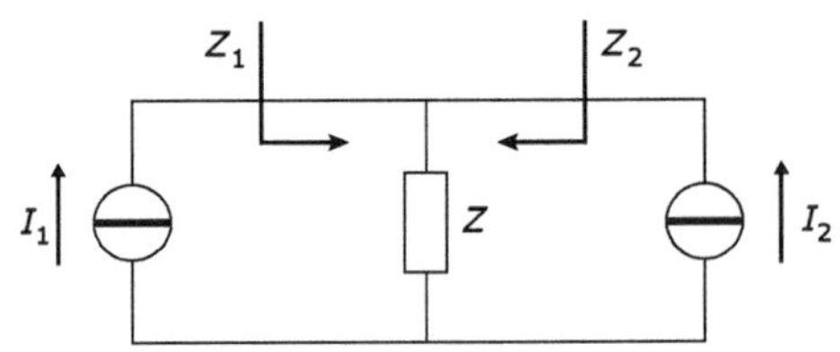

FIGURE 4.46

Active load modulation concept.

Computing the impedances seen by each current sources, it follows:

$$Z_1 = Z \cdot \left(1 + \frac{I_2}{I_1}\right)$$
$$Z_2 = Z \cdot \left(1 + \frac{I_1}{I_2}\right) \quad (4.68)$$

Thus, the idea is to combine a second device (the Auxiliary or Peaking), to provide the current (I_2 in Figure 4.46) in order to change the impedance seen by the Main device (Z_1). However, the direct connection does not work correctly, because when dealing with actual amplifiers, the voltage swing must be held constant while changing the output load if power performances are concerned. Moreover, the direct connection will results in a wrong load modulation, since the impedance seen by the Main Z_1 increases with I_2, while for the proper operation it should be decreased. Thus, an impedance inverter network (IIN) has to be posed between the Main and Auxiliary devices, to invert the impedance transformation.

The most simple implementation of such an IIN is represented by a $\lambda/4$ transformer, which of course introduces a phase shifting of 90 degrees. Thus, to compensate this phase shifting, a phase compensation network (eg, another $\lambda/4$), has to be inserted in front of the Auxiliary device.

The resulting typical DPA implementation is reported in Figure 4.47, and the behavior of the theoretical expected drain efficiencies of each device and of the overall DPA stage is depicted in Figure 4.48.

The complete analysis, including the simplified behavior and design guidelines for DPA, based on Class AB (Main) and Class C (Auxiliary) device biasing, is reported in [173,174] and summarized in the following.

Starting from the required output power level (namely $P_{\text{out,DPA}}$) and the signal PAPR or OBO (Output Back off), that is, the region where the efficiency is maintained at high level, defining the following parameter:

$$\alpha^2 = 10^{-\frac{OBO_{\text{dB}}}{10}} \quad (4.69)$$

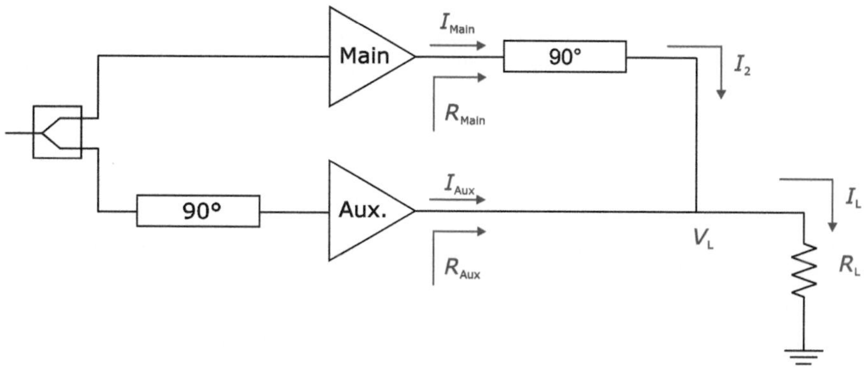

FIGURE 4.47

Typical DPA configuration.

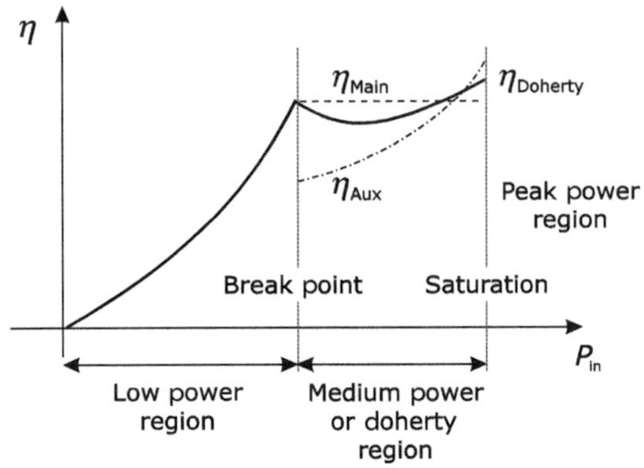

FIGURE 4.48

Theoretical drain efficiency behavior of DPA.

the main design figures can be inferred, as for example the optimum load impedance R_L at the common node, as well the impedance characteristic of the IIN $\lambda/4$, given, respectively, by:

$$R_L = \frac{(V_{DD} - V_k)^2}{2 \cdot P_{out,DPA}} \tag{4.70}$$

$$Z_0 = \frac{(V_{DD} - V_k)^2}{2 \cdot \alpha \cdot P_{out,DPA}} \tag{4.71}$$

where V_{DD} and V_k are the bias and knee voltage, respectively, of the selected device. For the active device dimensioning, it is required a Main device able to provide an output power given by:

$$P_{out,Main} = \alpha \cdot P_{out,DPA} \tag{4.72}$$

from which its maximum current (the amplitude of the fundamental current component, $I_{1,M}$) can be derived, depending on the selected bias point, while for the maximum current of Auxiliary device, its fundamental current component is given by:

$$I_{1,A} = \frac{1-\alpha}{\alpha} \cdot I_{1,M} \tag{4.73}$$

An example of realized Doherty amplifier is reported in Figure 4.49.

Also for the DPA, HT approaches can be adopted, mainly based on Class F design strategies for the Main device [175–180]. In Figure 4.50 is reported for example the photo of a Class F DPA, that is, based on a Class F approach for the Main device and tuned load for the Auxiliary one [175], while in Figure 4.51 are reported the measured comparisons between the tuned load DPA and the Class F DPA.

FIGURE 4.49

Photo of a DPA based on tuned load configuration [174].

FIGURE 4.50

Photo of a DPA based on Class F configuration for the Main device [175].

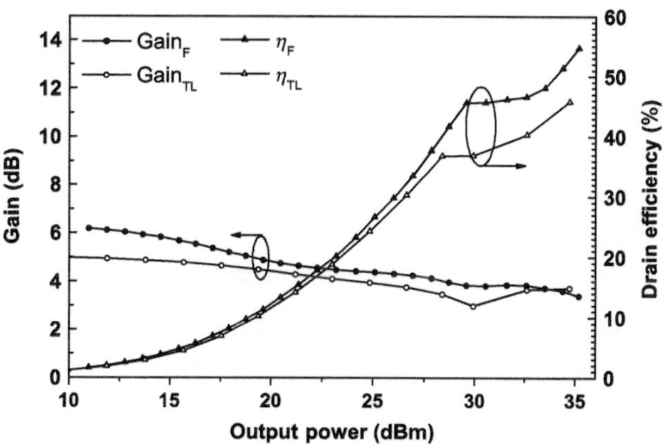

FIGURE 4.51

Comparisons between tuned load and Class F measurements [175].

The major frequency issues in a DPA design are related to the frequency limitation due to the output $\lambda/4$ transformer for the IIN, and the input (quadrature) splitter [181]. However, several approaches have been proposed and validated to extend the DPA frequency for multiband [104,182–184] or broadband [185–187] designs.

4.8 ANALYSIS ISSUES

The simulation of a PA has to account for its intrinsic nonlinear operating conditions. Specialized and dedicated textbooks are available [188–190], while in this section just a brief overview on this topic is addressed, introducing the principles behind each approach, focusing on those methods that are more frequently adopted in commercial CAD tools.

Because a PA is represented by a nonlinear model, while the circuit schematic usually includes both lumped and distributed elements, adequate analysis tools have to be adopted. Evidently, the circuit analysis implies the solution of a system of linear and nonlinear equations arising from the constitutive relationships of the circuit's elements and from the Kirchhoff laws [191]. When dealing with high-frequency circuits, as in the cases of microwave PA design, it has to be accounted for the widely different time constants in the circuit, due for instance to the large capacitance adopted for DC blocks or to the small parasitic associated to the active devices. Consequently, the time domain-based approaches (eg, SPICE-like program [192]) become unfeasible and very time consuming. Moreover, time domain descriptions of distributed elements are cumbersome and sometimes even impossible to directly extract, thus leading to the practical impossibility to analyze a real distributed nonlinear circuit.

The typical analysis approach used for PA simulation is the *harmonic balance* (HB) technique, which allows to solve nonlinear microwave circuits in steady-state condition [193]. It is crucial to understand that this technique, whatever is implemented, is based on the assumption that the solution of a nonlinear periodic circuit can be described by an infinite (practically finite due to computational limitation) number of basic functions, as for example the Fourier series:

$$x(t) = \sum_{n=-N}^{n=N} X_n \cdot e^{j2\pi f_n t} \tag{4.74}$$

where generically f_n is a frequency belonging to the set

$$\Lambda = \{0, f_1, f_2, \ldots, f_N\} \tag{4.75}$$

and X_n are the Fourier coefficient (complex values) that are assumed to be the unknown of the analysis problem [194]. Moreover, the following properties hold

$$f_n = -f_{-n} \qquad X_n = X^*_{-n} \tag{4.76}$$

The most simple (and used case) is when the input signal is periodic, with period T, then only one base frequency $f = 1/T$ exists and the frequency set Λ in Eq. (4.75) is formed with $f_n = n \cdot f$.

In the actual implementation of HB in CAD tools, named piecewise HB [193], the nonlinear circuit to be solved is split into two sub-networks, gathering all the

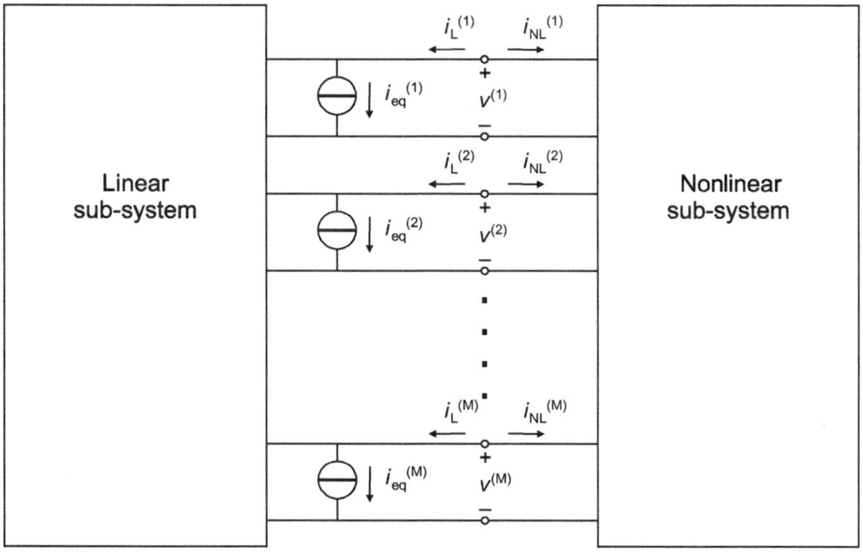

FIGURE 4.52

Example of circuit partition for the HB implementation in CAD tools.

nonlinear elements in one of them and the linear ones in the other, as shown in Figure 4.52. The two sub-circuits are connected through a number of ports M, partially related to the number of nonlinear elements. For the linear sub-circuit a frequency-domain technique is typically adopted (exploiting the superposition principle), while for the nonlinear sub-circuit a time-domain approach is selected. The results of these two different analyses are then combined through an appropriate algorithm, in order to minimize the balancing errors at the connecting ports [194–200]. To simplify computational efforts, different artifices can be adopted, sometimes known [200–205] but in general not disclosed for commercial competition among CAD vendors.

A pure HB analysis is based on the assumption that the nonlinear system is running in its steady-state condition, assuming a periodic excitation with frequency f. As a consequence, all the signals (and state variables) in the system can be represented through their Fourier series, even assuming a finite number N of harmonics. The main option demanded to the designer is the proper selection of the number N of the harmonics to be considered in the analysis. Clearly, a small value could result in a poorly approximated solution, while a large value for N could dramatically increase the computational time up to unacceptable levels.

From basic considerations derived by the results of Volterra series, an odd number of harmonics should be always considered, to account for the nonlinear phenomena which generate effects at fundamental frequency. From a practical point of view, usually $N = 5$ or $N = 7$ harmonics represent a good tradeoff for a fast and adequately accurate analysis during the design and the optimization of a typical nonlinear system as a PA. Obviously, at the end of the design procedure, the number N

could be reasonably increased to improve the accuracy of the results, while paying it with the increase in computational burden.

The single-tone HB analysis is usually satisfactory to provide enough basic information on the behavior of the nonlinear PA regarding large-signal gain and power compression phenomena. However, in order to evaluate other distortion phenomena, as for instance the intermodulation, a multi-tone HB technique can be used [198,206–218], which could be very cumbersome and time consuming, thus usually adopted when the PA design is almost completed.

When dealing with complex signals, likewise digitally modulated signals, as the ones actually adopted in modern communication applications, both time-domain analysis and HB techniques become ineffective. In this case, in fact, the former still exhibits drawback related to a long decaying transient analysis, while for the latter, the input signals cannot be easily represented as the combination of sinusoidal waveforms, especially if the number of required tones has to be kept reasonably high. In these cases, a different algorithm has to be adopted, based on an envelope analysis [219–223]. Unfortunately, this kind of analysis is very time consuming, rarely used at the end of the design phase and only if required to evaluate particular PA figures of merits (eg, ACPR or similar).

In some cases, however, the methods commonly adopted in commercial simulators fails to predict the actual behavior of the nonlinear PA circuit, due to the lack of validity of the underlying hypothesis. This is the case for instance of a wrong nonlinear model, or more critically related to the large-signal stability issues.

Stability issues are deeply discussed in Chapter 3; concerning a PA design procedure, the usual way to deal with its stability properties is to check its linear stability with standard and well-established approaches (eg, Rollet stability factor, geometric factors, etc.) [224]. Once the linear stability is guaranteed, the circuit is analyzed as a nonlinear one by the above described HB approach (or one of its variants depending on the stimulus). However, such analysis tools actually hide potential instability phenomena connected with the nonlinear nature of the circuit and to its topology. In fact, when the driving signal increases and becomes large enough that the small-signal assumption fails to be valid, parametric oscillations may rise, caused by large-signal instability. It means that some PA, stable under a small excitation, may evolve to oscillate when the stimulus becomes large enough. Such singularity, experienced on realized amplifier by many designers, is not predictable directly form standard microwave nonlinear analysis tools of the HB family.

Instead, dedicated test methods have been developed in the last decades, as the Ohtomo method [225], which, given a multi-device circuit, computes the Nyquist response at each possible loop present in the circuit, giving the designer the chance to prevent odd-mode oscillation by properly modifying its amplifier (generally with the insertion of parallel resistors). Nonetheless, this is still a linear analysis, which does not give information about possible parametric and large-signal oscillations. To this aim, ad hoc simulation tools are being developed, giving the possibility to perform a statistical (Monte Carlo) analysis of the circuit by varying several parameters (eg, input power, frequency, component values, etc.) and identifying the eventual presence of RHP (right-half plane poles), indication of possible parametric oscillation [226–229].

REFERENCES

[1] H. Sobol, K. Tomiyasu, Milestones of microwaves, IEEE Trans. Microw. Theory Techn. 50 (3) (2002) 594–611, ISSN 0018-9480, http://dx.doi.org/10.1109/22. 989945.

[2] R.N. Simons, E. Wintucky, J. Wilson, D. Force, Ultra-high power and efficiency space traveling-wave tube amplifier power combiner with reduced size and mass for NASA missions, IEEE Trans. Microw. Theory Techn. 57 (3) (2009) 582–588, ISSN 0018-9480, http://dx.doi.org/10.1109/TMTT.2008.2012298.

[3] W. Menninger, R. Benton, M. Choi, J. Feicht, U. Hallsten, H. Limburg, W. McGeary, X. Zhai, 70% Efficient Ku-band and C-band TWTs for satellite downlinks, IEEE Trans. Electron Devices 52 (5) (2005) 673–678, ISSN 0018-9383, http://dx.doi.org/10.1109/TED.2005.845840.

[4] K. Nakade, K. Seino, A. Tsuchiko, J. Kanaya, Development of 150W S-band GaN solid state power amplifier for satellite use, in: Asia Pacif. Microw. Conf., Yokohama, Japan, 2010, pp. 127–130.

[5] A. Wakejima, T. Asano, T. Hirano, M. Funabashi, K. Matsunaga, C-band GaAs FET power amplifiers with 70-W output power and 50% PAE for satellite communication use, IEEE J. Solid-State Circuits 40 (10) (2005) 2054–2060, ISSN 0018-9200, http://dx.doi.org/10.1109/JSSC.2005.854596.

[6] J. Shealy, M. Lefevre, B. Anderson, D. Runton, M. Poulton, J. Martin, Optimization of Gallium Nitride high power technology for commercial and military applications, in: IEEE Bipolar/BiCMOS Circ. Techn. Meeting (BCTM), Capri, Italy, ISSN 1088-9299, 2009, pp. 146–153, http://dx.doi.org/10.1109/BIPOL.2009.5314127.

[7] M. Skolnik, Role of radar in microwaves, IEEE Trans. Microw. Theory Techn. 50 (3) (2002) 625–632, ISSN 0018-9480, http://dx.doi.org/10.1109/22.989947.

[8] W. Keydel, Perspectives and visions for future SAR systems, IEE Proc. Radar, Son. Nav. 150 (3) (2003) 97–103, ISSN 1350-2395, http://dx.doi.org/10.1049/ip-rsn:20030494.

[9] B. Kopp, M. Borkowski, G. Jerinic, Transmit/receive modules, IEEE Trans. Microw. Theory Techn. 50 (3) (2002) 827–834, ISSN 0018-9480, http://dx.doi.org/10.1109/22.989966.

[10] J. Osepchuk, Microwave power applications, IEEE Trans. Microw. Theory Techn. 50 (3) (2002) 975–985, ISSN 0018-9480, http://dx.doi.org/10.1109/22.989980.

[11] Y. Meir, E. Jerby, Localized rapid heating by low-power solid-state microwave drill, IEEE Trans. Microw. Theory Techn. 60 (8) (2012) 2665–2672, ISSN 0018-9480.

[12] J. Osepchuk, A history of microwave heating applications, IEEE Trans. Microw. Theory Techn. 32 (9) (1984) 1200–1224, ISSN 0018-9480, http://dx.doi.org/10.1109/ TMTT.1984.1132831.

[13] R. Caverly, A. Paolella, F. Lexa, C. Grispino, High efficiency amplifier and signal source module for ISM-band applications, in: IEEE Top. Conf. Pow. Ampl. Wirel. Radio Appl. (PAWR), Phoenix, AZ, USA, 2011, pp. 13–16, http://dx.doi.org/10.1109/PAWR.2011.5725377.

[14] A. Rosen, M. Stuchly, A. Vander Vorst, Applications of RF/microwaves in medicine, IEEE Trans. Microw. Theory Techn. 50 (3) (2002) 963–974, ISSN 0018-9480, http://dx.doi.org/10.1109/22.989979.

[15] T. Copani, S. Min, S. Shashidharan, S. Chakraborty, M. Stevens, S. Kiaei, B. Bakkaloglu, A CMOS low-power transceiver with reconfigurable antenna interface

for medical implant applications, IEEE Trans. Microw. Theory Techn. 59 (5) (2011) 1369–1378, ISSN 0018-9480, http://dx.doi.org/10.1109/TMTT.2011.2116036.

[16] Q.-H. Liu, Z.Q. Zhang, T. Wang, J. Bryan, G. Ybarra, L. Nolte, W. Joines, Active microwave imaging. I. 2-D forward and inverse scattering methods, IEEE Trans. Microw. Theory Techn. 50 (1) (2002) 123–133, ISSN 0018-9480, http://dx.doi.org/10.1109/22.981256.

[17] M. Shahira Banu, S. Vanaja, S. Poonguzhali, UWB microwave detection of breast cancer using SAR, in: Int. Conf. Energy Eff. Techn. Sust. (ICEETS), Kanyakumari, Tamilnadu, India, 2013, pp. 113–118, http://dx.doi.org/10.1109/ICEETS.2013.6533366.

[18] A. Hassan, M. El-Shenawee, Review of electromagnetic techniques for breast cancer detection, IEEE Rev. Biomed. Eng. 4 (2011) 103–118, ISSN 1937-3333, http://dx.doi.org/10.1109/RBME.2011.2169780.

[19] G. Giakos, M. Pastorino, F. Russo, S. Chowdhury, N. Shah, W. Davros, Noninvasive imaging for the new century, IEEE Instrum. Meas. Mag. 2 (2) (1999) 32–35, ISSN 1094-6969, http://dx.doi.org/10.1109/5289.765967. 49.

[20] E. Fear, X. Li, S. Hagness, M. Stuchly, Confocal microwave imaging for breast cancer detection: localization of tumors in three dimensions, IEEE Trans. Bio. Eng. 49 (8) (2002) 812–822, ISSN 0018-9294, http://dx.doi.org/10.1109/TBME.2002.800759.

[21] I. Rekanos, A. Räisänen, Microwave imaging in the time domain of buried multiple scatterers by using an FDTD-based optimization technique, IEEE Trans. Mag. 39 (3) (2003) 1381–1384, ISSN 0018-9464, http://dx.doi.org/10.1109/TMAG.2003.810526.

[22] E. Fear, P. Meaney, M. Stuchly, Microwaves for breast cancer detection?, IEEE Potentials 22 (1) (2003) 12–18, ISSN 0278-6648, http://dx.doi.org/10.1109/MP.2003.1180933.

[23] O. Hammi, F. Ghannouchi, Comparative study of recent advances in power amplification devices and circuits for wireless communication infrastructure, in: IEEE Int. Conf. Electron. Circ. Syst. (ICECS), Yasmine Hammamet, Tunisia, 2009, pp. 379–382, http://dx.doi.org/10.1109/ICECS.2009.5410913.

[24] B. Kim, J. Kim, D. Kim, J. Son, Y. Cho, J. Kim, B. Park, Push the envelope: design concepts for envelope-tracking power amplifiers, IEEE Microw. Mag. 14 (3) (2013) 68–81, ISSN 1527-3342, http://dx.doi.org/10.1109/MMM.2013.2240851.

[25] A. Hussaini, I. Elfergani, J. Rodriguez, R. Abd-Alhameed, Efficient multi-stage load modulation radio frequency power amplifier for green radio frequency front end, IET Sci. Meas. Techonol. 6 (3) (2012) 117–124, ISSN 1751-8822, http://dx.doi.org/10.1049/iet-smt.2011.0127.

[26] C. Kim, P. Park, D.-Y. Kim, K.-H. Park, M. Park, M.-K. Cho, S.J. Lee, J.-G. Kim, Y.S. Eo, J. Park, D. Baek, J.-T. Oh, S. Hong, H.-K. Yu, A CMOS centric 77 GHz automotive radar architecture, in: IEEE Radio Freq. Integ. Circ. Symp. (RFIC), Montreal, QC, Canada, ISSN 1529-2517, 2012, pp. 131–134, http://dx.doi.org/10.1109/RFIC.2012.6242248.

[27] L. Larson, P. Asbeck, G. Hanington, E. Chen, A. Jayamaran, R. Langridge, X. Zhang, Device and circuit approaches for improved wireless communications transmitters, IEEE Pers. Commun. 6 (5) (1999) 18–23, ISSN 1070-9916, http://dx.doi.org/10.1109/98.799616.

[28] Y. Kobayashi, M. Hori, H. Noji, G. Fukuda, S. Kawasaki, The S-band GaN-based high power amplifier and rectenna for space energy transfer applications, in: IEEE Int. Microw. Workshop Series Inn. Wireless Power Trans. Techn. Syst. Appl. (IMWS), Kyoto, Japan, 2012, pp. 271–274, http://dx.doi.org/10.1109/IMWS.2012.6215809.

[29] B. Strassner, K. Chang, Microwave power transmission: historical milestones and system components, IEEE Proc. 101 (6) (2013) 1379–1396, ISSN 0018-9219, http://dx.doi.org/10.1109/JPROC.2013.2246132.

[30] P. Colantonio, F. Giannini, G. Leuzzi, E. Limiti, A fast tool for high-efficiency microwave power amplifier design, Microw. Eng. Eur. (May 1997) 33–41, ISSN 0960-667X.

[31] S.C. Cripps, RF Power Amplifiers for Wireless Communications, The Artech House Microwave Library, second ed., Artech House, Incorporated, 2006, ISBN 9781596930186.

[32] P. Colantonio, F. Giannini, E. Limiti, High Efficiency RF and Microwave Solid State Power Amplifiers, Microwave and Optical Engineering, Wiley, 2009, ISBN 9780470746554.

[33] J. Dilorenzo, W. Wisseman, GaAs power MESFET's: design, fabrication, and performance, IEEE Trans. Microw. Theory Techn. 27 (5) (1979) 367–378, ISSN 0018-9480, http://dx.doi.org/10.1109/TMTT.1979.1129635.

[34] D. Pavlidis, HBT vs. PHEMT vs. MESFET: what's best and why, Compound Semicond. 5 (5) (1999) 56–59.

[35] J. Walker, Handbook of RF and Microwave Power Amplifiers, Cambridge books online, Cambridge University Press, 2011, ISBN 9780521760102.

[36] G. Dambrine, A. Cappy, F. Heliodore, E. Playez, A new method for determining the FET small-signal equivalent circuit, IEEE Trans. Microw. Theory Techn. 36 (7) (1988) 1151–1159, ISSN 0018-9480, http://dx.doi.org/10.1109/22.3650.

[37] J. Wood, D. Root, Bias-dependent linear scalable millimeter-wave FET model, IEEE Trans. Microw. Theory Techn. 48 (12) (2000) 2352–2360, ISSN 0018-9480, http://dx.doi.org/10.1109/22.898984.

[38] G. Crupi, D. Xiao, D. Schreurs, E. Limiti, A. Caddemi, W. De Raedt, M. Germain, Accurate multibias equivalent-circuit extraction for GaN HEMTs, IEEE Trans. Microw. Theory Techn. 54 (10) (2006) 3616–3622, ISSN 0018-9480, http://dx.doi.org/10.1109/TMTT.2006.882403.

[39] G. Crupi, D. Schreurs, A. Raffo, A. Caddemi, G. Vannini, A new millimeter-wave small-signal modeling approach for pHEMTs accounting for the output conductance time delay, IEEE Trans. Microw. Theory Techn. 56 (4) (2008) 741–746, ISSN 0018-9480, http://dx.doi.org/10.1109/TMTT.2008.918147.

[40] A. Alt, D. Marti, C. Bolognesi, Transistor modeling: robust small-signal equivalent circuit extraction in various HEMT technologies, IEEE Microw. Mag. 14 (4) (2013) 83–101, ISSN 1527-3342, http://dx.doi.org/10.1109/MMM.2013.2248593.

[41] V. Vadalà, A. Raffo, S. Di Falco, G. Bosi, A. Nalli, G. Vannini, A load pull characterization technique accounting for harmonic tuning, IEEE Trans. Microw. Theory Techn. 61 (7) (2013) 2695–2704, ISSN 0018-9480, http://dx.doi.org/10.1109/TMTT.2013.2262803.

[42] A. Raffo, G. Bosi, V. Vadalà, G. Vannini, Behavioral modeling of GaN FETs: a load-line approach, IEEE Trans. Microw. Theory Techn. 62 (1) (2014) 73–82, ISSN 0018-9480, http://dx.doi.org/10.1109/TMTT.2013.2291710.

[43] M. Golio, RF and Microwave Semiconductor Device Handbook, The Electrical Engineering Handbook Series, Taylor & Francis, 2002, ISBN 9780849315626.

[44] J.C. Pedro, N.B. Carvalho, Intermodulation Distortion in Microwave and Wireless Circuits, Artech House, Boston, MA, 2003, ISBN 1580533566 (alk. paper).

[45] N. de Carvalho, J. Pedro, A comprehensive explanation of distortion sideband asymmetries, IEEE Trans. Microw. Theory Techn. 50 (9) (2002) 2090–2101, ISSN 0018-9480, http://dx.doi.org/10.1109/TMTT.2002.802321.

[46] P. Colantonio, F. Giannini, E. Limiti, A. Nanni, Distortion sideband asymmetries in power amplifiers through Volterra series, in: Int. Workshop on Integ. Nonlin. Microw. and Millim.-Wave Circ. (INMMIC), Rome, Italy, ISBN 88-88748-34-2, 2004, pp. 147–150.

[47] P. Colantonio, F. Giannini, E. Limiti, A. Nanni, Investigation of IMD asymmetry in microwave FETs via Volterra series, in: Eur. Gallium Arsenide and Other Semic. App. Symp. (GAAS), Paris, France, 2005, pp. 53–56.
[48] P. Colantonio, F. Giannini, E. Limiti, A. Nanni, V. Camarchia, V. Teppati, M. Pirola, Design approach to improve linearity and power performance of microwave FETs, Int. J. RF Microw. Comput. Aided Eng. 18 (6) (2008) 527–535, ISSN 1099-047X, http://dx.doi.org/10.1002/mmce.20328.
[49] S. Cripps, A theory for the prediction of GaAs FET load-pull power contours, in: IEEE MTT-S Int. Microw. Symp. Dig., Boston, MA, USA, ISSN 0149-645X, 1983, pp. 221–223, http://dx.doi.org/10.1109/MWSYM.1983.1130864.
[50] L. Kushner, Output performance of idealized microwave power amplifiers, Microw. J. 32 (1989) 103.
[51] H. Kondoh, FET power performance prediction using a linearized device model, in: IEEE MTT-S Int. Microw. Symp. Dig., vol. 2, Long Beach, CA, USA, 1989, pp. 569–572, http://dx.doi.org/10.1109/MWSYM.1989.38791.
[52] L. Kushner, Estimating power amplifier large signal gain, Microw. J. 33 (1990) 87.
[53] S.R. Mazumdert, A. Azizi, F. Gardiol, Improvement of a class-C transistor power amplifier by second-harmonic tuning, IEEE Trans. Microw. Theory Techn. 27 (5) (1979) 430–433, ISSN 0018-9480, http://dx.doi.org/10.1109/TMTT.1979.1129644.
[54] M. Maeda, H. Masato, H. Takehara, M. Nakamura, S. Morimoto, H. Fujimoto, Y. Ota, O. Ishikawa, Source second-harmonic control for high efficiency power amplifiers, IEEE Trans. Microw. Theory Techn. 43 (12) (1995) 2952–2957, ISSN 0018-9480, http://dx.doi.org/10.1109/22.475660.
[55] J. Staudinger, Multiharmonic load termination effects on GaAs MESFET power amplifiers, Microw. J. (1996) 60–67.
[56] P. Colantonio, F. Giannini, E. Limiti, G. Saggio, Experimental performances of 5 GHz harmonic-manipulated high efficiency microwave power amplifiers, Electron. Lett. 36 (9) (2000) 800–801, ISSN 0013-5194, http://dx.doi.org/10.1049/el:20000614.
[57] V. Vadalà, A. Raffo, P. Colantonio, E. Cipriani, F. Giannini, C. Lanzieri, A. Pantellini, A. Nalli, G. Bosi, G. Vannini, Evaluation of FET performance and restrictions by low-frequency measurements, in: International Workshop on Integrated Nonlinear Microwave and Millimetre-Wave Circuits (INMMiC), Leuven, Belgium, 2014, pp. 1–3, http://dx.doi.org/10.1109/INMMIC.2014.6815070.
[58] G. Bosi, E. Cipriani, P. Colantonio, F. Giannini, A. Raffo, G. Vannini, Gate waveform effects on high-efficiency PA design: an experimental validation, in: Eur. Microw. Int. Circ. Conf., Rome, Italy, 2014.
[59] B. Kopp, D. Heston, High-efficiency 5-watt power amplifier with harmonic tuning, in: IEEE MTT-S Int. Microw. Symp. Dig., vol. 2, New York, NY, USA, 1988, pp. 839–842, http://dx.doi.org/10.1109/MWSYM.1988.22163.
[60] M. Khatibzadeh, H. Tserng, Harmonic tuning of power FETs at X-band, in: IEEE MTT-S Int. Microw. Symp. Dig., vol. 3, Dallas, TX, USA, 1990, pp. 989–992, http://dx.doi.org/10.1109/MWSYM.1990.99745.
[61] S. Toyoda, High efficiency amplifiers, in: IEEE MTT-S Int. Microw. Symp. Dig., vol. 1, San Diego, CA, USA, ISSN 0149-645X, 1994, pp. 253–256, http://dx.doi.org/10.1109/MWSYM.1994.335324.
[62] T. Nojima, S. Nishiki, High efficiency microwave harmonic reaction amplifier, in: IEEE MTT-S Int. Microw. Symp. Dig., vol. 2, New York, NY, USA, 1988, pp. 1007–1010, http://dx.doi.org/10.1109/MWSYM.1988.22201.

[63] S. Nishiki, T. Nojima, Harmonic reaction amplifier—a novel high-efficiency and high-power microwave amplifier, in: IEEE MTT-S Int. Microw. Symp. Dig., vol. 2, Las Vegas, NV, USA, ISSN 0149-645X, 1987, pp. 963–966, http://dx.doi.org/10.1109/MWSYM.1987.1132581.

[64] A. Mallet, T. Peyretailade, R. Sommet, D. Floriot, S. Delage, J. Nebus, J. Obregon, A design method for high efficiency class F HBT amplifiers, in: IEEE MTT-S Int. Microw. Symp. Dig., vol. 2, San Franscisco, CA, USA, ISSN 0149-645X, 1996, pp. 855–858, http://dx.doi.org/10.1109/MWSYM.1996.511072.

[65] V. Tyler, A new high-efficiency high-power amplifier, Marconi Rev. 21 (130) (1958) 96–109.

[66] D. Snider, A theoretical analysis and experimental confirmation of the optimally loaded and overdriven RF power amplifier, IEEE Trans. Electron Devices 14 (12) (1967) 851–857, ISSN 0018-9383, http://dx.doi.org/10.1109/T-ED.1967.16120.

[67] F. Raab, Introduction to class-F power amplifiers, RF Design 19 (5) (1996) 79–84.

[68] C. Duvanaud, S. Dietsche, G. Pataut, J. Obregon, High-efficient class F GaAs FET amplifiers operating with very low bias voltages for use in mobile telephones at 1.75 GHz, IEEE Microwave Guided Wave Lett. 3 (8) (1993) 268–270, ISSN 1051-8207, http://dx.doi.org/10.1109/75.242219.

[69] S. Gao, High efficiency class-F RF/microwave power amplifiers, IEEE Microw. Mag. 7 (1) (2006) 40–48, ISSN 1527-3342, http://dx.doi.org/10.1109/MMW.2006.1614233.

[70] C. Trask, Class-F amplifier loading networks: a unified design approach, in: IEEE MTT-S Int. Microw. Symp. Dig., vol. 1, Anaheim, CA, USA, 1999, pp. 351–354, http://dx.doi.org/10.1109/MWSYM.1999.779491.

[71] A. Grebennikov, Circuit design technique for high efficiency Class F amplifiers, in: IEEE MTT-S Int. Microw. Symp. Dig., vol. 2, Boston, MA, USA, ISSN 0149-645X, 2000, pp. 771–774, http://dx.doi.org/10.1109/MWSYM.2000.863295.

[72] Y.Y. Woo, Y. Yang, B. Kim, Analysis and experiments for high-efficiency class-F and inverse class-F power amplifiers, IEEE Trans. Microw. Theory Techn. 54 (5) (2006) 1969–1974, ISSN 0018-9480, http://dx.doi.org/10.1109/TMTT.2006.872805.

[73] R. Negra, F. Ghannouchi, W. Bachtold, Study and design optimization of multiharmonic transmission-line load networks for class-E and class-F K-band MMIC power amplifiers, IEEE Trans. Microw. Theory Techn. 55 (6) (2007) 1390–1397, ISSN 0018-9480, http://dx.doi.org/10.1109/TMTT.2007.896769.

[74] A. Inoue, T. Heima, A. Ohta, R. Hattori, Y. Mitsui, Analysis of class-F and inverse class-F amplifiers, in: IEEE MTT-S Int. Microw. Symp. Dig., vol. 2, Boston, MA, USA, ISSN 0149-645X, 2000, pp. 775–778, http://dx.doi.org/10.1109/MWSYM.2000.863296.

[75] N. Sokal, A. Sokal, Class E—a new class of high-efficiency tuned single-ended switching power amplifiers, IEEE J. Solid-State Circuits 10 (3) (1975) 168–176, ISSN 0018-9200, http://dx.doi.org/10.1109/JSSC.1975.1050582.

[76] A. Grebennikov, H. Jaeger, Class E with parallel circuit—a new challenge for high-efficiency RF and microwave power amplifiers, in: IEEE MTT-S Int. Microw. Symp. Dig., vol. 3, Philadelphia, PA, USA, ISSN 0149-645X, 2002, pp. 1627–1630, http://dx.doi.org/10.1109/MWSYM.2002.1012169.

[77] A. Dani, M. Roberg, Z. Popovic, PA efficiency and linearity enhancement using external harmonic injection, IEEE Trans. Microw. Theory Techn. 60 (12) (2012) 4097–4106, ISSN 0018-9480, http://dx.doi.org/10.1109/TMTT.2012.2222918.

[78] A. AlMuhaisen, P. Wright, J. Lees, P. Tasker, S. Cripps, J. Benedikt, Novel wide band high-efficiency active harmonic injection power amplifier concept, in: IEEE MTT-S

Int. Microw. Symp. Dig., Piscataway, NJ, USA, ISSN 0149-645X, 2010, pp. 664–667, http://dx.doi.org/10.1109/MWSYM.2010.5514759.

[79] D.E. Dawson, Closed-form solutions for the design of optimum matching networks, IEEE Trans. Microw. Theory Techn. 57 (1) (2009) 121–129, ISSN 0018-9480, http://dx.doi.org/10.1109/TMTT.2008.2009041.

[80] W.L. Jung, J.-H. Chiu, Stable broadband microwave amplifier design using the simplified real frequency technique, IEEE Trans. Microw. Theory Techn. 41 (2) (1993) 336–340, ISSN 0018-9480, http://dx.doi.org/10.1109/22.216477.

[81] S. Masuda, A. Akasegawa, T. Ohki, K. Makiyama, N. Okamoto, K. Imanishi, T. Kikkawa, H. Shigematsu, Over 10W C-Ku band GaN MMIC non-uniform distributed power amplifier with broadband couplers, in: IEEE MTT-S Int. Microw. Symp. Dig., Piscataway, NJ, USA, ISSN 0149-645X, 2010, pp. 1388–1391, http://dx.doi.org/10.1109/MWSYM.2010.5517992.

[82] X. Zhou, L. Roy, R. Amaya, 1 W, highly efficient, ultra-broadband non-uniform distributed power amplifier in GaN, IEEE Microw. Wirel. Compon. Lett. 23 (4) (2013) 208–210, ISSN 1531-1309, http://dx.doi.org/10.1109/LMWC.2013.2250270.

[83] E. Cipriani, P. Colantonio, F. Giannini, The effect of 2nd harmonic control on power amplifiers performances, in: Eur. Microw. Int. Circ. Conf., Amsterdam, The Netherlands, 2012, pp. 345–348.

[84] S. Cripps, P. Tasker, A. Clarke, J. Lees, J. Benedikt, On the continuity of high efficiency modes in linear RF power amplifiers, IEEE Microw. Wirel. Compon. Lett. 19 (10) (2009) 665–667, ISSN 1531-1309, http://dx.doi.org/10.1109/LMWC.2009.2029754.

[85] P. Wright, J. Lees, J. Benedikt, P. Tasker, S. Cripps, A methodology for realizing high efficiency class-J in a linear and broadband PA, IEEE Trans. Microw. Theory Techn. 57 (12) (2009) 3196–3204, ISSN 0018-9480, http://dx.doi.org/10.1109/TMTT.2009.2033295.

[86] P. Tasker, V. Carrubba, P. Wright, J. Lees, J. Benedikt, S. Cripps, Wideband PA design: the "continuous" mode of operation, in: Comp. Semicond. Int. Circ. Symp. (CSICS), Piscataway, NJ, USA, ISSN 1550-8781, 2012, pp. 1–4, http://dx.doi.org/10.1109/CSICS.2012.6340118.

[87] C. Florian, R. Cignani, A. Santarelli, F. Filicori, Design of 40-W AlGaN/GaN MMIC high power amplifiers for C-band SAR applications, IEEE Trans. Microw. Theory Techn. 61 (12) (2013) 4492–4504, ISSN 0018-9480, http://dx.doi.org/10.1109/TMTT.2013.2286109.

[88] K. Horiguchi, S. Ishizaka, T. Okano, M. Nakayama, H. Ryoji, Y. Isota, T. Takagi, Efficiency enhancement of 250W Doherty power amplifiers using virtual open stub techniques for UHF-band OFDM applications, in: IEEE MTT-S Int. Microw. Symp. Dig., San Francisco, CA, USA, ISSN 0149-645X, 2006, pp. 1356–1359, http://dx.doi.org/10.1109/MWSYM.2006.249501.

[89] J. Yao, G. Shen, G. Zhang, J. Chen, A novel four stage 200W Doherty power amplifier for DVB-T transmitter, in: Int. Conf. Microw. Millim. Wave Techn. (ICMMT), vol. 1, Shenzhen, China2012, pp. 1–3, http://dx.doi.org/10.1109/ICMMT.2012.6230000.

[90] M. van der Heijden, M. Acar, J. Vromans, A compact 12-Watt high-efficiency 2.1-2.7 GHz class-E GaN HEMT power amplifier for base stations, in: IEEE MTT-S Int. Microw. Symp. Dig., Boston, MA, USA, ISSN 0149-645X, 2009, pp. 657–660, http://dx.doi.org/10.1109/MWSYM.2009.5165782.

[91] C. Hsia, D. Kimball, S. Lanfranco, P. Asbeck, Wideband high efficiency digitally-assisted envelope amplifier with dual switching stages for radio base-station envelope tracking power amplifiers, in: IEEE MTT-S Int. Microw. Symp. Dig.,

Piscataway, NJ, USA, ISSN 0149-645X, 2010, pp. 672–675, http://dx.doi.org/10.1109/MWSYM.2010.5517569.

[92] F. van Rijs, Status and trends of silicon LDMOS base station PA technologies to go beyond 2.5 GHz applications, in: IEEE Radio and Wirel. Symp., 2008, pp. 69–72, http://dx.doi.org/10.1109/RWS.2008.4463430.

[93] O. Hammi, F. Ghannouchi, Comparative study of recent advances in power amplification devices and circuits for wireless communication infrastructure, in: IEEE Int. Conf. on Electron. Circ. and Syst. (ICECS), 2009, pp. 379–382, http://dx.doi.org/10.1109/ICECS.2009.5410913.

[94] Y. Wei, J. Staudinger, M. Miller, High efficiency linear GaAs MMIC amplifier for wireless base station and femto cell applications, in: IEEE Top. Conf. Pow. Ampl. Wirel. Radio Appl. (PAWR), Santa Clara, CA, USA, 2012, pp. 49–52, http://dx.doi.org/10.1109/PAWR.2012.6174938.

[95] S. Seneviratne, M. Yagoub, R. Amaya, MMIC Doherty power amplifier for a 5W picocell base station in GaN HFET technology, in: Int. Conf. Microel. (ICM), Algiers, Algeria, 2012, pp. 1–4, http://dx.doi.org/10.1109/ICM.2012.6471411.

[96] X. Moronval, A. Dechansiaud, R. Abdoelgafoer, A 20W multi-band multi-mode MMIC power amplifier for base station applications, in: Eur. Microw. Int. Circ. Conf., Nuremberg, Germany, 2013, pp. 161–164.

[97] M. Franco, Mobile handset power amplifiers, IEEE Microw. Mag. 10 (7) (2009) 16–19, ISSN 1527-3342, http://dx.doi.org/10.1109/MMM.2009.934687.

[98] A. Hajimiri, Next-generation CMOS RF power amplifiers, IEEE Microw. Mag. 12 (1) (2011) 38–45, ISSN 1527-3342, http://dx.doi.org/10.1109/MMM.2010.939321.

[99] B. Franois, P. Reynaert, A fully integrated watt-level linear 900-MHz CMOS RF power amplifier for LTE-applications, IEEE Trans. Microw. Theory Techn. 60 (6) (2012) 1878–1885, ISSN 0018-9480, http://dx.doi.org/10.1109/TMTT.2012.2189411.

[100] P.H. Aaen, J.A. Pla, J. Wood, Modeling and Characterization of RF and Microwave Power FETs, The Cambridge RF and Microwave Engineering Series, Cambridge University Press, Cambridge, 2007, ISBN 9780521870665.

[101] R. Pengelly, S. Wood, J. Milligan, S. Sheppard, W. Pribble, A review of GaN on SiC high electron-mobility power transistors and MMICs, IEEE Trans. Microw. Theory Techn. 60 (6) (2012) 1764–1783, ISSN 0018-9480, http://dx.doi.org/10.1109/TMTT.2012.2187535.

[102] U.K. Mishra, L. Shen, T. Kazior, Y.-F. Wu, GaN-based RF power devices and amplifiers, IEEE Proc. 96 (2) (2008) 287–305, ISSN 0018-9219, http://dx.doi.org/10.1109/JPROC.2007.911060.

[103] P. Marchand, M. Diop, R. Lopes, J. Polian, F. Ribeiro, T. Ruan, High power (35 kW and 190 kW) 352 MHz solid state amplifiers for the soleil synchrotron, in: IEEE Part. Acc. Con., Knoxville, TN, USA, 2005, pp. 811–813, http://dx.doi.org/10.1109/PAC.2005.1590572.

[104] P. Saad, P. Colantonio, L. Piazzon, F. Giannini, K. Andersson, C. Fager, Design of a concurrent dual-band 1.8–2.4-GHz GaN-HEMT Doherty power amplifier, IEEE Trans. Microw. Theory Techn. 60 (6) (2012) 1840–1849, ISSN 0018-9480, http://dx.doi.org/10.1109/TMTT.2012.2189120.

[105] C. Florian, F. Scappaviva, R.P. Paganelli, I. Melczarsky, R. Cignani, M. Pirazzini, R. Giordani, M. Feudale, G. Vannini, F. Filicori, Hybrid high power amplifiers for L-band space application, in: IEEE MTT-S Int. Microw. Symp. Dig., Honolulu, HI, USA, ISSN 0149-645X, 2007, pp. 1589–1592, http://dx.doi.org/10.1109/MWSYM.2007.379988.

[106] R. Giofré, L. Piazzon, P. Colantonio, F. Giannini, A Doherty architecture with high feasibility and defined bandwidth behavior, IEEE Trans. Microw. Theory Techn. 61 (9) (2013) 3308–3317, ISSN 0018-9480, http://dx.doi.org/10.1109/TMTT.2013.2274432.
[107] E. Cipriani, P. Colantonio, F. Di Paolo, F. Giannini, R. Giofré, R. Diciomma, B. Orobello, M. Papi, A highly efficient octave bandwidth high power amplifier in GaN technology, in: Eur. Microw. Int. Circ. Conf., Manchester, UK, 2011, pp. 188–191.
[108] P. Colantonio, F. Giannini, R. Giofré, E. Limiti, A. Serino, M. Peroni, P. Romanini, C. Proietti, A C-band high-efficiency second-harmonic-tuned hybrid power amplifier in GaN technology, IEEE Trans. Microw. Theory Techn. 54 (6) (2006) 2713–2722, ISSN 0018-9480, http://dx.doi.org/10.1109/TMTT.2006.874872.
[109] A. Mediavilla, A. Tobia, P. Colantonio, 1 kW compact L-band pulsed power amplifier for radar applications, in: Microw., Radar and Rem. Sens. Symp. (MRRS), Kiev, Ukraine, 2011, pp. 43–46, http://dx.doi.org/10.1109/MRRS.2011.6053597.
[110] S. Theeuwen, J. Qureshi, LDMOS technology for RF power amplifiers, IEEE Trans. Microw. Theory Techn. 60 (6) (2012) 1755–1763, ISSN 0018-9480, http://dx.doi.org/10.1109/TMTT.2012.2193141.
[111] G.C. Barisich, S. Pavlidis, C.A.D. Morcillo, O.L. Chlieh, J. Papapolymerou, E. Gebara, An X-band GaN HEMT hybrid power amplifier with low-loss Wilkinson division on AlN substrate, in: IEEE Int. Conf. Microw. Comm. Ant. Elect. Syst. (COMCAS), Tel-Aviv, Israel, 2013, pp. 1–4, http://dx.doi.org/10.1109/COMCAS.2013.6685285.
[112] P. Schuh, R. Leberer, H. Sledzik, M. Oppermann, B. Adelseck, H. Brugger, R. Behtash, H. Leier, R. Quay, R. Kiefer, 20W GaN HPAs for next generation X-band T/R-modules, in: IEEE MTT-S Int. Microw. Symp. Dig., San Francisco, CA, USA, ISSN 0149-645X, 2006, pp. 726–729, http://dx.doi.org/10.1109/MWSYM.2006.249736.
[113] H. Chen, L.-J. Jiang, X.-F. Ji, Y.-X. Zhang, Design of an X-band pulsed SSPA based on a cascade technique, in: Int. Conf. Microw. Tech. Comp. Electrom., Beijing, China, 2011, pp. 152–155, http://dx.doi.org/10.1109/ICMTCE.2011.5915187.
[114] F. Canales, M. Abbasi, A 75–90 GHz high linearity MMIC power amplifier with integrated output power detector, in: IEEE MTT-S Int. Microw. Symp. Dig., Seattle, WA, USA, ISSN 0149-645X, 2013, pp. 1–4, http://dx.doi.org/10.1109/MWSYM.2013.6697600.
[115] C. Campbell, A. Balistreri, M. Kao, D. Dumka, J. Hitt, GaN takes the lead, IEEE Microw. Mag. 13 (6) (2012) 44–53, ISSN 1527-3342, http://dx.doi.org/10.1109/MMM.2012.2205829.
[116] Y.-F. Wu, D. Wapolnek, J. Ibbetson, P. Parikh, B. Keller, U. Mishra, 14-W GaN-based microwave power amplifiers, in: IEEE MTT-S Int. Microw. Symp. Dig., vol.2, Boston, MA, USA, ISSN 0149-645X, 2000, pp. 963–965, http://dx.doi.org/10.1109/MWSYM.2000.863517.
[117] S. Lee, B. Green, K. Chu, K. Webb, L. Eastman, Demonstration of a high efficiency nonuniform monolithic gallium-nitride distributed amplifier, in: IEEE MTT-S Int. Microw. Symp. Dig., vol 1, Boston, MA, USA, ISSN 0149-645X, 2000, pp. 549–552, http://dx.doi.org/10.1109/MWSYM.2000.861116.
[118] S. Sheppard, W. Pribble, D. Enerson, Z. Ring, R. Smith, S. Allen, J. Palmour, High power demonstration at 10 GHz with GaN-AlGaN HEMT hybrid amplifiers, in: Dev. Res. Conf. (DRC), Denver, CO, USA, 2000, pp. 37–38, http://dx.doi.org/10.1109/DRC.2000.877078.
[119] Y. Yoon, J. Kim, H. Kim, K.H. An, O. Lee, C.-H. Lee, J. Kenney, A dual-mode CMOS RF power amplifier with integrated tunable matching network, IEEE Trans. Microw. Theory Techn. 60 (1) (2012) 77–88, ISSN 0018-9480, http://dx.doi.org/10.1109/TMTT.2011.2175235.

[120] Q. Gu, Z. Xu, M.-C. Chang, Two-way current-combining W-band power amplifier in 65-nm CMOS, IEEE Trans. Microw. Theory Techn. 60 (5) (2012) 1365–1374, ISSN 0018-9480, http://dx.doi.org/10.1109/TMTT.2012.2187536.

[121] J.-W. Lee, B.-S. Kim, A K-band high-voltage four-way series-bias cascode power amplifier in 0.13 μm CMOS, IEEE Microw. Wirel. Compon. Lett. 20 (7) (2010) 408–410, ISSN 1531-1309, http://dx.doi.org/10.1109/LMWC.2010.2049442.

[122] J. Cressler, SiGe HBT technology: a new contender for Si-based RF and microwave circuit applications, IEEE Trans. Microw. Theory Techn. 46 (5) (1998) 572–589, ISSN 0018-9480, http://dx.doi.org/10.1109/22.668665.

[123] M. Kaynak, I. Tekin, Y. Gurbuz, Realisation of a single-chip, silicon germanium: C-based power amplifier for multi-band worldwide interoperability for microwave access applications, IET Microw. Antenna P. 4 (12) (2010) 2273–2280.

[124] U.R. Pfeiffer, S.K. Reynolds, B.A. Floyd, A 77 GHz SiGe power amplifier for potential applications in automotive radar systems, in: IEEE Radio Freq. Integ. Circ. Symp. (RFIC), Ft. Worth, TX, USA, 2004, pp. 91–94.

[125] V. Giammello, E. Ragonese, G. Palmisano, A 15-dBm SiGe BiCMOS PA for 77-GHz automotive radar, IEEE Trans. Microw. Theory Techn. 59 (11) (2011) 2910–2918, ISSN 0018-9480, http://dx.doi.org/10.1109/TMTT.2011.2166802.

[126] S. Nicolson, K. Yau, S. Pruvost, V. Danelon, P. Chevalier, P. Garcia, A. Chantre, B. Sautreuil, S. Voinigescu, A low-voltage SiGe BiCMOS 77-GHz automotive radar chipset, IEEE Trans. Microw. Theory Techn. 56 (5) (2008) 1092–1104, ISSN 0018-9480, http://dx.doi.org/10.1109/TMTT.2008.921268.

[127] P. Colantonio, F. Giannini, R. Giofré, E. Limiti, C. Lanzieri, S. Lavanga, A two stage high frequency class F power amplifier, in: Int. Workshop on Integ. Nonlin. Microw. and Millim.-Wave Circ. (INMMIC), Aveiro, Portugal, 2006, pp. 187–190, http://dx.doi.org/10.1109/INMMIC.2006.283543.

[128] I. Bahl, Fundamentals of RF and Microwave Transistor Amplifiers, Wiley, Hoboken, NJ, 2009, ISBN 9780470462317.

[129] R. Ulrich, W. Brown, R. Ulrich, Advanced Electronic Packaging, IEEE Press series on microelectronic systems, Wiley, 2006.

[130] B. Geller, J. Tyler, L. Holdeman, F. Phelleps, G. Laird, Method of packaging microwave semiconductor components and integrated circuits, 1991, US Patent 5,063,177.

[131] M. Mach, J. Muller, Thermal design considerations on wire-bond packages, in: Eur. Microel. Pack. Conf. (EMPC), Rimini, Italy, 2009, pp. 1–7.

[132] A. Bessemoulin, M. Parisot, M. Camiade, 1-Watt Ku-band power amplifier MMICs using low-cost quad-flat plastic package, in: IEEE MTT-S Int. Microw. Symp. Dig., vol 2, Ft. Worth, TX, USA, ISSN 0149-645X, 2004, pp. 473–476, http://dx.doi.org/10.1109/MWSYM.2004.1336015.

[133] R. Giordani, M. Amici, A. Barigelli, F. Conti, M. Del Marro, M. Feudale, M. Imparato, A. Suriani, Highly integrated and solderless LTCC based C-band T/R module, in: Eur. Microw. Int. Circ. Conf., Rome, Italy, 2009, pp. 407–410.

[134] P. Lavrador, T. Cunha, P. Cabral, J. Pedro, The linearity-efficiency compromise, IEEE Microw. Mag. 11 (5) (2010) 44–58, ISSN 1527-3342, http://dx.doi.org/10.1109/MMM.2010.937100.

[135] P. Kenington, High-Linearity RF Amplifier Design, Artech House microwave library, Artech House, 2000, ISBN 9781580531436.

[136] Y. Shmaliy, Continuous-Time Systems, Signals and Communication Technology, Springer, 2007, ISBN 9781402062728.

[137] V. Camarchia, F. Cappelluti, M. Pirola, S. Guerrieri, G. Ghione, Self-consistent electrothermal modeling of class A, AB, and B power GaN HEMTs under modulated RF excitation, IEEE Trans. Microw. Theory Techn. 55 (9) (2007) 1824–1831, ISSN 0018-9480, http://dx.doi.org/10.1109/TMTT.2007.903839.

[138] S. Nuttinck, E. Gebara, J. Laskar, H. Harris, Study of self-heating effects, temperature-dependent modeling, and pulsed load-pull measurements on GaN HEMTs, IEEE Trans. Microw. Theory Techn. 49 (12) (2001) 2413–2420, ISSN 0018-9480, http://dx.doi.org/10.1109/22.971629.

[139] S. Binari, K. Ikossi, J. Roussos, W. Kruppa, D. Park, H.B. Dietrich, D.D. Koleske, A. Wickenden, R.L. Henry, Trapping effects and microwave power performance in AlGaN/GaN HEMTs, IEEE Trans. Electron Devices 48 (3) (2001) 465–471, ISSN 0018-9383, http://dx.doi.org/10.1109/16.906437.

[140] J. Brinkhoff, A. Parker, M. Leung, Baseband impedance and linearization of FET circuits, IEEE Trans. Microw. Theory Techn. 51 (12) (2003) 2523–2530, ISSN 0018-9480, http://dx.doi.org/10.1109/TMTT.2003.819208.

[141] J. Brinkhoff, A. Parker, Effect of baseband impedance on FET intermodulation, IEEE Trans. Microw. Theory Techn. 51 (3) (2003) 1045–1051, ISSN 0018-9480, http://dx.doi.org/10.1109/TMTT.2003.808704.

[142] A. Richards, K. Morris, J. McGeehan, Removing the effects of baseband impedance on distortion in FET amplifiers, IEE Proc. Microw. Anten. P. 153 (5) (2006) 401–406, ISSN 1350-2417, http://dx.doi.org/10.1049/ip-map:20060005.

[143] J. Santiago, J. Portilla, T. Fernández, Study of the influence of bias and matching networks on the distortion and memory of FET-based power amplifiers, Int. J. RF Microw. Comput. Aided Eng. 18 (6) (2008) 517–526, ISSN 1096-4290, http://dx.doi.org/10.1002/mmce.v18:6.

[144] S. Maas, D. Neilson, Modeling MESFETs for intermodulation analysis of mixers and amplifiers, IEEE Trans. Microw. Theory Techn. 38 (12) (1990) 1964–1971, ISSN 0018-9480, http://dx.doi.org/10.1109/22.64581.

[145] B. OhAnnaidh, T. Brazil, An accurate nonlinear MOSFET model for intermodulation distortion analysis, IEEE Microw. Wirel. Compon. Lett. 14 (7) (2004) 352–354, ISSN 1531-1309, http://dx.doi.org/10.1109/LMWC.2004.829285.

[146] A. Raffo, G. Vannini, A. Santarelli, P. Traverso, M. Pagani, F. Palomba, F. Scappaviva, F. Filicori, Improvement of PHEMT intermodulation prediction through the accurate modelling of low-frequency dispersion effects, in: IEEE MTT-S Int. Microw. Symp. Dig., Long Beach, CA, USA, ISSN 01490-645X, 2005, pp. 4, http://dx.doi.org/10.1109/MWSYM.2005.1516630.

[147] J. Pedro, J. Perez, Accurate simulation of GaAs MESFET's intermodulation distortion using a new drain-source current model, IEEE Trans. Microw. Theory Techn. 42 (1) (1994) 25–33, ISSN 0018-9480, http://dx.doi.org/10.1109/22.265524.

[148] F. Giannini, G. Leuzzi, G. Orengo, P. Colantonio, Modeling power and intermodulation behavior of microwave transistors with unified small-signal/large-signal neural network models, Int. J. RF Microw. Comput. Aided Eng. 4 (13) (2003) 276–284, ISSN 1096-4290, http://dx.doi.org/10.1002/mmce.10091.

[149] J.A. Garcia, A. Tazon Puente, A. Mediavilla Sanchez, I. Santamaria, M. Lazaro, C.J. Pantaleon, J.C. Pedro, Modeling MESFETs and HEMTs intermodulation distortion behavior using a generalized radial basis function network, Int. J. RF Microw. Comput. Aided Eng. 9 (3) (1999) 261–276, ISSN 1099-047X, http://dx.doi.org/10.1002/(SICI)1099-047X(199905)9:3⟨261::AID-MMCE10⟩3.0.CO;2-P.

[150] J. Vuolevi, T. Rahkonen, Distortion in RF Power Amplifiers, Artech House microwave library, Artech House, 2003, ISBN 9781580536295.
[151] J. Bussgang, L. Ehrman, J. Graham, Analysis of nonlinear systems with multiple inputs, 62 (8) (1974) 1088–1119, ISSN 0018-9219, http://dx.doi.org/10.1109/PROC.1974.9572.
[152] R. Minasian, Intermodulation distortion analysis of MESFET amplifiers using the Volterra series representation, IEEE Trans. Microw. Theory Techn. 28 (1) (1980) 1–8, ISSN 0018-9480, http://dx.doi.org/10.1109/TMTT.1980.1129998.
[153] C.L. Law, C.S. Aitchison, Prediction of wide-band power performance of MESFET distributed amplifiers using the Volterra series representation, IEEE Trans. Microw. Theory Techn. 34 (12) (1986) 1308–1317, ISSN 0018-9480, http://dx.doi.org/10.1109/TMTT.1986.1133542.
[154] N. de Carvalho, J. Pedro, Large- and small-signal IMD behavior of microwave power amplifiers, IEEE Trans. Microw. Theory Techn. 47 (12) (1999) 2364–2374, ISSN 0018-9480, http://dx.doi.org/10.1109/22.808983.
[155] J. Garcia, M. De la Fuente, J. Pedro, N. Carvalho, Y. Newport, A. Mediavilla, A. Tazon, Time-varying Volterra-series analysis of spectral regrowth and noise power ratio in FET mixers, IEEE Trans. Microw. Theory Techn. 49 (3) (2001) 545–549, ISSN 0018-9480, http://dx.doi.org/10.1109/22.910560.
[156] P. Colantonio, F. Giannini, G. Leuzzi, E. Limiti, High-efficiency low-IM microwave PA design, in: IEEE MTT-S Int. Microw. Symp. Dig., vol. 1, Phoenix, AZ, USA, ISSN 0149-645X, 2001, pp. 511–514, http://dx.doi.org/10.1109/MWSYM.2001.966943.
[157] M. Meyer, D. Smely, J. Leeb, G. Magerl, Low intermodulation class-F power amplifier, in: Proc. of Radio and Wireless Conf., RAWCON 2003, Piscataway, NJ, USA, 2003, pp. 289–292, http://dx.doi.org/10.1109/RAWCON.2003.1227949.
[158] S. Goto, T. Kunii, T. Oue, K. Izawa, A. Inoue, M. Kohno, T. Oku, T. Ishikawa, A low distortion 25W class-F power amplifier using internally harmonic tuned FET architecture for 3.5 GHz OFDM applications, in: IEEE MTT-S Int. Microw. Symp. Dig., San Francisco, CA, USA, ISSN 0149-645X, 2006, pp. 1538–1541, http://dx.doi.org/10.1109/MWSYM.2006.249606.
[159] M.R. Moazzam, C. Aitchison, A low third order intermodulation amplifier with harmonic feedback circuitry, in: IEEE MTT-S Int. Microw. Symp. Dig., vol 2, San Franscisco, CA, USA, ISSN 0149-645X, 1996, pp. 827–830, http://dx.doi.org/10.1109/MWSYM.1996.511065.
[160] P. White, Effect of input harmonic terminations on high efficiency class-B and class-F operation of PHEMT devices, in: IEEE MTT-S Int. Microw. Symp. Dig., vol 3, New York, NY, USA, ISSN 0149-645X, 1998, pp. 1611–1614, http://dx.doi.org/10.1109/MWSYM.1998.700685.
[161] K.-I. Jeon, Y.-S. Kwon, S.-C. Hong, Input harmonics control using non-linear capacitor in GaAs FET power amplifier, in: IEEE MTT-S Int. Microw. Symp. Dig., vol. 2, Denver, CO, USA, ISSN 0149-645X, 1997, pp. 817–820, http://dx.doi.org/10.1109/MWSYM.1997.602919.
[162] S. Watanabe, S. Takatuka, K. Takagi, H. Kuroda, Y. Oda, Simulation and experimental results of source harmonic tuning on linearity of power GaAs FET under class AB operation, in: IEEE MTT-S Int. Microw. Symp. Dig., vol. 3, San Franscisco, CA, USA, ISSN 0149-645X, 1996, pp. 1771–1774, http://dx.doi.org/10.1109/MWSYM.1996.512286.
[163] P. Colantonio, F. Giannini, G. Leuzzi, E. Limiti, IMD performances of harmonically tuned microwave power amplifiers, Microw. Eng. Eur. (January) (2001) 49–55, ISSN 0960-667X.

[164] J. Sevic, K. Burger, M. Steer, A novel envelope-termination load-pull method for ACPR optimization of RF/microwave power amplifiers, in: IEEE MTT-S Int. Microw. Symp. Dig., vol. 2, New York, NY, USA, ISSN 0149-645X, 1998, pp. 723–726, http://dx.doi.org/10.1109/MWSYM.1998.705093.

[165] M. Steer, Beyond 3G, IEEE Microw. Mag. 8 (1) (2007) 76–82, ISSN 1527-3342, http://dx.doi.org/10.1109/MMW.2007.316254.

[166] F. Raab, P. Asbeck, S. Cripps, P. Kenington, Z. Popovic, N. Pothecary, J. Sevic, N. Sokal, Power amplifiers and transmitters for RF and microwave, IEEE Trans. Microw. Theory Techn. 50 (3) (2002) 814–826, ISSN 0018-9480, http://dx.doi.org/10.1109/22.989965.

[167] M. Ciappa, F. Carbognani, W. Fichtner, Lifetime prediction and design of reliability tests for high-power devices in automotive applications, IEEE Trans. Device Mat. Rel. 3 (4) (2003) 191–196, ISSN 1530-4388, http://dx.doi.org/10.1109/TDMR.2003.818148.

[168] L. Kahn, Single-sideband transmission by envelope elimination and restoration, Proc. IRE 40 (7) (1952) 803–806, ISSN 0096-8390, http://dx.doi.org/10.1109/JRPROC.1952.273844.

[169] C. Yu, A. Zhu, A single envelope modulator-based envelope-tracking structure for multiple-input and multiple-output wireless transmitters, IEEE Trans. Microw. Theory Techn. 60 (10) (2012) 3317–3327, ISSN 0018-9480, http://dx.doi.org/10.1109/TMTT.2012.2208653.

[170] D. Cox, Linear amplification with nonlinear components, 22 (12) (1974) 1942–1945, ISSN 0090-6778, http://dx.doi.org/10.1109/TCOM.1974.1092141.

[171] W. Doherty, A new high-efficiency power amplifier for modulated waves, AT&T Tech. J. 15 (3) (1936) 469–475, ISSN 0005-8580, http://dx.doi.org/10.1002/j.1538-7305.1936.tb03563.x.

[172] B. Kim, J. Kim, I. Kim, J. Cha, The Doherty power amplifier, IEEE Microw. Mag. 7 (5) (2006) 42–50, ISSN 1527-3342, http://dx.doi.org/10.1109/MW-M.2006.247914.

[173] P. Colantonio, F. Giannini, R. Giofré, L. Piazzon, The AB-C Doherty power amplifier. Part I: theory, Int. J. RF Microw. Comput. Aided Eng. 19 (3) (2009) 293–306, ISSN 1099-047X, http://dx.doi.org/10.1002/mmce.20350.

[174] P. Colantonio, F. Giannini, R. Giofré, L. Piazzon, The AB-C Doherty power amplifier. Part II: validation, Int. J. RF Microw. Comput. Aided Eng. 19 (3) (2009) 307–316, ISSN 1099-047X, http://dx.doi.org/10.1002/mmce.20351.

[175] P. Colantonio, F. Giannini, R. Giofré, L. Piazzon, Theory and experimental results of a class F AB-C Doherty power amplifier, IEEE Trans. Microw. Theory Techn. 57 (8) (2009) 1936–1947, ISSN 0018-9480, http://dx.doi.org/10.1109/TMTT.2009.2025433.

[176] C. Zhi Fan, X.-W. Zhu, J. Xia, L. Zhang, Efficiency enhanced class-F Doherty power amplifier at 3.5 GHz for LTE-advanced application, in: Asia Pacif. Microw. Conf., Coex, Seoul, Korea, 2013, pp. 707–709, http://dx.doi.org/10.1109/APMC.2013.6694963.

[177] Y.-S. Lee, M. Woo, Y.-H. Jeong, High-efficiency Doherty amplifier using GaN HEMT class-F cells for WCDMA applications, in: Int. Conf. on Microw. and Millim. Wave Techn. (ICMMT), vol. 1, Nanjing, P.R. China, 2008, pp. 270–273, http://dx.doi.org/10.1109/ICMMT.2008.4540359.

[178] Y. Suzuki, T. Hirota, T. Nojima, Highly efficient feed-forward amplifier using a class-F Doherty amplifier, in: IEEE MTT-S Int. Microw. Symp. Dig., vol. 1, Philadelphia,

PA, USA, ISSN 0149-645X, 2003, pp. 77–80, http://dx.doi.org/10.1109/MWSYM.2003.1210887.

[179] S. Goto, T. Kunii, A. Inoue, K. Izawa, T. Ishikawa, Y. Matsuda, Efficiency enhancement of Doherty amplifier with combination of class-F and inverse class-F schemes for S-band base station application, in: IEEE MTT-S Int. Microw. Symp. Dig., vol 2, Ft. Worth, TX, USA, ISSN 0149-645X, 2004, pp. 839–842, http://dx.doi.org/10.1109/MWSYM.2004.1339098.

[180] A. Grebennikov, S. Bulja, High-efficiency Doherty power amplifiers: historical aspect and modern trends, IEEE Proc. 100 (12) (2012) 3190–3219, ISSN 0018-9219, http://dx.doi.org/10.1109/JPROC.2012.2211091.

[181] K. Bathich, A. Markos, G. Boeck, Frequency response analysis and bandwidth extension of the Doherty amplifier, IEEE Trans. Microw. Theory Techn. 59 (4) (2011) 934–944, ISSN 0018-9480, http://dx.doi.org/10.1109/TMTT.2010.2098040.

[182] X. Chen, W. Chen, G. Su, F. Ghannouchi, Z. Feng, A concurrent dual-band 1.9; 2.6-GHz Doherty power amplifier with intermodulation impedance tuning, in: IEEE MTT-S Int. Microw. Symp. Dig., Tampa Bay, FL, USA, 2014, pp. 1–4, http://dx.doi.org/10.1109/MWSYM.2014.6848482.

[183] X.A. Nghiem, R. Negra, Design of a concurrent quad-band GaN-HEMT Doherty power amplifier for wireless applications, in: IEEE MTT-S Int. Microw. Symp. Dig., Seattle, WA, USA, ISSN 0149-645X, 2013, pp. 1–4, http://dx.doi.org/10.1109/MWSYM.2013.6697727.

[184] X.A. Nghiem, R. Negra, Novel design of a concurrent tri-band GaN-HEMT Doherty power amplifier, in: Asia Pacif. Microw. Conf., Kaohsiung, Taiwan, 2012, pp. 364–366, http://dx.doi.org/10.1109/APMC.2012.6421599.

[185] R. Giofré, L. Piazzon, P. Colantonio, F. Giannini, A closed-form design technique for ultra-wideband Doherty power amplifiers, IEEE Trans. Microw. Theory Techn. 62 (12) (2014) 3414–3424, ISSN 0018-9480, http://dx.doi.org/10.1109/TMTT.2014.2363851.

[186] R. Giofré, L. Piazzon, P. Colantonio, F. Giannini, An ultra-broadband GaN Doherty amplifier with 83% of fractional bandwidth, IEEE Microwave Wireless Compon. Lett. 24 (11) (2014) 775–777, ISSN 1531-1309, http://dx.doi.org/10.1109/LMWC.2014.2345193.

[187] L. Piazzon, R. Giofré, P. Colantonio, F. Giannini, A method for designing broadband Doherty power amplifiers, Prog. Electromagn. Res. 145 (2014) 319–331, http://dx.doi.org/10.2528/PIER14011301.

[188] F. Giannini, G. Leuzzi, Non-Linear Microwave Circuit Design, Wiley, 2004, ISBN 9780470847015.

[189] A. Suárez, R. Quéré, Stability Analysis of Nonlinear Microwave Circuits, Artech House Microwave Library, Artech House, 2002, ISBN 9781580535861.

[190] S. Maas, Nonlinear Microwave and RF Circuits, Artech House, 2003, ISBN 9781580536110.

[191] C. Alexander, M. Sadiku, Fundamentals of Electric Circuits, McGraw-Hill Higher Education, 2005, ISBN 9780073256436.

[192] M. Rashid, SPICE for Power Electronics and Electric Power, Electrical and Computer Engineering, third ed., Taylor & Francis, 2012, ISBN 9781439860465.

[193] M. Nakhla, J. Vlach, A piecewise harmonic balance technique for determination of periodic response of nonlinear systems, IEEE Trans. Circuits Syst. 23 (2) (1976) 85–91, ISSN 0098-4094, http://dx.doi.org/10.1109/TCS.1976.1084181.

[194] S. El-Rabaie, V. Fusco, C. Stewart, Harmonic balance evaluation of nonlinear microwave circuits—a tutorial approach, IEEE Trans. Electron Devices 31 (3) (1988) 181–192, ISSN 0018-9359, http://dx.doi.org/10.1109/13.2310.

[195] V. Rizzoli, A. Lipparini, E. Marazzi, A general-purpose program for nonlinear microwave circuit design, IEEE Trans. Microw. Theory Techn. 31 (9) (1983) 762–770, ISSN 0018-9480, http://dx.doi.org/10.1109/TMTT.1983.1131587.

[196] K. Kundert, A. Sangiovanni-Vincentelli, Simulation of nonlinear circuits in the frequency domain, IEEE Trans. Computer-Aided Design Integr. Circuits Syst. 5 (4) (1986) 521–535, ISSN 0278-0070, http://dx.doi.org/10.1109/TCAD.1986.1270223.

[197] V. Rizzoli, A. Neri, State of the art and present trends in nonlinear microwave CAD techniques, IEEE Trans. Microw. Theory Techn. 36 (2) (1988) 343–365, ISSN 0018-9480, http://dx.doi.org/10.1109/22.3524.

[198] K. Kundert, J. White, A. Sangiovanni-Vincentelli, Steady-State Methods for Simulating Analog and Microwave Circuits, The Springer International Series in Engineering and Computer Science, Springer US, 1990, ISBN 9780792390695.

[199] R.J. Gilmore, M.B. Steer, Nonlinear circuit analysis using the method of harmonic balance—a review of the art. Part I. Introductory concepts, Int. J. Microw. Millimeter-Wave CAE 1 (1) (1991) 22–37, ISSN 1522-6301, http://dx.doi.org/10.1002/mmce.4570010104.

[200] V. Rizzoli, A. Lipparini, A. Costanzo, F. Mastri, C. Cecchetti, A. Neri, D. Masotti, State-of-the-art harmonic-balance simulation of forced nonlinear microwave circuits by the piecewise technique, IEEE Trans. Microw. Theory Techn. 40 (1) (1992) 12–28, ISSN 0018-9480, http://dx.doi.org/10.1109/22.108318.

[201] F. Filicori, V. Monaco, C. Naldi, Simulation and design of microwave class-C amplifiers through harmonic analysis, IEEE Trans. Microw. Theory Techn. 27 (12) (1979) 1043–1051, ISSN 0018-9480, http://dx.doi.org/10.1109/TMTT.1979.1129789.

[202] H. Wacker, Continuation Methods: Proceedings of a Symposium at the University of Linz, Austria, October 3–4, 1977, Academic Press, 1978, ISBN 9780127292502.

[203] H.-G. Brachtendorf, G. Welsch, R. Laur, Fast simulation of the steady-state of circuits by the harmonic balance technique, in: IEEE Int. Sump. Circ. Syst., vol. 2, Seattle, WA, USA, 1995, pp. 1388–1391, http://dx.doi.org/10.1109/ISCAS.1995.520406.

[204] V. Rizzoli, F. Mastri, C. Cecchetti, F. Sgallari, Fast and robust inexact Newton approach to the harmonic-balance analysis of nonlinear microwave circuits, IEEE Microwave Guided Wave Lett. 7 (10) (1997) 359–361, ISSN 1051-8207, http://dx.doi.org/10.1109/75.631201.

[205] B. Troyanovsky, Z. Yu, R.W. Dutton, Physics-based simulation of nonlinear distortion in semiconductor devices using the harmonic balance method, Comput. Method Appl. M. 181 (4) (2000) 467–482, ISSN 0045-7825, http://dx.doi.org/10.1016/S0045-7825(99)00185-1.

[206] W. Curtice, Nonlinear analysis of GaAs MESFET amplifiers, mixers, and distributed amplifiers using the harmonic balance technique, IEEE Trans. Microw. Theory Techn. 35 (4) (1987) 441–447, ISSN 0018-9480, http://dx.doi.org/10.1109/TMTT.1987.1133668.

[207] K. Kundert, G. Sorkin, A. Sangiovanni-Vincentelli, Applying harmonic balance to almost-periodic circuits, IEEE Trans. Microw. Theory Techn. 36 (2) (1988) 366–378, ISSN 0018-9480, http://dx.doi.org/10.1109/22.3525.

[208] Z. Xiang-Dong, H. Xing-Nan, G. Bao-Xin, Accurate Fourier transform method for almost-periodic response simulation of microwave nonlinear circuits, Electron. Lett. 25 (6) (1989) 404–406, ISSN 0013-5194, http://dx.doi.org/10.1049/el:19890278.

[209] P. Heron, M. Steer, Jacobian calculation using the multidimensional fast Fourier transform in the harmonic balance analysis of nonlinear circuits, IEEE Trans. Microw. Theory Techn. 38 (4) (1990) 429–431, ISSN 0018-9480, http://dx.doi.org/10.1109/22.52585.

[210] P. Rodrigues, An orthogonal almost-periodic Fourier transform for use in nonlinear circuit simulation, IEEE Microwave Guided Wave Lett. 4 (3) (1994) 74–76, ISSN 1051-8207, http://dx.doi.org/10.1109/75.275585.

[211] V. Rizzoli, C. Cecchetti, A. Lipparini, A general-purpose program for the analysis of nonlinear microwave circuits under multitone excitation by multidimensional Fourier transform, in: Eur. Microw. Conf., Rome, Italy, 1987, pp. 635–640, http://dx.doi.org/10.1109/EUMA.1987.333678.

[212] L. Chua, A. Ushida, Algorithms for computing almost periodic steady-state response of nonlinear systems to multiple input frequencies, IEEE Trans. Circuits Syst. 28 (10) (1981) 953–971, ISSN 0098-4094, http://dx.doi.org/10.1109/TCS.1981.1084921.

[213] A. Ushida, L. Chua, Frequency-domain analysis of nonlinear circuits driven by multi-tone signals, IEEE Trans. Circuits Syst. 31 (9) (1984) 766–779, ISSN 0098-4094, http://dx.doi.org/10.1109/TCS.1984.1085584.

[214] E. Ngoya, J. Rousset, M. Gayral, R. Quere, J. Obregon, Efficient algorithms for spectra calculations in nonlinear microwave circuits simulators, IEEE Trans. Circuits Syst. 37 (11) (1990) 1339–1355, ISSN 0098-4094, http://dx.doi.org/10.1109/31.62409.

[215] D. Hente, R. Jansen, Frequency domain continuation method for the analysis and stability investigation of nonlinear microwave circuits, IEE Proc. Micow. Anten. P. 133 (5) (1986) 351–362, ISSN 0950-107X, http://dx.doi.org/10.1049/ip-h-2.1986.0064.

[216] P. Rodrigues, A general mapping technique for Fourier transform computation in non-linear circuit analysis, IEEE Microwave Guided Wave Lett. 7 (11) (1997) 374–376, ISSN 1051-8207, http://dx.doi.org/10.1109/75.641425.

[217] R. Gilmore, Nonlinear circuit design using the modified harmonic balance algorithm, IEEE Trans. Microw. Theory Techn. 34 (12) (1986) 1294–1307, ISSN 0018-9480, http://dx.doi.org/10.1109/TMTT.1986.1133541.

[218] F. Filicori, V. Monaco, G. Vannini, Computationally efficient multitone analysis of non-linear microwave circuits, in: Eur. Microw. Conf., vol. 2, Stuttgart, Germany, 1991, pp. 1550–1555, http://dx.doi.org/10.1109/EUMA.1991.336566.

[219] P. Feldmann, J. Roychowdhury, Computation of circuit waveform envelopes using an efficient, matrix-decomposed harmonic balance algorithm, in: IEEE/ACM Int. Conf. Comput. Aided Des., San Jose, CA, USA, 1996, pp. 295–300, http://dx.doi.org/10.1109/ICCAD.1996.569712.

[220] V. Borich, J. East, G. Haddad, The method of envelope currents for rapid simulation of weakly nonlinear communications circuits, in: IEEE MTT-S Int. Microw. Symp. Dig., vol. 3, Anaheim, CA, USA, 1999, pp. 981–984, http://dx.doi.org/10.1109/MWSYM.1999.779550.

[221] V. Rizzoli, A. Neri, F. Mastri, A. Lipparini, A Krylov-subspace technique for the simulation of integrated RF/microwave subsystems driven by digitally modulated carriers, Int. J. RF Microw. Comput. Aided Eng. 9 (6) (1999) 490–505, ISSN 1099-047X, http://dx.doi.org/10.1002/(SICI)1099-047X(199911)9:6⟨490::AID-MMCE6⟩3.0.CO;2-D.

[222] V. Rizzoli, A. Costanzo, F. Mastri, Efficient Krylov-subspace simulation of autonomous RF/microwave circuits driven by digitally modulated carriers, IEEE Microwave Wireless Compon. Lett. 11 (7) (2001) 308–310, ISSN 1531-1309, http://dx.doi.org/10.1109/7260.933780.

[223] V. Rizzoli, A. Lipparini, P. Ghigi, F. Mastri, C. Cecchetti, Pulsed-RF and transient analysis of nonlinear microwave circuits by harmonic-balance techniques, in: IEEE MTT-S Int. Microw. Symp. Dig., vol. 2, Boston, MA, USA, ISSN 0149-645X, 1991, pp. 607–610, http://dx.doi.org/10.1109/MWSYM.1991.147076.
[224] R. Jackson, Rollett proviso in the stability of linear microwave circuits—a tutorial, IEEE Trans. Microw. Theory Techn. 54 (3) (2006) 993–1000, ISSN 0018-9480, http://dx.doi.org/10.1109/TMTT.2006.869719.
[225] M. Ohtomo, Stability analysis and numerical simulation of multidevice amplifiers, IEEE Trans. Microw. Theory Techn. 41 (6) (1993) 983–991, ISSN 0018-9480, http://dx.doi.org/10.1109/22.238513.
[226] N. Ayllon, J.-M. Collantes, A. Anakabe, G. Soubercaze-Pun, S. Forestier, D. Langrez, Joint RF and large-signal stability optimization of MMIC power combining amplifiers, Int. J. Microwave Wireless Techn. 6 (2013) 683–688, http://dx.doi.org/10.1017/S1759078713000767.
[227] S. Dellier, R. Gourseyrol, G. Soubercaze-Pun, J.-M. Collantes, A. Anakabe, K. Narendra, Stability analysis of microwave circuits, in: Ann. Wirel. Microw. Techn. Conf., Cocoa Beach, FL, USA, 2012, pp. 1–5, http://dx.doi.org/10.1109/WAMICON.2012.6208480.
[228] A. Suarez, S. Jeon, D. Rutledge, Stability analysis and stabilization of power amplifiers, IEEE Microw. Mag. 7 (5) (2006) 51–65, ISSN 1527-3342, http://dx.doi.org/10.1109/MW-M.2006.247915.
[229] L. Pantoli, G. Leuzzi, A. Santarelli, F. Filicori, R. Giofré, Stabilisation approach for multi-device parallel power amplifiers under large-signal regime, in: Eur. Microw. Int. Circ. Conf., Manchester, UK, 2011, pp. 144–147.

CHAPTER

Technology design interaction: System driven technology choices

5

P.J. Zampardi*, K. Burger† and B. Moser‡

Qorvo, Inc, Newbury Park, CA, United States Qualcomm, San Diego, CA, United States†*
Qorvo, Inc, Greensboro, NC, United States‡

CHAPTER OUTLINE

5.1 INTRODUCTION

Power amplifiers (PAs) are one of the key components in any modern cellphone. One challenge for these circuits is that their development does not follow traditional Moore's law, so a great deal more effort is needed to determine what technology should be used for a given application. We start by discussing the application or system architecture considerations that have driven these technologies to where they are today. This is especially important for the development of III-V transistor

Microwave Wireless Communications. http://dx.doi.org/10.1016/B978-0-12-803894-9.00005-6

technologies. Once the applications are understood, we then discuss real-life product considerations that are generally not the topic of exciting research papers, but are critical for anyone targeting commercial applications. Things such as time to market, mask costs, and integration requirements are important. We next discuss technology features (active and passive device suites, as well as other features) arming designers with enough information to ask informed questions regarding any new processes they are considering using. More importantly, we explain WHY each of these has an impact at the product level for performance, size, and yield. Essential device characterization is presented and demonstrates that much of the concerns a designer has about a process can be garnered from relatively simple measurements. Finally, other considerations like packaging technology, circuit layout, and product yields are discussed and their impact on technology selection is presented.

5.1.1 SYSTEM ARCHITECTURE SELECTION

Before we can decide what technologies we should look at for a PA design, we first need to consider what the customer requirements are and what the PA needs to do. There has been some fragmentation in the market so that everything from "dumb" PAs to highly integrated front-end modules (FEMs) are important. Obviously, what the PA is supposed to do drives decisions about what technology should be used.

Many different technologies are used in a PA module. Here, we will focus on the PA chip (power amplification chain) only. There are trade-offs associated with all of these technologies used in PAs. However, there are standards for many of the parameters (such as gain, linearity, and ruggedness) and you only need to meet them (over bias, temperature, and voltage). Other parameters [like power-added efficiency (PAE), cost, and time to market] are key reasons for choosing a particular technology since they provide some benefit to the end user (and are not just pass/fail specifications).

Characteristics that distinguish good PA technologies are: low-cost (die size, reproducibility, and yield), good reliability and robustness, high power gain (device and assembly related), high breakdown voltage, good linearity, and positive turn-on voltage. Key specifications for the finished PA module include cost/size, ruggedness, gain, linearity (multiple channels), and efficiency.

For initial handset PAs, the device technologies of choice were laterally diffused metal-oxide semiconductor (LDMOS) devices and silicon (Si) bipolar since these were readily available from their use in other power applications and the supply voltages (5 V) aided in keeping these technologies competitive. The silicon bipolar designs consisted of discrete power devices integrated on module [1]. Gallium arsenide (GaAs) metal-semiconductor field effect transistors (MESFETs) were later used, but lost market presence due to negative gate voltage requiring a DC-DC converter (added cost), inferior size and yields compared to GaAs heterojunction bipolar transistors (HBTs). In the late 1990s, the supply voltage for PAs was lowered to a nominal of 3.4 V. This means that the PA needs to operate anywhere from 4.2 (on charger) to 2.8 V (battery almost dead). LDMOS and silicon BJT, possessing many of the important characteristics, began losing market share to GaAs HBTs based on

PA performance metrics such as gain, linearity, and PAE around this time. Currently, GaAs HBTs dominate both the linear and saturated PA markets due to superior performance. Handset manufacturers favor GaAs HBT PAs due to their better efficiency, especially at higher frequencies. In addition, the simpler process (compared to silicon), excellent yield, and high power density have all contributed to this market preference. Other basic considerations are now considered. D-mode MESFET and high-electron mobility transistor (HEMT or MODFETs) provide good PAE and linearity, good radio frequency (RF) integration capability, and provide a good option for RF switching. However, the negative supply voltage (as mentioned earlier) and manufacturing issues do not make them competitive for PAs. E-Mode MESFET/HEMTs, do not require a negative supply, but provide limited output power for PAs, and have manufacturability worse than D-mode, silicon, or bipolar technologies. Silicon or silicon-germanium (SiGe) bipolar require a positive supply, have excellent turn-on control, excellent manufacturability, but have noncompetitive PAE and gain for many PA applications. Silicon PAs can be competitive a lower frequencies (<1 GHz), but are not competitive at higher frequencies (1.9 GHz) [2] at higher power levels (28 dBm). RF CMOS/LDMOS does not require a negative supply, but the PAEs are noncompetitive at higher frequency bands [3], the lower power density leads to larger die, and inferior on-chip passives limit integration capability. In addition, the long turn-aroundtimes and mask costs lead to expensive development costs. III-V HBTs do not require a negative supply, have excellent turn-on control (although higher than silicon), competitive PAEs, and high power density. The simple process enables fast turn-around times (TATs) (time to market advantage) and minimizes mask costs. Also, a MESFET can be integrated into a III-V HBT process [4], adding design flexibility while maintaining a simple manufacturing process.

There is currently a great deal of development on silicon (BiCMOS, CMOS, and SOI-CMOS) based technologies since they offer the promise of "system-on-a-chip." However, these solutions will only be acceptable once the development time and cost of the PA is made to coincide with the rest of the development of the system. This will also limit the flexibility of power amplifier module (PAM) fabrication because the optimum technology for a given function of the PAM cannot necessarily be used. Other technologies [indium phosphide (InP), gallium nitride (GaN), and silicon carbide] will have to prove that they are capable of competing on cost, manufacturability, and demonstrate significant performance improvements to displace the incumbents in the handset market. As for which technology is "best" for PA applications, it is really more a question of how to use/partition all of the various technologies where they make sense, rather than selecting only one and compromising system performance [5,6].

Besides the PA chips, front-end solutions (even simple ones) have multiple technologies in them used for other functions. These other chips are predominantly, switches, controllers, and filters. Filters are primarily bulk acoustic wave (BAW) or surface acoustic wave (SAW) filters. Controllers are either complementary metal-oxide-semiconductor (CMOS) or bipolar with CMOS (BiCMOS) [RF front-end/working mobile industry processor interface (RFFE/MIPI), controllers low dropout regulators for global system for mobile communications (GSM)] and the switching function is currently dominated by silicon on insulator (SOI) based solutions. The components left that still have competition among different semiconductor technologies are the PAs and those will be the main focus of our discussion in this chapter.

Once some of the general considerations are understood, then the designer is ready to consider the technical aspects of the PA design. A list of requirements is shown below — these have a significant impact on technology selection.

- *Frequency — must operate across a band, not single point*
- *Supply voltage — a range* [*0.5* (*low power*) *to 6 V*]
- Bias supply voltage — a range (also low current draw)
- Output power minimum (−50 dBm) to maximum (27, 29, and 34 dBm)
- *Power gain* (*26–28 dB*)
- Gain variation over temp (−1 to 1 dB)
- *Adjacent channel power ratio* (*ACPR*) (first and second) and ACPR under mismatch. Adjacent channel leakage ratio (ACLR) and ACPR are used interchangeably.
- DC-DC compatibility
- Moisture sensitivity level
- Robustness [electro-static discharge (ESD) and voltage standing wave ratio]
- Max current
- Receive band noise
- Harmonics
- Leakage currents
- Switching time
- Stability, control slope
- PA duplexer configurations

The italicized specifications are the ones normally shown in publications — necessary, but not sufficient, for fabricating products. PAE, while not a specification, is a "money" spec in that if it is better, you make more money. A recent figure of merit that has gained popularity is PAE plus the absolute value of ACPR or (ACLR), some recent figures of merit (FOM) results are shown in Figure 5.1. This FOM is primarily driven by the PAE since the linearity is a fixed specification. More details on this FOM are discussed later in this chapter. The points shown in the figure are for wide-band code division multiple access (WCDMA) PAs and are not separated by frequency or what type of module they are integrated into. The dashed line is an FOM of 95 which is approximately the theoretical limit for a WCDMA amplifier with 6 dB power back-off.

5.1.2 BATTERY VOLTAGE CONSIDERATIONS

Changing battery or supply voltage has been a topic of discussion for some time [7]. However, there are different scenarios that drive completely different aspects of technology. The first potential change is a lowering, or removal of the reference

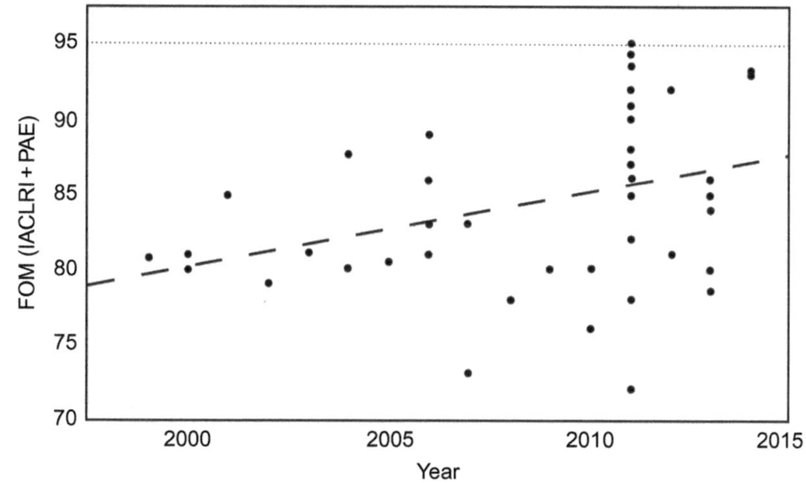

FIGURE 5.1

Linear power amplifier figure of merit (ACLR + PAE) by year. The dashed line is the trend line while the dotted line is the "theoretical limit" for WCDMA.

voltage supply. This is driven by the desire to use fewer regulated supplies in the phone. The second change in operating voltage, of concern for PA design, is driven by the digital content of the phone (ie, almost everything besides the PA) and is the lowering of the overall battery voltage (lowers digital current) of the phone. This would mean that the difference between the maximum (usually the charger) and minimum voltages would be the same, but the values would be lower than the current values. For example, the nominal voltage would shift from 3.4 to 2.4 V and the charger voltage (maximum voltage) and end-of-life (EOL) voltage (minimum) would change accordingly. Finally, the last possibility for voltage changes, and probably most imminent scenario, is that the EOL (lowest usable voltage) battery voltage will decrease to take advantage of newer battery technologies.

The first voltage related problem that PA designers need to deal with is the impact of a lower/no reference voltage on bias circuit design. At present, typical reference voltages supplied inside the phone have been from 2.85 to 2.75 V. Besides the voltage limitations, there are also stringent rules on how much current can be drawn from these supplies. At low reference voltages GaAs HBTs run into the well-known "two V_{be}" problem [8] which leads to degraded performance over temperature and bias voltage as the GaAs current-mirror circuit starves the PA of bias current. There are two technology choices that have been used effectively in alleviating this problem. The first solution is to use a material with a lower turn-on voltage. This has been demonstrated in commercial handset PAs by the implementation of gain (InGaAsN) based transistors. The most attractive feature of this solution is that it provides an easy "drop-in" solution that involves no added process complexity or re-tooling [9].

Unfortunately, it does have some limitations. For example, it still only provides designers with high turn-on, relative to silicon, npn transistors (making it very difficult to design high functionality bias circuits with the desired current over temperature and process control).

A more attractive long-term solution is the implementation of a BiFET technology that integrates a FET (field effect transistor) with the HBT allowing low and no-V_{ref} [10] circuits to be designed. Several examples of improved bias circuit design have been published. In addition to improved bias circuits, there are other benefits of HBT-FET integration related to mid-power efficiency (16 dBm output power) improvements (this FOM is discussed many times where the PA current is integrated over a range of output powers of urban or suburban operating). While other technologies, such as silicon, GaAs pseudomorphic HEMT (pHEMT), and InP HBTs can also address low reference voltage applications. There are various issues related to the RF performance that make them less attractive for some of these applications. For example, our experience has been that it is very easy to design no-V_{ref} circuits in silicon, but it is difficult to get good enough RF performance at the higher frequency bands for simple two-stage designs. There are some recent changes to technology that help mitigate this [11] but no reports, to date, have demonstrated a PA with complete performance comparable to GaAs HBT. As we will discuss later, the mask costs (34 masks vs. ~10 for III-V) make development expensive. Previous work [2] has shown that InP HBTs will likely have performance comparable to silicon, rather than GaAs at lower frequency bands (less than 5 GHz). Combined with substrate cost, this lack of RF performance advantage makes it a poor choice. In terms of FET/HBT integration, the characteristics of integrated FETs/InP HBTs were not very good from an analog stand-point [12] so designers would be back to designing with only higher turn-on npns. Finally, pHEMT is very capable of demonstrating excellent performance [13] (flatter gain curves look attractive for new envelope tracking (ET) requirements, however suffers practical limitations due to size [8] and poor PA yield due to V_p variation).

The attraction of using a lower EOL battery voltage is that there are new battery technologies available that have a higher power density [14], but have a discharge curve that is very different than the currently used batteries (see Figure 5.2). As a result, there is still energy to be used in the range of 2.4–2.7 V compared to older battery technologies, allowing the customer to go longer between charges. Also plotted in this figure is the required current (current draw determines battery operating life rated in ampere-hours) to provide a constant power of 29 dBm at an assumed efficiency of 40%. We observe that while we are taking advantage of the extra energy from the battery, we are also using it faster because of the voltage — taking away some of the perceived return. In addition, the higher current also creates issues due to the resistance of the battery itself.

From a PA output impedance, or Z_{load}, perspective a lower EOL voltage creates a multitude of challenges. First, the charger voltage (which is usually the highest applied voltage the PA will see and is the one that places limitations on ruggedness) remains unchanged and can be as high as 5.5 V. This also means that the PA must now perform

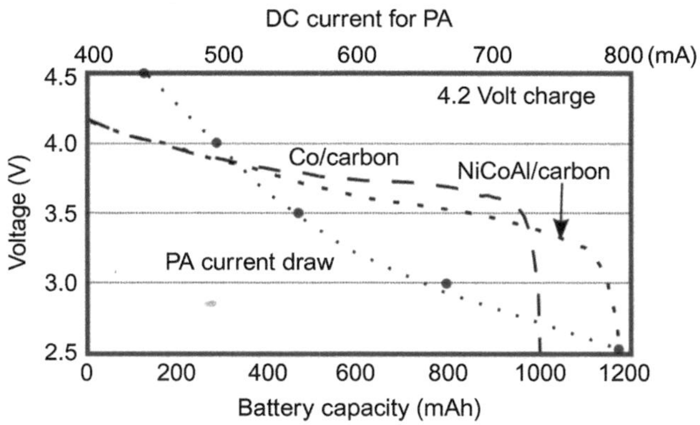

FIGURE 5.2

Battery discharge curves *(short and long dashed lines)* and required PA current assuming PAE=40% and an output power of 29 dBm *(dotted line)*. As the voltage decreases, the required current draw increases.

at least as well as before over a wider operating range — a very challenging task. This wider range creates similar issues for all potential PA technologies since it is very difficult to get optimum performance at all of the possible voltages (this wide range of operating conditions leads to many hallway conversations poking fun at comments like: "my PA is designed with a this or that load-line" known only at one freq, V_{supply} and temperature). The lower end of the voltage range, as shown in Figure 5.2, will need to supply much more current to achieve the same power and this will literally stress the technologies since many of the failure mechanisms for the active devices and wiring are current driven. Because this normally places a current density limitation, it would be desirable to alter array sizes (conceptually similar to [15]) to keep current densities within allowed limits for required foundry reliability limits. Another option is to drive the PA with a DC-DC converter [16] to step up the voltage once it goes below a certain level so that the PA only sees a low voltage state when a lower output power is required (so the current does not increase unnecessarily). Two concerns, for PA modules, are if the DC-DC converter is supplied from the phone manufacturer and how much loss or power the DC-DC converter itself uses (noise injection remains an integration challenge for convertor-based designs). Finally, another option is to re-tune the output network as a function of the desired output voltage, but this will be discussed in a later section on multimode, multiband challenges.

5.1.3 MID- AND LOW-POWER EFFICIENCY CONSIDERATIONS

Traditionally, improved mid-power efficiency has been handled using stage by-pass architectures, such as those in Figure 5.3. Figure 5.4 shows an example of PAE versus output power, as well as collector current for a stage by-pass architecture PAs such as in Figure 5.3A and B.

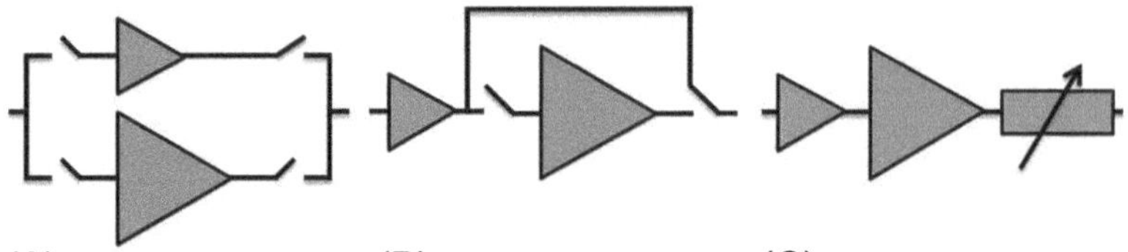

FIGURE 5.3

Examples of stage bypassing architectures or load-tuning. (A) Is a "quasi-Doherty" or parallel path PA, (B) is a stage by-pass architecture, and (C) is a PA with tunable output network.

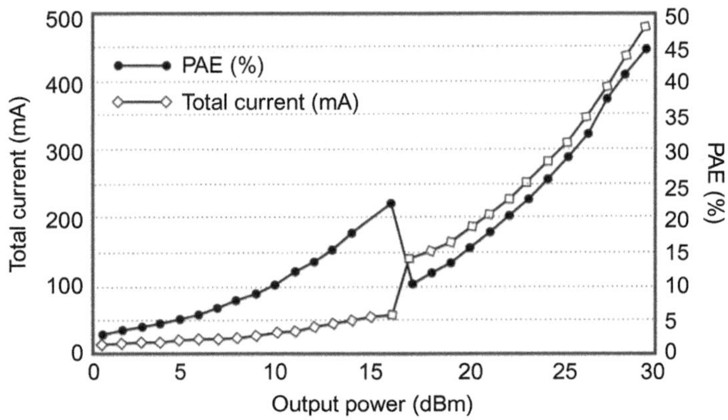

FIGURE 5.4

Example performance of a stage by-pass power amplifier showing mid-power PAE improvement at 16 dBm.

In contrast to saturated PAs (such as GSM) where the power is constantly being adjusted and controlled via the collector voltage, in "linear" PAs (eg, WCDMA), there has been a great deal of emphasis placed on techniques that allow for the lowest overall current consumption by the PA. In particular, the efficiency at mid-power (or integrated over a range of power) is a very important specification from some handset manufacturers. This is represented at the system level by the probability distribution function (PDF) that shows the likelihood of having to transmit a certain power. Figure 5.5A shows the CDMA (code division multiple access) standard (both urban and suburban environment). Figure 5.5B shows the PDF for WCDMA, including data. As we rely more and more on data, the probability shifts towards transmitting higher powers. This mid-power requirement is being addressed through both design topologies (quasi-Doherty [17], Figure 5.3A) and through technologies that allow bypassing of one or more PA stages [18], as in Figure 5.3B. For III-V semiconductor technologies, stage bypassing creates some interesting opportunities/challenges. Besides the technical challenges in size and network loss, cost, and

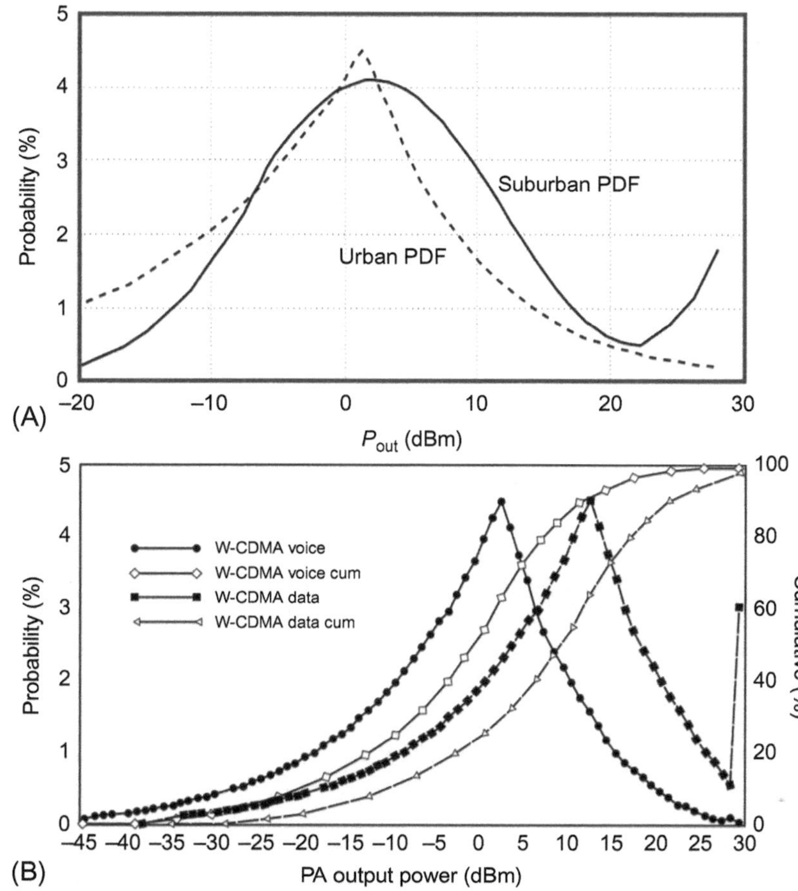

FIGURE 5.5

(A) Transmit probability curve. One would expect things to be optimized for 0 dBm, but normally specifications are at 16 dBm because that is what gets tested. (B) WCDMA probability and cumulative probability curves.

reproducibility are major factors. In the context of a PA module, Figure 5.6, when a switching function is desired, it also requires accompanying logic circuits. Thus, a technology that integrates the logic functions within the chip doing the switching (either a switch die or on the PA die) is desirable (and is already obtainable in silicon technologies) [19] and can also lead to lower overall size because of reduced board routing requirements. For III-V technologies, this means that an integrated HBT-FET technology should, ideally, include E/D (enhancement/depletion) mode FETs [20,21] to be competitive. One of the remaining challenges will be obtaining low enough current draw for the logic circuits to rival alternative silicon technologies. As discussed in the previous section, a DC-DC converter can also be

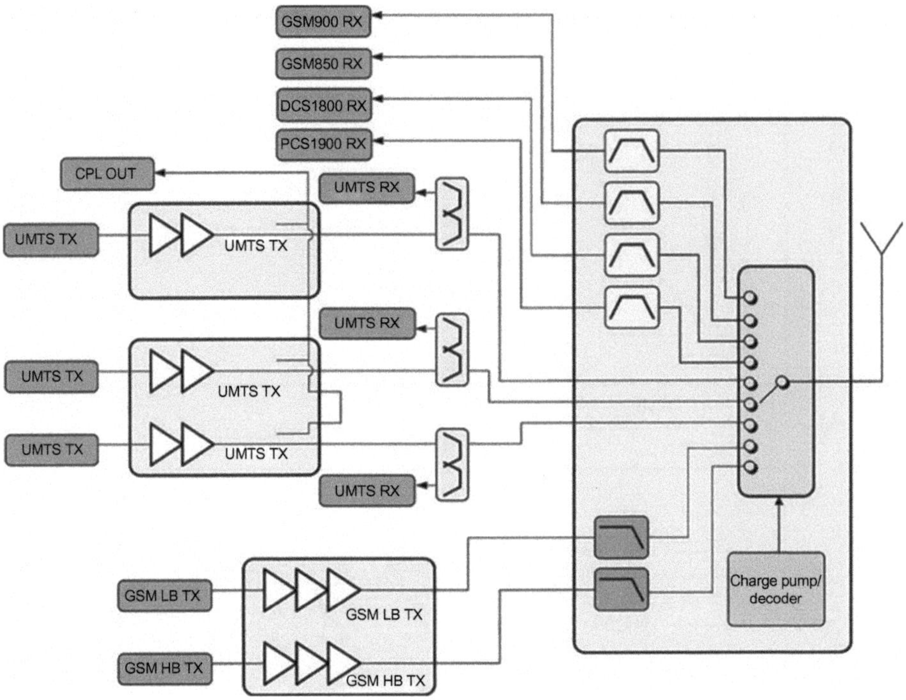

FIGURE 5.6

Example block diagram illustrating the diversity of technologies in a power amplifier module.

used to obtain similar efficiency improvements. A final concern is determining what technologies will be sufficient to do this switching if the power distribution curves (how often the PA transmits a given power) are modified [22].

5.1.4 CONSIDERATIONS FOR AVERAGE POWER TRACKING AND ET

With the advent of average power tracking (APT) or ET modules, which include a RFFE/MIPI controller, it is possible to return to HBT-only processes to reduce cost because control logic can be placed on the MIPI chip. Whether or not this is sufficient depends on the required integration level and design preference. For additional details on transistor operation in dynamic power supply transmitters, the reader is referred to Chapter 7. The MIPI RFFE is a standard that unifies the RF front-end in handsets allowing phone manufacturers to deal with the added complexity of multiple devices and multiple bands. It allows for a control bus that connects all the devices in a PA from low-noise amplifiers, PAs, switches, filters, etc. [23,24].

When a PA is run from a fixed supply voltage, the efficiency degrades significantly when the power is backed off from peak power. Unfortunately, this is where the PA spends most of its time transmitting based on the complementary cumulative distribution function curves [25]. This results in much less battery life. To increase battery life, there are several things that can be done at the system level. This is especially important for modulate's signals, such as long-term evolution (LTE), that have very high peak-to-average ratios. APT uses an efficient DC-DC converter connected between the battery and PA supply voltage (V_{cc}) and dynamically changes V_{cc} based on the PA average output power. When the PA output power is "backed off" from maximum output power, the PA supply voltage is reduced to improve PA efficiency. ET [26] forces the supply voltage of the amplifier to track the changes of the applied signal to the amplifier. An ET amplifier operates at maximum efficiency at all envelope levels. To a PA designer, the main focus is still on delivering the best PA at maximum output power. The main difference between APT and ET is that APT is a slowly varying envelope (updated each time slot) while ET attempts to instantaneously follow the envelope (see Figure 5.7). Studies on switcher design/integration [27] compared GaAs and GaN for high-frequency switcher applications and concluded that GaAs were technically superior for supply voltages $<\sim 18$ V. The faster the buck converter runs, the smaller the inductor and overall solution size. From a technology point of view, the use of APT or ET means that we would like transistor characteristics that do not vary (in particular, gain, PAE, etc.) as a function of the supply voltage (V_{cc} or V_{dd}). Ideally, the technology provides high peak PAE down to low voltages. This will also be discussed in detail in our discussion on active device characterization.

5.1.5 MULTIMODE, MULTIBAND PAs

Multiband (specifically quad-band) PAs (single chip) are already in existence (there are many quad-band GSM PAs already on the market). LTE-frequency division duplex (FDD)/time division duplex (TDD) bands 38, 39, 40, 41 show the need for operating in both modes within the same or overlapping frequency range. FDD is

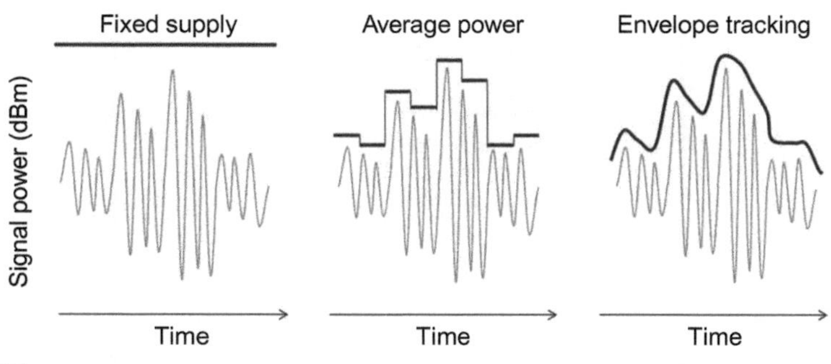

FIGURE 5.7

Comparison of V_{cc} bias — constant V_{cc}, average power tracking, and envelope tracking. The dark line represents the instantaneous supply voltage. Fixed supply does not change with signal, average power changes the supply every "slot" and envelope tracking continuously adjust the power supply to track the signal envelope.

very similar to WCDMA while TDD is more like time division synchronous code division multiple access or enhanced data GSM environment (GSM EDGE) in bias density and RF matching. A typical block diagram is shown in Figure 5.8. We observe that there are multiple PAs, matching networks, and switches that allow selection between the transmitting PAs and the receive ports. Figure 5.9 highlights the trade-off in complexity for various implementations for multimode, multiband PAs. There is a trade-off between performance and size/cost. Integrating "discrete" PAs gives the best performance, but takes the most space, and is more expensive. Fully integrating the PAs (to a single chip, even) is the lowest cost/size, but the reuse of the matching and additional switching will degrade performance. For a PA manufacturer, the dream scenario is multiple PAs and a single switch if the loss can be tolerated. Unfortunately, given size constraints, that is not a long-term practical solution as there is continuous pressure to shrink PA module size. This has been further exacerbated by the proliferation of bands required for phones. We have already

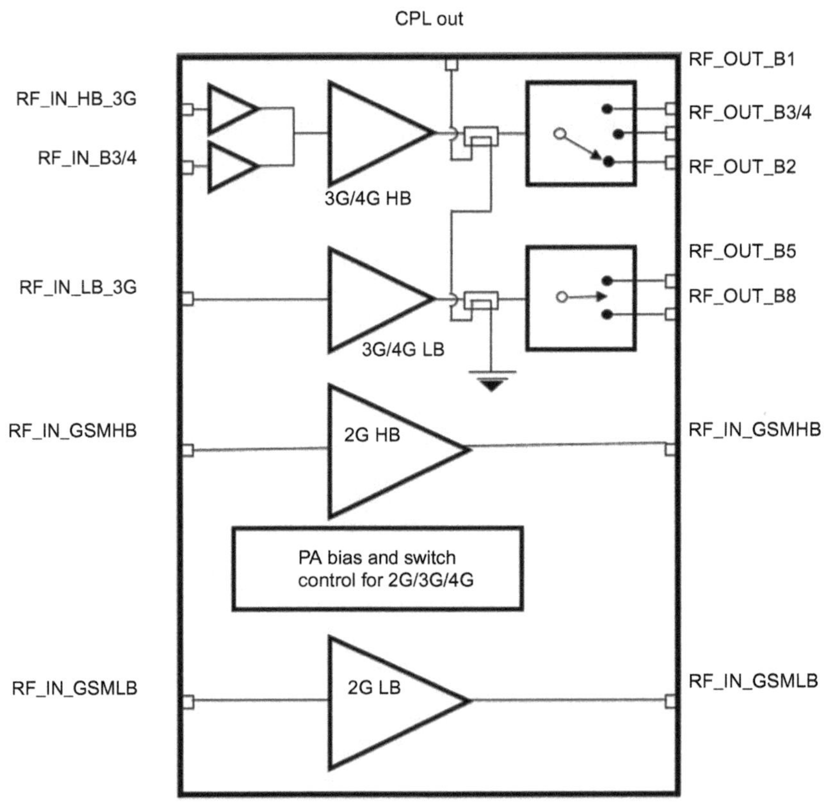

FIGURE 5.8

Example block diagram of multimode/multiband PA module. These contain PAs for different standards as well as couplers, switches, and a controller.

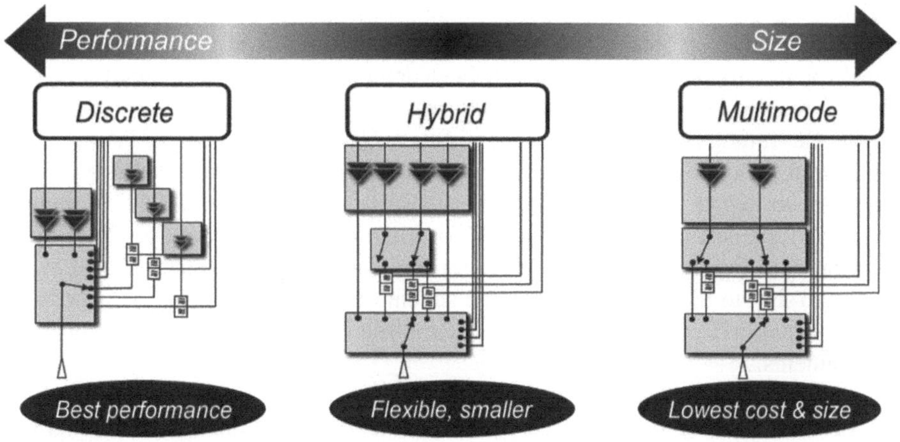

FIGURE 5.9

Example of complexity increase from discrete to multimode PA going from left to right. Discrete integration gives the best performance, but worst form factor. Hybrid still provides some flexibility in design and is smaller. Multimode is lowest cost and size and, unfortunately, lower performance.

discussed many of the important challenges that the individual PAs will face (based on current standards). One constant in PA technology development is that the material structures and transistor structures will continue to be refined to meet new requirements. It is expected that to meet new linearity demands of uplink carrier aggregation (UL CA) with 2×20 MHz channel bandwidths, this technology development will continue. Beyond that, however, true multimode, multiband operation will drive other, more substantial technology improvements as well. This area is one where integrability becomes very important. Now the technology needs to perform well in linear and saturated modes across a variety of peak-to-average ratios to minimize the number of PA dies.

To achieve a true multimode, multiband phone in a minimum size, switching or tunable output matching functions will become critical needs. Low-loss tunability for reconfigurable matching networks is in the same demand as a tunable SAW or BAW filter technology. For tunability, there are several different technologies for [microelectromechanical systems (MEMs) ferroelectric capacitors, and high-quality factor (high-Q) varactors]. MEMs has demonstrated a number of devices (switches, capacitors, etc.). While no one can dispute the impressive laboratory demonstration of these devices, there are still some practical questions related to using them in cellular handsets. Among these are cost, reliability, packaging, and the requirement for additional support circuitry. For ferroelectric capacitors, the advantages are: simple packaging and good power handling capability, and fast tuning time. The disadvantages are the trade-offs to get high power and high Q at the same time, relatively narrow tuning range, and high control voltages ESD robustness has also shown to be a weakness of BST (barium strontium titanate) ferroelectric caps. Finally, some novel varactor topologies have recently been demonstrated [28] which may also provide a suitable solution for tunable networks.

Besides tuning (load or inter-stage) and stage by-pass switching, for PA modules, the switch technology is still an area where GaAs faces strong challenges from silicon-based technologies depending on the throw count. While pHEMTs continue to provide the best device performance for low throw count, some of the features mentioned earlier, such as device suite for logic and size, have caused most of the market to shift to SOI switches. With the advent of carrier aggregation, it is possible that some pHEMT processes will make a comeback because of the stringent linearity requirements.

Finally, one of the biggest challenges facing module designers is how to partition the functions across technologies. Some of the PA functions can be integrated in silicon, while others still require GaAs. UL CA with 40 MHz bandwidth appears to be pushing GaAs PA capability to the lead position in the technology. This is especially true at higher frequencies [3]. As a result, the specifics of how the technologies are used to implement different functions will continue to drive the development and refinement of the various contending technologies.

For typical linear [time division multiple access, WCDMA (3G), LTE (4G)] PAs (without ET or APT), there are linearity (ACPR or ACLR) requirements and less stringent ruggedness requirements than saturated PAs (GSM). As we previously discussed, they often require other on-chip features such as stage by-pass architecture — super efficiency at low power, or high efficiency at low power PAs. In addition there is usually some level of bias control to manage the stage switching or to provide temperature compensation. APT or ET do not eliminate the linearity requirements unless there is digital pre-distortion in the system, in which case linearity is sacrificed for peak PAE [29]. LTE falls into the linear PA category.

The GSM is saturated, with no linearity requirement. This same chain is also used for linear (EDGE). The CMOS controller is usually used so there is no real need for on-chip logic, etc. There have been some publications that showed complete integration of controller and PA on the same chip [30]. GSM has generally avoided a real need for BiFET technologies other than for improved bias circuits using source followers. While GSM is being "phased out," it will still be sometime before it is no longer required in modules.

For multimode/multiband PAs, the number of bands has proliferated at incredible speed. The market is splitting in half so that some phone manufactures are very performance-driven and other emerging markets are far most cost-driven. This means there are distinct "low-cost" and "premium" market segments that drive very different design behaviors. For low-cost, cost is obviously the primary factor. PA specs are typically not as stringent (although this is changing a little and expected to change over time). Not really fancy and silicon (CMOS) actually plays okay in this space. For the premium market segment (mostly smart phones), performance is key. GaAs HBT (BiHEMT or BiFET) is still dominant technology here.

The recent integration trend — including duplexers or filters in the FEM — have muddied the water in the technology selection a little because the PA is becoming a smaller fraction of the overall cost and performance budget. For example, many PA designers have had the experience where a customer selects a cheaper filter and now requires the PA to deliver an extra dB or more power to compensate for this. Each unique handset design has different antenna path losses, leading to a wide range in required PA output powers. This makes it hard to deliver a single device to win multiple sockets. The path loss changes for MIMO configurations and front-end architecture choices become long, drawn-out roundtable discussions … in some cases with marketing trying to assert the required technologies, but not responsible for delivering the final high-yielding hardware at production.

5.1.6 WIRELESS LAN AMPLIFIERS

Wireless LAN (WLAN) PA standards do not seem to fluctuate as much as "handset" PA standards. This actually allows for some different design decisions. For most designs, the data rate is higher (up to 5.8 GHz), but the output requirements are lower. For the lower power/lower frequency band (2.4 GHz), SiGe PAs dominate the market. Even though SiGe is primarily used for the power-chain, some of the back-end limitations in silicon have led to the adoption of integrated passive devices (IPDs) — using passives on GaAs substrates — as the output matching network. The IPD can be designed and modified independent of the active devices (which allows a great time to market savings, as we will discuss later) and also allows the use of thicker metals than typically are available in active device technology processes.

For the higher power, higher frequency standard, pHEMT, or GaAs HBT is still used. While the power requirement is lower, the linearity requirements are quite stringent and complicated by the system level calibration.

The linearity requirement for WLAN (especially 802.11ac vs. 802.11N) is very stringent. The spec in terms of error vector magnitude (EVM), introduced in Chapter 2, is 1.8% versus 3%. The sources of EVM are also not necessarily traditional "linearity" sources. They include additive noise, nonlinear distortion, linear distortion, phase noise, spurious, and other errors. Some of these "other errors" may come from memory effects and changes in bias due to the calibration sequencing and thermal effects. The second challenge is dynamic operation because WiFi used TDD. The PA is pulsed on and off [31,32]. Much of this is managed by proper layout thermal behavior or bias circuit design [33]. The problem is that the data of the orthogonal frequency-division multiplexing signal is calibrated during the preamble burst. Unfortunately, the PA gain is usually not stabilized at this point, and this can lead to a dynamic EVM error. As with standard PAEs, it is required to get both good linearity and good PAE at the same time. This is more challenging since devices are usually more efficient in saturation, but that makes them less linear.

5.2 TECHNOLOGY SELECTION AND CHARACTERIZATION

As designers or engineers, we would all like to think that the only consideration for selecting a technology is performance. Unfortunately, some of the first considerations for PA technology selection have nothing to do with the technology performance, but are even more basic considerations such as:

- The required TAT, how fast in the competitive market is the solution needed? Winning at the sample down selection is key to remaining in business.
- What is the required integration level? Is control/logic integration required for the module?
- How much can the solution cost?
- Availability of the technology. If you are successful, can you fabricate enough parts?
- What sort of design support (design kits, models, applications engineering) is available?

From a designer's point of view, cycle time is different than just pure turn-time that a foundry normally quotes. For a designer, cycle time means design data in, to wafers out, as shown in Figure 5.10. That means it includes mask fabrication time as well (translating the data from electronic form to physical photomask plates used for

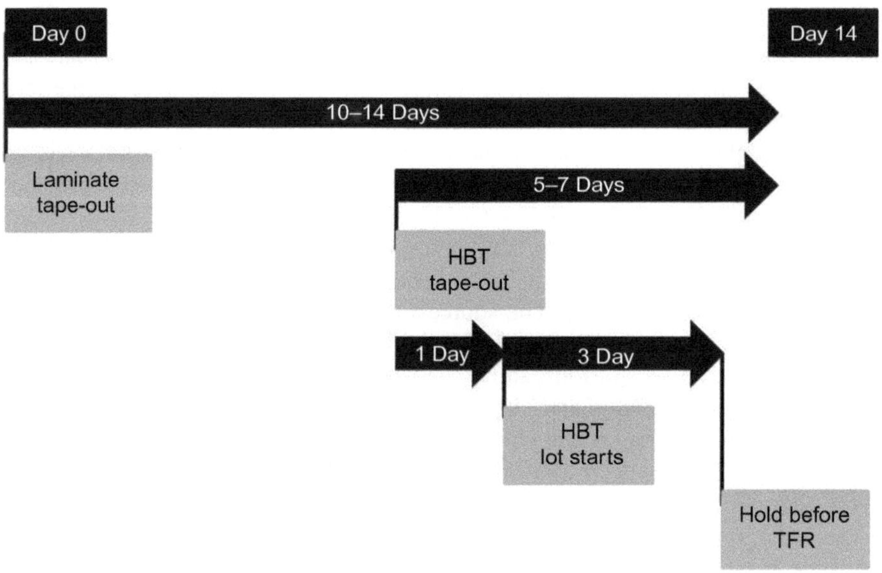

FIGURE 5.10

Example of a prototype design cycle. This highlights the designer view of turn-around time—from mask data submitted to testable parts back. This determines how many attempts a designer gets during a given schedule.

fabrication), not just wafer processing time. Mask preparation can vary from technology to technology based on complexity. For a module design, there is the additional part to the cycle time based on laminate and assembly time (also shown in Figure 5.10). Characterization time and the data collection for some customers require a small team itself to complete a compliance matrix on a handful of parts resulting in a data package to accompany the end devices. For cost, the cost per mm^2 is usually used. In practice, this is not the best figure of merit because the overall PRODUCT cost is what's important, but wafer cost is easily available and the relative costs of FEM technologies, relative to a base CMOS process is shown in Figure 5.11. Since the packaging cost will be nominally the same, the difference between product cost and wafer cost is the fact that different technologies lead to different die sizes. For example, Ref. [4] showed that there is a significant difference in die size between pHEMT and HBT designs (rated in power density as mW/mm). In addition, bond pad sizes, guard-rings (or chip seals), and scribe streets (pad-to-pad spacing) all have an impact on the overall die cost. Flip-chip (FC), backside vias (BSV), or through-silicon vias (TSV), etc. will impact die sizes and directly impact cost. So, anything impacting die size directly impacts cost.

5.2.1 HOW DO WE PICK A TECHNOLOGY?

Technology selection is driven by trading off a wide variety of factors such as product cost, performance, integration, module size, time to market, and technology availability.

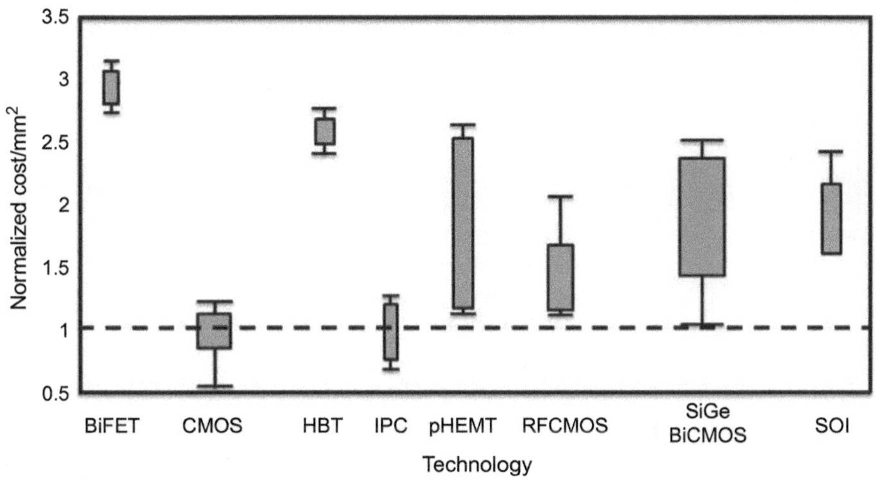

FIGURE 5.11

Technology cost per square millimeter relative to bulk CMOS. Variation is due to option selections for the various technologies.

While cost is a major factor in handset PAs, what is most important is the PRODUCT COST, not the wafer cost. There are multiple factors that influence the product cost that are independent of raw wafer cost. The most obvious is die size. For example, some silicon-based PA topologies rely heavily on inductors. Those inductors drive increased process cost and die size. Some second order features that impact the gross die per wafer are the size of the scribe streets and keep-out regions on the edge of wafers, number of die lost to process control monitor (PCM) structures or alignment marks, and various design rules for a given technology. PA cost breakdown typically shows that packaging, assembly, and testing are 50% of the cost of goods sold (COGS). Design for test, probe, and high-yielding assembly are key factors to financial success.

Table 5.1 highlights some of the important "high-level view" process considerations and compares them for silicon based processes and GaAs based processes. In a practical sense, the main differences are in voltage handling (in a handset this corresponds to ruggedness), thermal conductivity, mask costs, and turn-time. Turn-time, TAT (how long it takes from data in to chips back), is a major consideration for technology selection. Figure 5.12 shows the relative standard turn times (average) of FEM technologies relative to a simple GaAs HBT technology.

This is becoming a very important consideration because the time from initial discussion to production ramp has been significantly reduced since handsets first became a commodity. There is an expectation that samples are delivered to a customer within 6 months and that compliance to spec samples is delivered in less than 1 year. Production ramp is expected 2–6 months after that. Because of process simplicity, some processes provide much faster TAT than others. For example, typical GaAs processes (HBT or pHEMT) start with the device layers predefined by the starting material and take between 8 and 13 mask layers to complete. Compare that to a silicon process with upwards of 30 masks (BiCMOS 35 and SOI 22 depending on back-end of line options)

Table 5.1 High-Level Comparison of Silicon and III-V Technologies for Handset Power Applications

	Attribute	Silicon	III-V
Enablement	Voltage handling		✓
	Thermal conductivity	✓	
	Insulating substrate		✓
	Mechanical strength	✓	
Cost	Feature integration	✓	
	Mask costs		✓
	Unit costs	✓	
Support	Fab cycle time		✓
	Fab capacity	✓	✓
	Fab automation	✓	✓
	Models, PDK	✓	✓

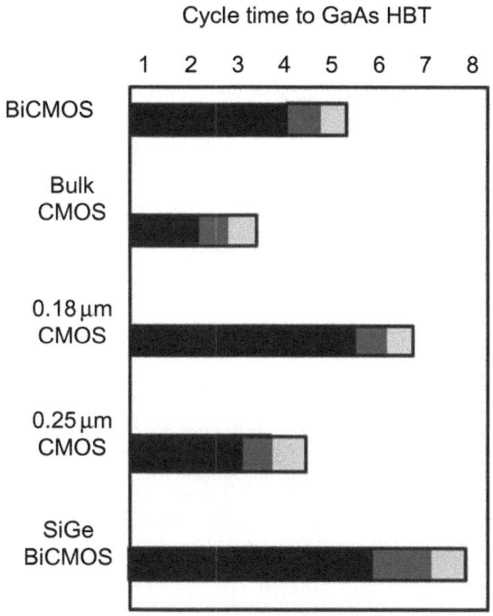

FIGURE 5.12

Relative cycle times of silicon technologies to GaAs HBT. Black bar is fastest from average, black + dark gray is the mean, and total bar is worst case. Higher values are worse.

and the fact that many of the steps are time-consuming anneals and it becomes understandable why III-Vs have been so popular in handset PAs. All PA designs (independent of technology) have similar issues in terms of system level modeling (the *perception* of better models for silicon processes does not really provide an advantage), so the one that turns around chips faster has a step up on the competitors. This is somewhat analogous to the integration of analog and digital portions of other chips — the analog portion always takes longer [34]. Staging wafers for silicon laminate mask changes can improve TAT. Another very important question, if a pure-play foundry is being used, is "what is your capacity for this process and how much of it can we have?" Once some of the logistical questions for a process are understood, then the next step is to look at what the process itself offers.

5.2.2 OVERALL PROCESS FEATURES

When a designer is evaluating a new process, they are often tempted to only pay attention to the active devices they will use for the power amplification chain. Unfortunately, for integrated circuits (ICs), many other technology features and devices are almost equally important for circuit size, performance, and yield. Each of these devices have their own FOM and, unfortunately, the figure of merit for a designer (or impact on circuit layout) may not agree with the figure of merit from the foundry! An example is seen in metal-insulator-metal (MIM) capacitors where foundries will

Table 5.2 Technology Features and Their Impact on Design

Feature	Design Impact
Device suite	Optimize transistor selection for application
Types of devices	Bias circuit, on-chip switching
Minimum device feature size	Small devices for low current logic. Low R_b devices.
Metal rules	Pitch/spacing impacts die size. Current handling impacts die size
Die thickness	Thermal and package keep-out size

quote the capacitance density (capacitance per area) for the film used to make the MIM. However, a designer really cares about the capacitance per area as defined by the footprint for the capacitor (taking into account any vias used to connect layers, as well as keep-out region). Not understanding these subtle differences can make a huge difference in die size and product cost.

As integration levels and the desire for die shrink increase, other factors — besides basic device performance also play a role in technology selection — a few of these are shown in Table 5.2. Some examples: what suite of devices (actives and passives) are offered? What are the minimum feature size devices (for logic)? What are the metal design rules and current handling capabilities? Is through-wafer via offered and what are the design rules and die thickness available or is the process compatible with FC?

A larger suite lets designers make trade-offs or optimize performance for different frequencies or performance requirements. For bipolar transistors, the power device unit-cell layout can be used to optimize the performance in different applications/frequencies. The layout can be modified by changing the emitter length, width, number of base fingers (or number of base fingers relative to emitter fingers). These changes allow important PA-related device parameters like base resistance, base-collector (BC) capacitance, and thermal resistance to be traded off. For example, at low frequency, very wide emitter fingers can be used to shrink the overall amplifier layout without compromising performance. For bias or logic functions, the minimum emitter size is important because it determines the necessary current draw for that function. In an amplifier, where PAE is important, minimizing any other currents is critical. For FETs (pHEMT or MESFET), there is less flexibility in the device design for power amplification since the minimum gate length is usually used. Gate width is really the only parameter that can be varied by the designer. For other functions (bias, logic, or switching) it is useful to have different gate lengths and widths supported (for things like loads), as well as different numbers of gates in a given channel (primarily for switching). Other active devices, such as diodes, should also be fully scalable. Unfortunately, many foundries only offer discrete selections for the active devices. This is usually limited by what the modeling teams are willing to support, rather than what the process is capable of.

The metal design rules (pitch and spacing) are very important as more functionality is added on a given chip because it will directly drive die size for dense ICs. An example is the circuit in [30], which would become very large if the metal-to-metal rules were increased. This may be an unfamiliar concept for designers used to using microstrip style layouts often used for millimeter wave ICs. The metal current carrying capability directly impacts die size because the output current for the PA is very high (usually requiring wide metal traces). Because the output current and the ground currents are very similar in magnitude, at least two metal layers should support high currents.

Whether or not the technology offers through-wafer vias (or if they are affordable), will drive packaging decisions. Details of through-wafer vias considerations are presented in the following section.

As you can see, all of these details need to be considered when looking at a process — designers do not design by power devices alone. Now we can more fully explore the considerations for the different technology elements such as capacitors, resistors, inductors, and through-wafer vias.

5.2.3 PASSIVE AND ACTIVE DEVICE CONCERNS

In this section, we begin by discussing passive devices and later discuss active device characteristics. Table 5.3 shows some of the passive elements used in a PA design, the parameters related to that device, and the impact that each element can have on the resulting design. For resistors, the basic parameters are the sheet resistance (R_{sh}), the minimum width or length (W_{min} or L_{min}), width variation (dW), and maximum current (I_{max}). For a base ballast (where the ballast resistor is placed on the base of the transistor) resistor (often used with HBTs where beta decreases with temperature), higher R_{sh}, smaller W_{min}, and higher I_{max} all result in smaller die. The variation in R_{sh} and in dW (especially for long, skinny resistors) can impact the yield. Bias resistors have the same dependence. For the emitter ballast, the resistor value is typically small and the minimum length is important instead of the overall length. In most III-V processes, the resistors are made from a thin-film resistor using either tantalum nitride or nichrome [35]. The variation of these resistors is ±5–10%.

Table 5.3 Passive Elements, Their Important Parameters, and Their Potential Impact on Die Size and Yield

Element	Parameter				Die Size				Yield	
Base ballast	R_{sh}	W_{min}	dW	I_{max}	X	X		X	X	X
Emitter ballast	R_{sh}	L_{min}	dW	I_{max}	X	X		X	X	X
Bias	R_{sh}	W_{min}	dW	I_{max}	X	X		X	X	X
Cap	C_a	W_{min}	dA	V_{max}	X			X	X	X
Metal	R_{sh}	W_{min}	S_{min}	I_{max}		X	X	X		

In silicon processes, semiconductor resistors are often used for these functions and have variations of around ±20–25% [36]. The thin-film resistor (at extra cost) in SiGe BiCMOS is approximately ±10% [37].

Most semiconductor processes offer multiple capacitors. In III-V technologies, these are almost exclusively based on the use of nitride materials as the insulator material [35]. The use of nitrides places some limitation on the value of capacitance achievable using two metal plates, so that "stacked" capacitors are popular in these technologies. The example we discussed earlier of the "fab" definition of density not matching the designer definition is exactly the case for the stacked capacitor since it requires vias to connect the various plates (this is also true in silicon processes as well). The key parameters for the capacitor are capacitance per footprint size (C_a), the minimum size cap (in case small caps are needed), dA is the variation in area, and is most important for small capacitor values. The V_{max} is the voltage the cap can be biased at. It will affect the die size because if it is not high enough two capacitors will be needed to be placed in series to survive the voltage. The V_{max}, on GaAs processes, is directly traded off with the cap density (from the insulator). On silicon processes, MIM caps can be higher density because of the use of other materials that provide both high dielectric constant (higher intrinsic capacitor density) and high breakdown voltage. Other caps, such as poly-poly caps are also available in silicon processes.

The important parameters for metallization are the sheet resistance for the various metals (this becomes more important towards the output of a PA where the currents are very high), the minimum width (W_{min}), and the minimum spacing, S_{min}. The impact was discussed in a previous section as was the maximum current handling capability.

5.2.3.1 Capacitors

As we just discussed, MIM capacitors are most important for matching in the circuit. The variation is very important to communicate to designers because it can impact the yield and performance. For matching applications (input, inter-stage, or output pre-matching), higher quality factor (Q) caps (like MIM) should be used. From a designer perspective, the capacitors should be compared based on the footprint they require in a layout, not just the value provided in the process specification. For evaluating technologies, it is also important to consider this layout impact and not just make assumptions about the Q. One example is two technologies where one has a 2 × capacitor density compared to the first, but also has 2 × the metal sheet resistance for one of the metal plates of the MIM. Designers automatically assume that the Q will be worse for the second process (based on the metal resistance). However, comparing caps of the same value, we find that the cap is ½ the length so that the Q is unaffected. The variation can differ by type, film thickness, and layout. Voltage dependence of the capacitor is not currently a great concern for PAs but could be in the future with carrier aggregation. In particular, the concern would be for linearity at high powers. For power supply bypassing, the Q is not as important, so the highest density cap that is available can be used. For example, a poly-poly cap (typically lower Q) in silicon can be used in this application. Some other considerations for comparing capacitors is when they are physically allowed to be placed in the circuit and how few masks are required to modify them in the process. The placement can be very important because they take up a large amount of the circuit area [38]. Can caps be placed under bond pads [39], over BSV, or under copper pillars? And also, how many mask layers need to be changed to modify the capacitor? The lesser number allows metal mask revisions that can really speed development time. These are important, but not typically talked about, features of capacitors.

5.2.3.2 Resistors

Resistors can be semiconductor or thin-film (as previously discussed). Sheet resistance is the primary information provided about a resistor, but there are several other important parameters that designers should know to make good choices for particular applications. The process variation for a resistor is a combination of the sheet resistance variation and the geometrical feature size variations that occur in all processes. As a result, it is important to know the total resistor variation as a function of the resistor geometry. Many otherwise good circuits have suffered massive yield hits from not considering this variation. The temperature coefficient of resistance (TCR) usually specified in parts per million per degree (ppm/°C) is also an important parameter. Most semiconductor resistors have a positive TCR while thin-film resistors can have a positive or negative TCR depending on the processing details. If resistors are available with opposite TCR's, a composite resistor can be made with very low variation over temperature. The maximum current rating (usually in mA/μm) is important — as mentioned earlier — for die size and reliability. For semiconductor resistors, there may be other layers under the resistor that cause a roll-off of the resistance with frequency. This means that the bandwidth of the resistor should be known and compared to the application frequency. The maximum voltage is another value that needs to be known for robustness.

For PA design, there are two classes of resistor necessary for building PAs: ballasting and bias resistors. For a ballast resistor, the important features are: small layout area, high current handling capability (especially if it is an emitter ballast resistor), positive TCR (so that there is more ballasting as the transistor gets hotter). If the resistor has an RF roll-off, this roll-off should be greater than the third harmonic of the desired signal. Besides the small layout area, it is also important to have a small "keep-out" region to other resistors or active devices so that a small array can be laid out. For emitter ballasting, low value resistors, in the 2–10 Ω range, are used. The limiting feature for these devices is usually the contact-to-contact spacing on the resistor (this limits how short the resistor can be). For semiconductor resistors, a second limitation on the length can be velocity saturation which begins to make the resistor nonlinear (so a minimum length restriction can be placed on the resistor to ensure the electric field is always in the linear region of the mobility). These resistors, because they need to support the emitter current, require high current handling capability. They have the same desired RF bandwidth consideration as base ballasting (>third harmonic). The desired layout width is limited, in practice by the desire to have the width of the resistor be about the same width as the cell it is ballasting for a compact array layout. A positive TCR and tight packing density are also desirable.

Finally, for other resistors used in biasing networks or other areas of the circuit (like detector circuits), we want small area resistors (this usually means higher sheet resistance) because these resistors can have quite large values (some on the order of 10s of kΩ). High current handling capability is desired and these resistors should have low TCR so that they do not contribute to bias point drift over temperature. The resistors can benefit from small keep-out regions and small resistor-to-resistor spacing because they are often serpentined. Since it is desirable to be able to "tune" resistors with metal mask changes, one should also consider how many mask layers need to be changed to modify the resistor value.

5.2.3.3 Inductor-like devices

Inductors, couplers, baluns, and transmission lines are fabricated from the metallization layers. Variation in the metal resistance, geometry, and inter-level dielectric (ILD) thickness can be important. The number of metal layers offered, the metal thicknesses (sheet resistance), and the thickness of the ILDs (affecting interlayer capacitance) determine what style inductors will be useful in a given process. Stacked (just stacking metal layers) and solenoid (winding one inductor on metal one, then winding second turn on second metal) [40] are most popular for GaAs. Other inductors are also usually possible, but not usually required for PA application (symmetric inductors, variable trace inductors, etc.). When looking to compare inductors across processes, a good way to evaluate the process capability is to plot the inductor Q as a function of the inductance for various layouts [41] at different fundamental frequencies. This allows for a direct comparison of the capability of the process, not just a few inductors. It should also be pointed out that not every inductor on a PA needs to be low-loss. Bias inductors, which feed the DC current to an array, do not require high Q. For that application, more emphasis is put on inductance per layout area to keep the chip size small. For PA modules, many of the inductors are actually fabricated in the laminate because very thick, low-loss metals are available.

5.2.3.4 Backside via (BSV) and metallization

Through-wafer vias (also called TSV, BSV, or substrate vias) are important for achieving low-inductance connections to ground [42]. The features associated with the BSV are important for die size. BSV size, keep-out region (spacing other elements must be away from it), pitch (spacing from via-to-via), and spacing to die edge are all important to consider when looking at a process. The geometry, combined with the wafer thickness, determines inductance. The major advantage of the BSV, compared to bonding is that the ground inductance is small and repeatable. It also does not require a pad that gets physically bonded to — saving die size. Typical III-V processes routinely offer BSV for 3 or 4 mil wafer thicknesses.

On silicon, the typical TSV is on a wafer thickness of 6 mils. Bond pads actually count as a device, but are usually not important until laying out the circuit. The size of the bond pad is limited by probe capability and wire-bonding capability, not by fabrication capability. The bond pad footprint is the key figure of merit and also needs to be considered for "ganged" or multibond pads. Since the bond pads will have an impact on the overall die size, there are many important features besides the pad size: pitch (bond pad to bond pad spacing), spacing to active circuitry, spacing to die edge, and can circuitry be placed under the pads? It is very typical in silicon processes to place elements like ESD circuits under pads. For GaAs processes that are non-planar, this is challenging.

We briefly mentioned the metallization while discussing inductors. Contrary to many MMICs (that use microstrip line), handset PAs tend to be dense circuits, so things such as minimum width/pitch, number of layers, metal thickness, inter-level dielectrics thickness are important. Ideally, one would like a small minimum width/pitch, the option to have multiple metal layers (if it shrinks the die, extra layers can be worth the cost addition), at least two thick metals and a thick, low dielectric constant dielectric between them. One interesting observation is that metallization in silicon processes use either aluminum-based metallization schemes or copper-based metallization. Most III-V (GaAs in particular) use gold as the interconnect metal. Gold does not typically show electro migration issues and is not typically de-rated versus temperature as metallization in silicon processes is. Figure 5.13A and B shows the rated current carrying capability as a function of metal thickness for Au, Al, and Cu wiring. The most surprising feature of this figure is that at 125°C, Cu is barely better than Al and is not nearly as good as gold. This makes copper a poor choice for metallization in power processes when the metal must be close to the heat source (ie, the device). This means that while Cu is attractive for cost reduction reasons, it may require larger die because of this inferior current handling. Having considered the passive elements available in a technology, we will now discuss active devices.

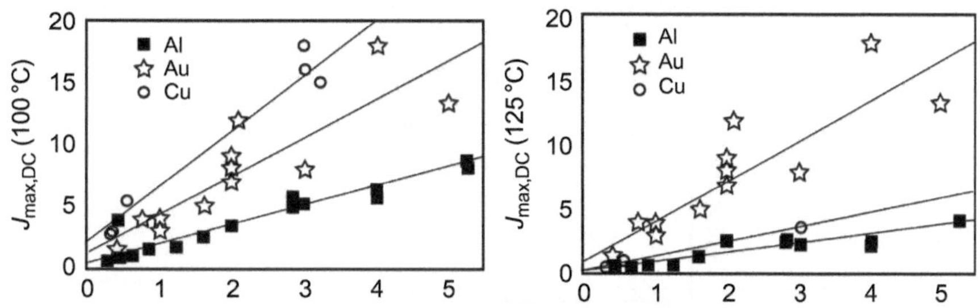

FIGURE 5.13

Current carrying capability of different metallization as a function of metal thickness at (A) 100°C and (B) 125°C. This capability can limit the ability to make small PA layouts.

5.2.4 DEVICE CHARACTERIZATION FOR PROCESS SELECTION

APT and ET PAs require high efficiency across a range of supply voltages. The performance of these amplifiers is usually characterized by a "waterfall curve" [43–45] — an example is shown in Figure 5.14. While circuit design choices can certainly be a factor in this roll-off of PAE with decreasing supply voltage, it is also necessary to evaluate device technologies to understand the trade-offs and opportunities they present for these applications. For technology development, it is advantageous to see if "simpler" measurements, such as RF knee (which is discussed in detail later) correlate to these waterfall measurements because they are easier to collect. The concept is to use these measurements to down select what devices can be further investigated with either load-pull structures or full PAs. We demonstrate the correlation of simpler measurements to the waterfall measurements and also compare the performance of different HBT material designs and other technologies.

The introduction of systems using APT and ET has created some new opportunities and requirements for PA technologies. One of the main requirements is less variation in RF gain with supply voltage and high PAE at low supply voltage. While load-pull is ultimately needed to evaluate these particular performance metrics (as discussed in Chapter 2 [46–48]), device development engineers need an efficient way to quickly screen new device designs or technologies before performing load-pull at either an array level or a circuit level. Because we are interested in understanding the device or process parameters that affect these curves, small-signal data is also useful in that it allows direct evaluation of device parameters such as parasitic resistances and capacitances, and this is discussed later for various technologies.

Because of these new system implementations, some of the questions that technologists get from design teams are "can you improve the low voltage PAE roll-off by improving the I_c-V_{ce} curve?," "do other technologies allow lower voltage operation?," "does re-matching dynamically provide better PAE?" To answer some of these questions, we investigated the correlations between the DC-IV curves, small-signal "RF knee" measurements, and waterfall curves measured by load-pull. First, we evaluated the impact of device design (layout) and epi-design on the I_c-V_{ce} output curve for various collector designs. We then did RF knee measurements on the

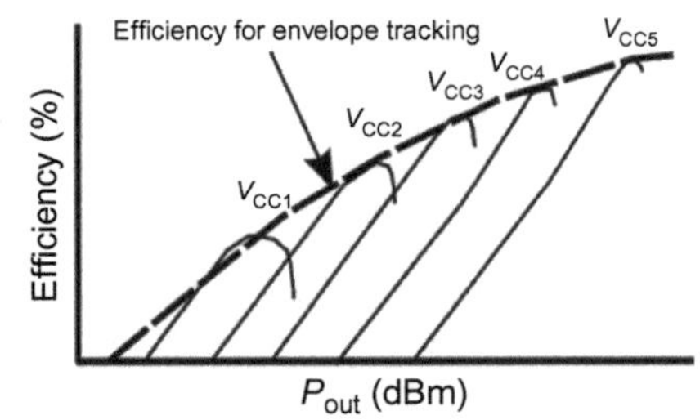

FIGURE 5.14

Example of efficiency versus P_{out} for ET applications. Ideally there would be no roll-off at low P_{out}.

different devices/materials. Down-selecting to a basic power cell, waterfall curves were measured. Waterfall curves are compared across different technologies (GaAs HBT, pHEMT, and SiGe) and this is discussed in the context of ET and APT PAs.

Many devices can readily be compared from simple DC and small-signal measurements. Table 5.4 shows typical measurements used for technology characterization and modeling. The first grouping is the "technology" characterization. The transmission line model is used to determine sheet resistances and contact resistances of the technology (important for device performance). Capacitance-voltage measurements are also performed on all junctions. This allows the doping profile to be monitored and the junction capacitance to be separated into area and perimeter terms (for modeling). Device specific characterization is performed (some more details for the power devices will be discussed shortly). All of this information should be made available to designers (and compared to modeled values) when using a process.

For DC measurements, forward and reverse, Gummel plots (Figure 5.15 shows example of forward Gummel), collector current versus collector-emitter voltage (I_c-V_{ce}) [forced base current, I_b, Figure 5.16, and forced base-emitter (BE) voltage, V_{be}], as well as the individual BE and BC diode curves are useful for understanding

Table 5.4 Example Measurements Used for Technology Characterization and Modeling

Feature/Device	Measurement
Technology	Emitter TLM
	Base TLM
	Collector TLM
	Channel TLM
HBTs	Gummel Plots (forward/reverse)
	I_c-V_{ce} (I_b and V_{be})[a]
	f_T versus J_c versus V_{ce}
	s-Parameters
	Load-pull
FETs	I_d versus V_{gs}
	Output curve (I_d-V_{ds})[a]
	f_T versus I_d versus V_{ds}
	s-Parameters
	Load-pull
Diodes	Forward IV
	Reverse IV
	Pulsed IV
	s-Parameters
Semiconductor Resistors	DC
	s-Parameters

[a]*Other special characterization, like pulsed IV or dispersion may also be needed.*

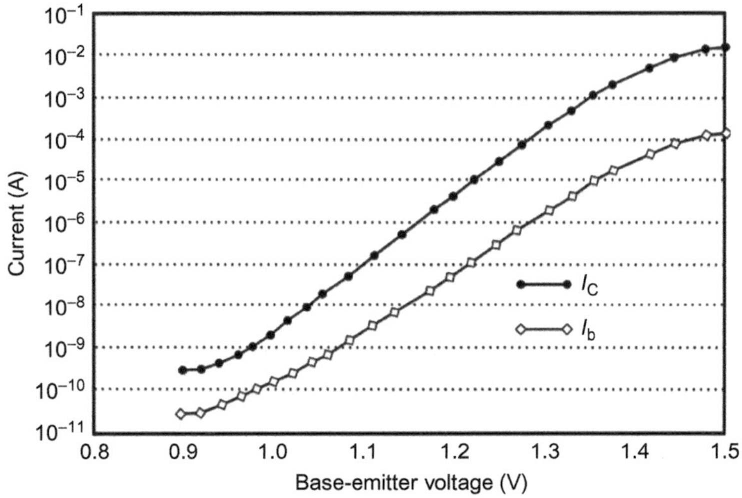

FIGURE 5.15

Example forward Gummel plot for HBT transistor. Many important device characteristics can be extracted from this data.

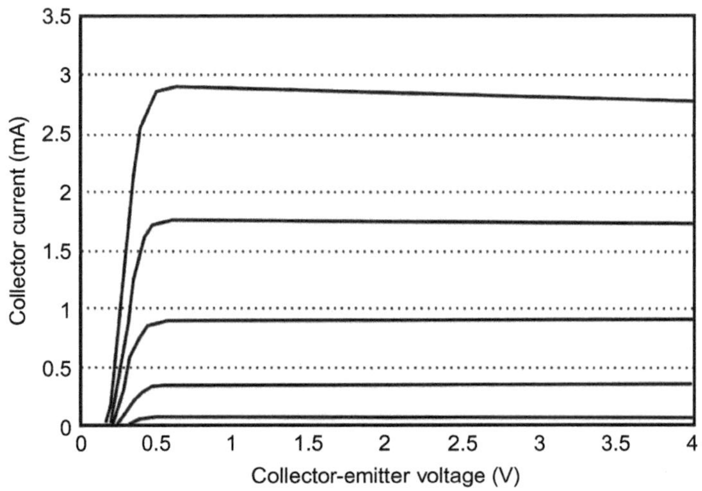

FIGURE 5.16

Example I_c-V_{ce} curve with fixed base current for HBT. For an ideal transistor, the current would go straight up from 0 V.

the basic device characteristics. From the Gummel plot (at different V_{ce}s), the basic turn-on characteristics, as well as emitter resistance, R_e, and thermal resistance, R_{th}, can be determined.

This approach is applicable to technologies that do not have other effects that change the DC or large-signal characteristics, such as low-frequency dispersion (due to traps [49]). Fortunately, most bipolars and LDMOS do not display significant dispersion. In addition, some pHEMT processes or GaN have significant trapping effects that require the use of pulsed IV just to obtain the effective DC-IV curve, besides the aforementioned impact on the large-signal performance. There are also more advanced load-pull systems [50] that can be used to perform these measurements, but are not always available in an industrial setting.

The small-signal characterization should consist of: unity gain cut-off frequency, f_T (usually single frequency) versus I_c (see Figure 5.17D) and V_{ce}, S-parameters at several currents versus V_{ce} (we normally transform these into RF knee [51] measurements), as well as S-parameters versus frequency at several bias points

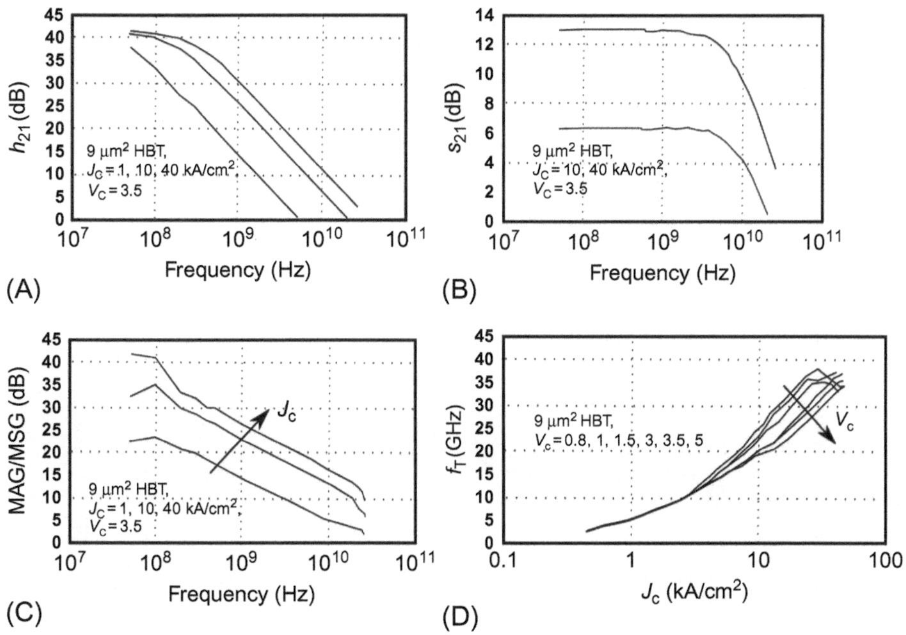

FIGURE 5.17

(A) h_{21} (RF current gain) versus frequency at three current densities, (B) S_{21} versus frequency at two bias currents, (C) MAG/MSG versus frequency at three bias currents, and (D) f_T versus J_c at 5 bias voltages. Example set of small-signal measurements. For power amplifier designers, (B) and (C) are the most important since they represent the bandwidth and power gain for the device.

(see Figure 5.17B) from which RF gain parameters such as h_{21} (see Figure 5.17A), maximum available gain (MAG)/maximum stable gain (MSG) (see Figure 5.17C) and Mason's unilateral gain, U.

Besides the RF gain just shown, a more useful form is to plot this over the "load-plane" as a function of I_c-V_{ce}. We call this the "RF knee" and this data provides a great deal of information about the transistors because this data can be converted to the input and output impedances — Z_{11} and Z_{22} (to see how much they are moving with bias) to basic device parameters such as BC capacitance, C_{bc} (using direct extraction techniques). While directly extracted parameters are not exactly the same as the model parameters, they provide qualitative insight into the device operation and identifying what device parameters in the technology can be improved. RF Knee data is found from the small-signal gain metric, MAG/MSG (MAG, $k > 1$, and MSG plotted when $k < 1$), measured with V_{ce} (V_{dd}) from 0.1 to 5 V and several target output current density from 1, 10, 20, 40 kA/cm^2 using a network analyzer. Plotting MAG/MSG versus V_{ce} allows the observation of RF knee — where the gain rolls off at low V_{ce} (V_{dd}). The RF knee is measured at multiple frequencies from 0.4 to 5.4 GHz to highlight which technologies have the least amount of gain and/or phase change over bias (as well as input/output impedances). Eqs. (5.1) and (5.2) show the link between the Z parameters and basic device parameters. Some examples are shown in Figures 5.18–5.20 for a high breakdown SiGe transistor, a high performance SiGe transistor, and a GaAs HBT. Figures 5.21–5.23 show Z_{11}. The measured Z_{11} corresponds to Eq. (5.1). This equation allows us to highlight some differences between the different technologies and ballast schemes. In Eq. (5.1), we can treat R_b as the intrinsic device R_b plus and extrinsic R_b added as ballast. The same

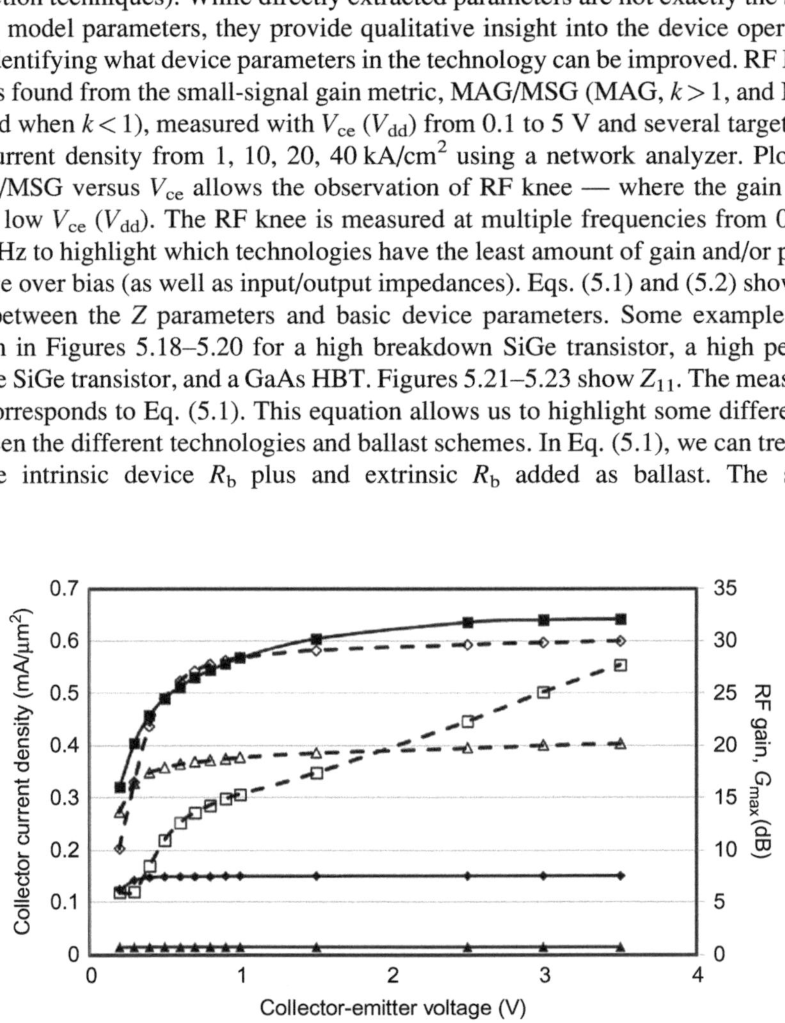

FIGURE 5.18

High breakdown SiGe. Note that the IV curves *(filled markers and solid lines)* and the RF *(open markers and dashed lines)* gain roughly track at low current, but at high current *(open boxes)*, the RF gain is degraded at low collector-emitter voltage due to Kirk effect limitations.

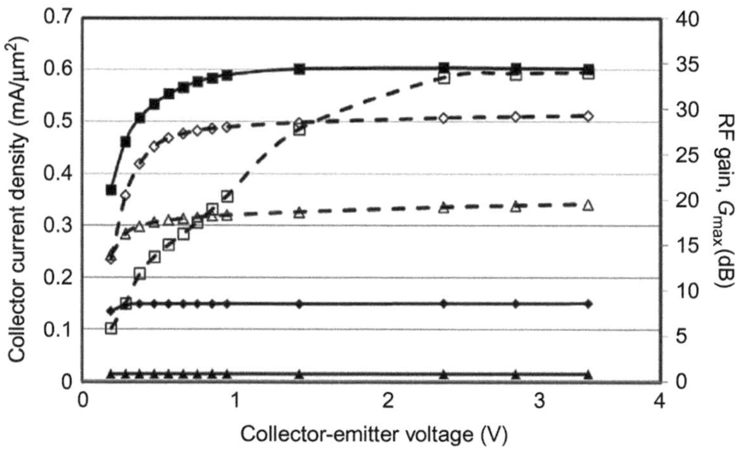

FIGURE 5.19

High performance SiGe. Note that the IV curves *(filled markers and solid lines)* and the RF *(open markers and dashed lines)* gain roughly track at low current, but at high current *(open boxes)*, the RF gain is degraded at low collector-emitter voltage due to Kirk effect limitations, but is better than the high breakdown device.

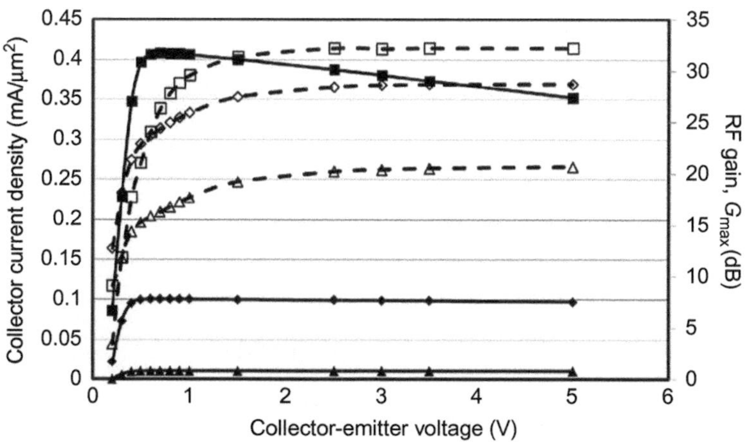

FIGURE 5.20

GaAs HBT. The IV curve is much sharper than the RF gain and the RF gain begins to roll-off sooner. The downward slope of the IV curve at high current is due to self-heating. There is no degradation at high current as was observed in the SiGe devices.

consideration can be made for R_e. The term $r_e||C_{be}$ is dominated by r_e, V_T/I_c, so, as a result, is very dependent on bias. This is NOT good if the transistor is the second stage of an amplifier (it effectively changes the load the first stage sees). A typical "trick" is to add some R_b in the RF path (provided the transistor has been designed to give adequate gain) so that this input impedance is dominated by the R_b, meaning it does not

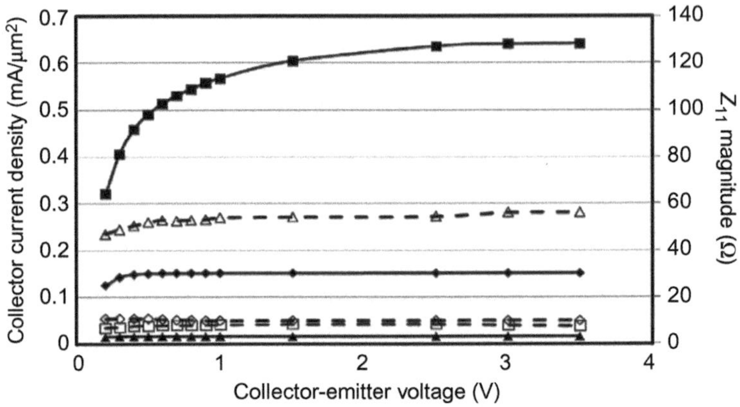

FIGURE 5.21

High breakdown SiGe — Z_{11} and collector current curves. The Z_{11} is strongly dependent on bias (IV).

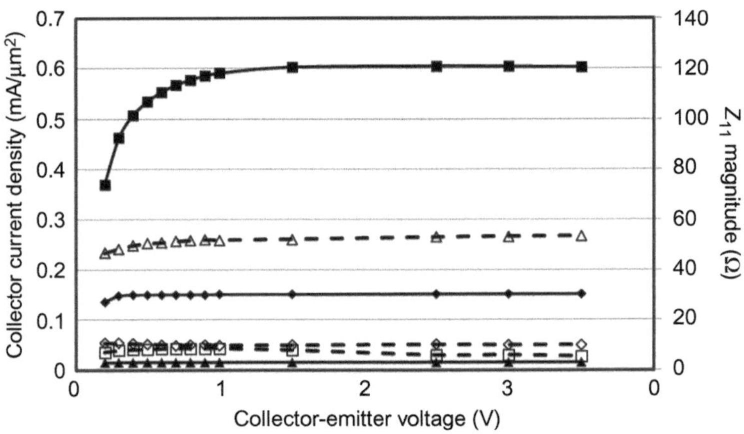

FIGURE 5.22

High performance SiGe — Z_{11} and collector current curves. The Z_{11} is strongly dependent on bias (IV).

move around with bias! The transistor design for bipolars allows many of the parameters in this equation to be traded off [52,53]. In particular, the usual device design trade-off is R_b versus C_{bc} (by changing the number of fingers or finger aspect ratio).

$$Z_{11} = R_b + R_{bx} + R_e + j\omega(L_b + L_e) + r_e||C_{be} - \frac{R_b(R_b + \beta)}{R_b + \frac{1}{j\omega C_{bcx}} + r_e||C_{be}} \tag{5.1}$$

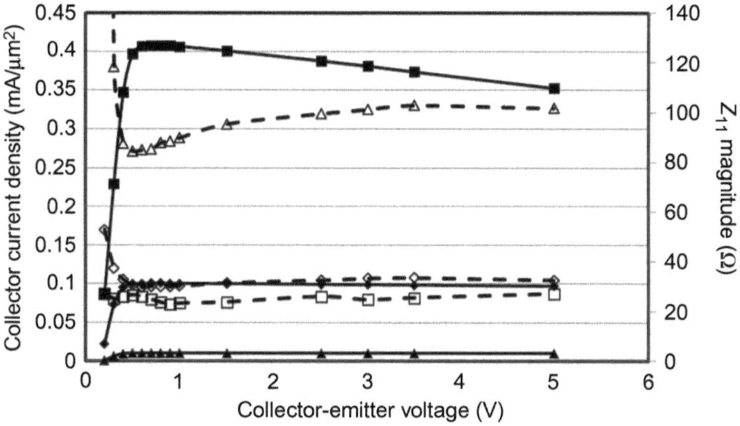

FIGURE 5.23

GaAs HBT — Z_{11} and collector current curves. The Z_{11} is strongly dependent on bias (IV). The impedance values are slightly higher than for the SiGe HBT devices.

β contains transit time information (not DC gain) and $r_c || C_{bc}$. R_b, R_{bx}, and R_e are the intrinsic base resistance, extrinsic base resistance, and emitter resistances. L_b and L_e are the parasitic inductances on the base and emitter, respectively. The last two terms are bias and frequency dependent. The real part dominates Z_{11}. As we can see, the HBT device parameters can be modified to change the input impedance (so layout can be important). Comparing the technologies, the input impedance being lower makes it harder to drive the stage so higher input impedance is preferred. Figures 5.24–5.26 show Z_{22}. Figure 5.2 represents the parameters that affect Z_{22}.

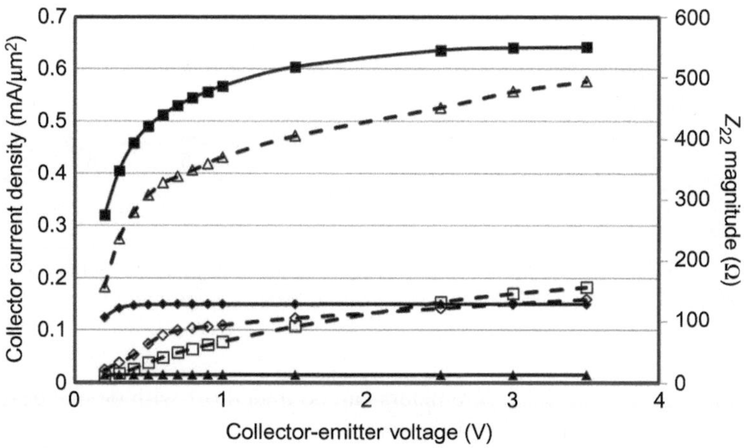

FIGURE 5.24

High breakdown SiGe — Z_{22} and collector current curves. The Z_{22} is strongly dependent on bias (I_c-V_{ce}).

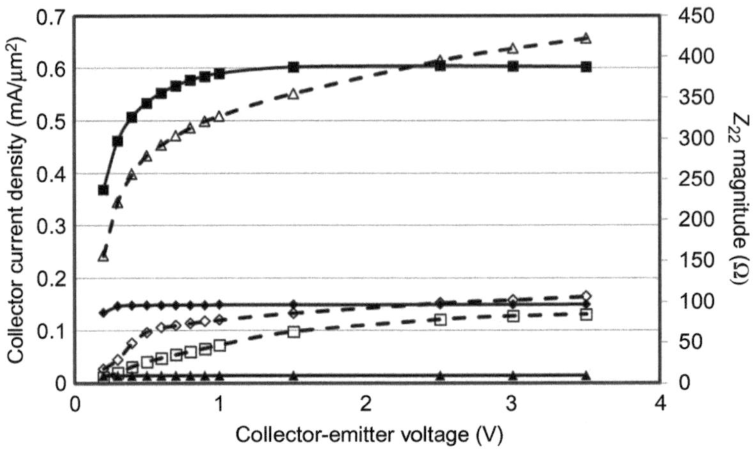

FIGURE 5.25

High performance SiGe — Z_{22} and collector current curves. The Z_{22} is strongly dependent on bias (IV).

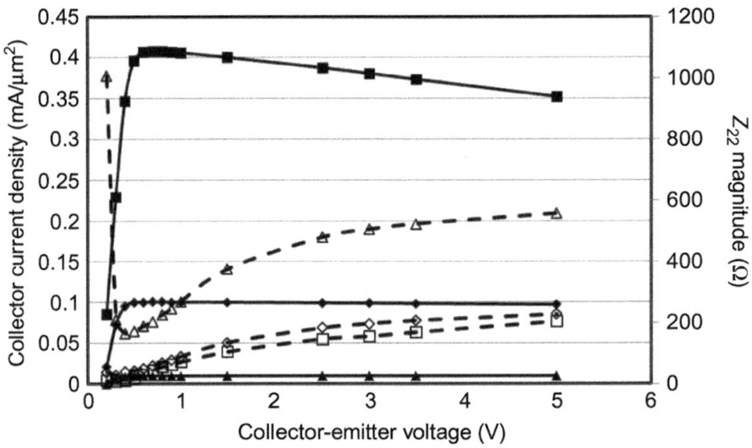

FIGURE 5.26

GaAs HBT — Z_{22} and collector current curves. The Z_{22} is strongly dependent on bias (IV). At higher voltage, the variation is similar to SiGe but at lower voltage there is more change.

Similar to Z_{11}, many of these parameters can be traded off with device design. The C_{bc} tends to dominate this part of the match.

$$Z_{22} = R_c + R_e + j\omega(L_c + L_e) + r_e||C_{be} + r_c||C_{bc} - \beta - \frac{r_{bc}||C_{bc}(\beta - r_{bc}|C_{bc})}{R_b + \dfrac{1}{j\omega C_{bcx}} + r_e||C_{be}} \tag{5.2}$$

The parameter definitions for 5.2 are the same as for Eq. (5.1). R_c is the total collector resistance and r_{bc} is the BC dynamic resistance. We will see later that FETs have parameters that vary far less over the load-plane so they should be simpler to match. However, they also have less flexibility to trade-off device geometry (or topology) and ballasting (because FETs typically do not require ballasting).

For FETs, drain current, I_d, versus gate-source voltage, V_{gs}, at different drain-source voltages, V_{ds} (see Figure 5.27), output curves (I_d vs. V_{gs} and V_{ds} — Figure 5.28) provide insight into how the device turns on, etc. The RF measurements would be similar to the HBT, only adjusted in the I_d versus V_{ds} plane. Some examples are shown in Figures 5.29 and 5.30. Figure 5.31 shows the RF gain measurement for a pHEMT device. Figure 5.32 — Z_{11} and this represent Eq. (5.3). Compared to bipolar devices, the FET input impedance does not depend on a many parameters. The parameters it does depend on, C_{gs} and C_{gd}, do not vary too much over the load-plane. Combined with the relatively higher input impedance (depending on width) it makes FETs easier to match and, in a two stage line-up, means that the first and second stages are generally well matched over power.

$$Z'_{11} = \frac{1}{\dfrac{sC_{gs}}{1 + sR_iC_{gs}} + (1 - A_v)sC_{gd}} \tag{5.3}$$

In Figure 5.33, Z_{22} is described by Eq. (5.4). Again, we see a relatively simple equation for FETs (compared to bipolar) and the variation is less than for the bipolars.

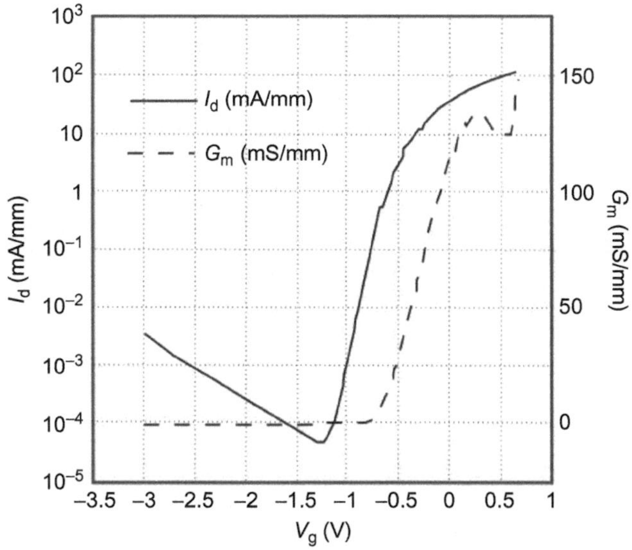

FIGURE 5.27

I_d versus V_{gs} curve and g_m versus V_{gs} curve for d-mode (turn-on less than 0 V) pHEMT device.

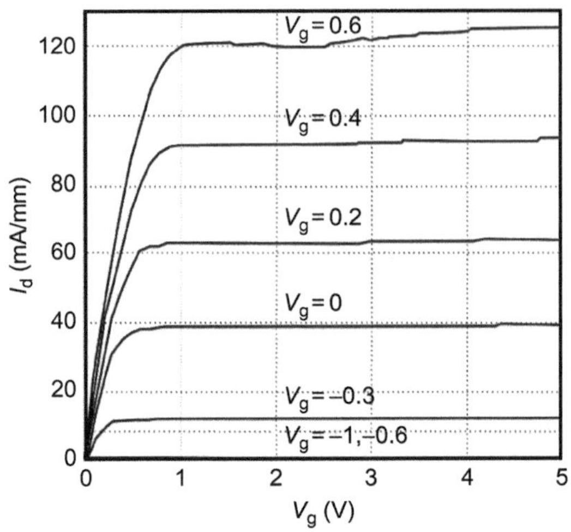

FIGURE 5.28

DC I_d versus V_{ds} at different gate voltages. For biasing, the DC curve is appropriate, for power or switching applications, the dispersion must be taken into account so pulsed IV or some other method will represent this under the "use" conditions.

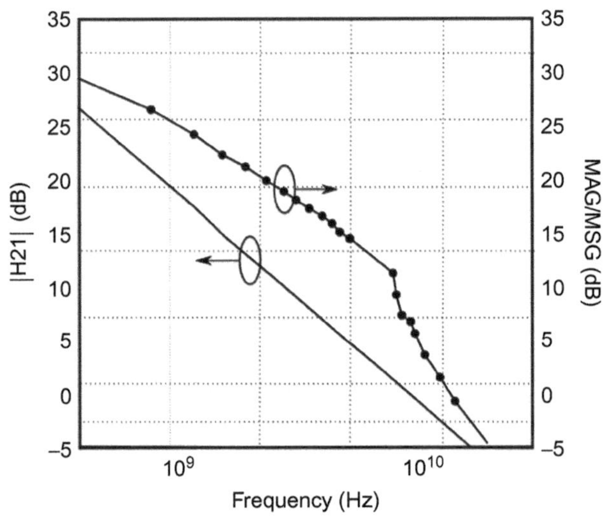

FIGURE 5.29

RF gain curve showing H_{21} and MAG/MSG versus frequency for pHEMT.

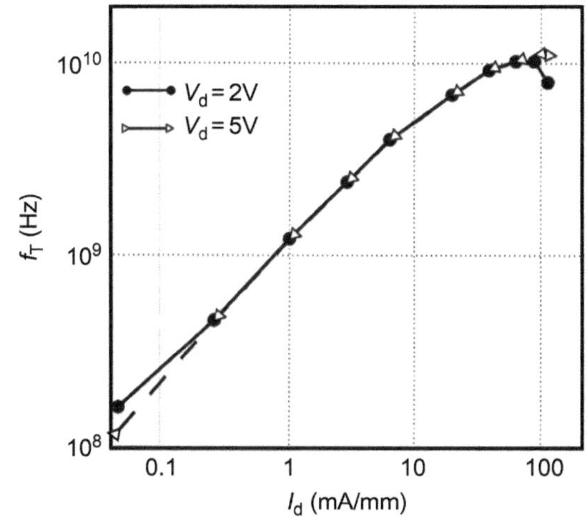

FIGURE 5.30

f_T versus I_c for pHEMT at two different V_{ds}.

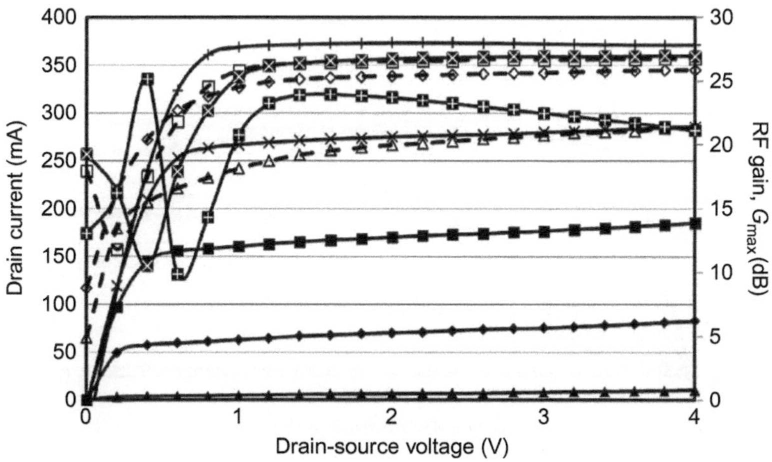

FIGURE 5.31

DC IV *(dark markers)* and RF gain curve *(white markers)* for (e-mode) pHEMT. Note that compared to bipolar transistors, there is not much change in the RF gain with voltage.

From a matching point of view, this means that the match is more "optimal" over the load-line.

$$Z'_{22} = \frac{1}{\left\{\left(1/R_{ds} + sC_{ds} + s\frac{C_{gs}C_{gd}}{C_{gs}+C_{gd}}\right) + g_m\frac{C_{gd}}{C_{gs}+C_{gd}}\right\}} \tag{5.4}$$

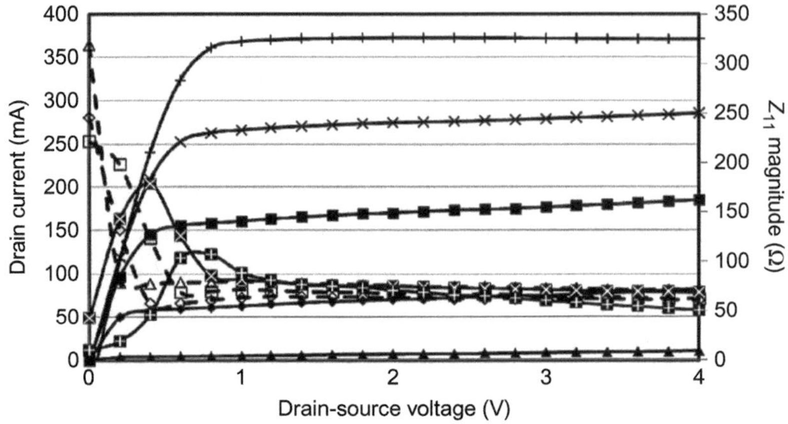

FIGURE 5.32

DC IV *(dark markers)* and Z_{11} *(white markers)* for e-mode pHEMT. Z_{11} is almost independent of the output current and does not show much change until low drain-source voltage. This makes matching pHEMT over bias simpler than for bipolars.

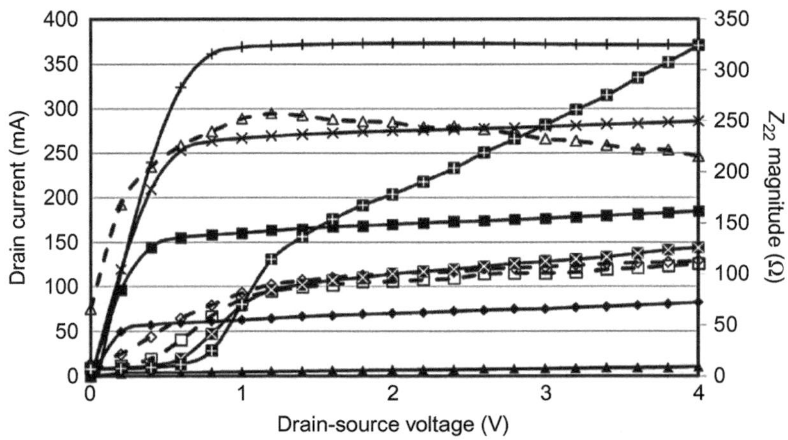

FIGURE 5.33

DC IV *(dark markers)* and Z_{22} *(white markers)*. The output impedance, Z_{22} shows much less variation than bipolars. The most variation occurs at highest current *(black boxes with white cross)*.

Combined with the evaluation of RF gain in the load-plane, pHEMT devices (from a purely technical point of view) should give superior results to bipolars in amplifiers.

The key observations from that RF knee (small-signal measurement) curves are that SiGe devices showed gain collapse at high currents and low V_{ce} (because of Kirk effect limitations). This is a surprising result because many people assume that SiGe can perform better at higher current (mostly based on observations for other types of circuits). The collapse is a result of the high bias current being on the falling edge of the f_T versus J_c curve. As the voltage is increased, the current density for collapse is increased. For all the HBTs, we note that the DC-IV curve has a lower "knee" than the RF gain. FETs showed gain down to lower voltages and had less variation in input/output impedance than bipolars. FETs also had better correlation of the RF knee and DC-IV curves than bipolars. This may provide some insight into the performance of FETs relative to HBTs in PA applications. These small-signal measurements will be compared to the "waterfall" measurements performed in the next section.

Finally, single device large-signal "waterfall curve" measurements on select unit cell geometries were performed at 0.9 GHz to examine large-signal performance trends with geometrical device factors. Unit cells are used because there is usually a design that can be traced back through historical technologies (at a given company), they can be measured in co-planar structures (removing effects from wafer thickness if BSV is used), and they remove the complication of bias circuit interactions. Some of these are, ultimately, evaluated at the circuit level. Here, waterfall measurements with tuning for max PAE are used. The transistors were biased with either I_b to achieve 0.1 mA/μm^2 collector current density or V_b (including a stabilization/ballast resistor) to achieve 0.011 mA/μm^2. FETs (pHEMTs) were biased to 13.3 mA/mm so that they roughly matched the biasing of the HBT in a PA application. The devices were then tuned for maximum PAE at $V_{ce} = 3.4$ V. The voltage was then swept. The measurements do not tune the harmonics so the theoretical maximum is 80% PAE [54]. The measurement consists of tuning the devices for best PAE or max P_{out} at a given bias current and 3.4 V. The bias current selected was one often used in amplifier designs. The P_{in} is then swept from a low value to high value and drives the device into compression. The highest PAE and the power level at which it occurs are recorded. This is repeated at successively lower V_{ce}s. It is called a waterfall curve because the entire plot of PAE versus P_{out} resembles a waterfall. A series of PAE versus P_{out} curves are shown for V_{ce} from 0.5 to 3.4 V for two different bias conditions: (1) fixed base current to achieve $J_{cq} = 10$ kA/cm^2 and (2) fixed base voltage using a resistor in series with the base to achieve $J_{cq} = 1.1$ kA/cm^2. The waterfall measurements are compared and contrasted with the small-signal RF knee to demonstrate that there is qualitative correlation and that there are significant device level differences for PAE versus supply. While this is not "exactly" how the device will get used in an ET application, it provides good guidance for technology developers.

Here, we measured waterfall curves on the same SiGe technology and similar pHEMT technologies. The results are shown in Figure 5.34. The roll-off in PAE for SiGe is similar to that observed in GaAs, even though the device DC turn-on (V_{be}) is much lower and has lower offset voltage. However, the peak for the SiGe occurs slightly below the nominal voltage (3.4 V). For pHEMT, we note that the PAE versus voltage remains high down to very low voltages (in agreement with [55] indicating that pHEMT may be a superior choice for ET and APT applications). We also know that for FETs the peak PAE is dependent on the device size, indicating that the knee (DC and RF) do impact the PAE for the FETs. These data compare quite

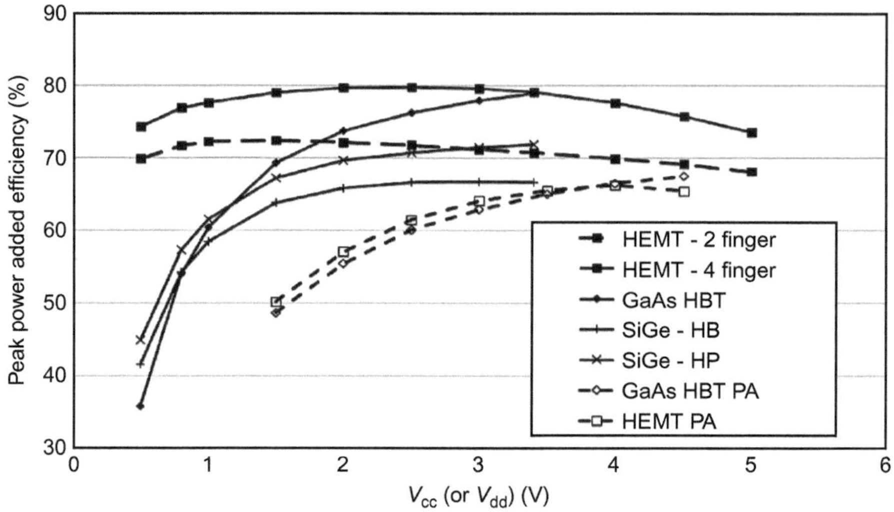

FIGURE 5.34

Waterfall curve for various technologies. We observe that bipolars (InGaP is GaAs HBT) have similar roll-offs, independent of the IV curves we saw before. FETs are much flatter with supply voltage. Research power amplifiers show similar roll-offs which indicates that design may still dictate the overall circuit performance.

well with the observations made from looking at the earlier RF knee plots (gain in the load-plane).

To further investigate and quantify the differences between these technologies, we fit the waterfall data to:

$$\mathrm{PAE}(V_{cc}) = \mathit{Eff}_{Ideal}(1 - V_{\mathrm{knee}}/V_{cc}) \tag{5.5}$$

Where *Eff_Ideal* is the peak/ideal efficiency and V_{knee} is a parameter that describes the PAE roll-off. PAE (V_{cc}) is the PAE at a given V_{cc}. The resulting parameters for GaAs HBT, SiGe, and pHEMT are shown in Table 5.5. The *Eff_Ideal* values are above the earlier mentioned limit of 80% (for class F) and are slightly higher than the observed values because it is a fitting parameter. The trends and clustering of

Table 5.5 Ideal Efficiency (Eff_Ideal) and Knee From Fitting PAE Waterfall for I_b and V_b Bias Conditions

Technology	Eff_Ideal, I_b	Knee, I_b (V)	Eff_Ideal, V_b (V)	Knee, V_b (V)
GaAs HBT	85.84	0.286	86.48	0.291
SiGe HB	81.96	0.315	72.49	0.205
SiGe HP	84.43	0.299	76.60	0.206
FET, 2 Finger	NA	NA	72.17	0.01
FET, 4 Finger	NA	NA	80.53	0.038

values follow the actual curves. For current biased devices, we note that the SiGe devices are competitive with the GaAs devices in terms of the maximum PAE. The extracted knee voltages for all of the bipolars are very similar with the high breakdown SiGe being slightly higher the GaAs HBT. This is correlated to the RF gain roll-off (Kirk limit) observed in the RF knee measurements. While the *Eff_Ideal* for the pHEMT shows a size dependence, the knee for the PAE is very low, indicating some promise of pHEMT for ET if the device is sized appropriately. This agrees with the circuit level comparisons shown in [43].

In the context of multistage PA design, using load-pull for device characterization gives essentially the same information as the small-signal data but requires a lot more care in the measurement. For example, the matching of the harmonics is critical for ultimate performance. Also, the base-band impedance is important (ballasting and bias circuit) [56,57]. Unfortunately, many circuit designers attempt to compare technologies ignoring, not only these terminations, but the ground inductance too. For APT and ET systems, where V_{cc} is varied to improve performance over power and improve battery life, understanding of waterfall curves is important. We find that the waterfall curves track the small-signal fairly well, but provide a good illustration of some of the matching/biasing differences. Figures show comparisons of GaAs and SiGe HBT as well as different types of FETs. We have observed that for FETs the device size does play a big role in the peak performance. Waterfall curves show that GaAs HBT has better performance than SiGe HBT. FETs show a flatter PAE versus P_{out}, but the PAE for silicon FETs is not nearly as good as III-V FETs. This can be adjusted, somewhat, by device size, but as the devices get larger, they become harder to match which will present a trade-off.

5.3 FIGURE OF MERIT, YIELD, AND COST

Linear PA RF figure of merit — the summation of PAE and ACPR1 — eg, PAE = 45% and ACPR1 = 45 dBc then FOM = 90. For product design, the natural question is "What is the upper bound of FOM for a WCDMA handset PA?" An upper bound can be estimated from the curves in [58]. In that work, the theoretical PAE is plotted for two different matches — class B and class F. Considering that the power must be backed off to account for the peak-to-average ratio (PAR), the highest theoretical efficiency for a class F match is 63% compared to 55% for class B. If we assume that the linearity requirement will be just barely met (32–35 dBc), then adding gives an ideal upper bound of 95–98. Of course, in reality there will be input matching loss as well as output matching network loss. For each 0.1 dB of output match loss, there is a 1% reduction in the PAE.

There are two types of yield to consider for PAs. The first is normally known as good die (KGD) probe which is on-wafer probing of bare amplifier die. This can be used to screen for defects as well as for quiescent currents. These tests consist of only DC parameters. More sophisticated foundries actually correlate these measurements back to PCM data to better understand how device parameters impact the circuit probe yield. The second is "final test" of the fully packaged PA(s) assembled into modules. Final test yield test not only the die(s), but also the assembly and packaging (laminate, components, and bond wire) contributions to part performance. In particular, laminate variation and surface mount device (SMD) component variation can have significant impact on yields.

Cost can be considered as the COGS or it can include the opportunity cost of winning the socket for high-volume production orders (so, having a great performing part and showing up late means you still lose). The ability to return on investment the team's efforts and profitability for winning the socket is measured in man months to return positive cash flow. Again, it is important to consider the total product cost when evaluating technologies, not just the raw wafer cost. Some of the features we described that influence die per wafer (including yield) must be factored in. A good example is that if you have a "cheaper" technology that barely passes final test (so yield is lower) but a slightly more expensive one that gives you better yield, then the slightly more expensive one may lower your overall cost. The yield consideration is especially important in highly integrated systems where one chip being bad means you must sacrifice many chips!

5.4 CIRCUIT LEVEL DESIGN

For PAs, there are several different "phases" of design. As we have been discussing up until now, defining what the amplifier needs to do (and what specifications it will have meet) is discussed at length with the customer before the actual design work begins. The simulation phase normally takes place without knowing the exact circuit layout (die or module). At this point, the three most important things for a designer to simulate are (1) the bias circuit (to go into detail on the bias circuits is well beyond the scope of this chapter, some good references are [59,60]), (2) the on-die RF chain (for handset PAs, this is usually handled using compact model representations of all the devices available in the technology) and (3) the output matching network which uses the module technology and SMD devices to perform the matching. Items (2) and (3) can be roughly simulated without the layout information. However, once layout is done, a good designer will repeat parts (2) and (3), including the real layout so that parasitics (coupling, ground inductance, etc.) can be accurately accounted for. How much time is spent on (2) and (3) also depends on how well thought out the initial layout (both die and module level) are.

5.4.1 GETTING STARTED AND FLOOR PLANNING

At the beginning of any product is the typical product pin-out discussion between customer and supplier. This takes place early in the platform definition phase for new consumer electronics such as mobile handset, laptops, and tablets. The goal from a system point of view is to specify one RF chain that meets many final applications and bill of materials. There will be a broad set of specifications that must be met to meet or exceed the end product type approval and consumer demand for performance such as talk time and connectivity. Carefully choosing the assembly scheme for product development hinges on the ability to accurately simulate and estimate performance and yield, respectively, in the allotted time frame. For example, when choosing a wire-bonded approach, speed and flexibility come before size and or yield. While a wire-bonded device can have very high yields, it often comes after many wire bond profile changes on the production floor. Getting samples to the customer first for evaluation trumps everything in this type of speed and flexible design approach. Often a design of this type can be shared between many different geographically located teams for reuse in different pin-outs and product definitions.

Let us consider the design of the output matching network for a PA. In this textbook starting point, the designer uses the nonlinear transistor model (the reader is referred to Chapter 1 for an extensive treatment of the subject [49,61–64]) to estimate the desired load impedance for achieving power and linearity goals. In doing this, the use of SMD capacitors and laminate inductors is a typical starting point. From the drain or collector of the active device the bias feed network and matching elements synthesis looks at harmonic rejection and bandwidth trade-offs, followed by a sensitivity study to yield a tight locus of impedance points which delivers the product specs. Care must be taken in this design step for the following factors: SMD ESR values and tolerance and the substrate chosen to create inductances. Low pass matching networks are common where the shunt SMD caps can serve both fundamental impedance matching and harmonic reduction at the output node. Systematizing the design of output matching networks has been accomplished through closed form equations to examine different matching topologies for insertion loss, yield, and tuning flexibility.

Laminate and die floor planning is discussed in [65]. One of the unfortunate things about the competitive PA landscape is that the designer does not always have control of the pin-outs for the die because that may have already decided by a competitor and later changed — the ability to quickly follow is key to winning in the end. For example, if you are a "secondary" supplier, you will have to make your module pin-for-pin compatible with the primary supplier. Important things for floorplan: bias circuits should be close to the array they provide bias for to avoid linearity degradation and "power settling" [66,67]. Output power pads/bond wires should be kept away from bias lines or from any path back to the input. For multimode, multiband parts, there is often more than a single PA on a given die. Ideally, these RF chains are kept at 90 degrees to one another (this simplifies wiring and also provides less coupling between the two RF chains) as shown in [13,30,68,69]. Keeping the arrays orthogonal is more challenging if FC packaging is used because of mold compound underfill requirements.

5.4.2 PACKAGING AND SYSTEM LEVEL IMPACTS

Some of the packaging requirements that are also driving wafer level process changes are "maximum stack height" considerations, package size (sizes are now down to 2.5 × 2.5 mm), etc. An example of a packaging stack-up is shown in Figure 5.35. Assembly considerations also become important for high-volume PAs. How the part will be assembled has ramifications for some of the technological requirements. There are two main assembly techniques currently used in high-volume PAs: wire-bond (WB) assembly (shown in Figure 5.30A) and FC assembly (shown in Figure 5.30B). Assembly drives: bond pad size (what size wire is used? — based on cost/performance trade-offs or WB vs. FC), guard-ring requirements around chip (chip-ring) will depend on how die are singulated — saw, scribe and break, or laser scribe. The singulation also drives the street (pad-to-pad spacing of adjacent devices on the mask) and impacts the cost through die per wafer. Following the lead of silicon, many multichip modules now employ some form of die-stacking. Wire bond: in concert with SMT component changes, can be quickly modified to meet ever-changing customer specifications. Die paddle/flag is usually slightly bigger than die so it allows more vias to be placed in the laminate, under the die. Typical modeling (PDK) is extracted with modeling structures in this configuration. Die can be thinned to 75 μm. With modern bonding equipment, the placement accuracy and bond wire length control are very good. FC: 40–65 μm thick copper pillars are used and allow for excellent thermal shunting across transistor arrays. Transistor orientations are limited by the flow of mold compound during assembly. This allows more direct thermal path from transistor to heat sink but, in practice, the thermal path is more confined as designers take advantage of being able to route beneath the die, as

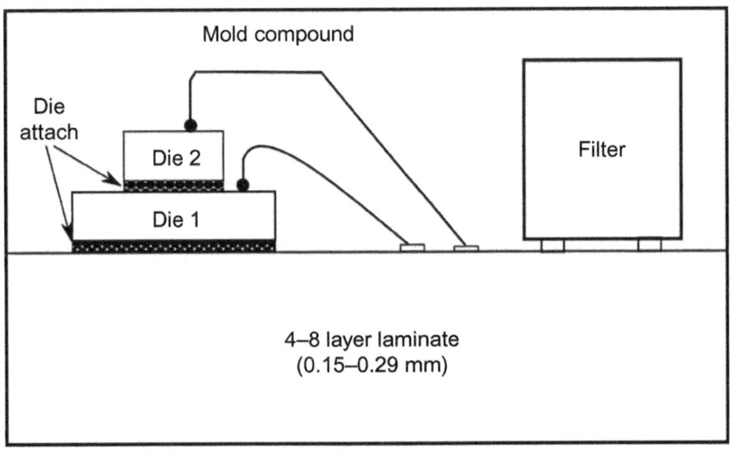

FIGURE 5.35

Example of stack height concerns. The total height is specified by the phone manufacturer and can dictate how modules are assembled. Shown in the figure are a normally attached die (Die 1), a stacked die (Die 2), and a flip-chipped filter.

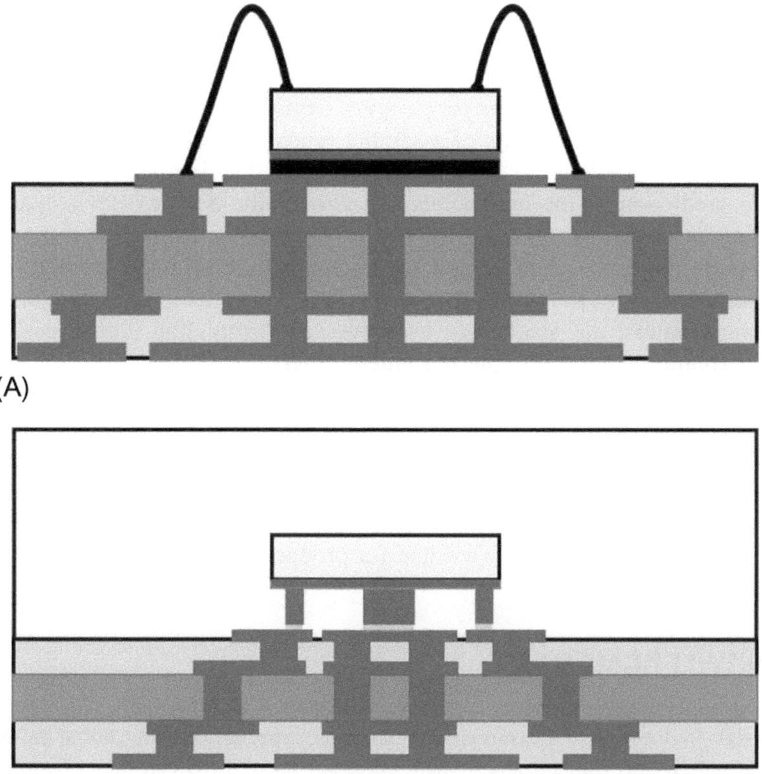

(A)

(B)

FIGURE 5.36

(A) Wire bond schematic versus (B) flip-chip. The die attach material in (A) is the thick black layer under the die. The first layer of the laminate (that extends out past the die) is the "ground paddle" or "die flag." There are strict rules on how far the bond wire must land from this area. In (B), a flip-chip die is shown. The flip-chip reduces the keep-out region, and the die flag is often shrunk to accommodate more routing.

shown in Figure 5.36. At higher frequencies, it can be easier to attain lower ground inductance. Generally, which technique is used for assembly is more a function of company supply-chain than electrical performance.

5.5 LARGE-SIGNAL MODELING AND VALIDATION AT THE CIRCUIT LEVEL

For validating modeling of the circuit, it is very important to progress from checking DC, then small-signal, then large-signal [70,71]. Because many PA performance parameters are related to the bias point, DC is critical. If DC is not correctly predicted, then it does not make sense to consider small- or large-signal. The most common reasons for poor DC fits are (1) non-scalability of transistor models, (2) thermal offsets between different parts of the circuit [33,67], (3) poor accounting for on-wafer critical dimensions for passives (resistors) or transistors, and (4) poor accounting for thermal impacts of assembly (epoxy) and laminate layouts. Small-signal, at the device level, is actually the most challenging to fit since it requires accurate prediction of the bias points and also of the detailed device parameters (including capacitances).

Our preferred approach to technology is actually looking at the circuit level large-signal results. One of the things that makes this so challenging to simulate is that everything matters: die level passives/actives and grounding, interconnect to the laminate (bump or wire), the die placement, SMD components, the laminate, and the evaluation board, all matter. In Ref. [72] the modeling requirements for each component were discussed in detail and the variation due to assembly/packaging was shown to be a very important consideration for showing agreement between measurement and simulation. As the FEMs become more complex, the ability to co-simulate all of the chips (PAs, filters, switches, controllers), as well as the package/interconnects, becomes critical for product success [73].

ACKNOWLEDGMENTS

We greatly acknowledge the input and conversations with all of our colleagues at Qorvo — from design engineering to measurement technicians, especially Kevin Choi of the Westlake Design Center, Denny Limanto and Richard Beresniewicz of the Greensboro Characterization Lab, and Marc Schulze Teneberge — now with Maury Microwave. We also are grateful for insightful discussions with Matt Ozalas (Keysight) and Bin Li (Teledyne Scientific).

REFERENCES

[1] R. Jos, Future developments and technology options in cellular phone power amplifiers: from power amplifiers to integrated RF front-end modules, in: 2000 Bipolar Circuits and Technology Conference Digest, Minneapolis, September, 2000, pp. 118–125.

[2] K. Nellis, P.J. Zampardi, A comparison of bipolar technologies for linear handset power amplifier applications, IEEE J. Solid State Circuits 39 (10) (2004) 1746–1754.

[3] C. Weitzel, RF power amplifiers for cellphones, in: GaAs Mantech Technical Digest, Scottsdale, Arizona, Paper 2.2, May, 2003.

[4] M.-C.F. Chang, P.M. Asbeck, R.L. Pierson, Planar HBT-FET Device, US Patent 5,250,826, October 3, 1993.

[5] M. Golio, The push towards low voltage transistors, IEEE Microw. Mag. 1 (1) (2000) 38–45.

[6] P. Bacon, C. Mohan, D. Fryklund, P.J. Zampardi, Semiconductor trends in wireless handsets, Microw. J. 48 (6) (2005) 22–33.

[7] M. Golio, Active devices for commercial subscriber power amplifier products, in: International Union of Radio Science, Lille, France, August, 1996.

[8] M. Bloom, Workshop WSB, in: IMS Symposium Short Course, Boston, 2000.

[9] P.J. Zampardi, M. Sun, L. Rushing, K. Nellis, K. Choi, J.C. Li, R. Welser, Demonstration of a low V_{ref} PA based on InGaAsN technology, in: 2006 Topical Workshop on Power Amplifiers (RWS), San Diego, California, Paper 4.4, January, 2006.

[10] A.G. Metzger, R. Ramanathan, J. Li, M. Sun, C. Cismaru, H.X. Shao, L. Rushing, K.P. Weller, C.J. Wei, Y. Zhu, A. Klimashov, G. Tkachenko, B. Li, P.J. Zampardi, An InGaP/GaAs merged HBT-FET (BiFET) technology and applications to the design of handset power amplifiers, J. Solid State Circuits 42 (10) (2007) 2137–2148.

[11] A. Joseph, L. Qizhi, W. Hodge, P. Gray, K. Stein, R. Previti-Kelly, P. Lindgren, E. Gebreselasie, B. Voegeli, P. Candra, D. Hershberger, R. Malladi, P.C. Wang, K. Watson, H. Zhong-Xiang, J. Dunn, A 0.35 μm SiGe BiCMOS technology for power amplifier applications, in: Digest of the 2007 Bipolar Circuits and Technology Meeting, Boston, Massachusetts, September, 2007, pp. 198–201.

[12] Y.K. Chen, D. Humphrey, D. Sivco, A.Y. Cho, J. Lothian, J.M. Kuo, F. Ren, J.S. Weiner, A. Tate, M.F. Change, R. Bernescut, Integrated AlInAs/InGaAs HEMT/HBT heterostructure grown by MBE, in: Conference Proceedings of the Seventh International Conference on Indium Phosphide and Related Materials, Sapporo, Japan, May, 1995, pp. 656–659.

[13] Y. Tkachenko, Performance comparison of HBTs and quasi E-mode PHEMTs for single supply high efficiency power amplifiers, in: 31st European Microwave Conference, London, England, September, 2001, pp. 1–4.

[14] M. Ito, New high capacity technology for Li-ion battery, Panasonic, in: IWPC Presentation, Dusseldorf, Germany, May 8, 2007.

[15] J. Deng, P. Gudem, L.E. Larson, P.M. Asbeck, A high average-efficiency SiGe HBT power amplifier for WCDMA handset applications, IEEE Trans. Microw. Theory Techn. 53 (2) (2005) 529–537.

[16] G. Hanington, P.F. Chen, V. Radisic, T. Itoh, P. Asbeck, Microwave power amplifier efficiency improvement with a 10 MHz HBT DC-DC converter, in: IEEE MTT-S Tech. Dig, Baltimore, MD, 1998, pp. 589–592.

[17] J. Kim, J. Kim, Y. Noh, Y. Kim, S. Kim, C. Park, An MMIC smart power amplifier of 21% PAE at 16 dBm power level for W-CDMA mobile communication terminals, in: IEEE GaAs IC Symp. Dig., Monterey, California, October, 2002, pp. 181–184.

[18] H. Patterson, Enhanced efficiency CDMA power amplifiers using gain stage switching, in: Technical Digest of the 2000 IEEE Topical Workshop on Power Amplifiers for Wireless Applications, San Diego, California, Paper 2.7, September, 2000.

[19] R. Novak, Making the call on 3G: finding the right switch solution for complex high-frequency bands, Microw. Prod. Digest (2005).

[20] T. Henderson, J. Middleton, J. Mahoney, S. Varma, T. Rivers, C. Jordan, B. Avrit, High-performance BiHEMT HBT/E-D pHEMT integration, in: Digest of GaAs Mantech, Austin, Texas, May, 2007, pp. 247–250.

[21] W.-J. Ho, J. Liu, H. Chou, C.S. Wu, T.C. Tsai, W.D. Chang, F. Chou, Y.-C. Wang, Manufacturing of GaAs MMICs for wireless communications applications, J. Semicond. Technol. Sci. 6 (3) (2006) 136–145.

[22] P. Wright, Power utilization curves and their implications for optimum power amplifier architecture, in: IWPC Presentation, Dusseldorf, Germany, May, 2007.

[23] F. Raab, P. Asbeck, S. Cripps, P. Kenington, Z. Popovic, N. Pothecary, J. Sevic, N. Sokal, Power amplifiers and transmitters for RF and microwave, IEEE Trans. Microw. Theory Techn. 50 (3) (2002) 814–826.

[24] E. McCune, Operating modes of dynamic power supply transmitter amplifiers, IEEE Trans. Microw. Theory Techn. 62 (11) (2014) 2511–2517.
[25] J. Sevic, Statistical characterization of RF power amplifier efficiency for CDMA wireless communication systems, in: Wireless Communication Conference, Boulder, Colorado, 1997, pp. 110–113.
[26] P. Asbeck, Efficiency enhancement design techniques for power amplifiers, in: IMS Symposium Short Course, Honolulu, Hawaii, 2007.
[27] V. Pala, H. Peng, P. Wright, M.M. Hella, T.P. Chow, Integrated high-frequency power converters based on GaAs pHEMT: technology characterization and design examples, IEEE Trans. Power Electron. 27 (5) (2012) 2644–2656.
[28] K. Buisman, L.C. de Vreede, L.E. Larson, M. Spirito, A. Akhnoukh, T.L. Scholtes, L. K. Nanver, Distortion-free varactor diode topologies for RF adaptivity, in: 2005 IEEE MTT-S International Microwave Symposium Digest, Long Beach, CA, June, 2005, pp. 157–160.
[29] C.D. Presti, A.G. Metzger, H.M. Banbrook, P.J. Zampardi, P.M. Asbeck, Efficiency improvement of a handset WCDMA PA module using adaptive digital predistortion, in: International Microwave Symposium, Anaheim, California, May, 2010, pp. 804–807.
[30] D.S. Ripley, A single chip HBT power amplifier with integrated power control, in: RFIC Symposium Digest, Seattle, Washington, June, 2013, pp. 337–340.
[31] P. Li, Recent developments in WiFi system and related challenges to power amplifier design, in: IEEE 11th International Conference on Solid-State Integrated Circuit Technology (ICSICST), Xi'an, China, October, 2012, pp. 1–4.
[32] A. Samelis, E. Whittaker, M. Ball, A. Bruce, J. Nisbet, L. Lam, W. Vaillancourt, Front-end modules with versatile dynamic EVM correction for 802.11 applications in the 2 GHz band, in: IEEE Topical Conference on Power Amplifiers for Wireless and Radio Applications (PAWR), Newport Beach, California, January, 2014, pp. 10–12.
[33] M. Ozalas, The impact of electro-thermal coupling on HBT power amplifiers, in: Compound Semiconductor Integrated Circuits Symposium (CSICS), San Diego, California, October, 2014, pp. 1–4.
[34] P. Viklund, Reducing RF design cycles through improved design methodologies, RF Des. 30 (2007) 40–43.
[35] S. Tiku, Front end of line (FEOL) device processing, in: Mantech Workshop Session 3, Portland, Oregon, May, 2010.
[36] A. Hastings, The Art of Analog Layout, Prentice-Hall, Upper Saddle River, NJ, 2001.
[37] R. Singh, M. Oprysko, D.L. Harame, Silicon Germanium: Technology, Modeling, and Design, Wiley-IEEE Press, Hoboken, NJ, 2003.
[38] M. Fresina, Trends in GaAs HBTs for wireless and RF, in: Proceedings of the IEEE Bipolar/BiCMOS Circuit and Technology Meeting, Atlanta, Georgia, September, 2012, pp. 150–153.
[39] J.P. Gambino, J. Malinowski, A. Cote, B. Guthrie, P. Chapman, A. Vize, W. Bowe, C. Griffin, E. Cooney, T. Aoki, Y. Chen, D. Wang, M. Jaffe, Reliability of circuits under pads for Au and Cu wire bonding, in: 19th IEEE International Symposium on Physical and Failure Analysis of Integrated Circuits (IPFA), Singapore, Singapore, July, 2012, pp. 1–6.
[40] I. Bahl, Lumped Elements for RF and Microwave Circuits, Artech House, Norwood, MA, 2003.
[41] F. Rotella, B. Bhattacharya, V. Blaschke, M. Matloubian, A. Brotman, Y. Cheng, R. Divecharia, D. Howard, K. Lampaert, P. Miliozzi, M. Racanelli, P. Singh,

P.J. Zampardi, A broad-band lumped element inductor model incorporating skin effect and substrate loss for technology performance assessment and RFIC design, IEEE Trans. Electron Devices 52 (2) (2005) 1429–1441.

[42] R.A. Pucel, Design considerations for monolithic microwave circuits, IEEE Trans. Microw. Theory Techn. 29 (6) (1981) 513–534.

[43] D.Y. Lie, J. Popp, A.H. Yang, J.F. Rowland, F. Wang, D. Kimball, Monolithic class-E SiGe power amplifier design with wideband high-efficiency and linearity, in: International Symposium on VLSI Design, Automation, and Test, Hsinchu, Taiwan, April, 2006, pp. 1–4.

[44] Y. Li, J. Lopez, D.Y. Lie, K. Chen, S. Wu, T.Y. Yang, Efficiency enhancement for linearity trade-offs for cascode vs. common-emitter SiGe power amplifiers for WiMAX polar transmitters, in: International Symposium on Circuits and Systems (ISCAS), Paris, France, May, 2010, pp. 1915–1918.

[45] F. Carrara, C.D. Presti, G. Palmisano, A. Scuderi, Power transistor guidelines and RF load-pull characterization of a 0.13-μm SOI CMOS technology, in: Proceedings of the European Solid State Device Research Conference, ESSDERC, Athens, Greece, September, 2009, pp. 444–447.

[46] P. Colantonio, F. Giannini, E. Limiti, V. Teppati, An approach to harmonic load-and-source-pull measurements for high-efficiency design, IEEE Trans. Microw. Theory Techn. 52 (1) (2004) 191–198.

[47] M. Marchetti, M.J. Pelk, K. Buisman, W. Neo, M. Spirito, L.C. de Vreede, Active harmonic load-pull with realistic wideband communication signals, IEEE Trans. Microw. Theory Techn. 56 (12) (2008) 2979–2988.

[48] A. Ferrero, M. Pirola, Harmonic load-pull techniques: an overview of modern systems, IEEE Microw. Mag. 14 (4) (2013) 116–123.

[49] A. Raffo, V. Vadala, D.M.-P. Schreurs, G. Crupi, G. Avolio, A. Caddemi, G. Vannini, Nonlinear dispersive modeling of electron devices oriented to GaN power amplifier design, IEEE Trans. Microw. Theory Techn. 58 (4) (2010) 710–718.

[50] A. Raffo, S. Di Falco, V. Vadalà, G. Vannin, Characterization of GaN HEMT low-frequency dispersion through a multiharmonic measurement system, IEEE Trans. Microw. Theory Techn. 58 (9) (2010) 2490–2496.

[51] P.J. Zampardi, B. Li, Power amplifier design and technology: what really impacts PAE and linearity? in: Workshop Design for Wireless Communications: From Transistor to System-Level Characterization, 80th ARFTG Microwave Measurement Conference, San Diego, California, November, 2012.

[52] P.J. Zampardi, J. Hu, C. Cismaru, V. Viswanathan, Transistor design considerations for power amplifier applications, in: IEEE Topical Workshop on Power Amplifiers for Wireless Applications, Orlando, Florida, Paper 6.2, 2008.

[53] M. Spirito, F.M. De Paola, V. d'Alessandro, K. Buisman, N. Rinaldi, Trade-offs in RF performance and electrothermal ruggedness of multifinger SiGe power cells, in: Proceedings of the 18th International Symposium on Power Semiconductor Devices & IC's, Naples, Italy, June, 2006, pp. 1–4.

[54] S. Kee, The Class E/F Family of Harmonic-Tuned Switching Power Amplifiers (Ph.D. Dissertation), California Institute of Technology, Pasadena, California, 2002.

[55] C. Clifton, Circuit and technology solutions to address the multi-mode mobile terminal front-end challenge, in: Joint ICT KTN and Radio SIG Event, New Materials, Devices, and Circuit Techniques for RF, Bath, England, October 10, 2013.

[56] M. Spirito, M.P. van der Heijden, M. Pelk, L.C. de Vreede, P.J. Zampardi, L.E. Larson, J.N. Burghartz, Experimental procedure to optimize out-of-band terminations for highly linear and power efficient bipolar class-AB RF amplifiers, in: Proceedings of the 2005 IEEE Bipolar/BiCMOS Circuits and Technology Meeting, Santa Barbara, California, October, 2005, pp. 112–115.

[57] V. Aparin, C. Persico, Effect of out-of-band harmonic terminations on intermodulation distortion in common-emitter circuits, in: IEEE MTT-S Digest, Anaheim, California, June, 1999.

[58] L.C. van den Oever, R.M. Heeres, H.F. Jos, Linearity efficiency load-pull measurements on GaAs HBT versus Si BJT for EDGE and OFDM modulation, in: Proceedings of the 2006 IEEE Bipolar/BiCMOS Circuits and Technology Meeting, Maastricht, The Netherlands, October, 2006.

[59] K. Yamamoto, InGaP-HBT power amplifiers, in: K. Iniewski (Ed.), Nano-Semiconductors: Devices and Technology (Devices, Circuits, and Systems), CRC Press, Boca Raton, FL, 2011, pp. 319–341 (Chapter 17).

[60] E. Jarvinen, S. Kalajo, M. Maitlainen, Bias circuits for GaAs HBT power amplifiers, in: IEEE MTT-S Digest, Phoenix, Arizona, May, 2001, pp. 507–510.

[61] C. Fager, J. Pedro, N. Carvalho, H. Zirath, Prediction of IMD in LDMOS transistor amplifiers using a new large-signal model, IEEE Trans. Microw. Theory Techn. 50 (12) (2002) 2834–2842.

[62] I. Angelov, N. Rorsman, J. Stenarson, M. Garcia, H. Zirath, An empirical table-based FET model, IEEE Trans. Microw. Theory Techn. 47 (12) (1999) 2350–2357.

[63] G. Crupi, D.M.-P. Schreurs, D. Xiao, A. Caddemi, B. Parvais, A. Mercha, S. Decoutere, Determination and validation of new non-ideal FinFET model based on lookup tables, IEEE Microw. Wirel. Componen. Lett. 17 (5) (2007) 361–363.

[64] H.-F. Wu, Q.-F. Cheng, S.-X. Yan, Q.-J. Zhang, J.-G. Ma, Transistor model for a microwave power heterojunction transistor, IEEE Microw. Mag. 16 (2) (2015) 85–92.

[65] J. Sifr, M. Ozalas, Integrated multi-technology physical RF design flow short-course, in: International Microwave Symposium Short Course SC1-1, Tampa, Florida, June, 2014.

[66] B. Bisla, RF system and circuit design challenges for WiMAX, Intel Technol. J. 18 (3) (2004) 189–200.

[67] P. Dai, P.J. Zampardi, K. Kwok, C. Cismaru, R. Ramanathan, Study of GaAs HBT thermal coupling effects by using simple current mirror circuits, in: 2005 CS MANTECH Digest, New Orleans, Louisiana, Paper 14.9, April, 2005.

[68] B. Vijayakumar, Packaging technology for FEMs-thermal design and modeling, in: International Microwave Symposium Short-Course WSO, Montreal, Quebec, Canada, June, 2012.

[69] S.-H. Tsai, Layout practices for die size reduction on InGaP/GaAs HBT MMICs for handset power amplifier applications, in: CS Mantech Conference Digest, New Orleans, Louisiana, May, 2013, pp. 307–310.

[70] P.J. Zampardi, Y. Yang, J. Hu, B. Li, M. Fredricksson, K. Kwok, H.X. Shao, Practical statistical simulation for efficient circuit design, in: Nonlinear Transistor Model Parameter Extraction Techniques, Cambridge University Press, New York, 2011, pp. 287–317.

[71] M. Tutt, GaAs based HBT large signal modeling using VBIC for linear power amplifier applications, in: IEEE BCTM Technical Digest, Minneapolis, Minnesota, October, 2000, pp. 58–61.

[72] P.J. Zampardi, Y. Yang, K. Kwok, B. Li, A.G. Metzger, C. Cismaru, H.X. Shao, W. Sun, M. Fredriksson, End-to-end modeling for handset power amplifiers — it's not just two transistors, in: Presented at Compact Modeling for RF (CMRF), Portland, Oregon, October, 2012.

[73] H.X. Shao, P.J. Zampard, Compound semiconductor process technologies, design kits, and enterprise design, in: CS Mantech Workshop, Indian Wells (Palm Springs), California, May, 2011.

CHAPTER

Radio frequency power amplifier for wireless communication

6

W. Chen*, Z. Wang†, X. Chen* and S. Zhang‡

Tsinghua University, Beijing, China Microsoft Corp., Beijing, China*†
Smarter Micro Inc., Shanghai, China‡

CHAPTER OUTLINE

6.1 INTRODUCTION

With the fast evolution of wireless communication, high-order quadrature amplitude modulation (QAM), and orthogonal frequency-division multiplexing (OFDM) are widely employed for capacity enhancement. These modulation schemes have more strict requirements on the radio frequency (RF), especially for device linearity and efficiency. In this situation, power amplifier (PA) devices and modules have attracted lots of attention and extensive study in the past decade. Compared with

Microwave Wireless Communications. http://dx.doi.org/10.1016/B978-0-12-803894-9.00006-8

passive components design, it is more difficult to model and design the nonlinear devices accurately, due to the limitation of nonlinear models and dynamic parasitic effects. In this chapter, we will do an overview the PA technologies from the transistor level to the architecture level, in order to provide a comprehensive study on the state-of-the-art PA design method.

A nonconstant envelope signal leads to a high peak-to-average power ratio (PAPR), which significantly decreases the efficiency of PA in order to meet the linearity requirements of wireless communication standards. As a result, the PA specifications in modern wireless communication systems are slightly different from the conventional continuous wave (CW) demand. First, the PA specification, including output capability, efficiency computation and linearity requirement, will be discussed.

Because the conventional single channel class-AB PA often leads to serious efficiency deterioration for high-PAPR signals application, several popular energy-efficiency PA architectures will be introduced, such as Doherty, linear amplification with nonlinear components (LINC), and envelope tracking (ET) technologies and the latest progress of these technologies will be also discussed.

Owing to the rapid advances of semiconductor technologies, many new devices such as gallium arsenide (GaAs), heterojunction bipolar transistor (HBT), and gallium nitride (GaN) further accelerate the development of PA technologies. These technologies have their respective advantages in diversified applications. For example, the GaN devices have strong potential for broadband and high-power base station application, while the GaAs HBT is quite suitable for cell phone application.

To meet the multiband and multimode demand, broadband and multiband high-efficiency PA design methods are also investigated in the final section. Given that common efficiency decreases in multiband PA, a new design method called inter-modulation tuning technique is also introduced for performance improvement in concurrent multi-standard wireless system.

6.2 PA SPECIFICATION

PA is an electronic amplification component that enlarges the signal amplitudes to increase the coverage area of communications. PA becomes one of the essential components in wireless communication systems, and it also has significant effects on the system linearity, efficiency, as well as cost. Along with the rapid evolution of wireless communications, the system brings severe requirements to the PAs. To satisfy these requirements, the PA cannot simply be characterized by conventional specifications, as new criteria should be introduced to evaluate the PA performance.

6.2.1 PA OUTPUT

The output power of PA mainly determines the signal density emitted from the transmitter and the communication distance due to the path loss, thus the output capability should be guaranteed to insure enough coverage range. In the conventional laterally

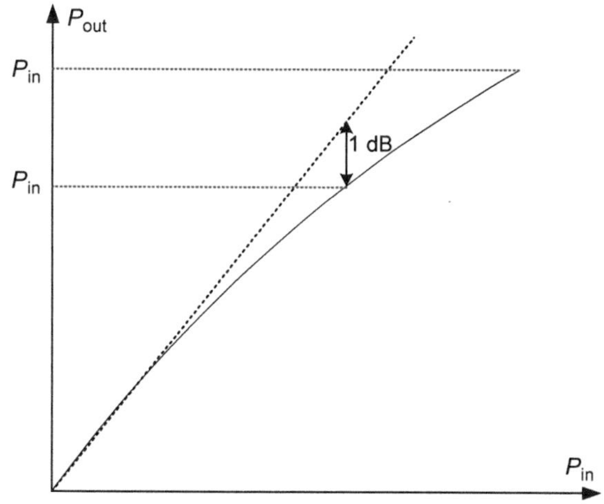

FIGURE 6.1

PA linear output capability.

diffused metal oxide semiconductor (LDMOS) transistor-based PA, P1dB point is usually regarded as the maximum linear output as shown in Figure 6.1. However, owing to the soft gain compression phenomenon in a GaN device, it is not appropriate to still use P1dB to evaluate the output power limitation, because the PA starts to depress its gain in the early stage. Benefiting from PA linearization techniques, such as digital predistortion (DPD), the PA can be corrected, even to be driven into saturation. In this situation, saturation output power is also often used as benchmark specification for new semiconductor devices.

Nonconstant envelope modulation schemes, such as high-order QAM and OFDM, have been widely used for high capacity in wireless communications. Due to the adoption of nonconstant envelope modification schemes, the PAPR is quite high for spectral efficiency enhancement. For high-PAPR application, the architecture designers pay more attention to the average output power instead of the peak power. Thereby, the PA output capability also can be characterized by the average output power under the modulated signals stimulus.

6.2.2 EFFICIENCY

With the exception of analog cellular wireless systems, most wireless systems exhibit a specific output power distribution [1]. The mean of this distribution of power, along with the associated RF PA efficiency at this power, determines the long-term time-average RF PA efficiency. As shown in Figure 6.2, the average efficiency is not determined by the maximum efficiency in saturation, it is actually an integration of the efficiency profile on the power distribution function. For distributions with small variances, this mean efficiency establishes the RF PA's impact on the phone talktime. This characterization approach is much more meaningful

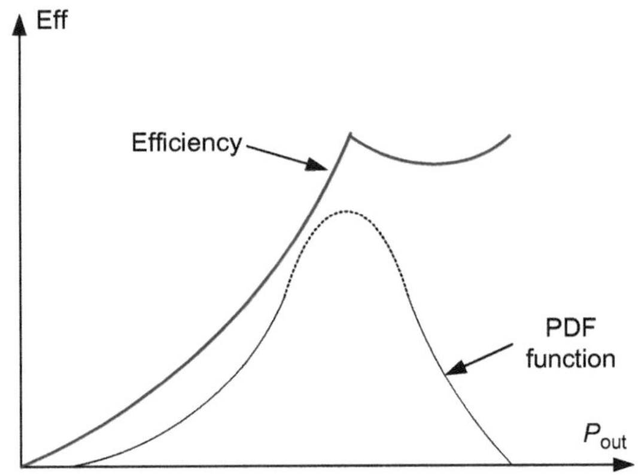

FIGURE 6.2

PA average efficiency with modulated signals.

because it includes system effects on the RF PA output power. Moreover, the power distribution function can be interpreted as a weighting function, showing that RF PA efficiency at maximum output power is significant only in special cases. Therefore, it is more essential for overall efficiency enhancement to design a PA which achieves high efficiency at average power point instead of maximum power point. Owing to this motivation, many advanced PA architectures such as Doherty, ET, and LINC techniques have been intensively investigated in modern wireless transmitters.

6.2.3 LINEARITY

In conventional transmitters, the intermodulation distortion (IMD) is often used to characterize the nonlinearities of PAs. However, such a characterization technique is not sufficient to fully evaluate the nonlinear systems with modulated signals, especially for broadband wireless applications. Alternatively, adjacent channel power ratio (ACPR) [ie, the ratio between the total power of adjacent channel (IMD signal) to the main channel's power (useful signal)], is utilized to evaluate the nonlinearities. There are two ways of measuring ACPR. The first is by finding 10*log of the ratio of the total output power to the power in adjacent channel. The second (and much more popular method) is to find the ratio of the output power in a smaller bandwidth around the center of the carrier to the power in the adjacent channel. The smaller bandwidth is equal to the bandwidth of the adjacent channel signal. The second way is more popular because it can be measured easily. ACPR is desired to be as low as possible: a high ACPR indicates that significant spectral spreading has occurred. As illustrated in Figure 6.3, an ACPR measurement template is presented in signal analyzer instruments.

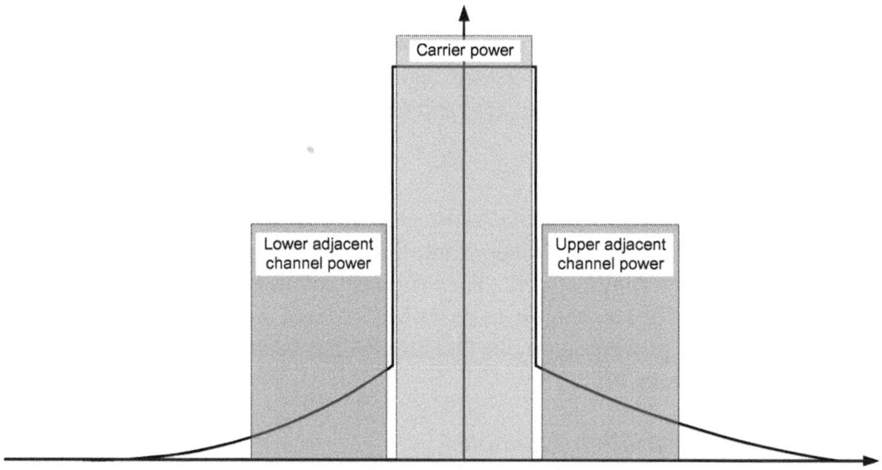

FIGURE 6.3

ACPR measurement template for modulated signals.

6.2.4 VIDEO BANDWIDTH

It has been reported that IMD asymmetries are caused by the termination impedance characteristics of the difference frequency and second harmonics. Difference frequency has been often terminated by a bias circuit of an amplifier, so the load impedance at the base band frequency will affect the PA memory effect significantly. To avoid serious asymmetries of IMD, it is essential to guarantee a flat and near to zero impedance for the transistor, which is referred to the video bandwidth (VBW) of the PA.

VBW is the primary limitation for the ability of modern DPD systems to correct spectral emission levels caused by high instantaneous bandwidth signals to meet regulatory standards. According to Ref. [2], VBW can be defined by using a CW two-tone test (with constant output power and swept tone-spacing, assuming the RF bandwidth of the PA is reasonably flat in the frequency range desired), as the tone-spacing at which the IMD3 products at $2f_2 - f_1$ and $2f_1 - f_2$ (where f_1 and f_2 are the frequencies of the individual tones) start to increase continuously in amplitude up to a point where the IMD3 power level is almost equal to that of the fundamental tones. It is understood that higher order intermodulation products (such as IMD5 and IMD7) must be considered for DPD correction as well. The fifth order products generally must be DPD corrected, but seventh order and above may fall below the level where DPD adds benefits due to the fact that their levels may fall at or below the noise floor of the system.

6.3 PA TOPOLOGIES FOR WIRELESS COMMUNICATION

With the rapid evolution of communication technologies, nonconstant envelope modulation schemes, such as high-order QAM and OFDM, have been widely used for high capacity in wireless communications. A nonconstant envelope signal leads to a high PAPR, which significantly decreases the efficiency of PA in order to meet the linearity requirements of wireless communication standards. To enhance the average efficiency of PAs with nonconstant envelope modulated signals, many advanced PA architectures are proposed and fully studied for high-efficiency RF transmitters including Doherty PA, ET, and LINC topologies.

6.3.1 DOHERTY PA

As introduced in Chapter 4, the Doherty amplifier was first proposed by W.H. Doherty in 1936. The original Doherty amplifier consisted of two tube amplifiers and an impedance inverting network. The efficiency of an RF PA is increased using the RF Doherty amplifier technique, as described in detail in Ref. [3]. To enhance the efficiency in the low-power region, the Doherty PA has been studied extensively and adopted in base stations [4].

6.3.1.1 Doherty operation principle

The circuit diagram of typical Doherty PAs is shown in Figure 6.4. It consists of a power splitter, two single-branch PAs, two offset lines, and two quarter-wavelength ($\lambda/4$) transmission lines. A simplified model of this Doherty PA, as shown in Figure 6.5, is used to describe the operation mechanism.

As shown in Ref. [4], the transistor outputs are modeled as current sources controlled by their input voltages; thus, the output voltages of the main amplifier and the load are calculated to be

$$\begin{cases} V_{\text{main}} = \dfrac{Z_{\text{T}}}{R_{\text{L}}} I_{\text{main}} - \text{j}Z_{\text{T}} I_{\text{peak}} \\ V_{\text{L}} = -\text{j}Z_{\text{T}} I_{\text{main}} \end{cases} \tag{6.1}$$

In Eq. (6.1), Z_{T} are the characteristic impedances of the $\lambda/4$ transformer. To illustrate the power backoff range induced by different characteristic impedances, the value of I_{main} when the main amplifier reaches saturation is defined as

$$I_{\text{th}} = \frac{1}{\gamma} I_{\text{main\_max}} \tag{6.2}$$

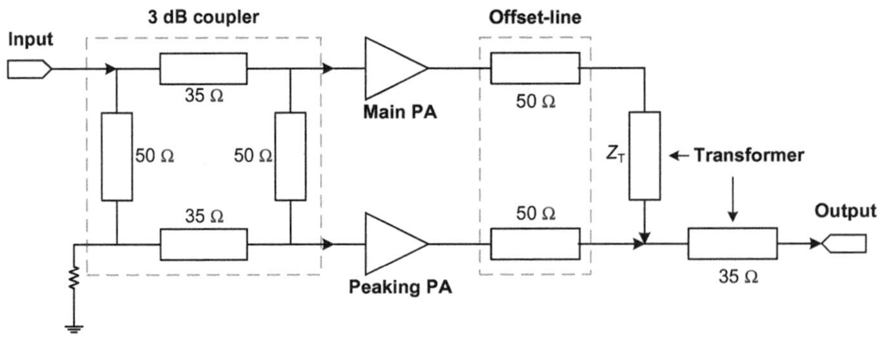

FIGURE 6.4

Circuit diagram of the Doherty PA.

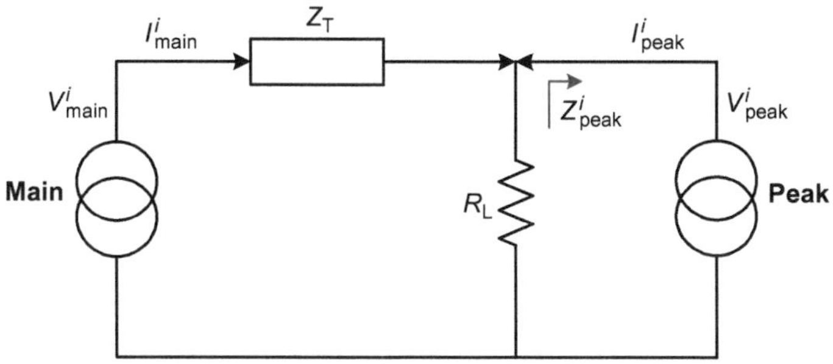

FIGURE 6.5

Equivalent circuit model of the Doherty PA.

where γ is a design parameter that determines the value of I_{th}, in relation to I_{main_max}. Consequently, the current in the peaking amplifier can be written as

$$I_{peak} = \begin{cases} 0, & I_{main} \prec I_{th} \\ -j\gamma(I_{main} - I_{th}), & I_{main} \geq I_{th} \end{cases} \tag{6.3}$$

By substituting Eqs. (6.2), (6.3) into Eq. (6.1) and solving for the Z_T that makes V_{main} constant, we can obtain

$$Z_T = \gamma R_L \tag{6.4}$$

In the low-power region, the peaking amplifier remains in the cut-off state, and the load impedance of the carrier amplifier is two times larger than that of the conventional amplifier. Thus, the carrier amplifier reaches the saturation state at the half input voltage because the maximum fundamental current swing is half and the maximum voltage swing reaches V_{dc}. As a result, the maximum power level is half of the carrier amplifier's allowable power level, and the efficiency of the amplifier is equal to the maximum efficiency of the carrier amplifier as shown in Figure 6.6.

In the high-power region, where the peaking amplifier is conducting, the current level of the peaking amplifier plays an important role in determining the load modulation of the Doherty amplifier. Assuming that g_m of the peaking amplifier is twice as large as that of the carrier amplifier, as illustrated in Figure 6.7A, the current and voltage swings of the peaking amplifier increase in proportion to the input voltage level and the voltage swing reaches the maximum voltage swing of V_{dc} only at the maximum input voltage. The load impedance of the carrier amplifier varies from $2Z_{opt}$ to Z_{opt}, and the peaking amplifier varies from ∞ to Z_{opt} according to the input voltage level as shown in Figure 6.7B. The efficiency of the Doherty amplifier at the maximum input voltage is equal to the maximum efficiency of the amplifiers. From the basic operation principle, we have explored the Doherty amplifier, which provides higher efficiency over whole power ranges compared to the conventional class-AB amplifiers. The resulting Doherty amplifier can solve the problem of maintaining a high efficiency for a large PAPR signal.

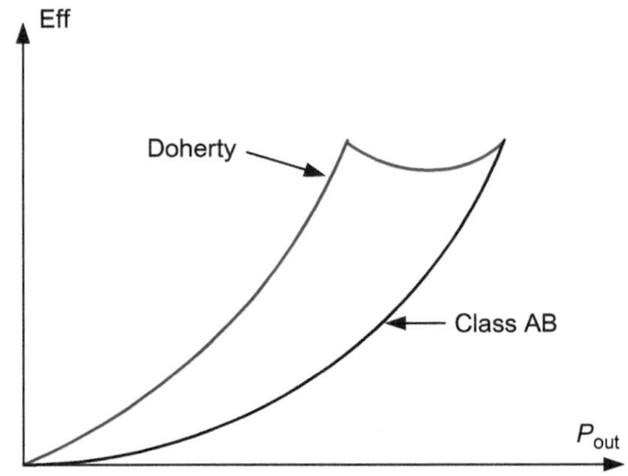

FIGURE 6.6

Efficiency profile of the Doherty PA.

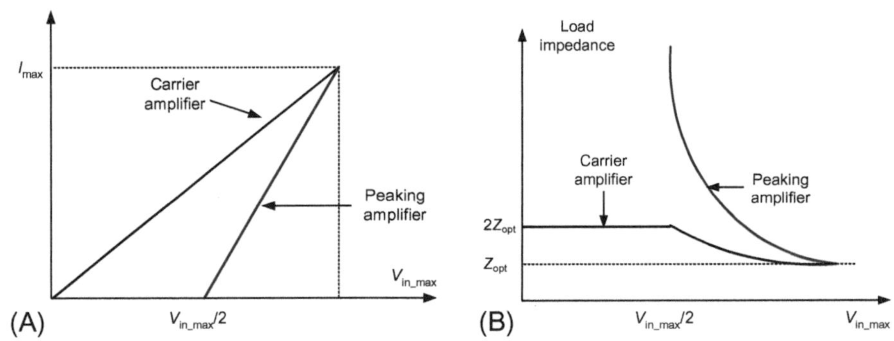

FIGURE 6.7

(A) The carrier cell and peaking cell currents of the Doherty PA, (B) the load impedances of the carrier and peaking cells of the Doherty PA.

6.3.1.2 Asymmetric Doherty PA

When the peaking and carrier amplifiers have the same size, which is normally the case, g_m of the two amplifiers are identical and the peaking amplifier cannot be fully turned on, so the power performance is degraded. Usually, both cells in conventional Doherty PA have identical size, matching circuits, and input drives. In addition, because the bias point of the peaking cell is lower than that of the carrier cell, the current of the peaking cell at maximum input drive reaches below the maximum allowable current level. To guarantee a full load modulation in Doherty PA, an asymmetrical Doherty PA was proposed to improve the performance [5], which consists of a class-AB biased carrier amplifier cell and a class-C biased peaking amplifier cell with larger output power.

To enhance the output power from the peaking cell, an asymmetrical Doherty PA with uneven power drive (ie, applying more power to the peaking cell) can be adopted. Figure 6.8A shows an operational diagram to analyze efficiency of the amplifier with an uneven drive times larger current source for the peaking cell. Figure 6.8B shows the fundamental currents from the two cells for even and uneven drives. As shown, in the case of an uneven power drive, an appropriate load modulation can be achieved, which avoids the disadvantage in conventional Doherty topology.

For equal output current operation by an uneven drive, the load line of the carrier cell follows the knee current without producing saturation due to proper load modulation. The voltage swing of the peaking cell increases in proportion to the power level and reaches the maximum voltage swing only at maximum power. At the maximum power level, both cells have optimum matching impedances. Therefore, the asymmetric PA with uneven drive operates more linearly and produces more power than an even drive.

6.3.1.3 Digital Doherty PA

The analog single-input Doherty architecture is widely employed in RF transmitters, but still suffers from inconsistent characteristics of two branches to the achievable power efficiency. First, at the system level, the difference in the class of operation of the carrier and peaking PAs results in gain imbalance between the two branches of Doherty PAs. At the transistor level, GaN devices have presented additional challenges to Doherty PA design, especially in terms of phase variation between the carrier and peaking branches. To improve the consistency between the two branches and the Doherty PA performance, a digitally driven dual-input Doherty PA was proposed, as shown in Figure 6.9 [6]. In this architecture, the input signal of each branch is digitally preprocessed and supplied separately to each branch of the Doherty PA in order to optimize the overall performance of the PA. Digital signal processing is applied to minimize the performance degradation due to phase impairment in Doherty PA's branches.

To minimize the adverse effects of phase imbalance between the carrier and peaking branches on the performance of GaN Doherty PAs, the dual-input digitally driven architecture is used, allowing for the adoption and implementation of a power adaptive phase-alignment mechanism.

To demonstrate the advantage of dual-input digitally driven Doherty PA, Figure 6.10 presents the simulated results of the load modulation of the dual-input Doherty PA with and without adaptive phase alignment. This figure clearly shows that the Doherty load modulation mechanism was significantly improved. Indeed, load impedance at the impedance inverter plane was modulated from 100 to nearly 50 Ω, as the input power increased. Similarly, load impedance at the impedance transformer plane was decreased gradually from a very high value to roughly 50 Ω.

As expected, by enforcing the constructive in-phase summation of the device output currents at all times, the power losses were considerably reduced, which evidently results in an increased total output power. In addition, it can be seen that the relative improvement in output power was maximal in the medium input drive levels.

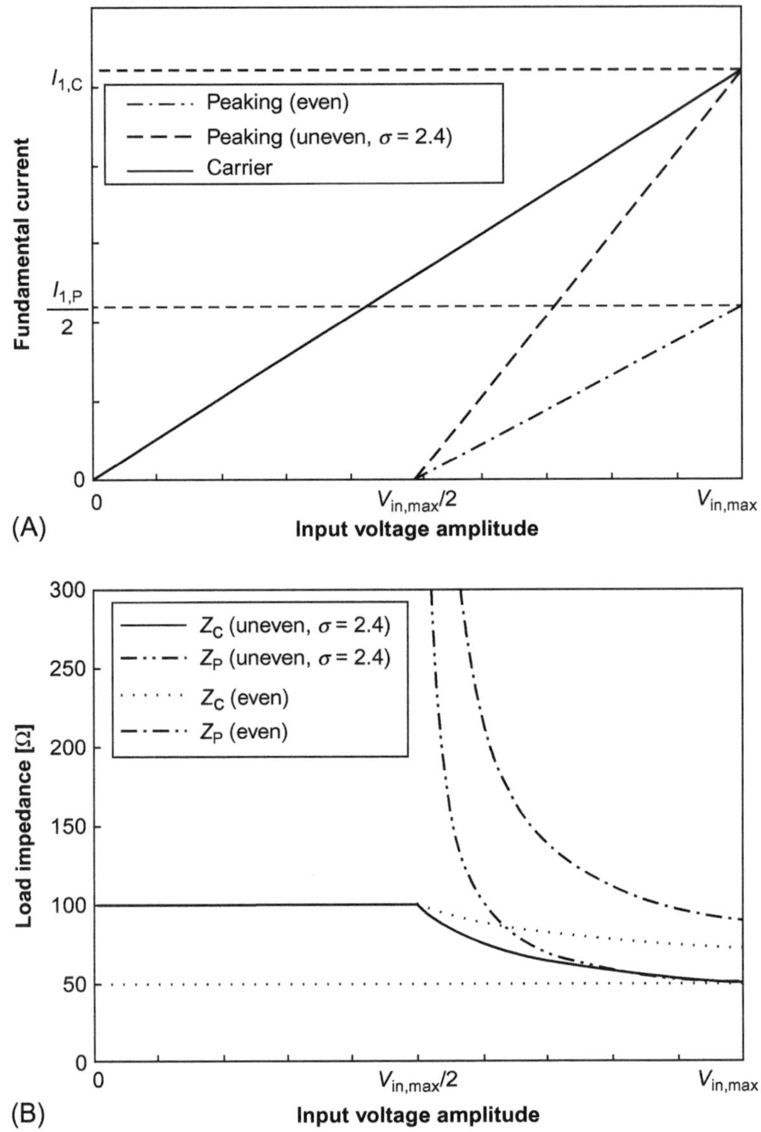

FIGURE 6.8

(A) The carrier and peaking cells currents of the asymmetrical Doherty PA, (B) the load impedances of the carrier and peaking cells of the asymmetrical Doherty PA.

From J. Kim, J. Cha, I. Kim, B. Kim, Optimum operation of asymmetrical-cells-based linear Doherty power amplifiers—uneven power drive and power matching, IEEE Trans. Microw. Theory Techn. 53 (5) (2005) 1802–1809. ©2009 IEEE. Reprinted with permission.

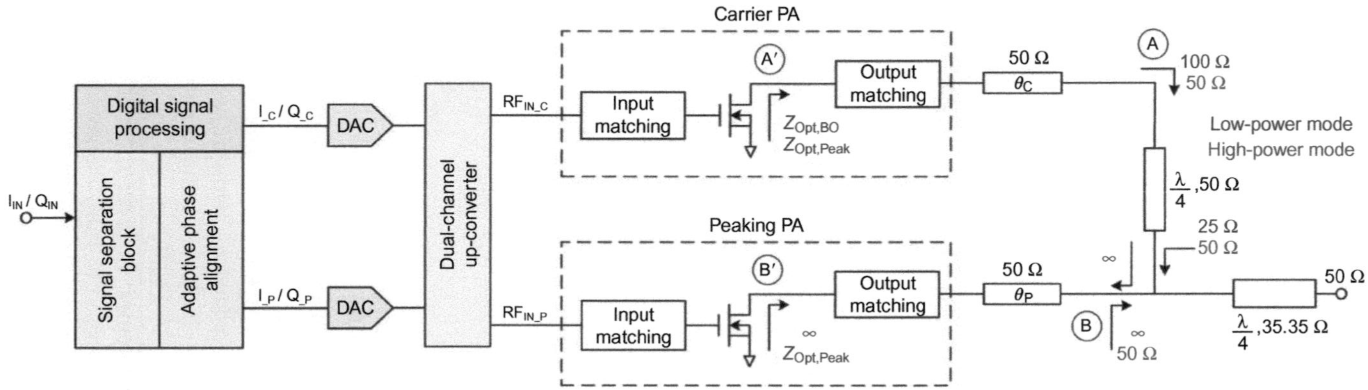

FIGURE 6.9

System diagram of digital Doherty PA architecture.

From R. Darraji, F.M. Ghannouchi, O. Hammi, A dual-input digitally driven Doherty amplifier architecture for performance enhancement of Doherty transmitters, IEEE Trans. Microw. Theory Techn. 59 (5) (2011) 1284–1293. ©2011 IEEE. Reprinted with permission.

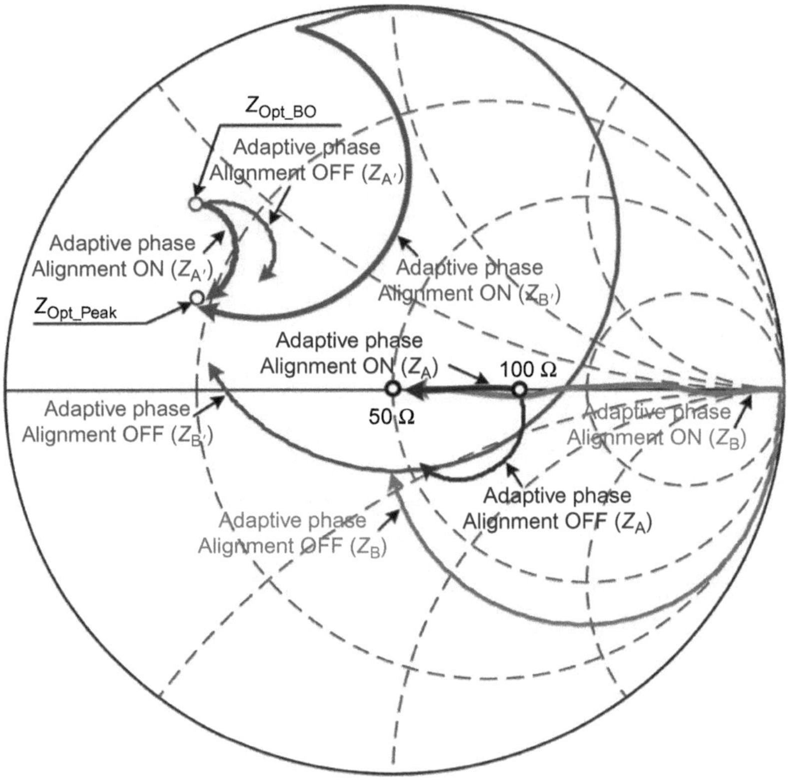

FIGURE 6.10

Load impedance improvement of digital Doherty PA.

From R. Darraji, F.M. Ghannouchi, O. Hammi, A dual-input digitally driven Doherty amplifier architecture for performance enhancement of Doherty transmitters, IEEE Trans. Microw. Theory Techn. 59 (5) (2011) 1284–1293.

6.3.1.4 Broadband Doherty PA

To meet the requirement of wideband wireless communications, broadband PAs are highly desirable for RF transmitters. Unfortunately, the limited bandwidth of Doherty PA is viewed as a bottleneck for broadband application. Moreover, a broadband amplifier can eliminate redundant hardware when the transmission of older wireless standards is needed for backward compatibility. As shown in Ref. [7], actually the bandwidth limitation is mainly caused by the quarter-wavelength transformers in the Doherty topologies. In order to remove this obstacle, a modified Doherty architecture was proposed to fulfill bandwidth enhancement as shown in Figure 6.11 [8].

For broadband Doherty architecture, the key idea is to guarantee a proper load modulation for the carrier cell over a broad bandwidth. However, the frequency-dependent

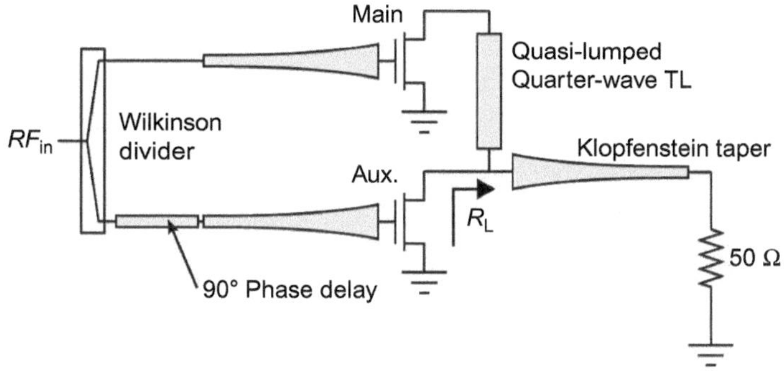

FIGURE 6.11

Circuit topology of the broadband Doherty amplifier.

From D.Y.-T. Wu, S. Boumaiza, A modified Doherty configuration for broadband amplification using symmetrical devices, IEEE Trans. Microw. Theory Techn. 60 (10) (2012) 3201–3213. ©2012 IEEE. Reprinted with permission.

characteristics of the quarter wavelength high deteriorate the bandwidth. It can be found that the Doherty amplifier can exhibit broadband behavior at a given power level if the load seen by the quarter-wave transmission line is equal to its characteristic impedance. For the conventional Doherty amplifier, this condition only occurs at the peak power. As a result, the high impedance can be maintained to lead a high average efficiency for broad bandwidth.

In the new architecture, the carrier and peaking cells current of the broadband Doherty PA are illustrated in Figure 6.12A. It can be seen that the current of the peaking cell is not identical to the carrier cell anymore, and the voltage of the peaking cell will be two times that of the carrier cell, which means that load impedance presented to the peaking cell is four times more than before. Figure 6.12B shows the load modulation in the proposed Doherty amplifier. At the center frequency, the impedance seen by the main device is identical to that of the conventional Doherty with perfect tracking of up to 6 dB of backoff power. Obviously, the load impedances presented to the carrier and peaking cells change little as the frequency deviates when compared to the conventional Doherty amplifier, which finally results in bandwidth enhancement significantly.

6.3.2 ET PA

6.3.2.1 EER and ET

Among the supply modulation PA architectures, there are two main techniques well known as envelope elimination and restoration (EER) and ET. EER was first invented by Leonard R. Kahn in 1952 and further developed later, being named Kahn transmitter and polar transmitter as well. EER exemplifies the use of constant amplitude nonlinear PAs with dynamically controlled supply voltage [9]. However, ET

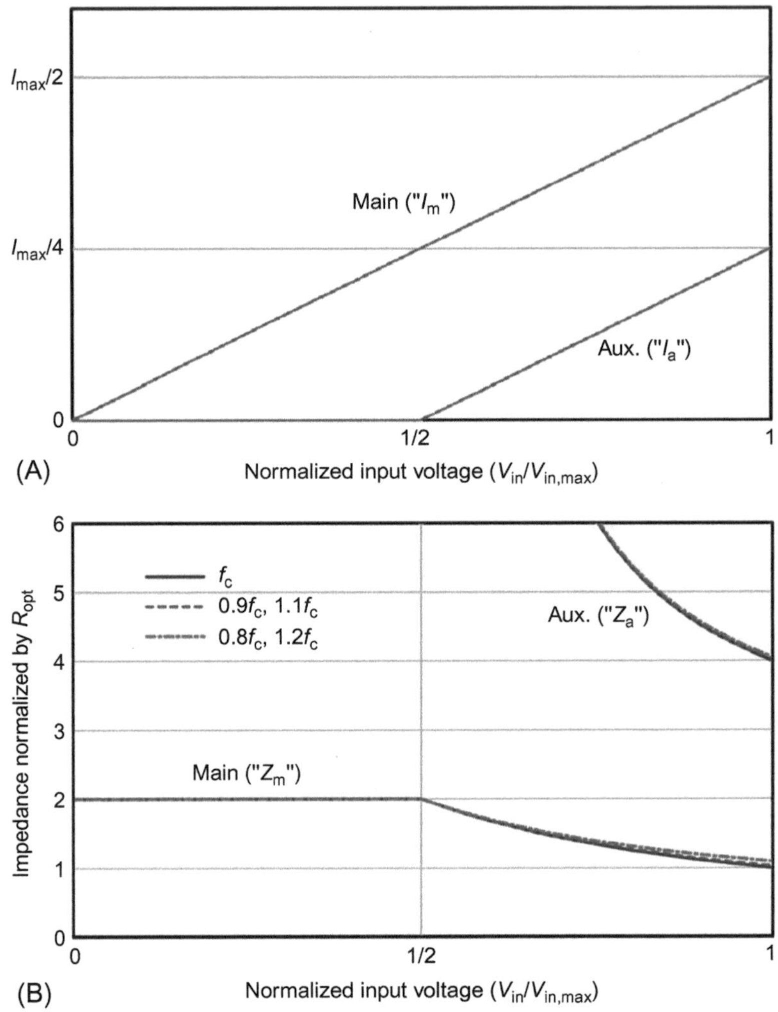

FIGURE 6.12

(A) The carrier and peaking cells currents of the broadband Doherty PA, (B) the load impedances of the carrier and peaking cells of the broadband Doherty PA.

From D.Y.-T. Wu, S. Boumaiza, A modified Doherty configuration for broadband amplification using symmetrical devices, IEEE Trans. Microw. Theory Techn. 60 (10) (2012) 3201–3213. ©2012 IEEE. Reprinted with permission.

utilizes linear PAs whose low efficiency weakness during relatively low-power periods can be mitigated by adjusting the supply voltage accordingly.

In Figure 6.13A of the EER transmitter, although the original input signal contains both the amplitude and the phase information, it is split and envelope-detected in the envelope generation block so that the envelope path is apart from the RF path

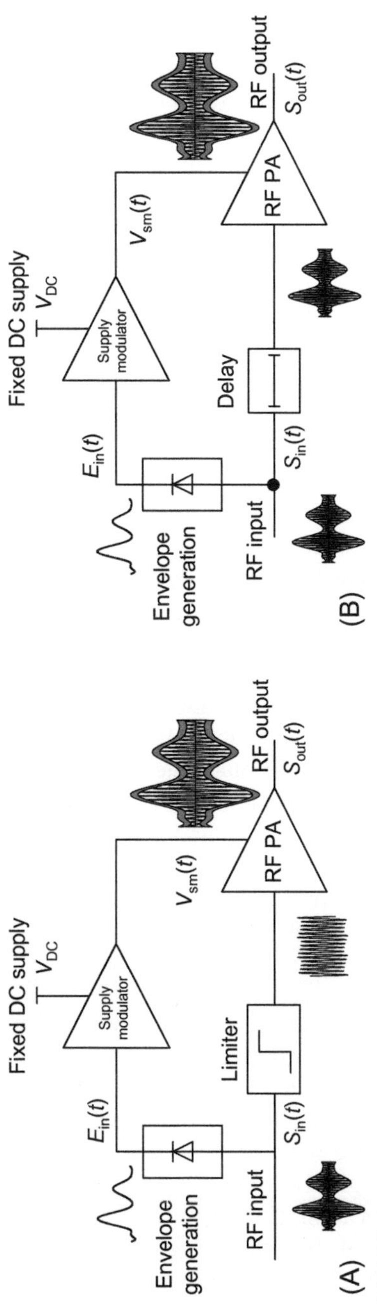

FIGURE 6.13

(A) EER transmitter with separated amplitude and phase modulation, (B) envelope tracking power amplifier.

From Z.C. Wang, Envelope Tracking Power Amplifier for Wireless Communications, Artech House, Norwood, MA, 2014. ©2014 Artech House. Reprinted with permission.

containing the constant envelope phase modulated signal. Obviously, significant disadvantages of EER are that the challenging broad phase bandwidth, stringent timing matching between envelope and signal paths, which render EER technique less attractive for fourth generation (4G) applications.

In contrast, an ET PA in Figure 6.13B comprises a linear RF PA and a supply modulator for providing a modulated supply voltage as a dynamic supply to the PA. The basic idea of the ET technique is to drive the linear PA in its high-efficiency region regardless of the magnitude of the input signal. This is accomplished by detecting the envelope of the input signal and using it to modulate the supply voltage of the linear PA.

Compared with EER transmitters, ET requires lower sensitivity to the timing mismatch between envelope and signal because phase bandwidth requirement of EER is much broader. ET also needs lower bandwidth requirement of the supply modulator because of allowing amplitude modulation input. ET can be more efficient than EER, because the EER driver permanently operates at maximum power. ET has higher gain at lower input power, because the supply modulator can keeps the RF PA nearly saturated with a specific constant compression envelope shaping function. ET has less feed-through signal in under-knee-voltage region to distort the transmitter output with applicable de-trough envelope shaping function. Benefiting from these merits, ET becomes more attractive for high-efficiency PA selection for 4G system and beyond, especially for mobile handsets.

However, there are also several major challenges for ET [10,11]. ET technique requires the use of a high-efficiency supply modulator in order to increase the overall efficiency. Efficiency of ET PAs is smaller at low envelope voltage than at high envelope voltage low-power region cannot be sufficiently compressed. The distortions in the supply modulator can be modulated into the output of RF PA in different forms. Therefore, linearity requirements force the use of new, but more complex, predistortion algorithms. Finally, although linearity and efficiency are tricky tradeoffs in traditional fixed DC supplied PAs, ET PAs can linearly amplify amplitude and phase modulated signals with much higher efficiency. Both linear and nonlinear PAs can be employed for ET by virtue of predistortion.

6.3.2.2 RF PA for ET

In ET PA design, the linkage between the supply modulator and RF PA should be examined carefully in order to maximize efficiency and linearity systematically. Here an exemplary design has been introduced for illustration, as shown in Figure 6.14.

In Figure 6.14A, the fundamental load and harmonic impedances induce both quasi half-sinusoidal voltage with an additional voltage peaking effect and quasi-rectangular current shaping. Also, there is almost zero drain voltage at the dip of the current, which means minimized power loss. Moreover, in the load line in Figure 6.14B it can be seen that the voltage has the minimum value during the on-state which reduces the losses. Sufficient margin can be observed by load line safely inside of area of safe operation region at junction temperature 200°C. Figure 6.14C shows the load impedance of 50 W GaN HEMT at 900 MHz on Smith

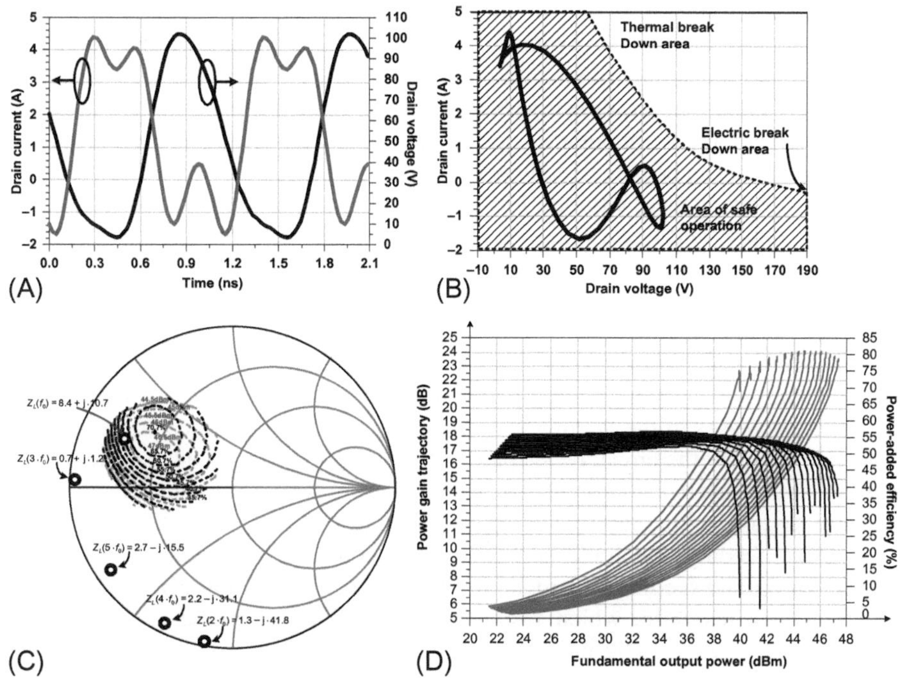

FIGURE 6.14

(A) Current and voltage waveforms observed at package plane and (B) load lines of PA up to the third harmonic, (C) fundamental load-pull contours of the output power and PAE, (D) gain-efficiency trajectory of the class-P PA.

From Z.C. Wang, Envelope Tracking Power Amplifier for Wireless Communications, Artech House, Norwood, MA, 2014. ©2014 Artech House. Reprinted with permission.

chart. The slightly inductive fundamental load impedance was designed as $8.4+j10.7\ \Omega$ at 900 MHz by load-pull analysis. The capacitive second harmonic impedance is also transformed to low impedance but capacitive $1.3-j41.8\ \Omega$. The third harmonic impedance is transformed into $(0.7-j1.2)$ as low impedance short. Therefore, under fixed DC supply conditions, power sweeping with various drain supply voltages can be employed to estimate the potential design for ET operation via both efficiency and gain trajectory as shown in Figure 6.14D. Since ET PA operates in compression over the majority of the envelope cycle, ET system linearization capability can be through the gain trajectory. The optimized envelope shaping function accordingly can be implemented to compromise efficiency and linearity.

Another important consideration of ET is the bias feeding line design. A conventional bias tee likewise makes use of a bank of large shunt capacitors on the drain supply side to present low impedance at the envelope bandwidth, enhancing low-frequency stability. However, in an ET mode, the supply modulator would be enforced to drive the bias tee capacitance in parallel with the RF transistor, leading to high capacitor power losses, and dramatically reduced supply modulator bandwidth. Therefore, the large capacitance must be removed in the ET mode operation. Remember that the supply modulator output impedance is ideally low over a bandwidth much larger than the envelope bandwidth, and thus presents low impedance to the RF transistor at low frequencies, just as the bank of capacitors had formerly. The stabilizing effect can be perceived when a PA operates under ET, with absence of the bias bypass capacitance, exhibits low-frequency oscillation when connected to a fixed DC supply, and nevertheless is stable when connected to supply modulator output.

As shown in Figure 6.15, there are the quarter-wave line and fundamental frequency resonant capacitor at the supply side. However, without the supply modulator present, additional shunt capacitance must be utilized again for stability and to prevent the drain voltage from drooping or ringing once a current appears. Unexpected resistance increases in the feeding line while the envelope saturates at the smallest value permitted by the envelope shaping function, which would cause some kind of resistance variation in load impedance, which further results in envelope feeding line response changes. Therefore, a RLC-snubber circuit located on the end of the symmetrical feeding line could be employed to control the impedance of the feed line. Inappropriate feeding line matching would lead to peaking effect on feeding line frequency response, which results in any residual envelope signal at those frequencies giving rise to the noise floor in RF spectrum at the same frequency offset around the center frequency.

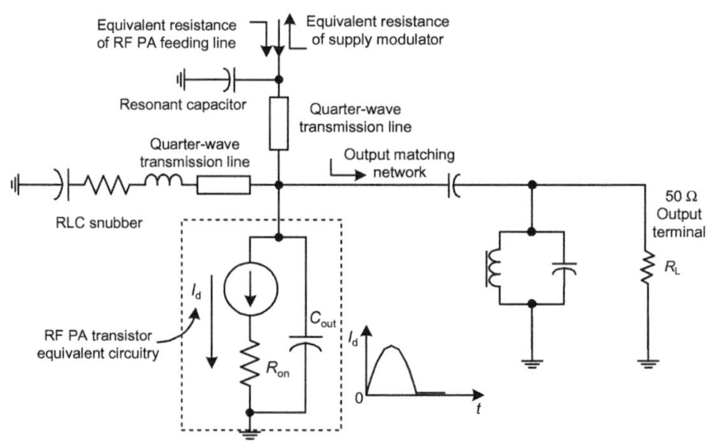

FIGURE 6.15

Envelope feeding line matching and RLC snubber for video bandwidth design.

From Z.C. Wang, Envelope Tracking Power Amplifier for Wireless Communications, Artech House, Norwood, MA, 2014. ©2014 Artech House. Reprinted with permission.

6.3.2.3 Supply modulator for ET

Because the overall efficiency of ET PA is to multiply the efficiency of the RF PA and that of the supply modulator, broadband and efficient supply modulator is indicative of the overall efficiency. Therefore, supply modulator design is considered as the centric technique of ET PA development. Typically, the category of the ET supply modulators can be classified as continuous and discrete types in general. However, there are actually subcategories under linear and discrete, either in standalone mode or hybrid combination, for the optimal performance and against the above-mentioned challenges of ET.

With the development of broadband high-PAPR signals, ever increasing the switching frequency decreases the efficiency of discrete modulators for ET [7]. Therefore, hybrid supply modulators emerge as the times require, taking advantages of both linear and discrete modulators, while minimizing their disadvantages. Hybrid supply modulator classification is flexible and designed for diverse application scenarios. For example, they can be either voltage-controlled or current-controlled; be connected in parallel or in series to the output. Figure 6.16 presents a serial hybrid supply modulator.

Serial hybrid topology can be simply understood as boosting supply modulator efficiency by means of "tracking envelope's envelope." It has combined the linear modulator with the discrete modulators to boost up efficiency of the linear modulator while maintaining the linearity. In Figure 6.16B, a multilevel/multiphase modulator supplies the linear modulator with discrete voltage levels that are as close as possible to the output voltage of the supply modulator. In this manner, the power losses on the linear modulator are minimal because they are directly proportional to the difference between its supply and output voltages. However, in order to guarantee correct operation of the linear modulator, the output voltage of the multilevel/multiphase modulator has to be higher than the output voltage of the linear modulator.

6.3.3 LINC PA

The outphasing technique consists of decomposing the magnitude and phase modulated signal into two phase modulated signals. Each of the signals can be amplified using highly nonlinear or switching mode PA. The power combiner adds the signals at the output of the PAs to recover the amplitude modulation of the original signal. In the literature, one can distinguish two major types of power combiners used with the outphasing technique: the isolated and the non-isolated combiners. The isolated combiners such as Wilkinson or hybrid combiners are matched at all input and output ports independently from the magnitude and phase of its input signals. Consequently, when input signals are out-of-phase, all the energy delivered to the isolated combiner is dissipated in the matching resistor as heat and the mean efficiency degrades considerably. The use of non-isolated power combiners, such as Chireix combiner, avoids the power loss. However, due to the nature of these non-isolated combiners, the loads presented by the combiner at its input ports vary as function of magnitudes

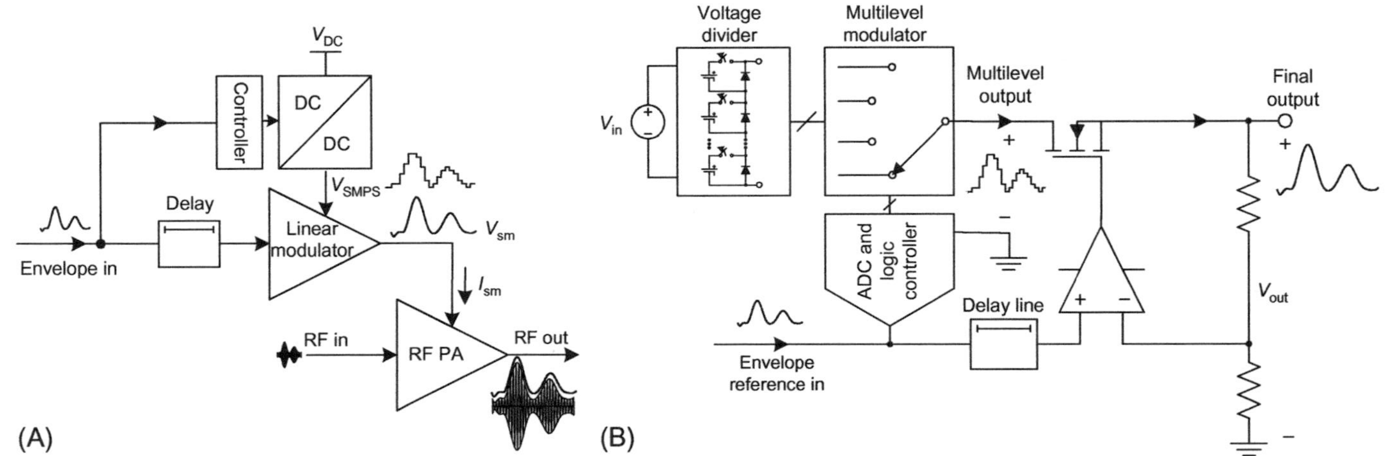

FIGURE 6.16

(A) Serial hybrid supply modulator, (B) exemplary serial hybrid architecture with stacked cells multilevel modulator in conjunction of a linear modulator.

From J.F. Sevic, Statistical characterization of RF power amplifier efficiency for WCDMA wireless communication systems, Wireless Communications Conference, Boulder, CO, USA, August 1997, 110–113. ©2014 Artech House. Reprinted with permission.

and phases of the input signals. Hence, a cross load modulation occurs between input ports, which introduces a variation in the PAs behavior such as the gain, the output power, and the DC consumption.

6.3.3.1 LINC principle

The principle of the LINC technique is to convert the amplitude modulation of the input signal $S(t)$ into phase modulations of two constant envelope signals $S_1(t)$ and $S_2(t)$, as the following equations [12]

$$S(t) = r(t) \cdot e^{j\varphi(t)} = S_1(t) + S_2(t) \tag{6.5}$$

$$r(t) = r_{max} \cdot \cos(\theta(t)) \tag{6.6}$$

$$S_i(t) = \frac{r_{max}}{2} \cdot e^{j(\varphi(t) \pm \theta(t))} \tag{6.7}$$

where r_{max} is the maximum of r, $\varphi(t)$ is the phase of the baseband signal, $\theta(t)$ is the additional phase modulation angle related to the amplitude of signal and given by

$$\theta(t) = \arccos\left(\frac{r(t)}{r_{max}}\right) \tag{6.8}$$

Because $S_1(t)$ and $S_2(t)$ have constant envelopes, they can be efficiently amplified by nonlinear amplifiers. The two amplified signals are then added to reproduce the original signal amplified. In this manner, the RF PAs can be operated at saturation, yielding maximum power efficiency; the amplified signals can then be summed precisely, yielding in principle, perfect linearity. Figure 6.17A illustrates the main building blocks of a LINC amplifier while Figure 6.17B shows a vector representation of the separated baseband components. Clearly, the three main building blocks of the LINC amplifier are the signal component separation, the nonlinear amplifiers, and the combiner.

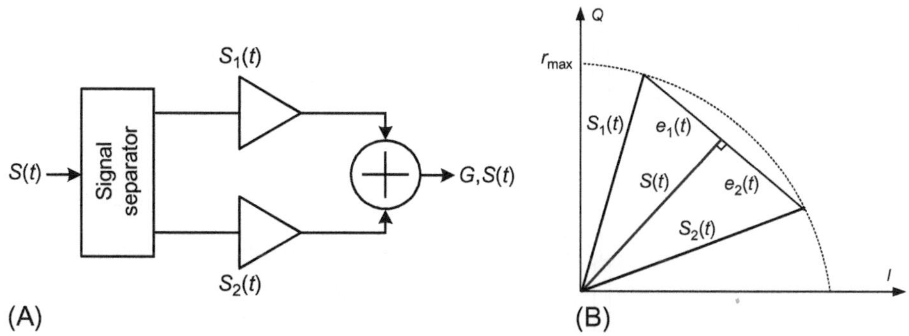

FIGURE 6.17

(A) Diagram of LINC PA, (B) baseband signal separation of LINC.

6.3.3.2 Combining structures for LINC

The use of two highly efficient PAs operated with constant envelope signals in the LINC transmitter is not sufficient to guarantee the high overall efficiency of the LINC amplifier. While the amplifiers may operate highly efficiently, it is the remaining available power at the output of the combiner that will determine the overall efficiency of the LINC system. It was shown that the average efficiency of the LINC system depends also on the power distribution of signal and the type of combiner used. The combining structure plays a key role in determining the overall power efficiency in conjunction with the PAs, as well as the linearity of the LINC PA, depending on how it affects the impedances seen by the amplifiers and how it combines the output signals. The combining structures can be divided in two broad classes which referred to the isolated combiners as in Figure 6.18 and the non-isolated combiners (also as Chireix combiners) as in Figure 6.19.

The hybrid combiner is a matched and lossy combining structure with high isolation between the combined paths. Typical structures of this type include the hybrid coupler and the Wilkinson combiner. Due to the isolation between the combined paths, these combiners yield perfect linearity at the output. Their efficiency tends

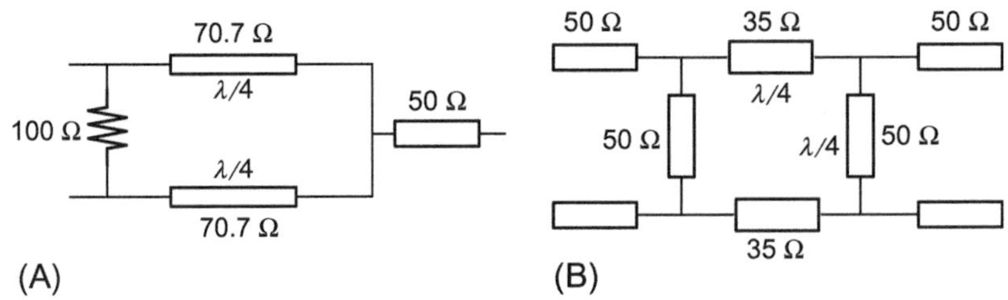

FIGURE 6.18

Isolated combining structures for LINC: (A) Wilkinson combiner, (B) hybrid coupler.

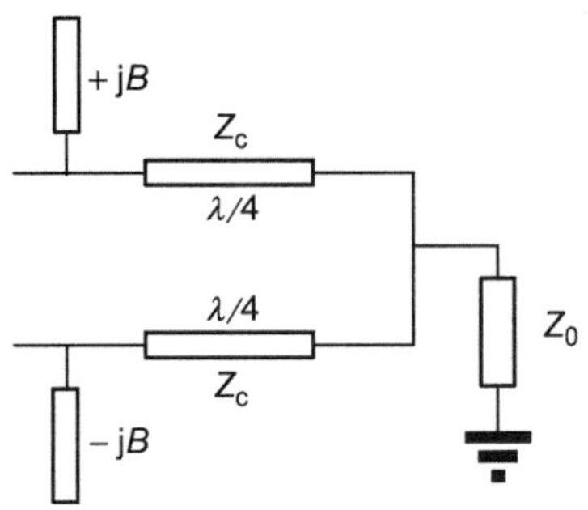

FIGURE 6.19

Non-isolated Chirex combining structure of LINC.

to degrade rapidly, which is due to the out-of-phase components of the combined signals ending up being delivered to the isolated port load. In this situation, a power recycling technique has been proposed and analyzed for efficiency-enhanced RF and microwave outphasing PAs for mobile wireless communications. With the use of a simple power recycling network, a considerable portion of the wasted power can be recovered back to the power supply.

The second class of combining structures is the non-isolated outphasing combiners. Figure 6.19 shows the Chireix-outphasing combiner structure with Z_c representing the characteristic impedance of the quarter-wavelength transmission line; jB is the magnitude of the susceptance of the shunt elements in the combiner by using transmission stubs. This susceptance is added to improve efficiency by cancelling the imaginary part of the load presented to the RF amplifier at certain power backoff.

6.3.3.3 Multilevel LINC

To eliminate the power consumption in an isolated combiner, instead of the traditional scaling technique using single level scale factor, multilevel LINC utilizes a multilevel scaling technique as a solution to achieve high efficiency [13]. Multilevel LINC adapts multilevel scale factor to shrink outphasing angle, which aims to enhance the Wilkinson combiner efficiency. Furthermore, asymmetrical multilevel LINC modulation decomposes a complex vector into two vectors such that the sum of the two vectors constructs the original complex vector with the minimum outphasing angle. As compared to the multilevel LINC technique, by making independent changes in the supply voltage for each of the two PAs, the asymmetrical multilevel LINC technique results in smaller outphasing angles so that higher efficiency can be achieved even in relatively high-PAPR standards [14], the principle and realized efficiency profiles are presented in Figure 6.20.

6.3.3.4 Mode-multiplexing LINC

To alleviate the rapid efficiency deterioration, as a function of the input signal level experienced in the LINC amplifier, mode-multiplexing LINC (MM-LINC) was proposed for efficiency improvement [15,16]. The main idea consists of using both class-B PAs in a balanced mode for high input signal levels and switching to the LINC mode at lower levels. Because the class-B efficiency drops slower than the LINC efficiency, the use of the balanced mode postpones the fast efficiency drop of the LINC system to a higher backoff. A high-efficiency value is maintained up to and about the mean power region of the signal excitation. When the gain of the class-B amplifier starts to drop for lower power values, the PAs switch to the LINC architecture mode. This allows the amplification system to maintain a high gain for the low-power levels. Moreover, a linear gain is maintained during the whole power range corresponding to the LINC operating mode. This preserves good performance in terms of signal quality over the whole dynamic range of the signal. In this way, the MM-LINC architecture benefits from the good efficiency of class-B

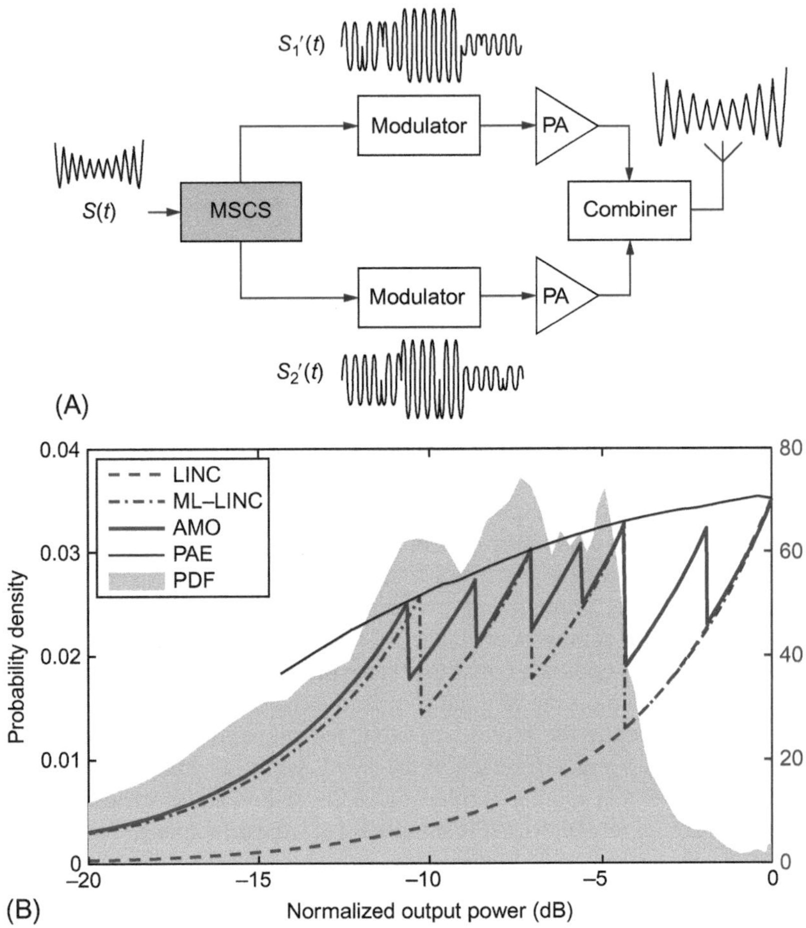

FIGURE 6.20

(A) Diagram of multilevel LINC, (B) efficiency profiles of multilevel LINC and asymmetrical multilevel LINC.

From K.-Y. Jheng, Y.-J. Chen, A.-Y. Wu, Multilevel LINC system designs for power efficiency enhancement of transmitters, IEEE J. Sel. Topics Signal Process. 3 (3) (2009) 523–532; S.W. Chung, P.A. Godoy, T.W. Barton, E.W. Huangy, D.J. Perreault, J.L. Dawson, Asymmetric multilevel outphasing architecture for multi-standard transmitters, IEEE Radio Frequency Integrated Circuits Symposium, June 2009, 1–4. ©2009 IEEE. Reprinted with permission.

amplifiers and the linearity of the LINC architecture, even when using highly nonlinear PAs. This approach achieves the same instantaneous efficiency of a class-B PA when the magnitude of the input signal is greater than the threshold. It admits a similar instantaneous efficiency to the LINC system for signal levels below the threshold value.

6.3.3.5 Power recycling LINC

An efficient power combining approach is necessary to preserve the high-power efficiency of the system. The purely reactive combiner such as Chireix combiner is a possible choice. However, this method suffers from a variation of the load impedance presented to the two PAs, which in turn creates the signal distortion. The load variation, which results from the time varying phase offset between the two signal components, can be eliminated with a conventional hybrid combiner. In the ideal case, the hybrid completely isolates the two PAs and presents each amplifier a 50 Ω load. With the isolated combiner, the desired output signal is obtained on the summing port, while the quadrature signal is obtained on the difference port. Unfortunately, the latter portion of the power is consumed at the resistive load and turns into waste heat, which degrades the overall power efficiency obviously.

The idea of the power recycling scheme is to replace the power-wasting resistive load with an RF-DC converter to recover as much of the wasted power as possible back to the power supply [17]. The schematic diagram of an implementation of this scheme is illustrated in Figure 6.21. A 180 hybrid combiner is configured as the power splitter to divide the wasted power into two 180 outphased portions. These two signals are then fed to a high-speed Schottky diode pair through an impedance matching network. The Schottky diodes rectify the RF waves and the DC components are withdrawn to the power supply. The RF choke loops provide a DC return path for the circuit. The matching network is to be adjusted to optimize the system performance. As it turns out, the efficiency of the recycling network is highly dependent on the power supply voltage, the characteristics of the diodes as well as the power delivered to the recycling network. Certainly, there still are challenges to implement high-efficiency RF-DC converter in real application scenario.

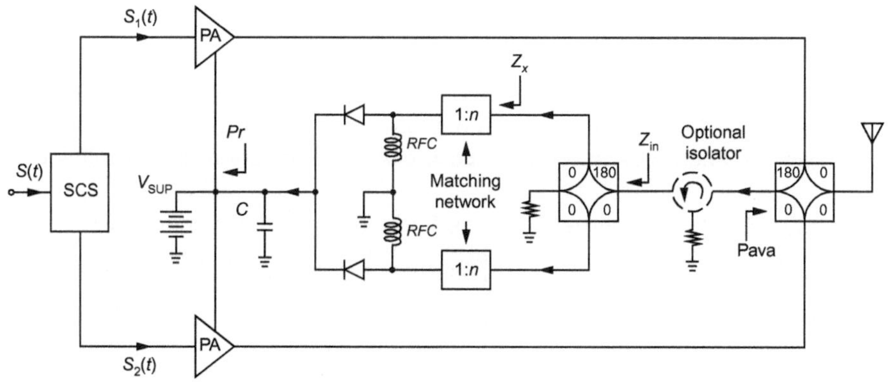

FIGURE 6.21

Power recycling technique for LINC.

From X. Zhang, L. E. Larson, P.M. Asbeck, R.A. Langridge, Analysis of power recycling techniques for RF and microwave outphasing power amplifiers, IEEE Trans. Circuits Syst. II, Analog Digit. Signal Process. 49 (5) (2002) 312–320. ©2002 IEEE. Reprinted with permission.

6.4 TRANSISTOR TECHNOLOGY FOR PA DESIGN

Herein some of the most popular transistor technologies in modern wireless communication will be introduced. Through analyzing the properties of these semiconductors and matching them to the real applications, we would like to provide a hint about the requirement of PA design on transistor features [18–21]. In this scenario, it is of fundamental importance to accurately characterize and model the transistor performance to target a successful circuit design [22–28]. For a detailed discussion on it, the reader is referred to Chapters 1 and 2. Before introducing the technologies, let's compare some of the critical properties of the commonly used semiconductors considering the described PA architectures. As already pointed out in Chapters 4 and 5, these properties are important to decide the roles of these semiconductors in the RF world.

It can be seen that SiC, for instance, has almost seven times the breakdown voltage of GaAs, which means that it could support much higher voltage with a small size, indicating a higher power density. At the same time, its thermal conductivity is eight times higher, so heat will not be a main factor to influence its high-power capability. On the other hand, the electron mobility of SiC is much lower which may limit the usage range of frequency.

In Table 6.1, the semiconductors can be divided into three categories. Silicon has the medium figures for almost all features. It is the base of the graceful electronic world, and attracts a great amount of attention, even in RF and microwave regions. GaAs and InP are the typical compound with high electron mobility, which makes them the first-class candidates for high-speed devices. SiC and GaN also attract a lot of attention, especially in the past decade, for their wide bandgap feature, making them the suitable material for high operation voltage applications, hence high-power density.

6.4.1 SILICON CMOS TECHNOLOGIES

The silicon devices are always attractive for RF and microwave designers because of their potential to be integrated with digital circuits on Si CMOS. The potential advantages are enormous in terms of cost, ease of fabrication, and perhaps even

Table 6.1 Material Properties of Some Common Used Semiconductors

	Si	GaAs	SiC	GaN
Bandgap (eV)	1.1	1.42	3.26	3.49
Electron mobility at 300 K ($cm^2/V\ s$)	1500	8500	700	1000–2000
Saturated (peak) electron velocity ($\times 10^7$ cm/s)	1.0 (1.0)	1.3 (2.1)	2.0 (2.0)	1.3 (2.1)
Critical breakdown voltage (MV/cm)	0.3	0.4	3.0	3.0
Thermal conductivity (W/cm K)	1.5	0.5	4.5	>1.5

Source: *From L.F. Eastman, U.K. Mishra, The toughest transistor yet, IEEE Spectrum, 39 (2002) 28–33.*

performance. By putting analog circuits on the same chip with the digital circuits, enormous convenience is provided for precise configuration and optimization, just like DPD and ET which have been discussed before.

In the past, industry and academics have focused many resources on the silicon processes in order to increase the integration scale and operation speed. As the minimum feature size of CMOS has been reduced to tens of nanometers, the speed capability of such systems has increased to above 100 GHz. Moreover, since the speed of many digital communication systems has reached several gigabits per second, the RF and microwave theory is also critical in high-speed digital circuit design.

For high-power RF designs, the lateral diffused MOSFET (LDMOS) has dominated the market for a long time. It is also fabricated in the CMOS process, but uses high resistivity silicon to achieve high breakdown voltages so as to handle high output power. One of the most evident differences from the regular CMOS is that the source of LDMOS is connected to ground with a "via," which is generated with a p+ sinker that runs vertically from the source contact at the surface through to the bottom of the substrate. These vias have much lower inductance compared to the bonding wires in CMOS, so the LDMOS could obtain higher gain in high frequency. LDMOS is a kind of enhancement mode FET, so it only needs a simple power supply circuit. It is also a quite rugged device and can withstand a high load mismatch.

The drawbacks of LDMOS are also evident. Its high input and output capacitance limit its usage frequency and make it difficult to design very broadband PAs. The harmonic modulations of the transistor output are also difficult to control, making a barrier for high-efficiency PA designs. In fact, LDMOS has been replaced by GaAs HBT in handheld applications, and probably will be replaced in favor of GaN HEMT in base stations, where such large amounts of energy are consumed per year.

While LDMOS is losing favor among PA designers, its big brother, CMOS, comes back to the battle. Because CMOS is so suitable for the low-cost trend of consumer market, many start-ups and even mainstream PA suppliers, like Skyworks, Avago, and RFMD, have taken part into the CMOS PA research and design. Until now, CMOS PAs have been proven a good balance of cost and performance in global system for mobile (GSM) communications uses, and tens of millions are shipped each month. Because of the difficulty implementing high-Q inductors on Si substrate, their efficiency and linearity is lower than GaAs PAs. However, since Qualcomm Inc. released their RF360 solution in 2013, which uses the multiband multimode CMOS PA together with ET module, the CMOS PA performance may be soon acceptable even in 3G/4G applications.

6.4.2 THE GaAs HBT

GaAs is the most common III-V compound semiconductors in the RF and microwave world. It became widely used in the 1980s in GaAs the metal-semiconductor field-effect transistor (MESFET) and then its heterojunction counter-part (HEMT). Since AlGaAs-GaAs HBT, sometimes called GaAs HBT for simplicity, was introduced

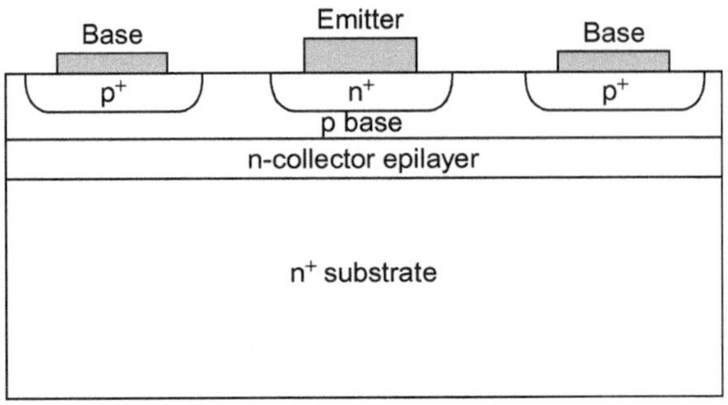

FIGURE 6.22

Simplified cross section of an NPN bipolar transistor.

From R. Gilmore, L. Besser, Practical RF Circuit Design for Modern Wireless Systems, Artech House, Boston, MA, 2003. ©2003 Artech. Reprinted with permission.

commercially in the 1990s, it has dominated most of the handheld markets for its good linearity and high efficiency.

GaAs HBT is similar in structure as Bipolar Junction Transistor (BJT), which is as Figure 6.22, but fabricated using p-type GaAs in the base sandwiched between an n-type GaAs layer in the collector and an n-type AlGaAs layer in the emitter, which is given the name "heterojunction." To obtain better performance, GaAs HBTs are also improved by using InGaP to replace AlGaAs in emitter layer. This could further decrease the turn-on voltage, so as to increase the power enhance efficiency when working under low collector voltage.

GaAs HBT has better linearity, lower turn-on voltage, and negligible output capacitance, rather than LDMOS, so it is a good choice for high-efficiency PA design, especially for handheld mobile applications, where battery voltage is low and communication quality and time are both critical. Another important advantage of GaAs HBT is that it is positive biased, so a specified negative voltage generator is not needed. This is different from MESFET and most HEMT, and very convenient for PA designers. As has been analyzed, GaAs is a kind of semiconductor that has fast electron transition, which makes it a good choice in high frequency use; and a poor thermal conductor, resulting in the domination of thermal effects in high/low temperature use.

6.4.3 THE GaN HIGH ELECTRON-MOBILITY TRANSISTOR

HEMT is a kind of horizontal structure device just like MESFET, but using a different compound semiconductor between the gate and substrate, which is very similar to the HBT compared to BJT. The GaN HEMT usually uses an AlGaN layer that forms the channel (called 2-dimension electron gas) between the gate and the GaN layer. This forms a heterojunction and results in a trapped layer of electrons having high electron velocity. In this way, it can obtain higher gain, extended frequency performance, and even better linearity, the same as the other HEMTs.

Unlike GaAs, InP and many other compounds, GaN is a wide bandgap semiconductor that can endure high breakdown voltage. As a result, the GaN HEMT can operate under very high drain-source voltage, and handle high-power density. In Sep. 2014, Cree, Inc. released a GaN HEMT, which can emit 320 W saturation power, 19 dB gain to 4 GHz. Most importantly, it could bear 150 V breakdown voltage, so it can operate with a drain bias voltage of 50 V. The high drain voltage lead to a relative high optimized resistance and prevent the output matching network from high resistance transforming ratio and narrow band (Figure 6.23).

Since the late 1990s and early 2000s, GaN HEMT has obtained intensive considerations by PA researchers and designers. In fact, it had a significant impact on the modern PA concepts and structures. Table 6.2 [18] compares the features of GaN HEMT and LDMOS (the "incumbent" technology for many applications), which are primary in different PA techniques. The main barrier that prevents GaN HEMT from real acceptance is its high cost. The lattice structure of GaN is not well matched to most substrate materials. At the present, SiC is the most commonly used, which makes them expensive to produce. As larger and larger SiC wafer have been reported to be fabricated (150 mm diameter by now, from Dow Corning), and the yield is being improved, the cost will probably soon decrease to an acceptable degree, and we may witness the great boom of GaN technology. Fortunately, GaN HEMTs can be realized on Si substrate, cheaper but worse performance than SiC.

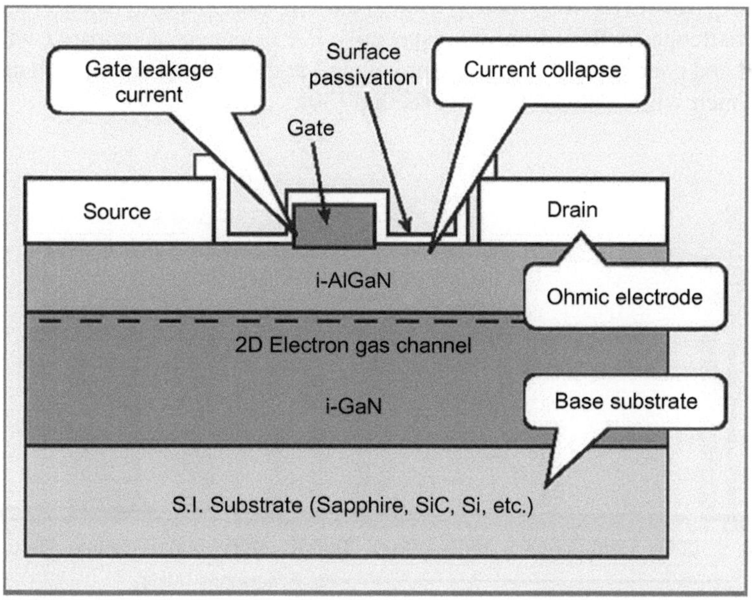

FIGURE 6.23

Schematic cross section of GaN HEMT device.

Table 6.2 Impact of GaN on PA Architectures

Concept	Silicon	GaN on SiC
Doherty	Low off-state impedance, high output capacitance	High off-state impedance, high output capacitance
LINC	Large nonlinear output capacitance	Small nonlinear output capacitance
EER/ET	Poor amplitude to phase modulation conversion, moderate bandwidth	Poor amplitude to phase modulation conversion, moderate bandwidth

Source: *From R. Pengelly, S. Wood, J. Milligan, S. Sheppard, W. Pribble, A review of GaN on SiC high electron-mobility power transistors and MMICs, IEEE Trans. Microw. Theory Techn. 60 (6) (2012) 1764–1783. ©2012 IEEE. Reprinted with permission.*

6.5 BROADBAND AND MULTIBAND PA

Along with the rapid evolution of the fourth and fifth generation (5G) communication systems, many communication standards have been proposed and put into operation, as shown in Figure 6.24. Such diversity of communication standards, along with the application of carrier aggregation, presents increasing requirements on flexibility of communication systems, especially for wireless communication infrastructures. One solution to provide such flexibility is the broadband transmitter, which features the ability to cover a wide continuous band of frequency. Another solution is the dual-band/multiband transmitter, with the ability of handling two separated bands [29]. However, constructing broadband and multiband transmitters presents a real challenge to RF engineers, especially PA designers. Compared with their narrow-band counterparts, designing broadband or multiband PAs is much more difficult, which will be covered in this section [30].

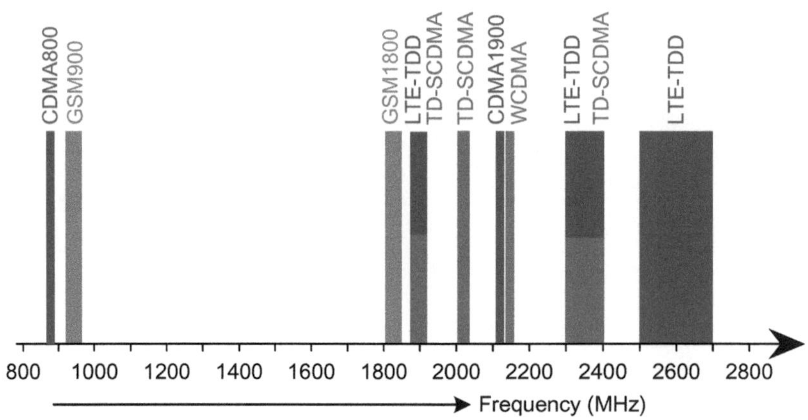

FIGURE 6.24

Typical frequency configuration of mobile communication systems in a city.

6.5.1 BROADBAND PA DESIGN

6.5.1.1 Broadband impedance matching networks

Similar to narrow-band PA, the startpoint of designing broadband PAs is acquiring the desired source and load impedances of the target power transistors, which can be performed based on load-pull technique for optimized output power, efficiency, linearity, or compromise among those performances. The optimized load impedance of a typical 10 W at 1.9–2.6 GHz frequency band is shown in Figure 6.25.

With the optimized impedance acquired, output matching networks should be designed to perform the impedance matching over the target frequency band. Generally speaking, we have two candidate techniques to perform the broadband impedance matching. The first one is the network synthesis method, based on network synthesis method such as real frequency technique. The second one is the empirical method based on practical topology and optimization of the value of components. For the transistor shown in Figure 6.25, two matching networks are shown in Figures 6.26 and 6.27, based on real frequency technique and empirical L-sections, respectively, the impedance response of which are shown and compared in Figure 6.25.

As can be seen from Figure 6.25, both network synthesis and empirical method can give satisfactory performances. Obviously, the network synthesis method is

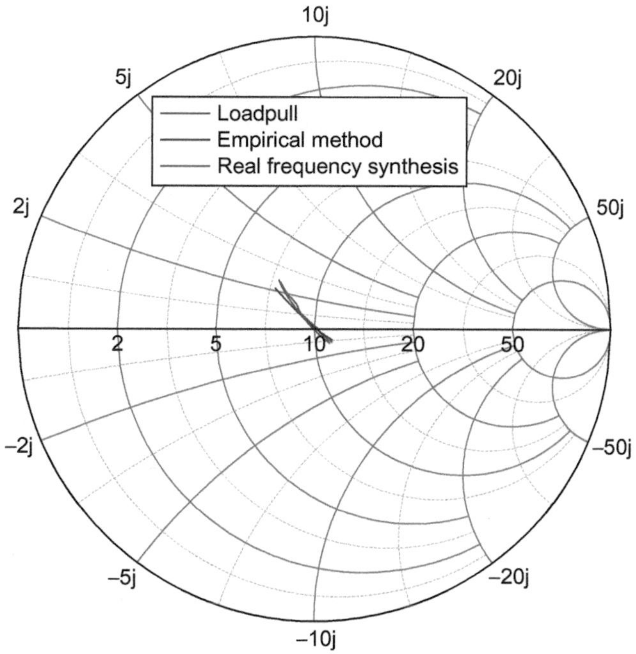

FIGURE 6.25

Impedance comparison of real frequency method and empirical method over the 1.9–2.6 GHz LTE band, shown on a 10-Ω Smith chart.

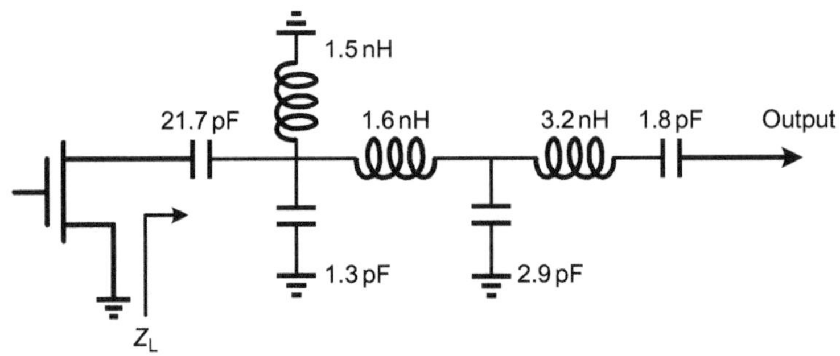

FIGURE 6.26

Broadband output matching network synthesized via real frequency technique.

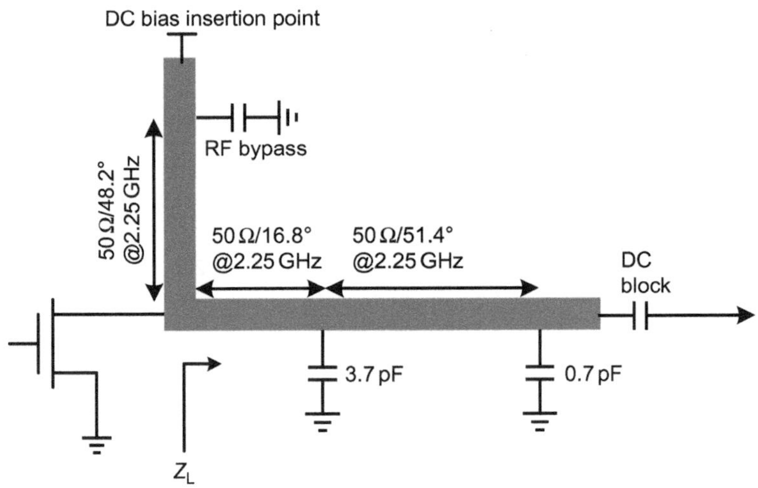

FIGURE 6.27

Empirical broadband output matching network based on multistage L-sections.

much more elegant. However, when talking about PAs (instead of other impedance matching problems), network synthesis method exhibits its inherent weaknesses. Firstly and most significantly, PAs' performance are highly sensitive to harmonics impedances, however, existing network synthesis focuses on fundamental impedance while it cannot take harmonics impedance into account, resulting in uncontrolled harmonics impedances. Secondly, the synthesized networks are not always suitable for PA circuit implementation (eg, the network shown in Figure 6.26 does not provide suitable DC bias insertion point). Finally, network synthesis technique generally give lumped circuits containing lumped inductors and capacitors, while in typical communication frequencies, distributed circuits are easier to implement. Accordingly, in PA design, empirical methods are used more widely than network synthesis method. However, network synthesis technique still finds its way into PA design, especially as the guide to select empirical topologies and to provide a starting point of optimization.

6.5.1.2 Broadband bias networks

Generally speaking, the bias networks are inserted between the bias insertion point of the matching network and the DC power supply. The most significant function of the bias network is to separate RF circuit from DC power supply circuit, or in other words, to prevent the RF circuit from being affected by the unpredictable RF behavior of DC power supply circuit.

In narrow-band PAs, the single-section bias networks as shown in Figure 6.28 are widely used, in which the RF isolation is provided by RF bypass capacitor while video components (as known as envelope components) isolation is provided by video bypass capacitor. Typically, high-Q chip capacitors are selected as the RF bypass capacitors for their low equivalent series inductance (ESL) performance and consequently high self-resonant frequency (SRF). In narrow-band PAs, the RF bypass capacitor's values are selected to settle its series SRF at the working frequency to provide lowest impedance and best isolation. However, for broadband PAs, the simple bias network shown in Figure 6.28 cannot provide satisfactory RF isolation. Instead, the multistage bias network can be employed in broadband PAs to provide sufficient RF isolation, as shown in Figure 6.29.

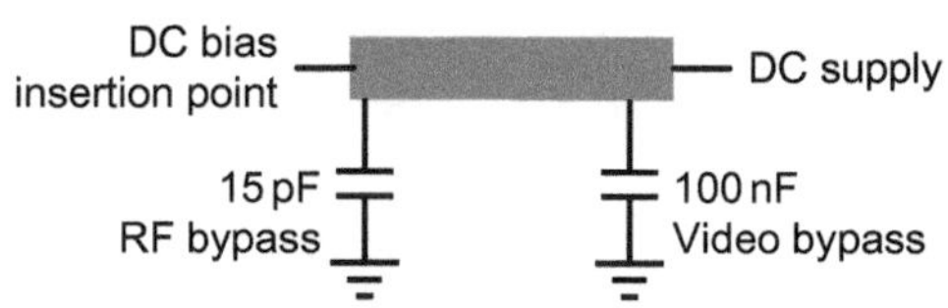

FIGURE 6.28

Narrow-band bias network working at 2.25 GHz.

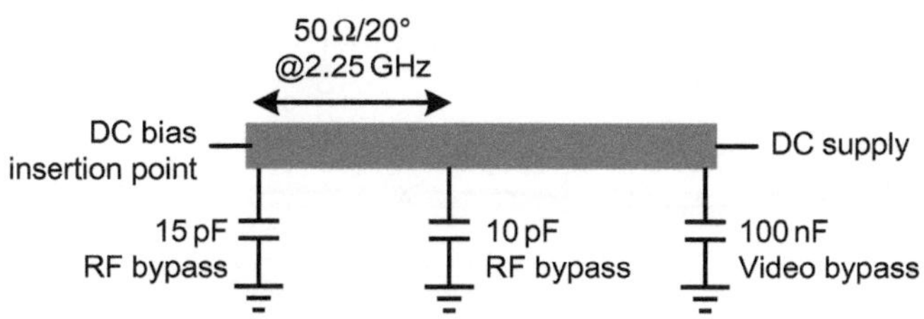

FIGURE 6.29

Broadband bias network working at 1.9–2.6 GHz.

6.5.2 MULTIBAND PA DESIGN

6.5.2.1 Multiband PAs

PAs that can work at two or more separated frequency bands, which is known as dual-band/multiband PAs, have received wide concern in recent years. However, the term "multiband PAs" is somewhat ambiguous and can be identified into three different classes of PAs, that are reconfigurable multiband PAs, nonconcurrent multiband PAs, and concurrent multiband PAs, as shown in Figures 6.30–6.32, respectively.

1. Reconfigurable Multiband PAs

   These PAs can work at two or more separated bands, with the reconfiguration of matching components, bias voltages, or other parts of the PA. The reconfiguration is generally controlled by a band select signal.
2. Nonconcurrent Multiband PAs

   Similar to reconfigurable multiband PAs, noncurrent multiband PAs reduce the complexity of the system by eliminating the requirement of the reconfiguration. The elimination is typically achieved by deliberate passive networks design of matching networks and/or bias networks. Though such PAs can work at two or more separated bands without reconfiguration of the circuit, it cannot handle the two bands concurrently, which is known as nonconcurrent multiband PAs.

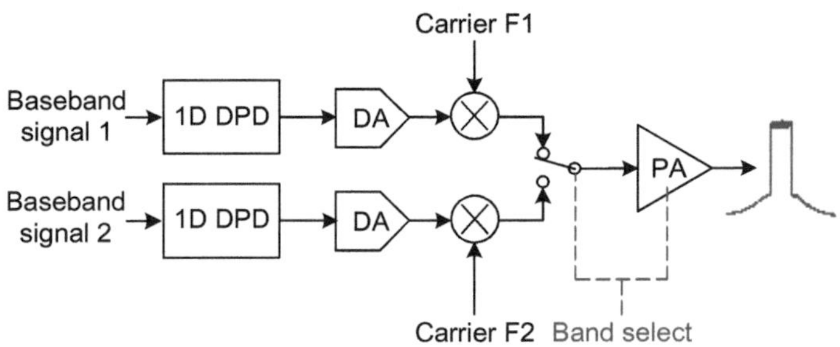

FIGURE 6.30

Reconfigurable multiband PAs.

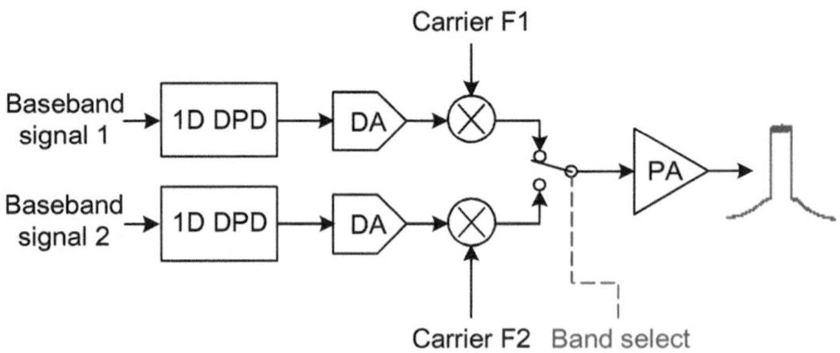

FIGURE 6.31

Nonconcurrent multiband PAs.

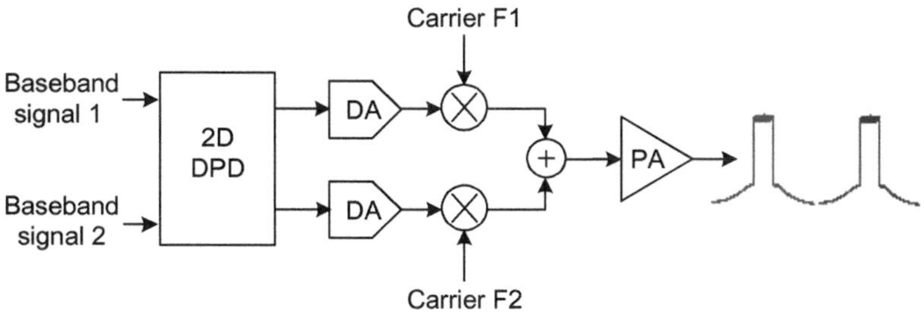

FIGURE 6.32

Concurrent multiband PAs.

3. Concurrent Multiband PAs

Concurrent multiband PAs also eliminate the need of reconfiguration. Furthermore, these PAs can work under the stimulus of the two bands concurrently. Typically, the ability of working under concurrent stimuli is achieved by manipulating the operation of the active device.

6.5.2.2 Analysis of concurrent multiband PAs

Obviously, among the three classes of multiband PAs, the concurrent multiband PAs can introduce more flexibility to the system and dramatically reduce the complexity of the system. Especially, concurrent multiband PAs can eliminate the need of cavity multiplexer and consequently reduce the cost and size of the system. Accordingly, concurrent multiband PAs are more attractive to communication operators and equipment manufacturers, compared with reconfigurable and nonconcurrent multiband PAs. With the stimulus of two carriers, the behavior of concurrent multiband PAs is much more complex than their single-band counterparts, which will be analyzed theoretically in this section. The analysis is based on the concurrent CW analysis, involving stimulating the PA using two carriers with varied amplitudes and phases and then observing the response of the device (Figure 6.33).

The concurrent CW stimulus signals can be written mathematically as

$$v_g(t) = V_{s1}\cos(\omega_1 t + \theta_1) + V_{s2}\cos(\omega_2 t + \theta_2) \tag{6.9}$$

Without losing generality, it is assumed that the transistor is biased at ideal class-B mode. In this case, the drain current can be decided according to the gate voltage stimulus. Two sets of typical waveforms are shown in Figures 6.34 and 6.35, respectively.

Obviously, the stimulus and response waveforms are highly dependent on the frequencies, amplitudes, and phases of the two carriers, as can be shown from Figures 6.34 and 6.35. However, it can be derived that the spectrum of the drain current is only related to the magnitudes of the two carries while nearly independent from frequencies and phases. The drain current spectrum is calculated and shown in Figure 6.35 when $K_s = V_{s2}/V_{s1} = 1$, $\omega_2/\omega_1 = 4/3$.

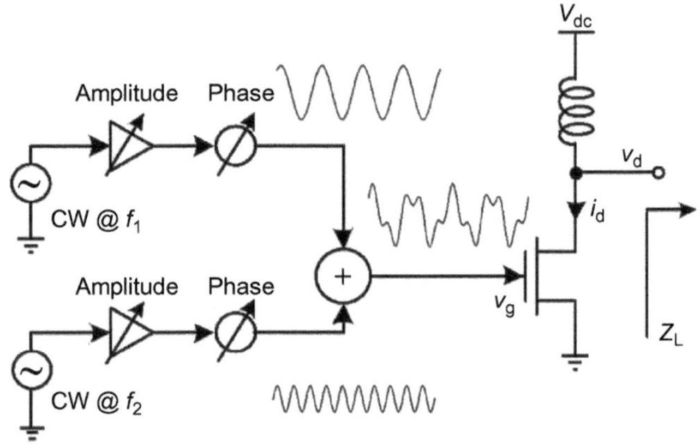

FIGURE 6.33

Concurrent CW analysis.

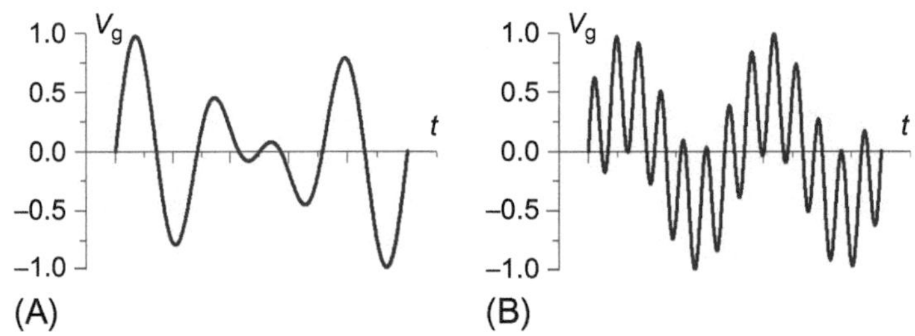

FIGURE 6.34

Concurrent CW signals with various parameters. (A) $K_s = V_{s2}/V_{s1} = 1$, $\omega_2/\omega_1 = 4/3$, $\Delta\theta = 0$. (B) $K_s = V_{s2}/V_{s1} = 1$, $\omega_2/\omega_1 = 13/2$, $\Delta\theta = 0$.

From X. Chen, W. Chen, F.M. Ghannouchi, Z. Feng, Y. Liu, Enhanced analysis and design method of concurrent dual-band power amplifiers with intermodulation impedance tuning, IEEE Trans. Microw. Theory Techn. 61 (12) (2013) (Part 2) 4544–4558. ©2013 IEEE. Reprinted with permission.

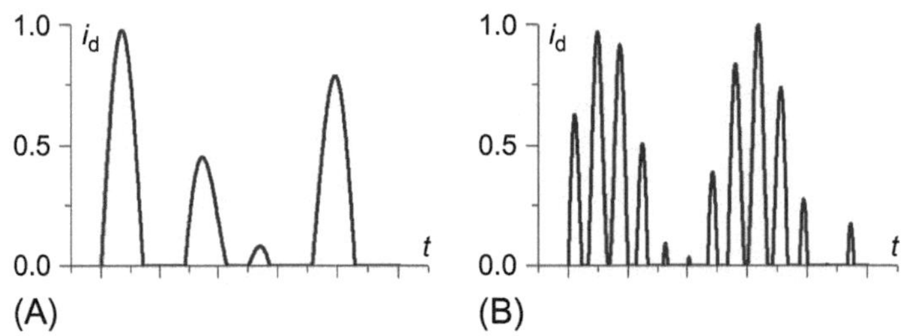

FIGURE 6.35

Normalized drain current waveforms for class-B. (A) $K_s = V_{s2}/V_{s1} = 1$, $\omega_2/\omega_1 = 4/3$, $\Delta\theta = 0$. (B) $K_s = V_{s2}/V_{s1} = 1$, $\omega_2/\omega_1 = 13/2$, $\Delta\theta = 0$.

From X. Chen, W. Chen, F.M. Ghannouchi, Z. Feng, Y. Liu, Enhanced analysis and design method of concurrent dual-band power amplifiers with intermodulation impedance tuning, IEEE Trans. Microw. Theory Techn. 61 (12) (2013) (Part 2) 4544–4558. ©2013 IEEE. Reprinted with permission.

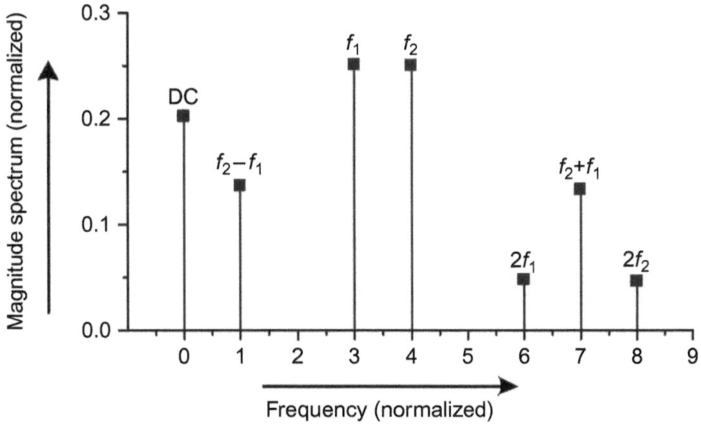

FIGURE 6.36

Spectrum of class-B drain current ($V_{s2}/V_{s1}=1$, $\omega_2/\omega_1=4/3$).

From X. Chen, W. Chen, F.M. Ghannouchi, Z. Feng, Y. Liu, Enhanced analysis and design method of concurrent dual-band power amplifiers with intermodulation impedance tuning, IEEE Trans. Microw. Theory Techn. 61 (12) (2013) (Part 2) 4544–4558. ©2013 IEEE. Reprinted with permission.

6.5.2.3 Intermodulation impedance matching

According to Figure 6.36, in addition to fundamental components and harmonics components, the drain current also contains intermodulation (IM) components, which directly lead to the concept of IM tuning. The basic idea of the IM tuning is that the IM components in drain current must be presented proper load impedances to reduce the drain voltage swing and consequently improve the PA's performance, which is similar to the harmonic tuning in classical single-band PA. Obviously, the simplest way of IM tuning is simply shorting them out by exhibiting zero load impedances to the drain at all IM frequencies. With perfect IM shorting, accompanied with harmonics shorting, the drain voltage only contains fundamental components plus a DC bias.

Noteworthy here, in the conventional design of dual-band PA, a concept of IM suppression is sometimes mentioned. Though it also can be considered as one special type of IM tuning, IM suppression is a completely different concept. The intention of IM suppression is suppressing the IM components in the output signal, so that no IM power is transferred to the load. Therefore, from the perspective of load impedance, IM suppression equals presenting an arbitrary reactive impedance to the transistor. While with proposed IM tuning, the IM impedance should be specified as zero, which is more restrictive than IM suppression. The comparison between the proposed IM tuning and conventional IM suppression is shown in Figure 6.37. According to Figure 6.37, with IM suppression, the load impedance is located on the $|\Gamma|=1$ circle, other than specified. Such a random load reactance will cause the PA's concurrent performance unpredictable and have the potential to deteriorate the performance evidently.

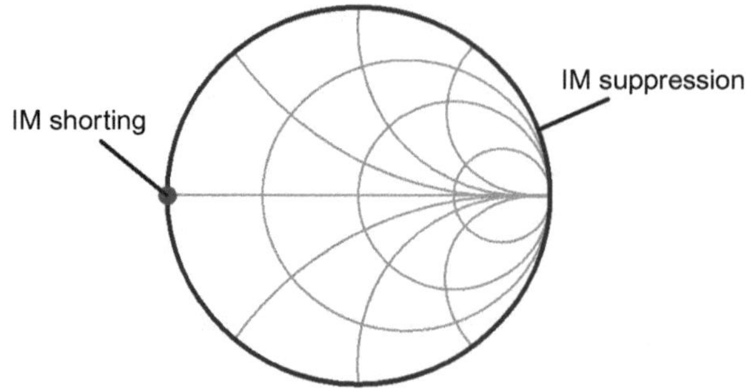

FIGURE 6.37

The desired load impedance suggested by the proposed IM tuning (IM shorting in this case) and conventional IM suppression.

From X. Chen, W. Chen, F.M. Ghannouchi, Z. Feng, Y. Liu, Enhanced analysis and design method of concurrent dual-band power amplifiers with intermodulation impedance tuning, IEEE Trans. Microw. Theory Techn. 61 (12) (2013) (Part 2) 4544–4558. ©2013 IEEE. Reprinted with permission.

The desired load impedance for concurrent modes are summarized in Figure 6.38A, showing perfect harmonic shorting and IM shorting. It can be proven that among all the IM components, the impedance matching at the lower second-order IM frequency (lower IM2) $\omega_2 - \omega_1$ is most important. The simulated performance versus varied lower IM2 load impedance is shown in Figure 6.38B for a 1.9–2.6 GHz dual-band PA, showing the significance of IM tuning.

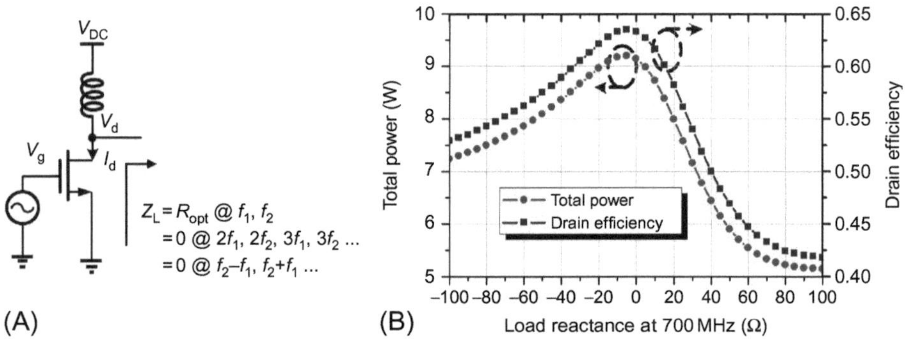

FIGURE 6.38

(A) Desired load impedances for concurrent reduced conduction angle modes, (B) simulated concurrent performance versus the load reactance at lower second-order IM frequency ($f_2 - f_1 = 700$ MHz).

From X. Chen, W. Chen, F.M. Ghannouchi, Z. Feng, Y. Liu, Enhanced analysis and design method of concurrent dual-band power amplifiers with intermodulation impedance tuning, IEEE Trans. Microw. Theory Techn. 61 (12) (2013) (Part 2) 4544–4558. ©2013 IEEE. Reprinted with permission.

REFERENCES

[1] J.F. Sevic, Statistical characterization of RF power amplifier efficiency for WCDMA wireless communication systems, in: Wireless Communications Conference, Boulder, CO, USA, August, 1997, pp. 110–113.

[2] H. Ladhani, J. Jones, G. Bouisse, Improvements in the instantaneous bandwidth capability of RF power transistors using in-package high-k capacitors, in: IEEE MTT-S International Microwave Symposium Digest, Baltimore, MD, USA, June, 2011, pp. 1–4.

[3] F.H. Raab, Efficiency of Doherty RF power amplifier system, IEEE Trans. Broadcast. BC-33 (3) (1983) 77–83.

[4] B. Kim, J. Kim, I. Kim, J. Cha, The Doherty amplifier, IEEE Microw. Mag. 7 (5) (2006) 42–50.

[5] J. Kim, J. Cha, I. Kim, B. Kim, Optimum operation of asymmetrical-cells-based linear Doherty power amplifiers—uneven power drive and power matching, IEEE Trans. Microw. Theory Techn. 53 (5) (2005) 1802–1809.

[6] R. Darraji, F.M. Ghannouchi, O. Hammi, A dual-input digitally driven Doherty amplifier architecture for performance enhancement of Doherty transmitters, IEEE Trans. Microw. Theory Techn. 59 (5) (2011) 1284–1293.

[7] K. Bathich, A.Z. Markos, G. Boeck, Frequency response analysis and bandwidth extension of the Doherty amplifier, IEEE Trans. Microw. Theory Techn. 59 (4) (2011) 934–944.

[8] D.Y.-T. Wu, S. Boumaiza, A modified Doherty configuration for broadband amplification using symmetrical devices, IEEE Trans. Microw. Theory Techn. 60 (10) (2012) 3201–3213.

[9] F.H. Raab, Intermodulation distortion in Kahn technique transmitters, IEEE Trans. Microw. Theory Techn. 44 (12) (1996) 2220–2225.

[10] B. Kim, J. Moon, I. Kim, Efficiently amplified, IEEE Microw. Mag. 11 (5) (2010) 87–100.

[11] Y. Li, System and circuits investigation of wideband RF polar transmitters using envelope tracking for mobile WiMAX/WiBRO applications, Master of Science Thesis, Electrical Engineering, Texas Tech University, Lubbock, TX, 2009.

[12] A. Birafane, M. El-Asmar, A.B. Kouki, M. Helaoui, F.M. Ghannouchi, Analyzing LINC systems, IEEE Microw. Mag. 11 (5) (2010) 59–71.

[13] K.-Y. Jheng, Y.-J. Chen, A.-Y. Wu, Multilevel LINC system designs for power efficiency enhancement of transmitters, IEEE J. Sel. Topics Signal Process. 3 (3) (2009) 523–532.

[14] S.W. Chung, P.A. Godoy, T.W. Barton, E.W. Huangy, D.J. Perreault, J.L. Dawson, Asymmetric multilevel outphasing architecture for multi-standard transmitters, in: IEEE Radio Frequency Integrated Circuits Symposium, June, 2009, pp. 1–4.

[15] M. Helaoui, S. Boumaiza, F.M. Ghannouchi, A.B. Kouki, A. Ghazel, A new mode-multiplexing LINC architecture to boost the efficiency of WiMAX up-link transmitters, IEEE Trans. Microw. Theory Techn. 55 (2) (2007) 248–253.

[16] J.H. Qureshi, M.J. Pelk, M. Marchetti, W.C.E. Neo, J.R. Gajadharsing, M.P. van der Heijden, L.C.N. de Vreede, A 90-W peak power GaN outphasing amplifier with optimum input signal conditioning, IEEE Trans. Microw. Theory Techn. 57 (8) (2009) 1925–1935.

[17] X. Zhang, L.E. Larson, P.M. Asbeck, R.A. Langridge, Analysis of power recycling techniques for RF and microwave outphasing power amplifiers, IEEE Trans. Circuits Syst. II, Analog Digit. Signal Process. 49 (5) (2002) 312–320.

[18] R. Pengelly, S. Wood, J. Milligan, S. Sheppard, W. Pribble, A review of GaN on SiC high electron-mobility power transistors and MMICs, IEEE Trans. Microw. Theory Techn. 60 (6) (2012) 1764–1783.

[19] R. Gilmore, L. Besser, Practical RF Circuit Design for Modern Wireless Systems, Artech House, Boston, MA, 2003.

[20] M. Golio, RF and Microwave Semiconductor Device Handbook, CRC Press, Boca Raton, FL, 2003.

[21] L.F. Eastman, U.K. Mishra, The toughest transistor yet, IEEE Spectrum 39 (2002) 28–33.

[22] D. Schreurs, J. Wood, N. Tufillaro, L. Barford, D. Root, Construction of behavioral models for microwave devices from time domain large-signal measurements to speed up high-level design simulations, Int. J. RF Microw. Comput. Aided. Eng. 13 (1) (2003) 54–61.

[23] P. Colantonio, F. Giannini, E. Limiti, V. Teppati, An approach to harmonic load- and source-pull measurements for high-efficiency PA design, IEEE Trans. Microw. Theory Techn. 52 (1) (2004) 191–198.

[24] O. Jardel, F. De Groote, T. Reveyrand, J.-C. Jacquet, C. Charbonniaud, J.-P. Teyssier, D. Floriot, R. Quéré, An electrothermal model for AlGaN/GaN power HEMTs including trapping effects to improve large-signal simulation results on high VSWR, IEEE Trans. Microw. Theory Techn. 55 (12) (2007) 2660–2669.

[25] A. Ferrero, M. Pirola, Harmonic load-pull techniques: an overview of modern systems, IEEE Microw. Mag. 14 (4) (2013) 116–123.

[26] G. Bosi, G. Crupi, V. Vadalà, A. Raffo, A. Giovannelli, G. Vannini, Nonlinear modeling of LDMOS transistors for high-power FM transmitters, Int. J. Numer. Model. Electron. Netw. Dev. Fields 27 (5–6) (2014) 780–791.

[27] H. Jang, P. Roblin, Z. Xie, Model-based nonlinear embedding for power-amplifier design, IEEE Trans. Microw. Theory Techn. 62 (9) (2014) 1986–2002.

[28] G. Avolio, A. Raffo, I. Angelov, V. Vadalà, G. Crupi, A. Caddemi, G. Vannini, D.M.M.-P. Schreurs, Millimetre-wave FET nonlinear modelling based on the dynamic-bias measurement technique, IEEE Trans. Microw. Theory Techn. 62 (11) (2014) 2526–2537.

[29] W. Chen, S.A. Bassam, X. Li, Y. Liu, K. Rawat, M. Helaoui, F.M. Ghannouchi, Z. Feng, Design and linearization of concurrent dual-band Doherty power amplifier with frequency-dependent power ranges, IEEE Trans. Microw. Theory Techn. 59 (10) (2011) 2537–2546. Part 1.

[30] X. Chen, W. Chen, F.M. Ghannouchi, Z. Feng, Y. Liu, Enhanced analysis and design method of concurrent dual-band power amplifiers with intermodulation impedance tuning, IEEE Trans. Microw. Theory Techn. 61 (12) (2013) 4544–4558. Part 2.

[31] Z.C. Wang, Envelope Tracking Power Amplifier for Wireless Communications, Artech House, Norwood, MA, 2014.

CHAPTER

Nonlinear applications at the transmitter system level

7

E. McCune

RF Communications Consulting, Santa Clara, CA, United States

CHAPTER OUTLINE

7.1 INTRODUCTION

There is a long history of trading off linearity for better energy efficiency in transmitter design. This tradeoff is fundamental: if either the linearity or efficiency specification is extremely important, then optimizing one always forces a minimum of the other. Transistor characteristic curves and Ohm's law set the parameters of this tradeoff. Fortunately, the transmitter design problem allows the use of circuit and system nonlinearity, which provides some options for simultaneously achieving high efficiency and output signal accuracy. It is this output signal accuracy that truly is important, and it is not necessary to have circuit and system linearity to achieve it.

Microwave Wireless Communications. http://dx.doi.org/10.1016/B978-0-12-803894-9.00007-X

In this chapter, the specific physics of this tradeoff between linearity and efficiency are illustrated, using representative transistor characteristic curves. Accessing the available high-efficiency options is described, from high-voltage linear design to ultimately maximizing available efficiency which requires strongly nonlinear operation. Use of dynamic power supply (DPS) operation as a middle ground is explored, requiring characterization of power amplifiers as three ports. From this new 3-port characterization, multiple power amplifier (PA) operating modes are identified. Their individual characterization guides designs later in this Chapter. Matching available transistor types to these operating modes leads to establishing design preferences based on the overall transmitter performance specifications.

7.2 POWER DISSIPATION VERSUS LINEARITY

When an amplifier is evaluated for efficiency, it is necessary to be unambiguous in what this evaluation is actually doing. There are more than seven separate uses for the word "efficiency" in the literature with regard to amplifiers [1], which causes a significant amount of ambiguity and likely misunderstanding of the authors' intentions.

One fundamental conservation principle in physics is the conservation of energy (CoE). But when the efficiency of an amplifier is evaluated, a power conservation method is used, as shown in Figure 7.1A. Though energy and power are related, but they are not the same. Can we really get away with this?

To get this answer, we begin with the power conservation equation from Figure 7.1A, including the units for each term,

$$P_{\text{IN}}\ (\text{J/s}) + P_{\text{DC}}\ (\text{J/s}) + P_{\text{OUT}}\ (\text{J/s}) + P_{\text{DISS}}\ (\text{J/s}) = 0 \tag{7.1}$$

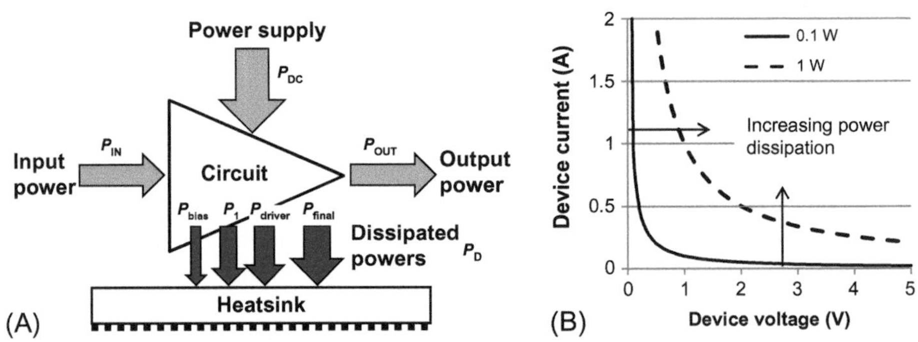

FIGURE 7.1

Energy efficiency of any circuit is dependent on the circuit dissipated power: (A) sum of powers block diagram, (B) dissipated power increases as the instantaneous operating point moves away from either axis on the *I*-*V* plane.

Knowing that watt = joule per second, as long as the seconds are considered equally, then that unit can be algebraically removed. This leaves what appears as a CoE relationship, except that the numbers are all wrong because they are not energies, they are still powers. Yet, this shows that this widespread calculation method does achieve its intended purpose: doing a conservation of power evaluation of an amplifier does provide a correct result in alignment — but not direct numeric agreement — with the physical CoE principle.

This means that the appropriate name for this PA efficiency is *energy efficiency*. This is more appropriate than the term "power efficiency" for two reasons. First, the term "power efficiency" is already defined as a communication link metric [2] relating the power from the transmitter needed to achieve a particular receiver bit-error rate or demodulator input signal-to-noise ratio. This means that giving something else the name "power efficiency" is an immediate ambiguity and therefore should not be done. Second, the actual physical principle that the efficiency calculation is based on is CoE — there is no such principle as conservation of power.

Evaluation of Eq. (7.1) for the fundamental relation of energy efficiency gives the result

$$\frac{P_{OUT}}{P_{DC}} = 1 - \frac{P_{DISS} - P_{IN}}{P_{DC}} \tag{7.2}$$

Two important realizations are apparent from Eq. (7.2). One is that the input power, particularly for the usually designed common-source or common-emitter circuits, is dissipated and does *not* pass through to the output power as envisioned by the concept of power-added efficiency. Second is that to maximize the efficiency it is equivalent to minimize the dissipated power. Looking at Figure 7.1B, it is equivalent to minimize the power dissipation by operating the transistor close to either axis of the *I*-*V* plane.

Overlaying power dissipation contours from Ohm's law on a typical set of transistor characteristic curves is shown in Figure 7.2. The controlled current source (CCS) region of transistor operation is located away from the plane axes, so the power dissipation in CCS operation, necessary for linear amplifier operation, is fundamentally higher.

7.2.1 POWER ALONG THE CHARACTERISTIC CURVES

When a load resistance is placed on a transistor, load line analysis techniques as shown in Figure 7.3A are appropriate and widely used. Linear operation only occurs when the transistor controlled-port voltage (V_{DS} for field effect transistors (FETs), V_{CE} for bipolars) exceeds the "knee voltage", which is the boundary of the transistor CCS region. This knee voltage V_k is a property of the transistor and is not a constant value. Instead, the knee voltage follows a profile over the characteristic curves like that shown in Figure 7.3A.

The bias point in Figure 7.3B, at the mid-voltage between V_k and the supply, is very near the peak of the transistor power dissipation profile. If the amplifier is

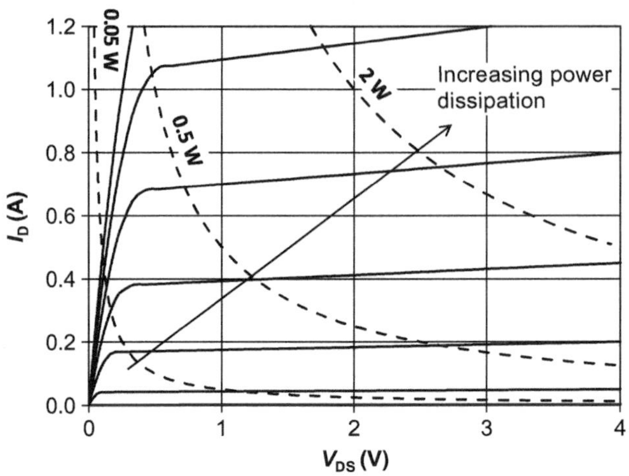

FIGURE 7.2

Overlay of transistor characteristic curves and constant power dissipation contours. Controlled current source operation only exists where power dissipation is relatively high.

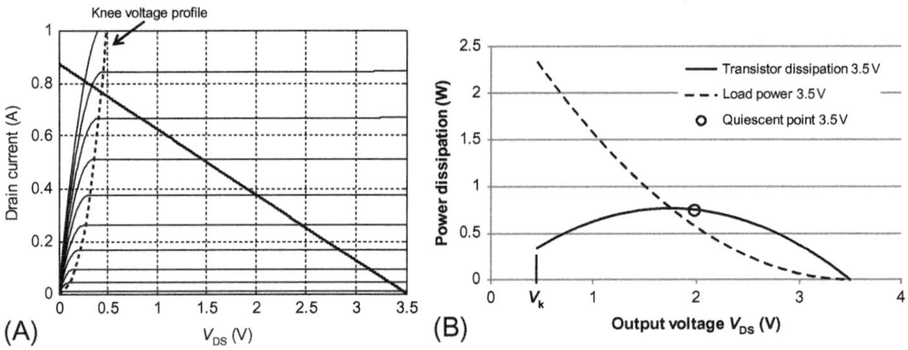

FIGURE 7.3

Load line analysis of an amplifier: (A) load line overlay on transistor characteristic curves; (B) power dissipated in the transistor and the load at all accessible operating points along the load line.

putting out a small signal, the power dissipation will stay very near to this point. This corresponds to a power dissipation maximum, and according to Eq. (7.2) is also an amplifier efficiency minimum.

To improve amplifier efficiency at small output signals, this power dissipation must be reduced. But once the supply voltage and load line are selected, the power dissipation profile of Figure 7.3B is fixed and immutable. As discussed in Chapter 4, the only way to reduce small signal power dissipation is to reduce the bias, as shown in Figure 7.4. This choice has an important cost.

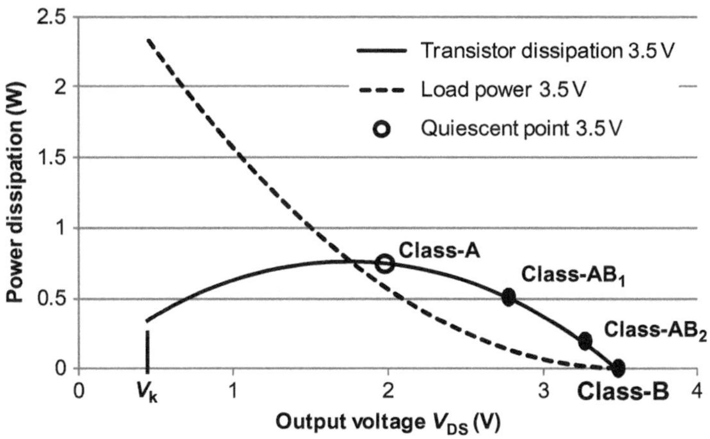

FIGURE 7.4

Power dissipation decreases away from class-A operation, because nonlinear cutoff operation has no power dissipation. This corresponds to the efficiency increase as the bias moves away from class-A.

The cost for this reduced bias is an inherent loss of amplifier linearity. As the bias moves from continuous current in class-A to transistor cutoff at class-B, the transistor output spends an increasing amount of time within each radio frequency (RF) cycle at zero current. This is a very fundamental nonlinearity, and it is present at the transistor output waveform. The use of class-AB and class-B bias conditions enters the energy efficiency versus amplifier linearity tradeoff. This tradeoff is fundamental: transistor linearity occurs with only class-A bias, the condition of minimum energy efficiency. Any improvement in energy efficiency requires the transistor to spend time in cut-off operation, which is fundamentally nonlinear.

We are left with this tradeoff: linear amplification or energy efficient operation, and must choose one.

Class-AB is considered a linear amplifier technique only because of resonant loads that restore the sinusoidal waveform at the output by bandpass (or lowpass) filtering. The entire amplifier appears to behave linearly, even though the transistor itself is not.

7.2.2 KNEE VOLTAGE PROFILES

The knee voltage is a property of the transistor, as seen in Figure 7.3A. By definition this is the boundary of CCS operation, which makes the knee voltage the minimum V_{DS} value that satisfies the condition

$$\frac{dI_D}{dV_{DS}} = 0 \tag{7.3}$$

There are no transistors that exactly match this condition (Eq. (7.3)), due to conditions like the early voltage in bipolar transistors or like channel length modulation in FETs. This means that a practical definition for knee voltage evaluate from a nonzero value for Eq. (7.3). One practical definition is the minimum V_{DS} value that satisfies the condition

$$\frac{dI_D}{dV_{DS}} = 0.01 \tag{7.4}$$

Definition (Eq. (7.4)) leads to the concept of power supply rejection (PSR) provided by the transistor

$$\mathrm{PSR} \equiv 20\log_{10}\left(\frac{dI_D}{dV_{DS}}\right) \tag{7.5}$$

Profiles of locations along the transistor characteristic curves that share the same PSR values are shown in Figure 7.5. Not only is the CCS boundary definition (Eq. (7.4)) evaluated, other profiles are also shown for reduced values of PSR. Each of the profiles with reduced PSR has lower voltage values, at the price of reduced linearity.

The conclusion is that the knee voltage can be "reduced" only by moving away from transistor CCS operation, which corresponds to reduced linearity (more nonlinearity) in the amplifier. Using Eqs. (7.4) and (7.5) allows the specific increase in nonlinearity to be calculated. Similarly, any transistor type can be evaluated for the operating conditions necessary to assure linear operation by remaining in CCS operation.

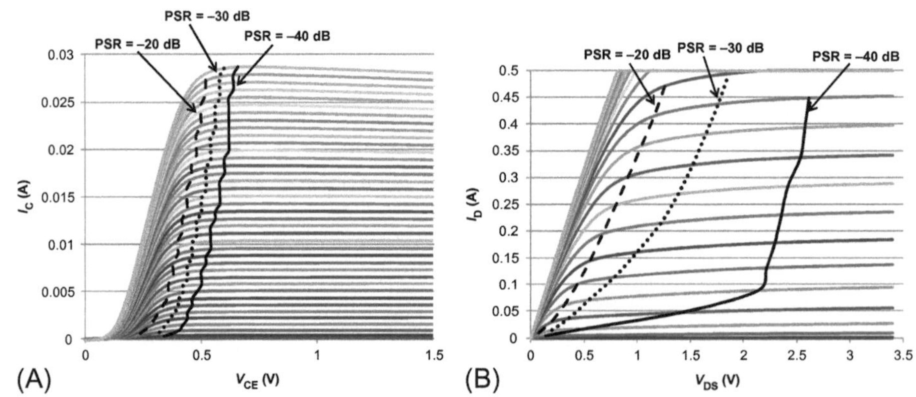

FIGURE 7.5

Knee voltage is a parameter of the transistor itself: (A) bipolar (GaAs heterojunction bipolar transistor (HBT)); (B) FET (GaN high electron mobility transistor (HEMT)). (Lack of smoothness in these curves is due to coarsely quantized data in their evaluation.)

7.2.3 LOAD LINE SELECTION FOR EFFICIENCY

Even though Ohm's law is symmetrical for power dissipation, amplifier operation is not symmetrical. In any real amplifier it is far more efficient to generate signal power from the high voltage and low current case. The presence of the knee voltage adds a loss mechanism that is much worse for the low voltage and high current case of this power-based signal process. It is important to note that this behavior is the exact opposite of the energy-based signal process used by complementary metal-oxide semiconductor (CMOS) digital circuitry, where power dissipation is minimized at low supply voltage.

Figure 7.6A and B demonstrates this difference. Beginning with the low supply voltage and high current case, the corresponding load line is presented in Figure 7.6A. The effective knee voltage of this transistor is around 0.5 V for this design. Operating waveforms are shown in Figure 7.6B. Of particular interest is the instantaneous power dissipation waveform. This waveform goes to zero only at the point of transistor cutoff operation. Power dissipation when the output waveform drops to the knee voltage is just slightly lower than the peak power dissipation of this load line.

At the other extreme is the high voltage and low current case presented in Figure 7.7. To achieve the same output power, the higher supply voltage works together with a higher load resistance value, and the transistor current falls as shown in Figure 7.7A. At this lower current value, the knee voltage is also smaller. The transistor power dissipation with this lower knee voltage and lower device current is dramatically lower as seen in Figure 7.7B. The scale on both parts of Figure 7.7 is identical to those in both parts of Figure 7.6, allowing direct comparison between these two design approaches.

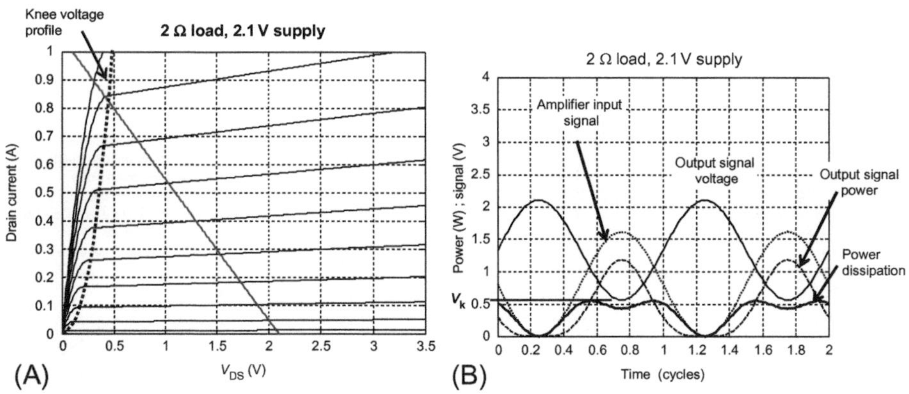

FIGURE 7.6

Low voltage and high current generation of output signal power: (A) the load line intersects the knee voltage profile at a relatively high value for V_k; (B) amplifier waveforms showing the knee-voltage-limited linear amplifier output signal voltage waveform, along with the waveform of instantaneous power dissipation.

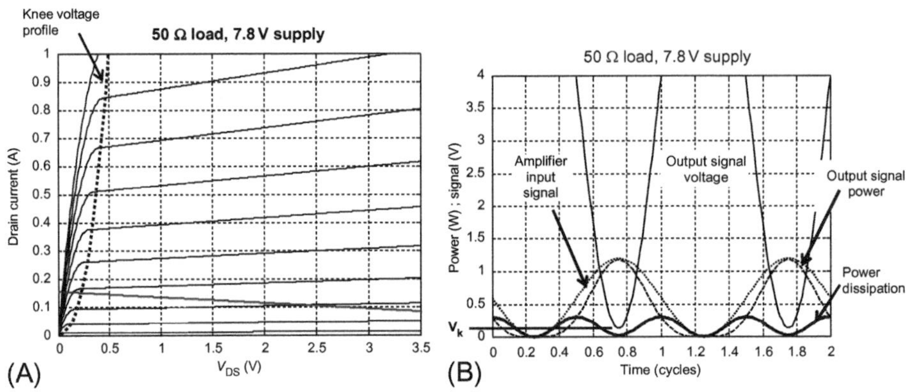

FIGURE 7.7

High voltage and low current generation of output signal power, at the identical output signal power from Figure 7.6: (A) the load line intersects the knee voltage profile at a relatively low value for V_k; (B) amplifier waveforms showing the knee-voltage-limited linear amplifier output signal voltage waveform, along with the waveform of instantaneous power dissipation. Total power dissipation is far below that of Figure 7.6.

7.2.4 VARIABLE POWER SUPPLY OPTION

One additional design option is to allow the supply voltage to vary based on circuit operating requirements. Using one value of load resistance — when the supply voltage varies — the load line shifts horizontally along with it. This is seen in Figure 7.8A. On each of the drawn load lines, solid circles are placed at each end of linear operation along that load line. At the midpoint is an open circle that

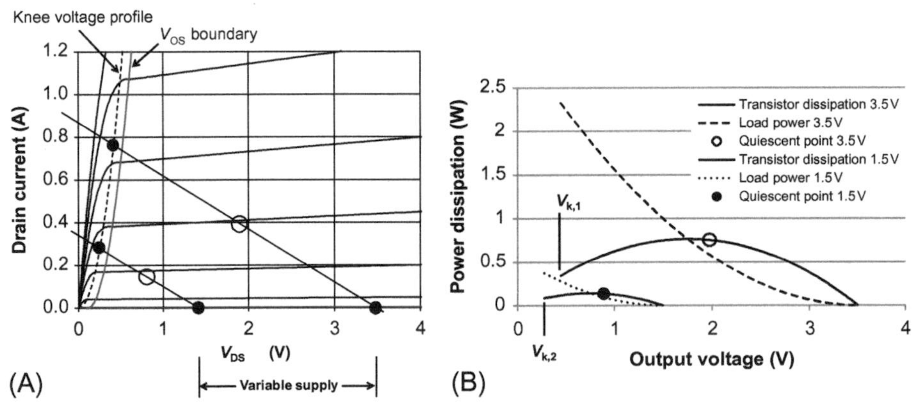

FIGURE 7.8

Variable supply operation of an amplifier: (A) load line shifts along the characteristic curves; (B) power dissipation characteristics at two different supply voltage values.

represents the ideal bias operating point. Smaller supply voltage values correspond to smaller output signal capability. Interestingly, the bias points also move along with the supply variations.

How using variable supply operation impacts the amplifier power dissipation is illustrated in Figure 7.8B. Here the power dissipation along each load line is shown, along the same plan used in Figure 7.3B. Each of these load lines experiences a different knee voltage value. Overall power dissipation along these two load lines is much lower at the smaller supply voltage. In accordance with Eq. (7.2) this means that the operating energy efficiency for a smaller output signal is much higher when all the conditions for lower supply voltage operation are met.

7.3 PA OPERATING MODES WITH A VARIABLE SUPPLY VOLTAGE

When the power supply is allowed to vary its value, then any amplifier must be evaluated as the three-port circuit that it is, and has always been. The supply voltage is very much a circuit input as the amplifier input signal is. Historically the supply voltage is considered as a fixed value, and then an amplifier can be reduced to a two-port circuit for design and analysis. This reduction is no longer valid when the fixed supply voltage value assumption is no longer true.

When the amplifier operates as a three port, there are two inputs (supply voltage, RF input) and one output, the RF output. A full characterization requires both inputs to be considered as independent variables, and the effects on the output signal are measured to determine the individual and joint transfer functions to variations on either input.

7.3.1 BOOTH CHART FUNDAMENTALS

The Booth Chart [3] is a family of circuit power transfer function measurements made at different values of supply voltage. One example is shown in Figure 7.9A. The main usefulness of this chart is not simply in the data curves it shows, but in the many relationships that readily follow from analyzing the data shown. Much as the Bode plot is an extremely useful tool for design and analysis of control loops, the Booth Chart is extremely useful in design and analysis of amplifiers operated as three ports.

The most immediate result visible on a Booth Chart is the existence of three separate operating modes that a three-port amplifier exhibits. Shown in Figure 7.9B, these modes are: L-mode, where the output power is sensitive to the input power and independent of the supply voltage value; C-mode, where the output power is sensitive to the supply voltage value and independent of the input power; and P-mode where the output power is sensitive to both the supply voltage value and the input power. These mode names correspond to linear (L), compressed (C), and product (P) operation of the amplifier.

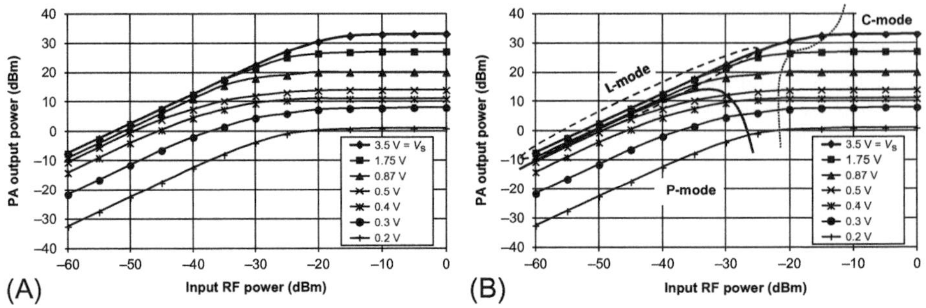

FIGURE 7.9

Booth charts are families of amplifier power transfer functions taken at different supply voltage values: (A) example from a multistage HBT amplifier simulation; (B) identification of the three PA operating modes on the Booth chart.

7.3.2 L-MODE OPERATION

L-mode is the traditional linear operation of the amplifier. This follows from operating along the load line when the transistor is operated in its CCS region, as shown in Figure 7.10A. Here two load lines are drawn corresponding to two different supply voltage values. The dotted ellipses show where the amplifier load current does not change when the supply voltage is varied, but the load current changes significantly with changes of the input signal. These precisely match the observed properties of L-mode. The load line analysis also shows us that to have L-mode

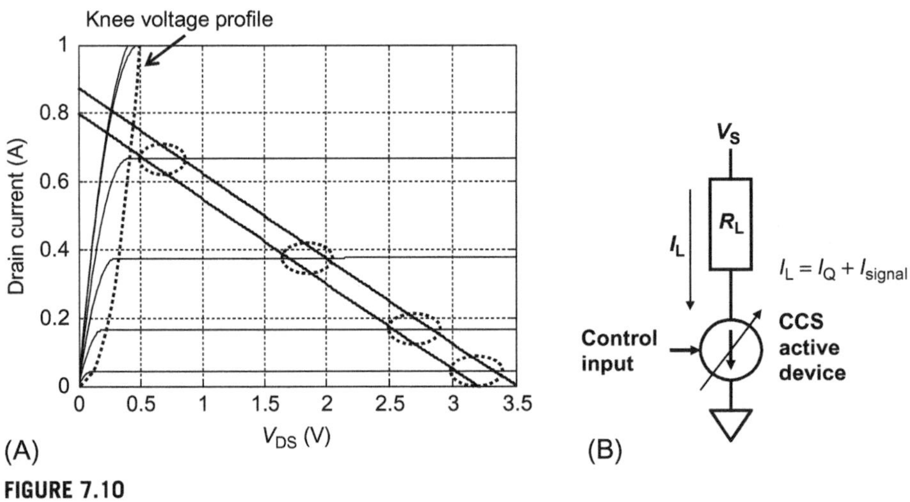

FIGURE 7.10

Conventional linear operation happens within the L-mode region: (A) the L-mode characteristic of load current being independent of the supply voltage; (B) block level circuit model for L-mode.

operation, the voltage at the transistor controlled port (the output voltage waveform) must always match or exceed the knee voltage profile.

The corresponding circuit model for L-mode is the resistively loaded CCS, presented in Figure 7.10B. This very conventional and widely used circuit model represents that the transistor is regulating the current flow through the load at all times.

7.3.3 C-MODE OPERATION

C-mode operation is very different from L-mode operation. Again looking at the transistor characteristic curves with the same load lines, Figure 7.11A shows where the transistor must operate in order to exhibit the properties of C-mode: no sensitivity to changes in the input signal, and a large sensitivity to changes in the supply voltage value. These characteristics are completely opposite to the L-mode properties.

To get into C-mode, the input drive to the transistor must increase sufficiently to push the transistor out of its CCS region and into its resistive region. This means that the voltage across the transistor is now well below the knee voltage, at least at its maximum value of device current. In order to remain consistent with the observed properties of C-mode, the minimum value of the input signal also cannot operate the transistor in its CCS region, which means that the transistor must transition all the way to cutoff operation. The transistor is either a low value resistance, or it is an open circuit. It therefore operates as a switch, alternating between cutoff (OFF) and resistive (ON) operation on each half of the input RF signal cycle.

The corresponding circuit model for C-mode operation changes to that presented in Figure 7.11B. Now the transistor is shown as an ideal switch, in series with a resistor

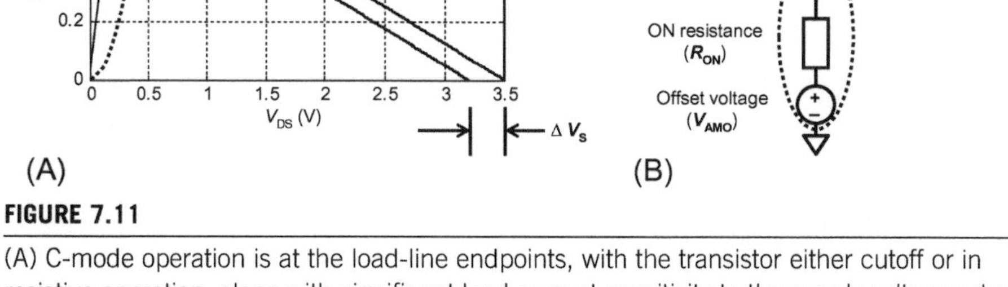

FIGURE 7.11

(A) C-mode operation is at the load-line endpoints, with the transistor either cutoff or in resistive operation, along with significant load current sensitivity to the supply voltage value; (B) block level circuit model for C-mode that emphasizes the switching operation.

representing the ON resistance. This model also includes an offset voltage V_{AMO} for completeness, which is generally zero for FETs and nonzero for bipolar transistors [4]. Unlike in L-mode, here the transistor does not regulate current flowing through the load. Rather, the load current is found from Ohm's law to be the total voltage across the resistances divided by the total resistance of the ON circuit. For intentional C-mode operation, taking the two new design conditions of $V_S \gg V_{AMO}$, and $R_L \gg R_{ON}$, then the load current becomes dependent only on V_S and R_L and is independent of the transistor details. This leads to inherent stabilities of a C-mode circuit, exhibiting stable output power independent of temperature, manufacturing variations in the transistors, and freedom from oscillation [5].

Proper switching power amplifier operation also assumes that the time spent in the ON and OFF states dominates the RF signal period. This means that transition times between OFF and ON, in either direction, must be short compared to the signal period. One quantification of what this means is provided in Figure 7.12, which evaluates how much higher the transistor f_T must be above the operating frequency, with the assumption that f_T is sufficient to describe the transition time of the transistor.

This assumption is partially true [6]. While not being universally precise, two very important results are illustrated in Figure 7.12. First, the transistor f_T is preferably 50 × to 100 × higher than the operating frequency for switching (C-mode) operation. It certainly must exceed a 10:1 ratio. Second, the available efficiency of a C-mode stage increases as the ratio of load resistance R_L over ON resistance R_{ON} increases. This is consistent with Figure 7.11B, because we do want power dissipated in R_L and not in R_{ON}.

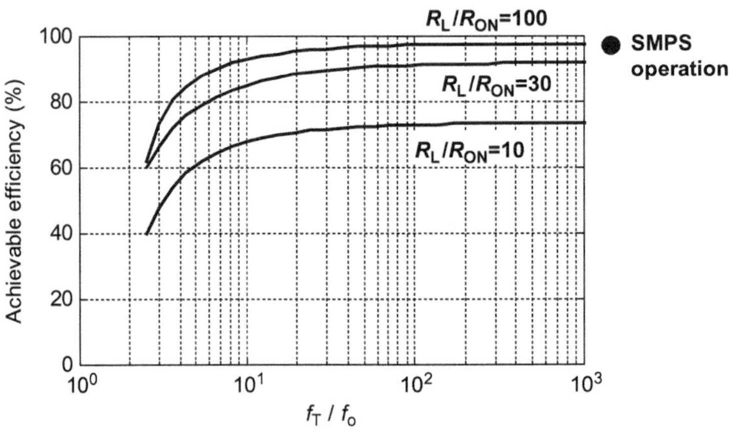

FIGURE 7.12

Assuring switching operation requires the transistor f_T to be significantly higher than the operating frequency, preferably 100:1. Higher load resistance values further improve available efficiency. A typical switching (switch-mode) power supply DC-DC converter operates at the point shown.

It is also good to note that this conclusion is consistent with the results of Section 7.2.3. Further, using this same analysis for the known high-efficiency circuitry of a switch-mode power supply (SMPS), the results are consistent. Shown in Figure 7.12 is an additional point called "SMPS operation" for a typical design with 96% conversion efficiency. These transistors are operated at a frequency that is less than 0.1% of the switching transistor f_T, and at a load to ON resistance ratio of 100:1.

7.3.4 P-MODE OPERATION

The third three-port amplifier operating mode is P-mode, which is sensitive to both the input signal value and the supply voltage value. Looking at the Booth chart in Figure 7.9B it is observed that P-mode only appears when the amplifier supply voltage has a low value, here below 0.9 V. Again, basing this evaluation on the load line analysis, Figure 7.13A shows that P-mode occurs when the transistor is not driven sufficiently to either be considered ON or OFF. In this region the transistor behaves as a controlled variable resistance. Operation remains below the knee voltage, like for C-mode, so this is not a linear amplifier. The transistor is not regulating current through the load.

A third circuit model is required for P-mode operation, and is presented in Figure 7.13B. Here the transistor is shown as a variable resistance, and keeps the possible presence of the offset voltage V_{AMO} for generality. With these three different circuit models, it is vital that the circuit analysis keep track of the transistor operating point to know which of these models is appropriate. This model selection holds for each instant in time, making the job of three-port amplifier analysis much more complicated than is historically needed for traditional two-port operation.

Literature for envelope tracking (ET) mentions an effect called "gain collapse" when the ET supply voltage gets below the knee voltage. The physical reason for this

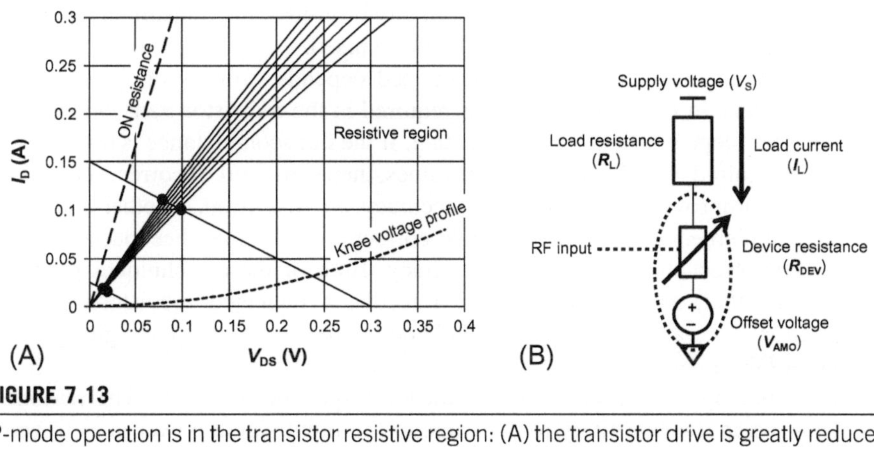

FIGURE 7.13

P-mode operation is in the transistor resistive region: (A) the transistor drive is greatly reduced to avoid cutoff, and has a significant load current sensitivity to the supply voltage value; (B) block level circuit model for P-mode that emphasizes the variable resistance operation.

reported rapid drop in gain is that the transistor enters P-mode operation. Once the transistor leaves CCS operation by crossing the knee voltage, its characteristic curves get closer together which signifies a drop in transconductance. P-mode is an operating region that must be avoided in ET designs.

7.4 SIGNAL LINEARITY AND ACCURACY REQUIREMENTS

Two measures are in wide use to determine the output signal accuracy from a transmitter. These measures are error-vector magnitude (EVM) and adjacent-channel power (ACP). Both of these measures are needed because they measure different things. EVM is an in-band measure of signal accuracy, with the intent to assure that the signal state (magnitude and phase pair) is accurately placed near the ideal value. ACP is a measure of spectral occupancy, primarily out-of-band, which is more sensitive to the transitions between signal states instead of the values of the states themselves [1]. In general, circuit design metrics such as intermodulation distortion (IMD) and harmonic distortion are not uniquely tied to these modulated signal metrics. IMD, in particular, does impact both EVM and ACP performance. Yet specifically how this impact occurs is distinct to each modulation type, and at this moment must be separately evaluated for each individual modulation type.

When reporting the signal accuracy from any transmitter design, it is essential to provide both the ACP and EVM performance measures, because they are essentially independent: one measure can be excellent, and the other can be simultaneously horrible. Only when both measures are good is there assurance that the intended output signal accuracy is acceptable.

Transmitter design is fundamentally an a priori problem, meaning that all information is available to the transmitter regarding what the output signal needs to be. This means that nonlinear circuitry can certainly be used in the transmitter as long as the accuracy of the output signal is not compromised.

How the output signal accuracy is attained depends strongly on the amplifier operating mode. In L-mode, accuracy is required in the transistor transconductance g_m so that the slope gain $g_m R_L$ [7] is constant. If the transconductance is not uniform across the required input signal variation values, then some sort of correction method is needed to improve the output signal accuracy. Predistortion, whether analog or digital in implementation, is one possible correction technique. Feedback can also be used to correct the output signal accuracy. Both of these techniques can also be used together, essentially realizing a feedback managed predistortion. This combination is presently very popular, and goes by the name adaptive digital predistortion (A-DPD).

In C-mode operation, all of these prior techniques do not work. The reason is simple — when the sensitivity to the value of the input signal power goes to zero, changes to that input signal power from any predistortion technique have no effect on the output signal. Different techniques are needed to provide any needed output signal accuracy corrections with C-mode operation.

The active transfer function in C-mode is from the supply voltage to output RF magnitude. Responsibility for the output signal magnitude, once the new design conditions from Section 7.3.3 are satisfied, is transferred to the supply voltage. Signal accuracy corrections are made by modifying the supply voltage, and not the input signal magnitude.

During P-mode operation there are two active transfer functions, greatly complicating any signal accuracy process. If needed, output signal accuracy corrections are practically done by treating the two sensitivities as separate processes [3], where the sensitivity to input RF power is used only for output power control, and signal modulation is provided only by the supply voltage variation. This division of purpose for the two transfer functions allows the C-mode correction system to also apply during P-mode operation. For this P-mode technique to work, all envelope modulation must be removed from the RF input signal, leaving it with only phase modulation.

7.5 DPS TRANSMITTER PRINCIPLES

When the supply voltage is intended to be varied on an amplifier, this enters the class of DPS transmitters. DPS transmitters can operate in any of the three modes described in Section 7.3. These lead to very different transmitter operating and behavioral characteristics, and have long had unambiguous definitions for these different operations [8,9]. There are two extremes, ET on one end, and direct polar (DP) on the other. A concise summary of their individual definitions is provided in Figure 7.14. An older technique called average power tracking operates under the same rules as ET.

7.5.1 ET: L-MODE ONLY

The primary feature of ET is that the exact value of the supply voltage is not important. The amplifier operates linearly, meaning that the RF input signal defines what the amplifier output signal is (given the amplifier gain). Of the three PA operating modes that exist, only L-mode meets this feature. Therefore any ET design can only use L-mode operation.

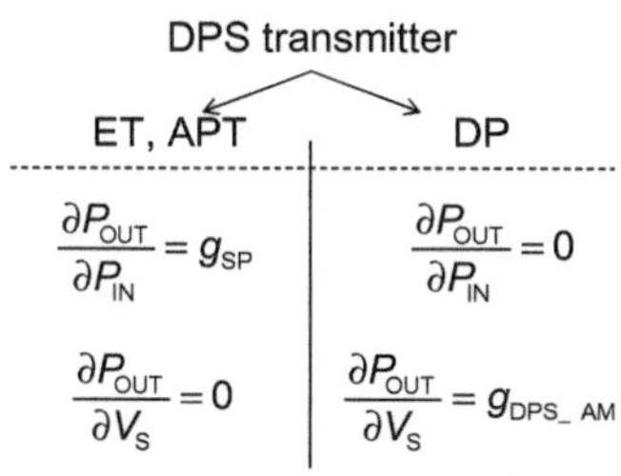

FIGURE 7.14

Envelope tracking and polar operation of dynamic power supply transmitters have opposite characteristics and very different transfer functions.

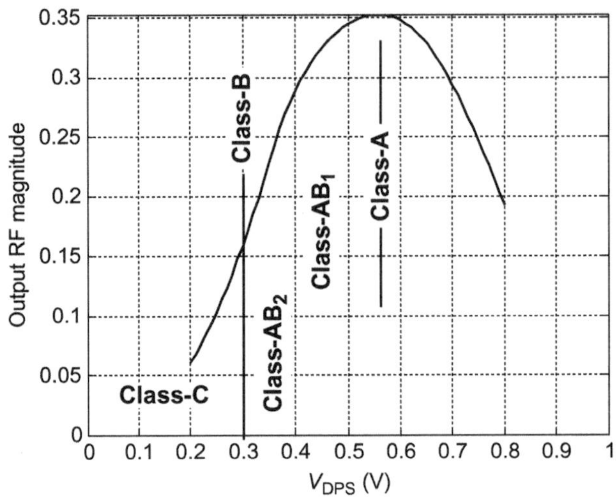

FIGURE 7.15

For a fixed input signal magnitude, the output signal magnitude changes as the amplifier bias is varied. This corresponds to amplifier gain, which peaks at class-A.

It is common to attempt ET operation on any PA designed for linear operation. This approach usually ends up as a disaster because there are several important design differences needed for successful amplifier operation as a three port. The first major change required is to the bias strategy. For a fixed magnitude input RF signal, the magnitude of the RF output signal varies with the amplifier bias as shown in Figure 7.15. Output signal magnitude (and therefore gain) peaks with class-A bias. At any other bias, the output signal magnitude drops. When the bias heads toward cutoff (as shown in Figure 7.4) the transistor output signal drops because of the time the transistor spends cutoff and therefore not generating any output signal current. A related effect happens when the bias is shifted toward saturation, but this region is not interesting because it dissipates even more power than class-A. Output power continues to fall as the bias progresses ever deeper into class-C. Energy efficiency continues to improve as these bias levels drop, eventually approaching 100% efficiency as the output signal goes to zero magnitude — another uninteresting condition.

Why this is a new design requirement for linear amplifiers intended for ET use is seen in Figure 7.16. If the bias remains constant while the supply voltage varies, then the amplifier operating class must vary. And when the operating class varies, Figure 7.15 shows that the gain will likely vary. This is a huge problem. As Figure 7.16 shows, if the amplifier is biased for class-A operation at the minimum supply voltage, then that same bias corresponds to class-AB_2 when the supply voltage is near maximum. If the design strategy is to bias for class-A (maximum gain) at the highest supply voltage, then as the supply voltage is reduced, the bias point quickly reaches the knee voltage and severely restricts the control dynamic range.

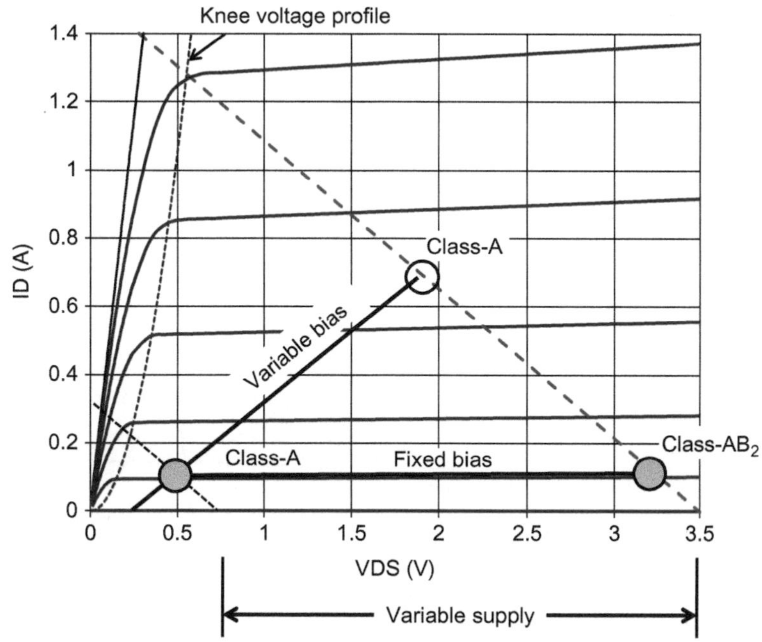

FIGURE 7.16

When the supply voltage is varied, holding a fixed bias causes the amplifier operating class to change. To maintain a constant amplifier class while the supply voltage changes, the bias must also change in unison with the supply.

Maximizing control dynamic range and gain jointly happens when the amplifier is re-biased to class-A at each value of the supply voltage. This requires a variable bias that is highly correlated to the DPS voltage. Moreover, this bias must vary at the same speed as the DPS variations. This is not an easy design to do.

Equally important is knowing what the supply voltage value needs to exceed in order to maintain linear amplifier operation. This calculation depends only on knowing two things: (1) the transistor characteristic curves and (2) the value of the load resistance. From each value of supply voltage, a load line is drawn on the characteristic curves, noting end points at intersections of each load line with the voltage axis and at an offset voltage V_{OS} slightly above the knee voltage. This process is illustrated in Figure 7.17.

Then further assuming maximum gain operation, which corresponds to class-A bias, the available output signal envelope for each power supply value is determined. The results are plotted together on a chart like that shown in Figure 7.18. The ET supply voltage profile increases above the required output magnitude as the output envelope increases, which is required by the increasing knee voltage profile in order to maintain linear amplifier operation. At the lower end is a supply voltage limit necessary to avoid entering P-mode operation.

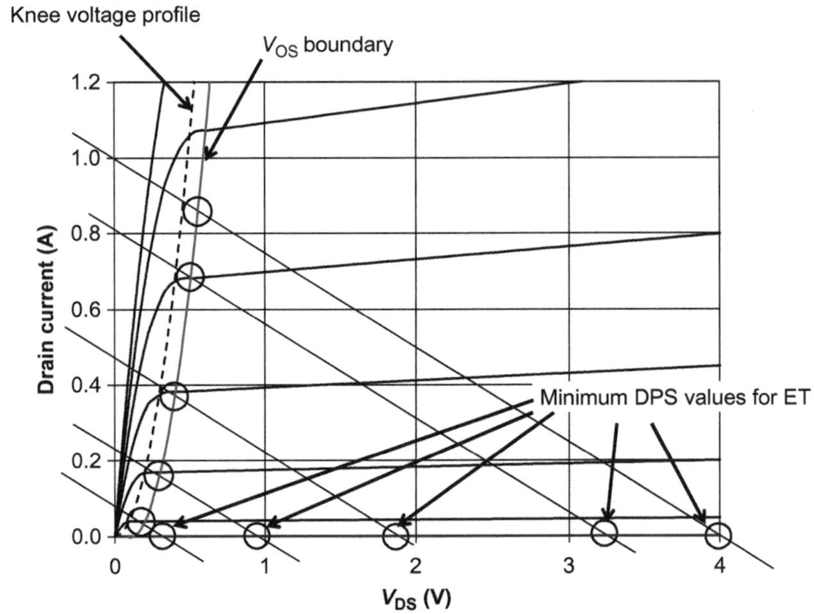

FIGURE 7.17

Knowing the transistor knee voltage profile and the design load-line value, the necessary minimum supply voltage for envelope tracking at each output envelope value is uniquely determined.

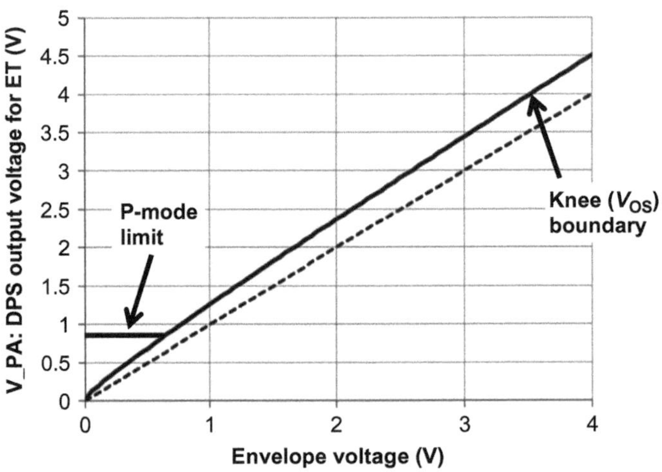

FIGURE 7.18

The envelope tracking supply voltage must increase above the output envelope value in order to maintain amplifier linearity. P-mode operation must be avoided, which sets a minimum value for the supply voltage.

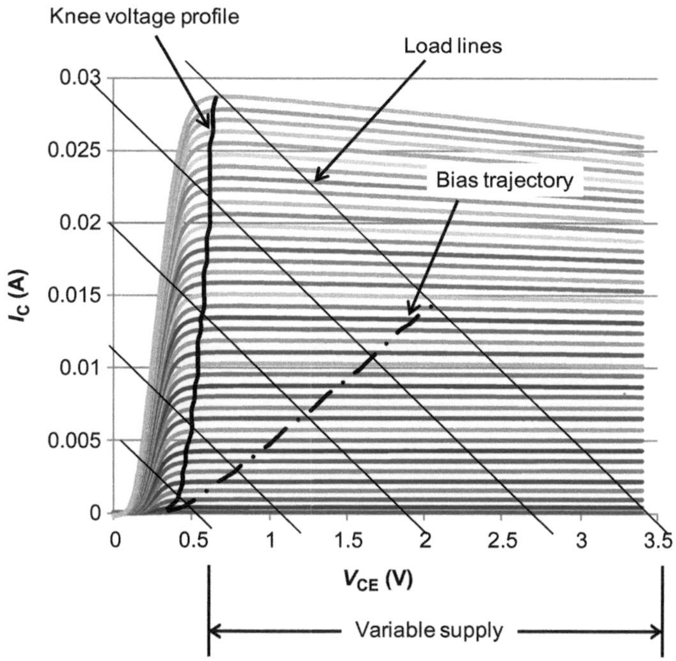

FIGURE 7.19

Envelope tracking (ET) operating parameters for an amplifier based on a GaAs HBT. The offset of the knee voltage at low supply voltages demonstrates why the ET supply voltage has a minimum value for this transistor type.

Closed-form equations exist for the knee voltage profile of a FET. Similar equations do not exist for a bipolar transistor, forcing the evaluation of knee voltage and the ET supply profile by calculation as described earlier. One example is provided in Figure 7.19, which also includes the path of the dynamic bias trajectory for this value of load resistance. Note that this calculation algorithm naturally provides a voltage offset for this GaAs HBT device.

The point of ET operation is to increase the PA energy efficiency for any envelope varying signal. The mechanism for how this efficiency improves is shown in Figure 7.8B. At maximum supply, the energy efficiency is the same as for the linear amplifier at that supply voltage. As the output signal power decreases, the falling supply voltage drops the PA power dissipation which helps maintain PA efficiency as illustrated in Figure 7.20A. The maximum available energy efficiency from any ET transmitter is limited by the same physics that limit class-A amplifiers. This means that ET designs are limited to 50% energy efficiency, and, as discussed in Chapter 4, can only approach that limit when the transistor knee voltage $V_k = 0$ V [1].

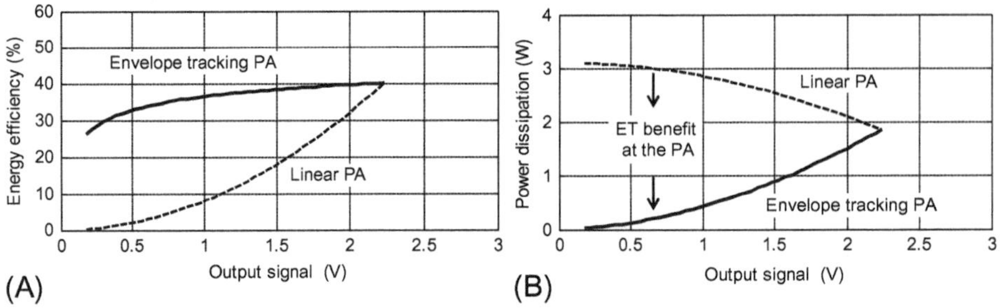

FIGURE 7.20

Energy efficiency improves dramatically with envelope tracking (ET) operation: (A) PA energy efficiency is largely constant from the amplifier maximum linear output efficiency; (B) ET power dissipation in the amplifier goes down to provide this efficiency increase.

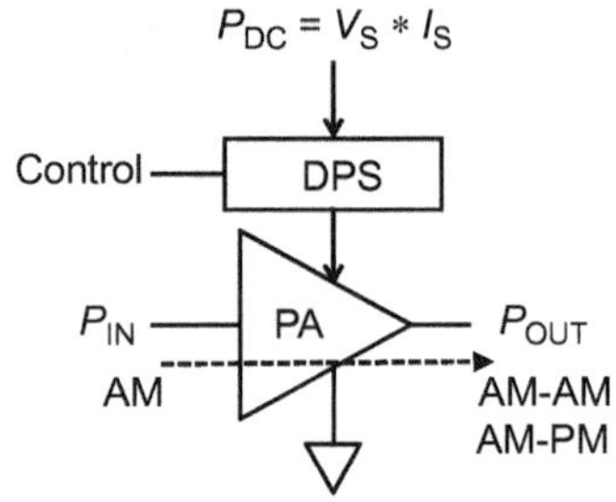

FIGURE 7.21

Nonlinear distortions in an envelope tracking transmitter are the same physical processes experienced in linear amplifiers (though now a surface w/varying supply voltage).

In accordance with Eq. (7.2) this PA efficiency improvement is achieved by reducing the PA transistor power dissipation. The transistor power dissipation comparison between a class-A linear amplifier and an envelope tracked linear amplifier with correct dynamic bias is shown in Figure 7.20B. While the linear PA dissipation actually increases as the output signal magnitude drops (as shown in Figure 7.8B), the power dissipation in the ET PA monotonically decreases. The difference between these two curves is the energy efficiency benefit of ET.

Because an ET PA always operates linearly, the primary distortion mechanisms are the same as for conventional operation: amplitude modulation (AM)-AM distortion which represents any deviation from constant gain, and AM-phase modulation (PM) distortion which represents any phase shift that occurs based on the input (and output) signal magnitude. These distortions are properties of the amplifier and the RF signal path, as shown in Figure 7.21. With ET operation, these distortions likely change for each supply voltage value, and therefore are not profiles but are best represented as surfaces. Managing these distortions, and correcting for them, becomes a multidimensional problem.

7.5.2 DPS CHARACTERIZATION

Adding the DPS also generates a new interface into any DPS design. This interface is the connection between the DPS and the PA, shown in Figure 7.22. The behavior of this interface depends on detailed characteristics of both circuit blocks.

The voltage at the interface is set by the DPS. But the current through this interface is set by the supply port impedance characteristics of the PA. The DPS has no direct control of its output current. This means that knowing the DPS output impedance — an S_{22} measurement in the language of RF PA design — is a vital parameter. Since the DPS acts as a voltage regulator, a typical wideband output impedance measurement for a fast operational amplifier is needed. One typical measurement is shown in Figure 7.23.

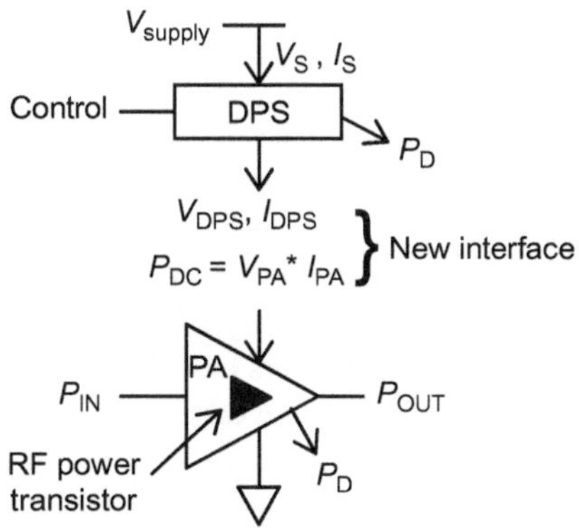

FIGURE 7.22

A new interface exists in an envelope tracking transmitter, where the PA connects to the dynamic power supply (DPS) output. The DPS sets the interface voltage and the PA supply input impedance sets the interface current.

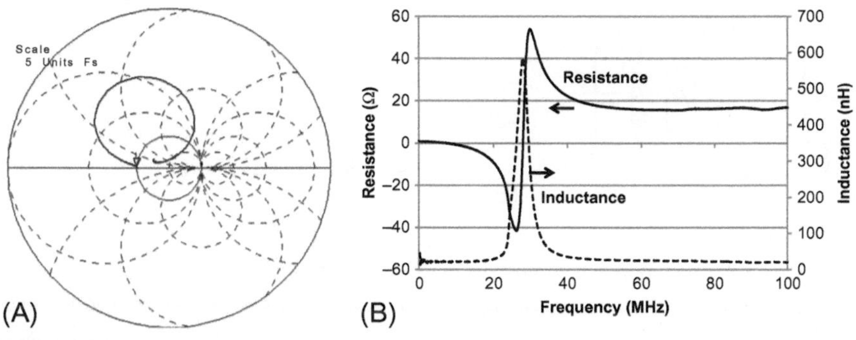

FIGURE 7.23

Measured output impedance of a linear regulator stage is inductive at all frequencies, and goes negative resistive outside the operating bandwidth of the regulator: (A) Smith chart measurement; (B) equivalent resistance and inductance values across frequency for these measurements.

The output impedance measurement in Figure 7.23A is plot on a Smith chart. It shows that at low output frequency the regulator output impedance is very close to a short circuit, as expected. However it is not purely resistive, but is also inductive. A little thought shows why this inductance makes sense — any feedback system has a finite delay in its response to output perturbations. This delay time means that the current experienced by the output probing voltage source is delayed from the applied voltage. Voltage phase-leading current is an inductance property.

A bigger surprise is what happens as the test frequency moves outside the closed loop bandwidth of the op-amp regulator. This measurement shows that the output impedance stays inductive but also becomes a negative resistance. In a negative resistance, the current goes down as the voltage goes up, which happens if the phase of the feedback current response exceeds a lag of 90 degrees. This is a natural occurrence in any feedback system.

Translating the Smith chart measurements to an impedance format yields the impedance chart in Figure 7.23B. The effective inductance is not constant, and peaks at the frequency where the Smith chart trace returns to positive resistance. The region of negative resistance is present over a finite band of frequencies.

This measurement illustrates a fundamental instability in this new interface. If the DPS output impedance is negative-resistive and inductive, then with a capacitor added to the DPS output there is a possibility of oscillation. The circuit for this is shown in Figure 7.24A. All power amplifiers have bypass capacitors on their supply inputs, so it is not possible to avoid having a capacitance at the interface node. How tolerant this particular DPS is to capacitance on its output is shown in Figure 7.24B. No oscillation occurs if the capacitance is less than 82 pF. All higher values of

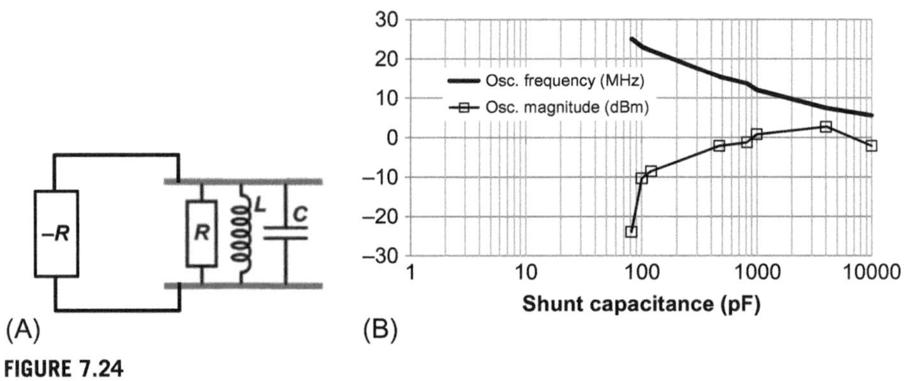

FIGURE 7.24

With the output impedance of the dynamic power supply (DPS) being inductive with negative resistance, adding an output capacitance can form an oscillator: (A) effective schematic when the PA with RF bypass capacitor(s) is attached; (B) resulting interface oscillation frequency and magnitude depends on the DPS output impedance and the attached capacitance value.

capacitance resulted in sinusoidal oscillation at the interface. As the capacitor value increases, the oscillation frequency decreases, but the oscillation magnitude increases until the value exceeds 5 nF. This range includes the typical range of PA bypass capacitance, confirming this fundamental interface instability.

This result shows that it may be more important to keep the PA supply pin capacitance small for reasons of DPS stability than for the more widely recognized power loss from repeated charging and discharging current flows necessary to vary the capacitance voltage — current flows that must occur but do not flow through the PA RF transistor and therefore cannot contribute to the RF output signal. The responsibility to solve this issue is shared between the designers of the DPS and the PA. The DPS designer must ensure that the output impedance of the DPS never presents a negative resistance component at any frequency. The PA designer must keep the pin capacitance at the supply voltage input below 100 pF anyway to minimize reactive current flows from the DPS and correspondingly improve overall DC to RF conversion efficiency.

7.5.3 POLAR MODULATION: C-MODE AND P-MODE ONLY

Polar modulation is the fundamental result of any three-port circuit where a control input sets a polar coordinate of the output signal. There are two types of polar modulation possible, one where the signal process controls the output magnitude and the other which controls the output signal phase. These polar processes are illustrated in Figure 7.25. Of particular importance is that the RF input to a magnitude polar modulator does not have to be constant envelope, as long as the magnitude control has complete control of the output magnitude.

In Section 7.3.3 it is shown that when the amplifier operates in C-mode, the current through the load resistance is fully controlled by the value of the supply voltage.

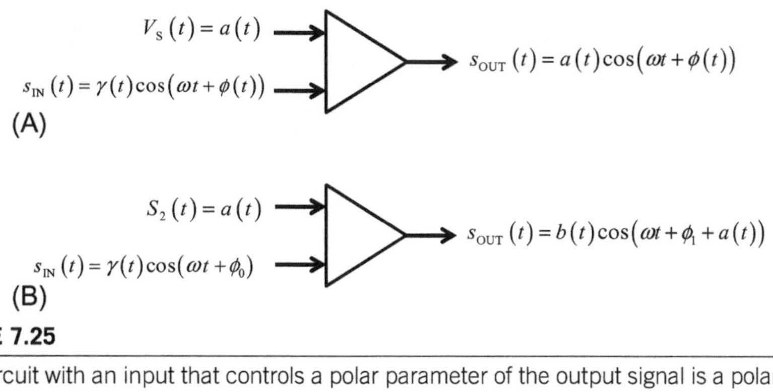

FIGURE 7.25

Any circuit with an input that controls a polar parameter of the output signal is a polar signal process: (A) magnitude control; (B) phase control.

Combining this result with the Booth chart of Figure 7.9B which shows that in C-mode, the sensitivity to input signal magnitude is zero, there can easily be input RF signal magnitude variations while still doing DP operation.

Revisiting the characteristic curves and load line, Figure 7.26 shows that increasing the input signal magnitude beyond the limit of ET operation pushes the maximum current operating point below the knee voltage, and also to a higher current through the load from the same supply voltage. Also, the power dissipation in the PA transistor falls by nearly a factor of 2, improving its energy efficiency. Gain must drop from its value in ET operation because of the larger input signal.

Comparison of polar and ET operation is shown in Figure 7.27, which is an expansion of Figure 7.20. Reduced power dissipation resulting from operating the polar transmitter well below the knee voltage corresponds to the higher energy efficiency seen in Figure 7.27A. Also visible in Figure 7.27A is the higher output signal power available from the C-mode operation of the polar transmitter for the same supply voltage value. The corresponding power dissipations in Figure 7.27B confirm that in C-mode the transistor power dissipation is between one-half and one-third of the dissipation from ET.

Accuracy of the output signal from polar modulation is readily estimated by drawing a triangle on the characteristic curves, with the load line at the highest

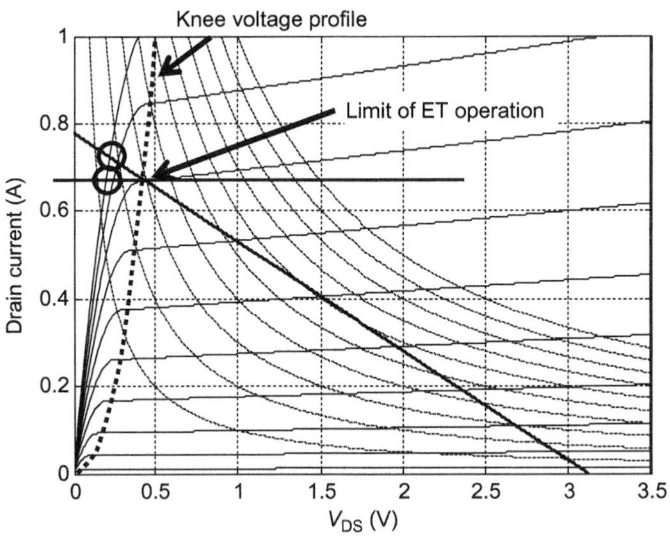

FIGURE 7.26

Polar operation requires operating the PA in C-mode. Input drive must increase, lowering PA gain. Output power increases for the same supply voltage. To hold the same output power from linear amplifier operation, the supply voltage must decrease. In all cases the amplifier power dissipation decreases compared to linear operation.

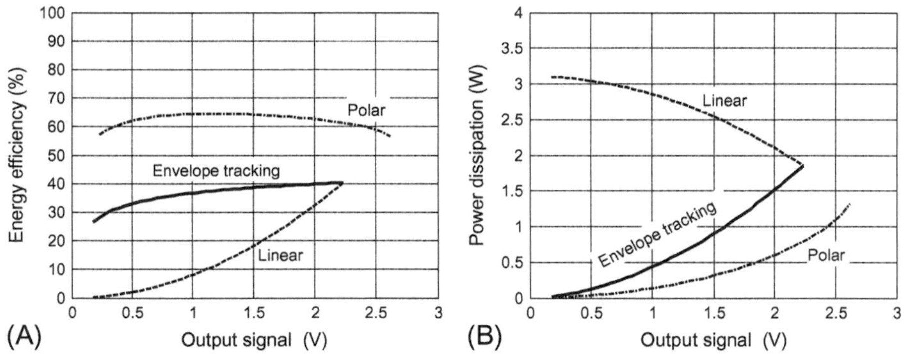

FIGURE 7.27

Higher output power and reduced dissipation combine to provide higher PA energy efficiency: (A) comparison with envelope tracking operation; (B) corresponding power dissipation relationships.

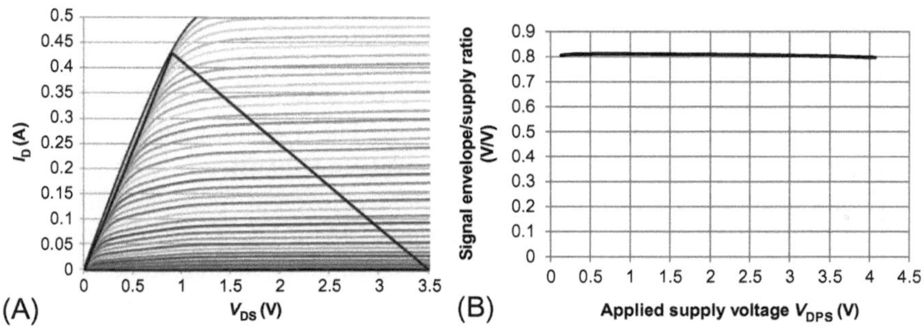

FIGURE 7.28

Estimating modulation accuracy for a polar transmitter: (A) a polar modulation triangle drawn on the characteristic curves of a CMOS PA; (B) measured polar modulation accuracy for this CMOS PA showing very low polar distortion.

supply voltage as one side, the voltage axis as a second side, and touching the *I-V* plane origin. The third side of this triangle is between the *I-V* plane origin and the intersection of the load line with the highest characteristic curve. An example is shown in Figure 7.28A, overlaying on a set of CMOS characteristic curves. As the supply voltage is varied, the load line side of the modulation triangle shifts left to right. The base side of the modulation triangle stays along the voltage axis. Accuracy of the resulting polar modulation depends on whether the third side changes slope as the modulation triangle is rescaled in size. The more consistent the slope of this third side remains as the supply voltage varies, the more accurate the polar modulation is. Correspondingly, such a polar transmitter will appear to be more linear with respect to the transfer function between the supply voltage value and the RF envelope value.

To the extent that this third side aligns with the transistor characteristic curves in their resistive region, polar modulation using this transistor is expected to have good accuracy. The measured results from this amplifier are provided in Figure 7.28B, demonstrating that this PA does provide accurate polar modulation across a wide dynamic range. Here the evaluated dynamic range is 30 dB.

Polar modulation cares nothing about the knee voltage profile. By operating in C-mode all circuit linearity is suppressed, so even the characteristics of the transistor CCS region are of no concern. But the detailed transistor characteristics of its resistive region are a primary concern, as seen in Figure 7.29A. Drawing the polar modulation triangle on this set of GaAs HBT characteristic curves shows that there is little alignment between this third side and the characteristic curves. Actual polar modulation measurements from this amplifier are provided in Figure 7.29B which show very poor accuracy. The modulation error is on the lower side, which does correspond to the characteristic curves being below the modulation triangle third side. One may consider that polar modulation of a bipolar transistor is hopeless. This is not quite true. Polar performance can be significantly improved using the following technique.

When the modulation triangle vertex at the origin is moved until the third side is best-fit to the bipolar characteristic curves the polar modulation accuracy improves dramatically. The resulting offset triangle is shown in Figure 7.30A, and the modified polar modulation is shown in Figure 7.30B. The dynamic range of this polar modulator is greatly improved, but still remains below 20 dB.

Primary distortion mechanisms are completely different from those in ET and linear amplifier transmitters. The conventional AM-AM and AM-PM distortion processes are not active in a polar transmitter because all input magnitude variations, including predistortions, are removed by the self-limiting of C-mode operation.

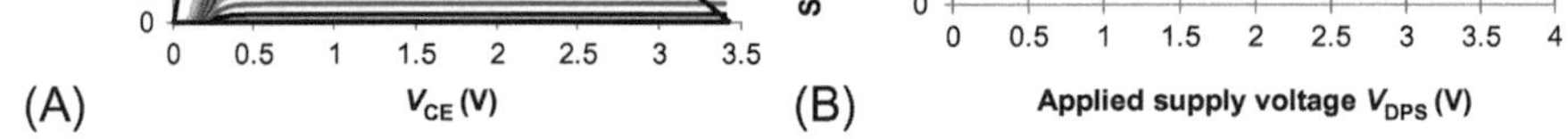

FIGURE 7.29

Problems with polar modulation and bipolar transistors: (A) modulation triangle does not align well with HBT characteristic curves; (B) resulting polar modulation has poor accuracy, with lower than expected output at both ends of the modulation range: polar distortion is severe.

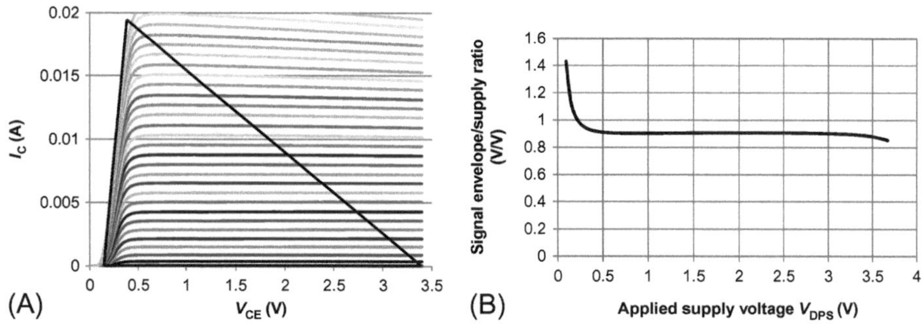

FIGURE 7.30

Offsetting the modulation triangle from the *I-V* plane origin improves polar modulation accuracy: (A) resulting modulation triangle for a GaAs HBT; (B) polar distortions have different signs at each end of the linearized polar modulation characteristic.

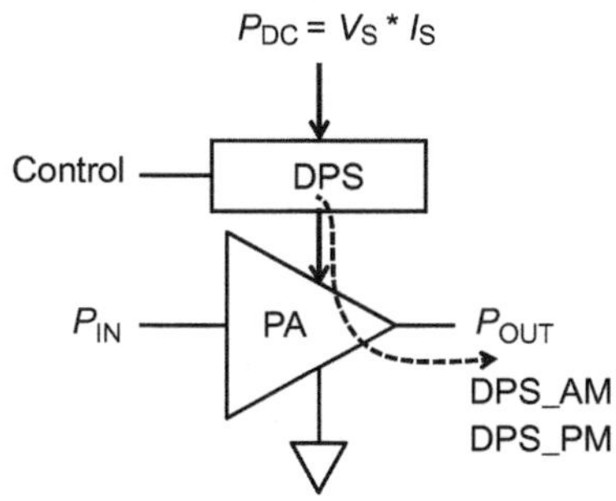

FIGURE 7.31

Polar modulation transfer function is very different from envelope tracking and linear PA operation. Nonlinearities in polar modulation therefore also have different physical manifestations.

Replacing these distortion mechanisms are new processes that are associated with the new transfer function from the DPS output to the RF output magnitude. These distortion processes are called DPS_AM and DPS_PM and are defined in Figure 7.31. Any accuracy improvement algorithms must not predistort the input to the PA, but rather modify the control input to the DPS in a polar transmitter.

Tools are now in place to evaluate hybrid designs using both polar and ET operation. Hybrid operation is actually very common. One manifestation of this hybrid operation is shown in Figure 7.32, which shows the effect of selecting a DPS profile called "de-troughed" and plot with a dashed curve. At low output powers, this transmitter operates with a linear (essentially fixed supply) power amplifier. When the output envelope exceeds 0.5 V, the transmitter becomes ET, until around 1.6 V at the envelope. Here the power supply crosses the minimum voltage needed to maintain linear operation, forcing the transistor voltage at signal peak current to be below

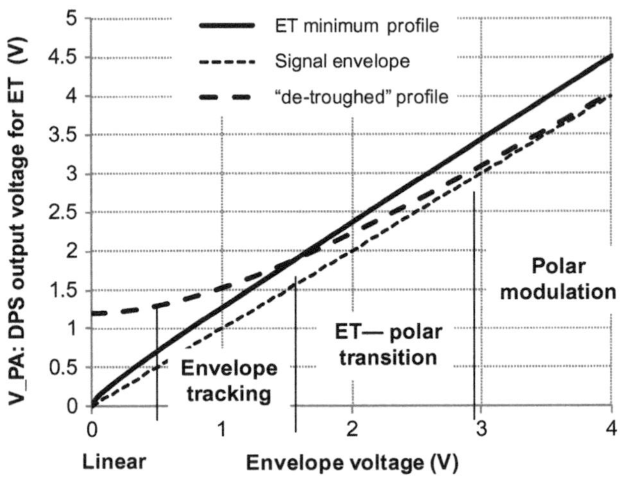

FIGURE 7.32

Hybrid operation between linear, envelope tracking, and polar modulation is readily invoked by the selection of the dynamic power supply profile compared to the required RF output signal envelope.

the knee voltage. This begins a transition to polar modulation, which ends when the envelope voltage exceeds 3 V. At higher envelope voltages, this amplifier is transitioned to C-mode and all circuit linearity is suppressed.

Gain cannot be constant with this profile. At low envelope values the amplifier operation is linear with the high gain of class-A. At envelope peaks the gain is lower due to the compression and nonlinearity required of C-mode operation. It is not uncommon to have this gain spread exceed 12 dB. Managing output signal accuracy with this type of DPS profile is a difficult challenge.

There are four primary techniques available to maintain the required output signal accuracy. None of these techniques is wholly satisfactory, so tradeoffs remain for each specific transmitter design project. Brief summaries of each primary technique are provided below, based on greater details for each provided in Ref. [6]:

- Pure ET: when class-A bias is dynamically maintained across the supply voltage variation, gain flatness is good and linearization may not be necessary. There is a tradeoff in the necessary supply voltage headroom necessitates a slight efficiency drop below its maximum available value.
- Pure DP: polar operation provides very good envelope accuracy as long as C-mode amplifier operation is maintained, which corresponds to achieving the maximum available efficiency. The governing transfer function is the supply voltage to the output envelope, and the normal gain of the RF amplifier (change in output envelope for a change in the input envelope) is zero. Conventional DPD no longer operates when the normal gain of the amplifier is zero. The major tradeoff is difficulty achieving accurate low envelope values.

- "constant gain ET": this technique is misnamed, and actually is a constant-ratio joint transfer function because in order to achieve the accurate ratio of output signal power to input signal power there must be a corresponding precise control of the supply voltage. This technique is a highly correlated "dance" between input power and power supply, and therefore does not meet the fundamental definition of what a "gain" is. This technique is neither ET nor DP, but remains in the hybrid region between these two primary techniques.
- ET with DPD: following the trajectory of Figure 7.32 requires managing the large gain shift between compressed polar operation at the signal peaks and linear operation at the envelope valleys. This distortion is primarily when the PA operates linearly, so use of conventional DPD can work. DPD does not work at the signal peaks, but the polar operation can be inherently precise when the proper transistor technology is adopted. The trade-off here is the added complexity and power consumption needed by the DPD and any feedback receiver it requires.

7.5.4 TRANSISTOR TYPES WITH BEST PERFORMANCE

As discussed earlier in this chapter, the selected RF transistors are key components for the transmitter system. Therefore, the design of any transmitter is strongly based on the prerequisite of an accurate characterization and resulting modeling of microwave transistors [10–14]. For a detailed treatment of the subject, the reader is referred to Chapters 1 and 2. Furthermore, as outlined in Chapters 3–6, the transistor performance heavily depends on the specific technology used.

Not all RF power transistor types perform equally with respect to these nonlinear transmitter operations. This is particularly true for the high-efficiency DPS transmitter family when hybrid operation like that shown in Figure 7.32 is desired. As seen in Figure 7.32, any transmitter following the hybrid operation profile must perform well as both a linear L-mode amplifier at low output levels, and as a switching C-mode "amplifier" (really a modulator) at high output levels. It therefore is very important to evaluate transistor applicability for use in DPS transmitter designs.

Examining these two endpoints, we adopt metrics taken from Figure 7.14. At the linear end, the transistor must not be sensitive to supply voltage value variations. This is identical to saying that the amplifier provides significant rejection of any supply voltage variations, which includes noise on the supply. Measurements of 11 different transistor technologies and types are presented in Figure 7.33A, which shows that only bipolar transistors and the tested GaN HEMT meet the low sensitivity to power supply variations requirement of Eq. (7.4). All other measured FET technologies do not satisfy Eq. (7.4) and therefore never qualify for L-mode operation. Instead, these transistors never leave P-mode operation, even at high supply voltages.

At the other end of the hybrid operation profile is polar modulation, whereby definition and practicality there is no possibility to reject noise on the supply voltage. The transistor evaluation metric changes then to the measured dynamic range of accurate polar modulation like those shown in Figures 7.28B, 7.29B, and 7.30B.

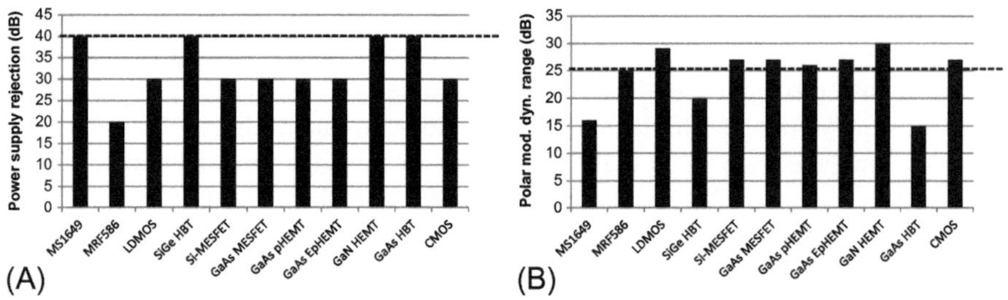

FIGURE 7.33

Transistor performance comparison by type, for transistor applicability in envelope tracking and polar transmitters: (A) for good envelope tracking operation, high power supply rejection is best, which is true for bipolar transistor types; (B) best polar modulation dynamic range is exhibited by FET transistor types.

Across these same transistor types, the values of these accurate polar modulation dynamic range measurements are provided in Figure 7.33B. Here we see that the FET technologies perform best.

At the present time, then, there is no transistor type that is well suited for high-efficiency transmitter use in hybrid DPS architectures like that shown in Figure 7.32 (although the GaN HEMT comes closest). Hopefully this situation will change in the future with transistor improvements in these performance parameters. Right now, though, this means that most hybrid DPS transmitter designs will not meet their performance expectations. The best bipolar transistors for PA power supply variation rejection are the worst performers for polar modulation.

In order to determine whether a transistor can achieve switching power amplifier operation as identified in Figure 7.12, we identify a transistor metric that is proportional to the ability of a transistor to transition from OFF to ON, which is a time constant of the transistor ON resistance and its output capacitance

$$\tau_{ON} = R_{ON} \cdot C_{OUT} \tag{7.6}$$

This time constant is a transistor metric in that it only contains parameters of the transistor, and that this is a best case performance because it does not account for any capacitance in the output circuit. The results of this transistor technology survey that includes τ_{ON} are provided in Figure 7.34.

Here the nominal values of ON resistance and output capacitance for transistors of various technologies are plot against each other, and overlaid with contours for constant values of τ_{ON}. Because transistor technologies tend to scale well with different device sizing (smaller ON resistance designs have larger capacitance, and smaller capacitance designs have higher ON resistance) the position of a particular technology point here is not as important as its relative position to a τ_{ON} contour. The ability of a particular transistor type to support switching operation at the operating frequencies listed across the top of this chart depends on which contour it is close to.

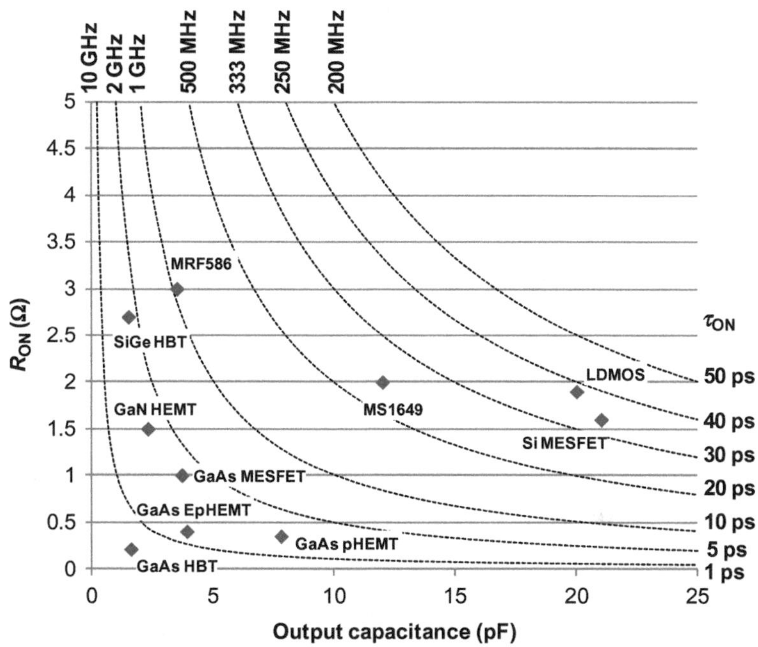

FIGURE 7.34

Comparison of transistor ON resistance and output terminal capacitance along with contours for constant τ_{ON}.

Nonlinear techniques at the circuit and system level are required to improve energy efficiency performance of transmitters with envelope varying signals. Maximizing circuit linearity inherently minimizes energy efficiency, which forces transmitter designers to adopt nonlinear transistor operation. This is already widely done, though it is also rarely recognized as adopting nonlinearity.

One of the important nonlinear applications is to operate the transmitter amplifiers as the three-port systems that they actually are, and always have been. Now with both the RF signal input and the supply voltage input as independent variables, amplifiers are seen to exhibit three separate operating modes. One of these modes is conventional linear operation, and the other two are nonlinear modes that are needed to improve transmitter energy efficiency. Yet even using the linear L-mode, overall transmitter energy efficiency is seen to significantly improve when full three-port operation is adopted.

In addition to greatly improving transmitter energy efficiency, nonlinear applications provide system-level advantages of inherent stability in the circuitry, and uniform performance across temperature and across manufacturing variations. Design conditions are shown to change, and when the new conditions are followed these stabilities are assured.

REFERENCES

[1] E. McCune, Practical Digital Wireless Signals, Cambridge University Press, Cambridge, 2010.

[2] H. Meyr, M. Moenclaey, S.A. Fechtel, Digital Communication Receivers, Wiley Interscience/John Wiley & Sons, New York, NY, 1998.

[3] E. McCune, Operating modes of dynamic power supply transmitter amplifiers, IEEE Trans. Microw. Theory Techn. 62 (11) (2014) 2511–2517.

[4] E. McCune, Polar modulation and bipolar RF power devices, in: Proc. of the 2005 Bipolar Circuits and Technology Meeting (BCTM), Santa Barbara, CA, USA, October 2005, pp. 1–5.

[5] E. McCune, High-efficiency multi-mode, multi-band terminal power amplifiers, IEEE Microw. Mag. 6 (1) (2005) 44–55.

[6] E. McCune, Dynamic Power Supply Transmitters, Cambridge University Press, Cambridge, 2015 (Chapter 8).

[7] E. McCune, Gain: Changed meanings for compressed amplifiers, in: Proc. of the 2013 Midwest Symp. on Circuits and Systems (MWSCAS), Columbus OH, USA, 2013.

[8] Z. Wang, Envelope Tracking Power Amplifiers for Wireless Communications, Artech House, Boston, MA, 2014 (Section 2.2).

[9] F. Raab, P. Asbeck, S. Cripps, P. Kenington, Z. Popovic, N. Pothecary, J. Sevic, N. Sokal, Power amplifiers and transmitters for RF and microwave, IEEE Trans. Microw. Theory Techn. 50 (3) (2002) 814–826.

[10] W. Van Moer, Y. Rolain, A large-signal network analyzer: why is it needed? IEEE Microw. Mag. 7 (6) (2006) 46–62.

[11] A. Ferrero, M. Pirola, Harmonic load-pull techniques: an overview of modern systems, IEEE Microw. Mag. 14 (4) (2013) 116–123.

[12] I. Angelov, H. Zirath, N. Rorsman, A new empirical nonlinear model for HEMT and MESFET devices, IEEE Trans. Microw. Theory Techn. 40 (12) (1992) 2258–2266.

[13] A. Raffo, V. Vadalà, D.M.M.-P. Schreurs, G. Crupi, G. Avolio, A. Caddemi, G. Vannini, Nonlinear dispersive modeling of electron devices oriented to GaN power amplifier design, IEEE Trans. Microw. Theory Techn. 58 (4) (2010) 710–718.

[14] G. Avolio, A. Raffo, I. Angelov, V. Vadalà, G. Crupi, A. Caddemi, G. Vannini, D.M.M.-P. Schreurs, Millimetre-wave FET nonlinear modelling based on the dynamic-bias measurement technique, IEEE Trans. Microw. Theory Techn. 62 (11) (2014) 2526–2537.

CHAPTER

System-level nonideality characterization for compensation

8

S. Farsi*,† and D.M.M.-P. Schreurs*

KU Leuven, Leuven, Belgium Qualcomm Inc., San Diego, CA, United States*†

CHAPTER OUTLINE

8.1 INTRODUCTION

The design of a radio frequency (RF) system typically starts with a very simplistic case. At that stage, it is assumed that the building blocks are ideal, meaning that they

Microwave Wireless Communications. http://dx.doi.org/10.1016/B978-0-12-803894-9.00008-1

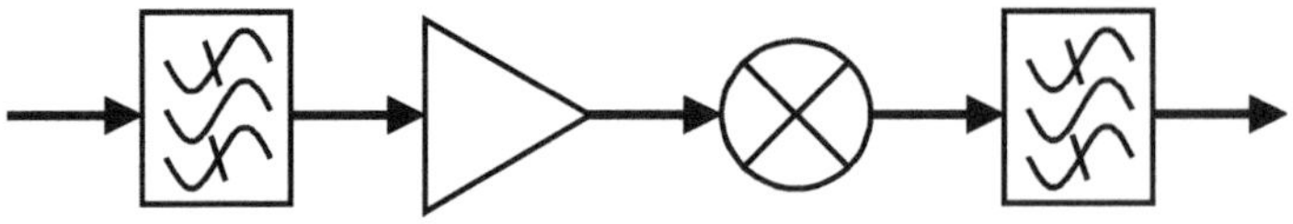

FIGURE 8.1

Simplistic design of a receiver.

work exactly the way they are meant for. However, when such a design is realized, those assumptions are not valid anymore. Therefore, if the assumption violations are not considered in the next stage of the design, the performance is very likely to be unacceptable. These violations are called nonidealities or nonideal effects, and should be properly characterized and compensated in order to have a successful design.

As an example, assume that a simple receiver is required to be designed. The simplistic conceptual design can be like Figure 8.1. In this design, the filters have a rectangular-shaped frequency-domain transfer function, the amplifier is linear, and the mixer changes the center frequency of the signal without changing its spectrum shape. However, this receiver has many nonideal effects in reality. For instance, the filter response is not rectangular, and the amplifier and the mixer do not have linear transfer functions. The amplifier and mixer can even have some frequency-dependent behavior. If these effects, along with many other nonidealities are not considered, the receiver output can easily have large error.

With the advent of next-generation RF systems with higher sensitivity to nonidealities, particularly telecommunication systems, the need for characterization and compensation of nonideal effects has great importance, since it can drastically enhance RF systems' performance. These challenges are the subject of this chapter.

The chapter starts with a brief review on the baseband representation and its applications in Section 8.2. Next, the main nonidealities in RF systems are described in detail in Section 8.3, followed by an overview of their characterization in Section 8.4. In Section 8.5, a recent characterization approach, offset multisine characterization, is highlighted and described in detail. The final Section 8.6 covers a practical application, namely the characterization and modeling of transmitter emission into the receive band.

8.2 BASEBAND CHARACTERIZATION AND MODELING

RF signals used in practice are band-limited. Limited bandwidth leads to the definition of a type of signal representation, called baseband, which makes analysis simpler. In fact, by using baseband representation, we skip considering the RF carrier in analysis which does not contain any information, but is used only for carrying information. The baseband representation is widely used in system-level simulation and analysis, and for modeling and linearization purposes, as will be shown in this

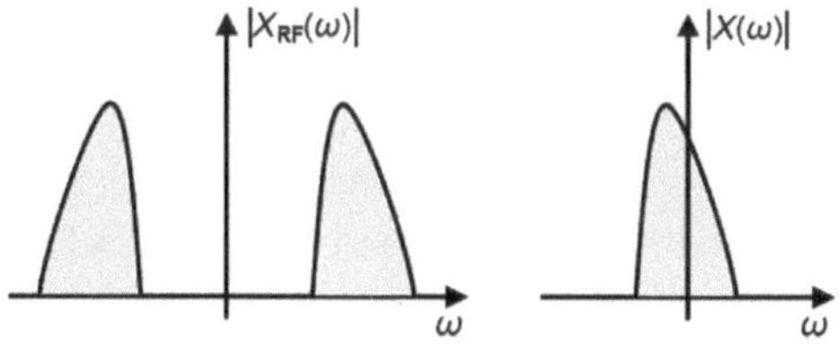

FIGURE 8.2

Relationship between $X_{RF}(\omega)$ and $X(\omega)$ spectra.

chapter. If a signal $x_{RF}(t)$ is band-limited around the center (or carrier) RF frequency f_c, it can be described by:

$$x_{RF}(t) = Re\{x(t)\exp(j2f_c t)\} \tag{8.1}$$

where $x(t)$ is a complex low-pass signal. The low-pass $x(t)$ is the baseband form of $x_{RF}(t)$, and has a spectrum shape $(X(\omega))$ which is the same as that of $x_{RF}(t)$ $(X_{RF}(\omega))$, as shown in Figure 8.2. As can be seen in the figure, $X(\omega)$ is generally asymmetric, which means that $x(t)$ is complex-valued.

8.3 SYSTEM-LEVEL NONIDEALITY

In this section three important types of nonideal effects that usually occur in RF systems will be explained: nonlinearity, memory effects, and in-phase and quadrature (IQ) imbalance. However, it should be noted that there are many more nonideal effects in RF systems, such as phase noise, local oscillator (LO) leakage, LO pulling, and linear distortions, including magnitude and group delay distortion.

8.3.1 NONLINEARITY

A linear system is the one that has a linear relationship between its input $x(t)$ and output $y(t)$ that can be illustrated as a function $f\{.\}$:

$$y(t) = f\{x(t-\tau)\} \tag{8.2}$$

$$f\{a_1x_1(t) + a_2x_2(t)\} = a_1f\{x_1(t)\} + a_2f\{x_2(t)\} \tag{8.3}$$

where τ is a constant output delay with respect to the input, and a_1 and a_2 are constant coefficients. It can be shown that in a time-invariant system, a linear function has the form:

$$f\{x(t)\} = ax(t) + b \tag{8.4}$$

in which a and b are constant coefficients. In RF systems, the coefficient a is usually referred to as gain. The coefficient b determines the bias point, or operating point, around which the output is varying. This coefficient usually adds only a constant value to the output and does not bear any variation of the response. Hence, b is usually disregarded in the analysis of RF systems, since it is usually assumed that the operating point is fixed. The simplified relationship in discrete-time is then:

$$y_n = a x_n \tag{8.5}$$

ignoring the constant delay difference between the input and output. If the delay difference $n0$ is to be considered, x_n will be replaced with x_{n-n0}. The signals y_n and x_n denote the output and input values at the sample time n, $y(n)$ and $x(n)$, respectively. Note that in baseband representation, the gain a can be complex-valued. The phase of this complex value represents the phase difference between the input and output.

8.3.1.1 Nonlinearity in baseband

Contrary to the linear system, the nonlinear system has a nonconstant gain $g(t)$ which depends on the amplitude of the input, and not its phase:

$$y_n = g(|x_n|)x_n \tag{8.6}$$

Note that this equation with the absolute value of x_n is valid in the baseband domain, where the signals are complex. However, in the real time domain or RF time domain, the gain is a function of the signal itself. This discrepancy is explained as follows.

Assume that the input-output relationship in the RF time domain is

$$y_{RF} = a x_{RF} + b x_{RF}^2 + c x_{RF}^3 \tag{8.7}$$

where it is assumed that the nonlinearities are second- and third-order only, though the explanation can be easily generalized in the same way to include higher orders. The baseband signals x and y can be derived from

$$x_{RF} = Re\{x\exp(j\omega_c t)\} = \frac{1}{2}\left(x e^{j\omega_c t} + x^* e^{-j\omega_c t}\right) \tag{8.8}$$

$$y_{RF} = Re\{y\exp(j\omega_c t)\} = \frac{1}{2}\left(y e^{j\omega_c t} + y^* e^{-j\omega_c t}\right) \tag{8.9}$$

where $\omega = 2\pi f_c$. Now we insert Eqs. (8.8) and (8.9) into Eq. (8.7)

$$\frac{1}{2}\left(y e^{j\omega_c t} + y^* e^{-j\omega_c t}\right) = \frac{a}{2}\left(x e^{j\omega_c t} + x^* e^{-j\omega_c t}\right) + \frac{b}{4}\left(x e^{j\omega_c t} + x^* e^{-j\omega_c t}\right)^2 + \frac{c}{8}\left(x e^{j\omega_c t} + x^* e^{-j\omega_c t}\right)^3 \tag{8.10}$$

Since a baseband equation is desired, we keep only the terms that are around the fundamental frequency, which means the terms that have the exponent of $e^{j\omega_c t}$. Therefore, from Eq. (8.10) it can be seen that the second order does not have any contribution to the baseband signal, because it does not include the exponent $e^{j\omega_c t}$. On the contrary, the third order does have contribution to the baseband signal through the multiplication of

$$3\left(x e^{j\omega_c t}\right)^2 \cdot x^* e^{-j\omega_c t} \tag{8.11}$$

After removing the irrelevant terms and keeping the contributors to the baseband signal, the baseband input-output relationship becomes

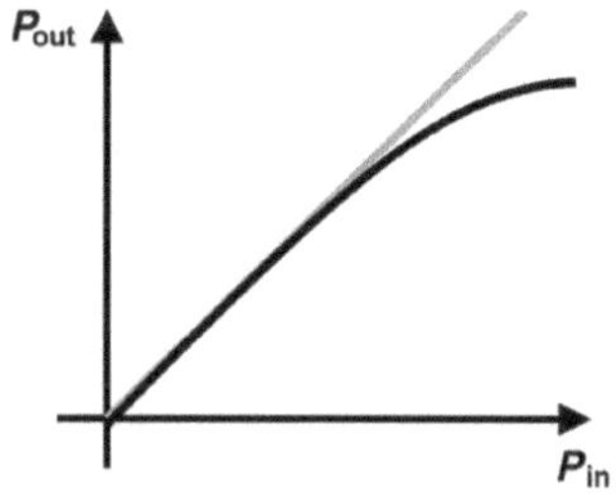

FIGURE 8.3

Typical output power versus input power in linear *(gray)* and nonlinear *(black)* systems.

$$y = ax + \frac{3c}{4}|x|^2 x = g(|x|)x \tag{8.12}$$

which shows that the gain function in the baseband domain is a function of only the input magnitude $|x|$ (compare with Eq. (8.6)). It also shows that the even orders do not have any contribution to the baseband signal. The even orders in Eq. (8.7) after expansion like Eq. (8.10) result in products located at $e^{j2k\omega_c t}$ with k being an integer number including zero. With $k=0$ the product lies at zero frequency. Therefore the even orders always have contributions to the DC component of a nonlinearity. In other words, DC components are generated by even orders because they mix each frequency components with its frequency image, so that the resulting frequency becomes zero.

Figure 8.3 shows the typical plots for the output power versus the input power in linear and nonlinear systems.

8.3.1.2 Weak versus hard nonlinearity

Generally, there are two types of nonlinearity: hard nonlinearity and mild nonlinearity [1,2]. The former includes hard clipping of the waveform due to some mechanism such as cutoff, breakdown, or saturation. However, the latter, also known as weak nonlinearity, includes weak nonlinearities that exist even with small signal variation around the operating point. Figure 8.4 shows these two types of nonlinearity in a typical characteristic curve of an amplifier. Observe that the curve is not straight, even in its middle. This weak deviation from being a straight line (dotted line) is the mild nonlinearity, whereas the clippings at the two ends of the curve are hard nonlinearities. Note that the mild nonlinearity may also be referred to as the weak nonlinearity just before the hard clipping at the high power levels.

8.3.1.3 Harmonic generation and intermodulation

Among the most important consequences of nonlinearity is the generation of harmonics and intermodulation (IM). Considering a tone at the frequency f_c, mathematically expressed by $\exp(j2\pi f_c t)$, when a nonlinear block of the form $a_1 x_{RF} + a_2 x_{RF}^2 + a_3 x_{RF}^3 + \cdots$ is excited by such a tone, the output will be

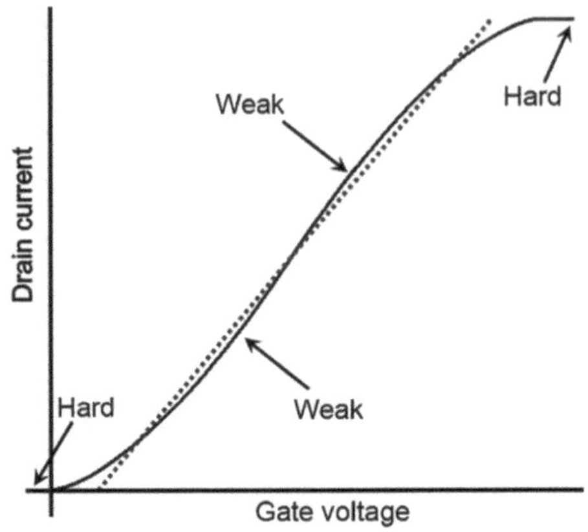

FIGURE 8.4

Demonstration of weak and hard nonlinearities in a characteristic plot of a field-effect transistor.

$a_1 \exp(2\pi(1f_c)t) + a_2 \exp(j2\pi(2f_c)t) + a_3 \exp(j2\pi(3f_c)t) + \cdots$. In other words, the harmonics of the input frequency f_c, namely $2f_c$ and $3f_c$ and $\ldots$, are generated.

On the other hand, the phenomenon of IM happens when two or more frequency components of the input signal mix together due to nonlinearity. Such mixing products can lay around the carrier frequency, zero frequency (DC), or higher harmonics. Those products around DC and around the carrier frequency are of special interest, as will be seen later in this chapter. Figure 8.5 illustrates the harmonic and IM products

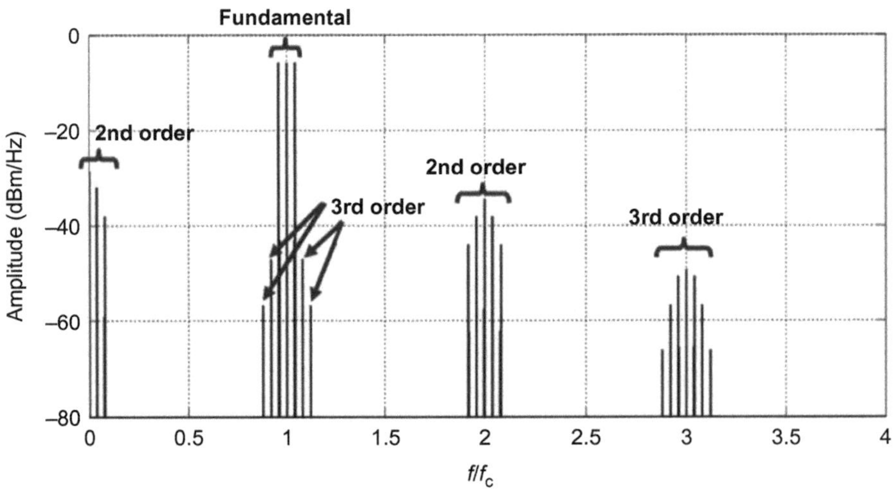

FIGURE 8.5

Generation of harmonics and IM products due to nonlinearity excited by a three-tone signal.

of a nonlinear block with linear, second-order and third-order terms, excited by a three-tone excitation. The figure shows the nonlinear order that each harmonic or IM product comes from. The generation of IM products around the carrier frequency is usually referred to as spectral regrowth. In Figure 8.5 spectral regrowth is caused by third-order nonlinearity only. Spectral regrowth is generally created by odd-order nonlinearities, since even-order mixing products do not lie around the carrier frequency, as shown in Section 8.3.1.

A special case of IM is of particular interest, when two separate bands are simultaneously present at the input of a nonlinear block. In this case, not only the above-mentioned frequency components of each band are intermodulated together, but the frequency components of the different bands are intermodulated as well. This type of mixing is called cross-modulation, and is shown in Figure 8.6. In this figure, IM products of each band along with the cross-modulation products can be seen. It is clear that the frequency location of cross-modulation products can be different from that of IM products, as is the case in Figure 8.6.

8.3.2 MEMORY EFFECTS

In general, the input-output relationship of a system could be static or dynamic. Here, the static or memoryless relationship is the one that has the form:

$$y_n = f(|x_n|) \tag{8.13}$$

In other words, in the static case, the output is a function of only the present input. Note that in this section we assume that the input and output samples are time-aligned

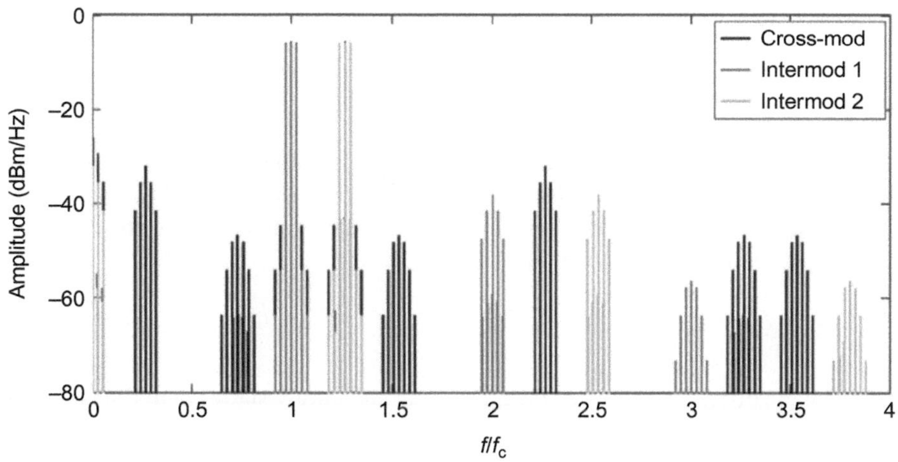

FIGURE 8.6

Generation of IM and cross-modulation products when two separate bands are simultaneously present at the input of the nonlinearity.

so that we can assume that without the loss of generality we do not consider any delay in Eq. (8.13) between the input and output.

Opposite of the memoryless relationship, the dynamic or memory-including relationship is the one that relates the output to not only the present input, but also to the previous values of the input, that is,

$$y_n = f(x_n, x_{n-1}, x_{n-2}, \ldots) \tag{8.14}$$

Under this assumption, the system is said to have memory. Memory in the frequency domain is translated to frequency dependency or dispersion. The simplest type of memory is linear memory or linear distortion. The output of a system with linear memory has the form:

$$y_n = h \cdot x_n = \sum_m h_m \cdot x_{n-m} \tag{8.15}$$

in the time domain. The output is the convolution of the impulse response with the input $x(.)$. In the frequency domain, it is the multiplication of the Fourier transformations of $h(.)$ and $x(.)$, $H(\omega)$ and $X(\omega)$, given by:

$$Y(\omega) = H(\omega)X(\omega) \tag{8.16}$$

The above-mentioned distortion is linear, because the coefficients of x_{n-m} in Eq. (8.15) do not depend on the input value. As opposed to this type of memory, there is nonlinear memory. In such a case, the coefficients are dependent on the input values. For example:

$$\begin{aligned} y_n &= \sum_m g_m(x_{n-m})x_{n-m} \\ g_m(x_n) &= \sum_k |x_n|^k \end{aligned} \tag{8.17}$$

in which the coefficients $g_m(\cdot)$ are themselves functions of the input, has nonlinear memory.

There is an intermediate condition between the memoryless and memory-including cases, called the quasistatic or quasimemoryless case [3]. In such a condition, both the output amplitude and output phase are assumed to be a static nonlinear function of the input amplitude, given by [3,4]:

$$|y_n| = f(|x_n|) \tag{8.18}$$

$$\angle y_n = \varphi(|x_n|) \tag{8.19}$$

Note that the difference between the quasistatic and purely static condition is that, in the former case, the phase is also dependent on the input amplitude, whereas in the latter case, only the amplitude is dependent. In other words, in the static case, we have only amplitude modulation to amplitude modulation (AM/AM) conversion, or only Eq. (8.18). Eq. (8.19) is called amplitude modulation to phase modulation (AM/PM). More explanation will be provided in Section 8.4.2.

Memory effects may have small or large time constants, called short-term or long-term memory effects, respectively [5,6]. Short-term memory effects are usually caused by frequency dependence of device internal parasitics, matching networks, and/or mismatching between different parts in the RF path. They are often in the scale of nanoseconds, and can be usually diagnosed as the amplitude variation of the main-band output spectrum. As discussed in Chapter 1, long-term memory effects are often in the scale of milliseconds or microseconds, and are known to have three major sources: nonideal bias circuit [7,8], thermal effects [7–10], and trapping effects [10,11]. The thermal and trapping effects are usually observed in the transistor level, and their degrading effects on the RF systems are basically through their effects on the transistor performance [12–15].

Long-term memory effects in a power amplifier (PA) can be described by a feedback mechanism [16], as shown in Figure 8.7. In this mechanism, an envelope signal is created by even-order nonlinearities of $F_{NL}(\omega)$, which represents the static nonlinearity of the device. This low-pass signal then manipulates either the thermal or the electrical DC operating point. The manipulation mixes the generated envelope signal with the existing signal, resulting in some nonideal effects around the fundamental frequency and its harmonics. Note that long-term memory effects usually are referred to as the effects around the fundamental frequency, not the harmonics. Since the thermal or electrical manipulation is frequency-dependent, reflected in $F_{FB}(\omega)$, it leads to memory effects. Considering that the thermal and DC operating point variation have large time constants compared to the RF signal, these memory effects are long-term. The two input and output blocks, $H(\omega)$ and $O(\omega)$, are representatives of linear distortion effects at the input and output of the PA circuit.

As mentioned above, along with nonideal bias circuits and thermal effects, trapping effects can also cause long-term memory [6]. Such effects are related to the trapping or releasing of electrons or holes in surface states and/or bulk material due to imperfections in the semiconductor layer structure. As the time constants of the trapping and releasing processes are relatively large, this physical phenomenon mainly affects low-frequency signals, and thus corresponds to long-term memory effect.

8.3.3 IQ IMBALANCE

Many RF systems use in-phase (I) and quadrature (Q) modulator/demodulator for up or downconversion [17]. In these demodulators/demodulators, it is assumed that the signals in the I and Q paths are equal in amplitude and have 90 degrees difference

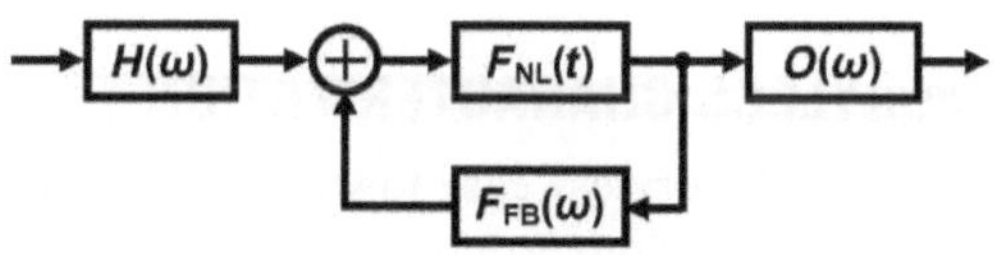

FIGURE 8.7

Feedback mechanism describing long-term memory effects in a PA circuit.

in-phase. However, these assumptions are not perfectly fulfilled in practice, and there are always deviations from these two assumptions [18–20]. These deviations are called IQ imbalance. In other words, the I and Q paths can have unequal amplitudes (amplitude imbalance) and/or a phase difference other than 90 degrees (phase imbalance). IQ imbalance has degrading effects in the system performance and should be compensated in many applications [18–20].

8.4 CHARACTERIZATION APPROACHES

Since the nonidealities of the RF systems can have degrading effects on the system performance, they should be suitably identified and measured. The appropriate identification and measurement of nonideal effects is called characterization. In the following subsections some of the widely used characterization approaches are briefly explained.

8.4.1 MEMORYLESS CHARACTERIZATION

Being memoryless is a plausible assumption in many RF systems. Under this assumption, the input-output relationship can be simply represented by their magnitudes only, that is, by a AM/AM curve. Here we show the rationale behind this fact.

Assume that a memoryless system is represented with Eq. (8.7) in the RF domain, that is,

$$y_{\mathrm{RF}} = ax_{\mathrm{RF}} + bx_{\mathrm{RF}}^2 + cx_{\mathrm{RF}}^3 \tag{8.20}$$

Note that in this equation the coefficients a, b, and c are real numbers, since it is in the RF time domain. In Section 8.3.1 it is already shown that the baseband equivalent equation is

$$y = ax + \frac{3c}{4}|x|^2 x = \left(a + \frac{3c}{4}|x|^2\right)x = g(|x|)x \tag{8.21}$$

In this equation, since a, c, and $|x|$ are real, $g(|x|)$ is also real. Hence, according to Eq. (8.21) the relationship between y and x can be simply shown with a gain $g(.)$ which is real-valued and a nonlinear function of the input magnitude. This means that an AM/AM curve is sufficient to show the input-output relationship in a memoryless system (ignoring the delay). In this case, the AM/AM curve is in fact the plot of $g(.)$ versus the input magnitude $|x|$.

8.4.2 QUASIMEMORYLESS CHARACTERIZATION

A small step forward from memoryless case towards one with memory is the quasimemoryless case. In this case the input-output relationship is not completely characterized by the AM/AM curve only, and an additional AM/PM curve is also required. Here we show in an example how the AM/PM curve can be an implication

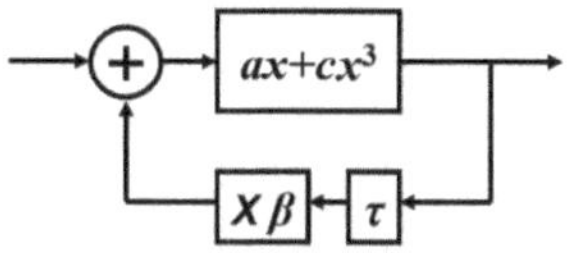

FIGURE 8.8

A simple delayed feedback mechanism.

of partial memory. Assume that the system has a delayed feedback from the output to the input, as show in Figure 8.8. This delayed feedback imposes memory to the system, since the system output is now a function of not only the present input, but also the delayed (previous) input because of the feedback.

Also assume that the delay is very short compared to the baseband variation (or equivalently, envelope variation). In other words, the delay is so short that:

$$x_{RF}(t-\tau) = Re\{x(t-\tau)\exp(j\omega_c(t-\tau))\} \approx Re\{x(t)\exp(j\omega_c(t-\tau))\} \tag{8.22}$$

Taking $\exp(j\omega_c\tau) = e^{j\Phi}$:

$$x_{RF}(t-\tau) \approx Re\{x(t)e^{j\Phi}\exp(j\omega_c t)\} \tag{8.23}$$

Eq. (8.23) means that the equivalent baseband signal of a short delay of $x_{RF}(t)$ is $x(t)e^{j\Phi}$. We use this property in the following. Assume that the nonlinear block of Figure 8.8 is represented by a third-order equation:

$$y_{RF}(t) = ax_{RF}(t) + cx_{RF}^3(t) \tag{8.24}$$

If the feedback factor β is small enough such that $a\cdot\beta \ll 1$, the system output can be found by first-order approximation (only one iteration):

$$y_{RF}(t) = a\left(x_{RF}(t) + \beta\left(ax_{RF}(t-\tau) + cx_{RF}^3(t-\tau)\right)\right) + c\left(x_{RF}(t) + \beta\left(ax_{RF}(t-\tau) + cx_{RF}^3(t-\tau)\right)\right)^3 \tag{8.25}$$

Assuming that

$$a\cdot\beta \ll 1, \quad c\cdot\beta\cdot x^2 \ll 1 \tag{8.26}$$

the system output becomes

$$\begin{aligned} y_{RF}(t) \approx\ & ax_{RF}(t) + \beta a^2 x_{RF}(t-\tau) + \beta ac x_{RF}^3(t-\tau) + cx_{RF}^3(t) + 3\beta ac x_{RF}^2(t)\,x_{RF}(t-\tau) \\ & + 3\beta c^2 x_{RF}^2(t) x_{RF}^3(t-\tau) \end{aligned} \tag{8.27}$$

Following the method presented in Section 8.3.1.1 for deriving baseband representation, and considering that, as explained earlier in this subsection, the baseband form of the delayed input $x_{RF}(t\text{-}\tau)$ is $x(t)e^{j\Phi}$, the baseband equivalent of Eq. (8.27) is:

$$\begin{aligned} y \approx\ & ax + \beta a^2 x e^{j\Phi} + \frac{3}{4}\beta ac|x|^2 x e^{j\Phi} + \frac{3}{4}c|x|^2 x + \frac{3}{4}\beta ac|x|^2 x\left(e^{-j\Phi} + 2e^{j\Phi}\right) \\ & + \frac{9}{16}\beta c^2 |x|^4 x\left(e^{-j\Phi} + 2e^{j\Phi} + e^{j3\Phi}\right) \\ =\ & a\left(1 + \beta a e^{j\Phi}\right)x + \frac{3}{4}c\left(1 + 3\beta a e^{j\Phi} + \beta a e^{-j\Phi}\right)|x|^2 x + \frac{9}{4}\beta c^2 \cos^2\varphi e^{j\Phi}|x|^4 x \end{aligned} \tag{8.28}$$

After some resorting, the equation becomes

$$y=\left(a\left(1+\beta a e^{j\Phi}\right)+\frac{3}{4}c\left(1+3\beta a e^{j\Phi}+\beta a e^{-j\Phi}\right)|x|^2+\frac{9}{4}\beta c^2\cos^2\varphi e^{j\Phi}|x|^4\right)x=g(|x|)x \quad (8.29)$$

The coefficients a and c are real, but the gain function $g(.)$ is complex now. Therefore

$$\angle y=\angle g(|x|)+\angle x \quad (8.30)$$

This equation shows that the phase difference between the input and output is not constant anymore and is a function of the input magnitude which means AM to PM conversion. Note that although in Eq. (8.29) the terms with the coefficient β are small, they become important when the power increases, because in that case $c|x|^2x$ becomes comparable with ax, but with the opposite phase. Hence these two terms add destructively, and the remaining terms which are all with the coefficient β get more important.

In summary, we showed how AM/PM arises in a quasimemoryless case in which the delay of the feedback which caused memory was short enough such that it did not affect the baseband signal, namely, $x(t-\tau)\approx x(t)$. This case is not memoryless because it has AM to PM conversion, but it is neither like memory-including cases because the output is still represented by the present input only (Eq. (8.29)). That is why it is called quasimemoryless.

AM/AM and AM/PM conversions are typically illustrated by a plot, as in Figure 8.9. If the system was memoryless (static case), only the AM/AM curve would be defined and the phase would be independent of the input power level, as shown in Figure 8.9. In other words, the AM/PM curve would be a horizontal straight line. As can be seen in the AM/PM curve, at low input magnitudes, the nonlinear effects on the output phase are negligible. However, as the input magnitude increases, the AM to PM conversion comes into play and makes the AM/PM curve dependent on the input magnitude.

One of the most well-known AM/AM and AM/PM models is the Saleh model. It was originally used for traveling-wave tube. It relates the gain and phase of the transfer function to the input amplitude, through rational polynomial functions [21,22].

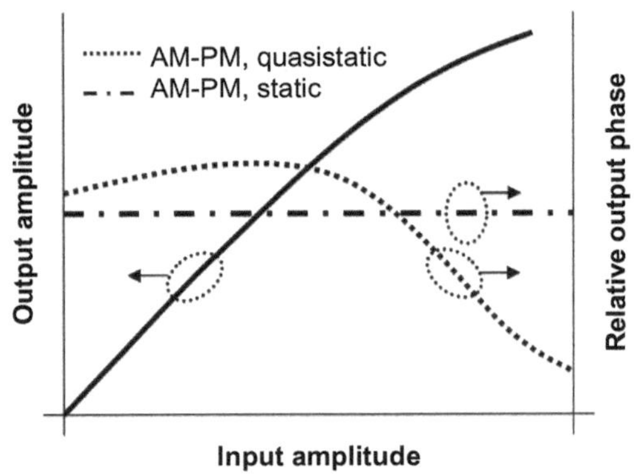

FIGURE 8.9

Typical AM/AM and AM/PM curves of a quasistatic system.

Although the AM/AM and AM/PM curves are basically for quasimemoryless cases where time variation of the parameters are not considered, extensions can be applied to such curves to incorporate memory effects. For example, it was shown in Ref. [23] that if dependence of AM/AM and AM/PM curves on the circuit variables such as supply voltage and temperature are considered and measured, the long-term memory effects caused by variation of supply voltage and operating temperature can be well estimated.

8.4.3 CHARACTERIZATION WITH VOLTERRA MODELS

What we explained in Sections 8.4.1 and 8.4.2 was nonideality characterization that can be summarized in the form of one or two curves (AM/AM and AM/PM curves). In a more general case, we can summarize the nonideal behavior of the system in the form of a mathematical equation. We refer to all these summarizations, either curves or equations, as behavioral models. One of the most common and strongest modeling approaches is Volterra-based modeling, which is the topic of this subsection.

Sections 8.4.1 and 8.4.2 were about memoryless and quasimemoryless cases. More general cases that include more memory may be characterized with Volterra-based models. For example, if the output is represented as

$$y_n = g(|x_n|, |x_{n-1}|, \ldots, |x_{n-m}|)x_n \tag{8.31}$$

where $g(.)$ is a nonlinear gain function, the system has a nonlinear memory with memory depth of m. Another representation of systems with nonlinear memory is the famous memory polynomial, which has been successfully used in modeling many systems, given by:

$$\begin{aligned} y_n &= \sum_{m=0}^{M} g_m(|x_{n-m}|)x_{n-m} \\ g_m(|x_{n-m}|) &= \sum_{k=1}^{K} a_{mk}|x_{n-m}|^{2k} \end{aligned} \tag{8.32}$$

In other words, the output is the weighted sum of input samples at different times, with weights being nonlinear functions of the magnitude of the same input sample.

Memory polynomial is a special case of a more general formulation called Volterra series. Volterra series and its derivations are widely used for behavioral modeling. Its baseband discrete-time form can be represented by [2]:

$$y_n = \sum_{p=1}^{P} \sum_{m_1} \cdots \sum_{m_{2p+1}} \left[h_{2p+1}(m_1, \ldots, m_{2p+1}) \prod_{i=1}^{p+1} x_{n-m_i} \prod_{j=p+2}^{2p+1} x^*_{n-m_j} \right], \quad k = 2p+1 \tag{8.33}$$

where $h_k(\cdot)$ is the kth-order Volterra kernel. Take the kth-order term for instance, represented as:

$$y_{nk} = \sum_{m_1} \sum_{m_2} \cdots \sum_{m_k} \left[h_k(m_1, \ldots, m_{2p+1}) x_{n-m} \cdots x_{n-m_{p+1}} x^*_{n-m_{p+2}} \cdots x^*_{n-m_{2p+1}} \right], \quad k = 2p+1 \tag{8.34}$$

Considering what is explained in Section 8.3.2, this term includes nonlinear memory effects. If $k=1$, the term is the linear convolution, and thus describes the linear behavior of the system, including the linear distortion. The baseband representation of Volterra series contains only odd orders, since the even orders do not lead to any frequency product around the fundamental frequency as shown in Section 8.3.1. Also the series is truncated at the order $2p+1$ for practical implementation. It is assumed that the contributions of higher orders are insignificant. The Volterra-based models are used to model mild and close-to-hard nonlinearities, and usually fail to have acceptable performance for very hard nonlinearities [2–21].

Volterra series can also be represented in the frequency domain. Taking the Fourier transform of the Volterra series that is written in the continuous-time domain leads to the frequency-domain representation [21]. The result is

$$\begin{aligned} Y(\Omega) = {} & H_1(\Omega)X(\Omega) + H_3(\omega_1,\omega_2,\omega_3)X(\omega_1)X(\omega_2)X^*(\omega_3) \\ & + H_5(\omega_1,\omega_2,\omega_3,\omega_4,\omega_5)X(\omega_1)X(\omega_2)X(\omega_3)X^*(\omega_4)X^*(\omega_5) + \cdots \\ \Omega = {} & \omega_1 + \omega_2 - \omega_3 \end{aligned} \tag{8.35}$$

where $X(.)$ and $Y(.)$ are the spectra of $x(t)$ and $y(t)$, respectively. The weights of $H_{2k+1}(.)$ are the frequency-dependent coefficients called frequency-domain Volterra kernels. The first-order kernel $H_1(.)$ is simply the transfer function of a linear system. Eq. (8.35) nicely shows the frequency translation mechanism caused by nonlinearity. For instance, the third-order nonlinearity contributes to the output baseband via the term $H_3(\omega_1,\omega_2,\omega_3)X(\omega_1)X(\omega_2)X^*(\omega_3)$. This term illustrates that the input frequency components $X(\omega_1)$, $X(\omega_2)$, and $X(\omega_3)$ which are at frequencies ω_1, ω_2 and ω_3 are the contributors of the nonlinearity that lies at $\omega_1+\omega_2-\omega_3$. The kernel $H_3(.)$ is generally frequency-dependent, depending on the three frequencies ω_1, ω_2, and ω_3. If the system is memoryless, this Volterra kernel will be constant, namely frequency-independent. Considering that ω_1, ω_2, and ω_3 are close to each other, we conclude that the term in Eq. (8.35) with the order of $2k+1$ generate products in $k+1$ regions. For example, the third-order term generates products around the main band via $\omega_1+\omega_2-\omega_3$, and in the third-harmonic region via $\omega_1+\omega_2+\omega_3$. Likewise, the fifth-order contributes to the fundamental region via $\omega_1+\omega_2+\omega_3-\omega_4-\omega_5$, the third-harmonic region via $\omega_1+\omega_2+\omega_3+\omega_4-\omega_5$, and the fifth harmonic region via $\omega_1+\omega_2+\omega_3+\omega_4+\omega_5$. Therefore, it is seen that all the odd-order terms contribute to the fundamental region. Some of the generated products lay exactly at the frequencies of the excitation frequency components, for example $\omega_1+\omega_2-\omega_3$ in the third order which lays at ω_1. Although this product has the same frequency as $X(\omega_1)$, it is not correlated with $X(\omega_1)$, because its value is $H_3(\omega_1,\omega_2,\omega_3)X(\omega_1)X(\omega_2)X^*(\omega_3) = H_3(\omega_1,\omega_1,\omega_1)|X(\omega_1)|^2X(\omega_1)$: the amplitude of the excitation component at ω_1 is one, while that of the third order is $H_3(\omega_1,\omega_1,\omega_1)|X(\omega_1)|^2$, thus they are uncorrelated. Such uncorrelated products cause in-band distortion and degradation of the error-vector magnitude (EVM), introduced in Chapter 2, while the products of odd orders which are in the fundamental region, but outside the excitation band cause spectral regrowth and degradation of the adjacent-channel leakage ratio (ACLR). Figure 8.10 shows such

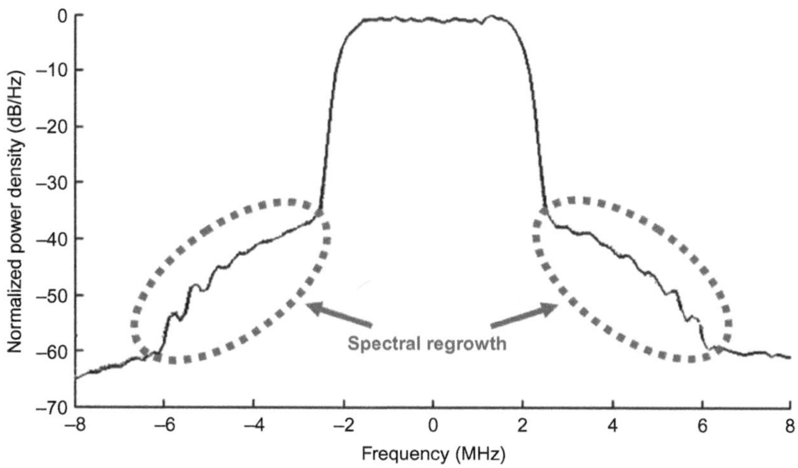

FIGURE 8.10

Spectral regrowth at the output of the nonlinearity.

products. In other words, the third-order IM (IM3) products which are generated by $H_3(\omega_1,\omega_2,\omega_3)X(\omega_1)X(\omega_2)X^*(\omega_3)$ affect both EVM and ACLR, Since the in-band third-order products are normally out-of-phase as compared with the fundamental component (first-order component), and hence add to them destructively, they cause compression at the nonlinearity output. A classic metric for such compression is P1dB, as discussed in previous chapters. Note that IM3 is not the only nonlinearity contributing to P1dB, ACLR, or EVM. All higher-order nonlinearities have an impact on those metrics in the same way. However they can have different effects in the sense of being constructive or destructive. For example, the fifth-order product can be in-phase with the fundamental, and thus, can be added to it constructively.

8.4.3.1 Identification of volterra-based models

The unknown parameters of Eq. (8.33) are the coefficients $h_{2p+1}(m_1, \ldots, m_{2p+1})$. All these coefficients have first order, which leads to a linear equation with respect to the unknown parameters. Therefore, to find the unknown values, a sufficient number (more than the number of unknown values) of Eq. (8.33) for different n values should be considered. This system of linear equations (8.33) with varying n can be written as:

$$Y = Xb \tag{8.36}$$

where Y is the vector of output samples y_n, b is the vector of unknown coefficients $h_{2p+1}(m_1, \ldots, m_{2p+1})$, and X is the input matrix consisting of elements $\prod_{i=1}^{p+1} x_{n-m_i} \prod_{j=p+2}^{2p+1} x^*_{n-m_j}$, with different n values. Such a system of linear equations can be solved by using least square methods [24].

The number of coefficients in Eq. (8.33) increases exponentially with increasing the truncation order $K=2p+1$. In many applications, the orders up to at least five or seven are required to have acceptable estimation of the system nonlinearity. This requirement leads to a huge number of coefficients for computation. Apart from increased complexity, the large number of coefficients may lead to overfitting.

A solution to this complexity problem is to make simplifications in the Volterra series. There are an enormous number of papers in the literature reporting different variations and simplifications of the Volterra series, for example [25–32]. In this subsection, only some of the most widely used and most accurate ones with suitable compromise between accuracy and simplicity will be addressed.

8.4.3.2 Effect of cross terms

Some of the terms in Eq. (8.33) are multiplication products of the same input sample, that is, $h_k(m,m,\ldots,m)$. The other terms are the multiplication products of different input samples. These terms are termed as cross terms. Cross terms have been proven to cause improvement in modeling capability as compared to the Volterra-based models without cross terms [32]. This improvement is due to the fact that cross terms actually imply that different input samples mixing may occur due to, for example, feedback or multipath.

8.4.3.3 Including even-order terms

Even-order terms in baseband have the form

$$h_{2p}(m_1,\ldots,m_{2p})\prod_{i=1}^{2p-1}|x_{n-m_i}|\cdot x_{n-m_j} \tag{8.37}$$

where p can be any integer number. Starting from the RF representation of the Volterra series, that is,

$$y_n=\sum_{k=1}^{K}\sum_{m_1}\cdots\sum_{m_k}\left[h_k(m_1,\ldots,m_k)\prod_{i=1}^{k}x_{n-m_i}\right] \tag{8.38}$$

and keeping only the terms that generate frequency components around the fundamental frequency, Eq. (8.33) can be derived [33]. Such a derivation does not include even-order terms as also shown in Section 8.3.1. In other words, even-order terms cannot be derived from the Cartesian representation of the Volterra series. However, it was shown in Ref. [8] that starting from the polar representation on the Volterra series and with some assumption regarding the nonlinear mechanisms in the circuit, even-order terms can be included in the formulation. The assumed mechanisms in Ref. [8] came from long-term memory effects. It has been shown in the literature that including even-order terms in Volterra-based models improves the modeling accuracy in many applications [34,35].

8.4.4 CHARACTERIZATION WITH VARIOUS EXCITATIONS

AM/AM and AM/PM curves are traditionally obtained with power sweep of a continuous-wave (CW) excitation. However, these curves are usually used to characterize RF systems that are going to operate in real applications. In real applications, the signals and excitations used in the systems are in many cases modulated signals. Therefore the envelope of the signal changes over time, and a CW signal with constant envelop is not a fair emulation of such a signal. For example, as discussed in Chapter 6, the peak-to-average power ratio (PAPR) of a modulated signal is larger than that of a CW signal. Thus, with the same average power, the modulated signal experiences more nonlinearity than the CW signal, since it goes more into the nonlinear region because of its higher peak value. Another dissimilarity between the two cases with different excitations is that a modulated signal can excite the memory of the system to a much higher extent because of its varying amplitude or frequency. In such conditions the AM/AM and AM/PM model obtained by CW sweep cannot be an accurate model of the system. Therefore, characterization by using signals with nonconstant envelope is a more reliable way. Envelope-varying waveforms used for characterization are usually multisine or real modulation excitations, as explained in the following.

8.4.4.1 Two-tone characterization

The simplest modulated or envelope-varying waveform is a two-tone waveform. It is extensively deployed in characterization of many RF systems. It is very easy to generate and well-defined. It is composed of two sine waves that, in general, have independent amplitude and phase. The shape and the variation speed of the envelope can be adjusted by adjusting the amplitudes frequencies and phases of the two sines (tones). However, in two-tone waveforms that are usually used, the two tones have the same amplitude and phase.

One of the interesting applications of the two-tone waveform is studying IM and memory effects. Traditionally, the IM3 product is characterized in nonlinear circuits as a measure of nonlinearity. This IM3 characterization is usually done with a two-tone excitation. The IM3 products appear at the output spectrum as two additional tones on the two sides of the main two tones. The power ratio of these IM3 products over the main tones is a measure of nonlinearity.

Two-tone excitation can also be used for memory effects characterization [7]. As was shown in Section 8.3.2, memory effects have frequency-dependent behavior. It is shown in the literature [7] that the RF systems may show memory via responding differently to two-tone signals with different tone spacings. Figure 8.11 illustrates an example of such memory effects as compared to the memoryless case.

8.4.4.2 Multisine characterization

More than two tones can be used to get a waveform with a more complicated envelope. Multisine waveforms have also been widely investigated in the literature and have proven their capabilities for characterization [37,38]. Since there are more than

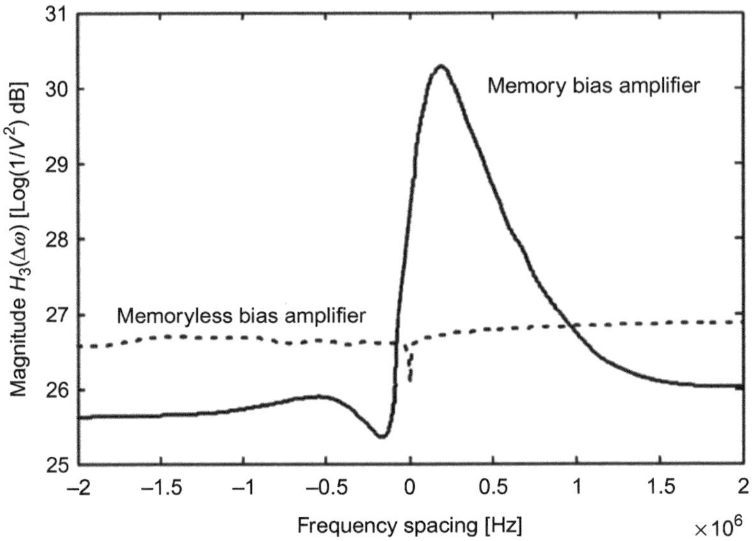

FIGURE 8.11

Dependence of memory effects on tone spacing [36].

two tones in this case, there are more degrees of freedom to make various envelope shapes. In this way, multisine excitations can be designed as to mimic realistic modulations [39].

8.4.4.3 Characterization with real modulation

As discussed in Chapter 2, real modulated waveforms can also be used in characterization to get the most reliable results from characterization. However, this kind of test is not as straightforward as multisine characterization, because, for example, there are usually several parameters and conditions that should be well-determined in such a modulated signal, and also these waveforms are practically nonperiodic. An easier and more feasible way is to use modulated waveforms that are simplified to be cyclic. In this way, they are periodic and a repeated behavior can be more easily and accurately measured.

8.4.5 CHARACTERIZATION WITH X-PARAMETERS

X-parameters, also known as S-functions, are the generalization of S-parameters [40]. They characterize nonlinear circuits in terms of their response at fundamental and harmonic frequencies, whereas S-parameters are used to characterize linear circuits in the frequency domain.

X-parameters are steady-state models that model nonlinear devices or circuits excited by a large input tone which determines the operating point. The other tones available at the input and output at fundamental and harmonic frequencies are assumed to be much smaller than that large tone. These additional small tones can be generated by nonlinearity and reflections. Under these assumptions, a multi-harmonic linearization approximation around the operating point is made: the center value of each tone is determined by the large tone, and the additional generated small tones are perturbations that change the values linearly [41]. The model relates A-waves and B-waves, where A- and B-waves are the incident and reflected traveling power waves [42] defined as:

$$A = \frac{V + Z_c I}{2} \tag{8.39}$$

$$B = \frac{V - Z_c I}{2} \tag{8.40}$$

where V, I, and Z_c are the voltage, current, and the characteristic impedance of the port, respectively. The X-parameter model is mathematically illustrated by [40]:

$$\begin{aligned} B_{pm}(|A_{11}|, f) = &\sum_q \sum_n Sf_{pmqn}(|A_{11}|, f) \cdot A_{qn} \cdot \exp(\mathrm{j}(m-n) \cdot \phi(A_{11})) \\ &+ \sum_q \sum_n Sfc_{pmqn}(|A_{11}|, f) \cdot A^*_{qn} \cdot \exp(\mathrm{j}(m+n) \cdot \phi(A_{11})) \end{aligned} \tag{8.41}$$

where A_{pm} and B_{pm} are the A- and B-waves. Sf_{pmqn} and Sfc_{pmqn} are the X-parameters which are the complex coefficients whose values depend on the magnitude and frequency of the large tone. The parameter p denotes the port number at which the reflected wave is to be calculated, and m is the harmonic number of the reflected wave. The parameter q denotes the port number to which the A wave is incident, and n is the harmonic number of the incident wave. Eq. (8.41) assumes that the reflected wave is a nonlinear function of A_{11}, but is linear with respect to the other A-waves. This assumption is valid as long as these waves are much smaller than A_{11} so that they do not change the operating point. The equation also shows dependency of reflected waves on not only the incident waves, but also their conjugate values A_{qn}, modeling the nonanalyticity of the describing function [40]. Another point is the dependency of the reflected waves on the absolute phase value of A_{11}, which is the direct result of time invariance assumption [43].

Because of its form, the X-parameter model is well-suited for harmonic balance simulation. The formulation already provides the required relationships between the tones at the fundamental and harmonic frequencies. Furthermore, the X-parameter model provides high-accuracy estimation for different nonlinear characteristics, including compression, AM/PM, harmonics, load-pull, and time-domain waveforms [40]. The model can accurately predict the reflected B-wave in nonlinear systems with mismatches at both the fundamental and harmonics. This capability enables the accurate total distortion estimation of circuits composed of a cascade of nonlinear components with known X-parameters for each of them.

On the other hand, X-parameter models are accurate as long as the assumption of the dominant large tone is valid. If one or more of the A-waves other than A_{11} are not much smaller than A_{11}, the assumption is violated and the accuracy is not guaranteed. The violation conditions include large mismatches. Failure in the case of large mismatches is common with other behavioral models as well. However, there is another failure with modulated excitations which is not the case in many other models. X-parameters can be used for modulated excitation conditions under the assumption of a quasistatic situation. This assumption holds as long as the time of change in the modulating signal is much larger than the physical time constants that are inherent in the system. However, this is not the situation in many wideband applications, for example when long-term memory effects are not negligible, or when the wideband excitation causes different memory effects. There are efforts to include long-term memory in the X-parameter model [44,45]. Nevertheless, the derived models are complicated and still not general enough to be used in many applications.

8.5 CHARACTERIZATION WITH OFFSET MULTISINE EXCITATION

As mentioned in Section 8.4.4, the envelope shape of multisine excitations can be easily controlled by adjusting the amplitude and phase of their comprising sine waveforms. However, another degree of freedom for controlling the properties of the multisine excitation corresponds to manipulating the tone frequencies, to be able to detect different nonlinearity orders in the frequency domain individually. There are different approaches based on this manipulation. One approach is to use a number of tones over a wide range of frequencies, and to turn off some of these tones to let higher-order IM products, which are smaller than the main tones, present themselves in the response [46]. This technique has been extensively adopted [47,48] to determine whether the nonlinear IM product is even-order or odd-order. The second approach presented in Ref. [49] focused on the second-order nonlinearity, and tried to distinguish between different second-order products by engineering the frequency locations of the tones so that each of the products lies at a unique frequency. This technique was again used for a wide frequency range to characterize the second-order frequency-domain Volterra kernel. However, higher-order products were not considered in this work and the Volterra model was used in real-time domain for a very low-frequency application.

Another approach, which will be explained more in this section, is to manipulate the multisine excitation to discriminate between different individual IM products, as explained in Ref. [50]. By adopting this approach, the IM products are located at separate frequency locations, each different from the location of the other IM products. In this way, each single IM product can be characterized separately in a single measurement. The fundamental differences between the approach of Ref. [50] and the one used in Ref. [49] are that (1) the approach of Ref. [50] considers nonlinearity orders of three and higher; (2) through its efficient frequency manipulation it can characterize memory effects and nonlinearity order levels; (3) because of its efficient frequency allocation of the tones, the tone spacings are not much different from each other; and (4) it is dedicated to modulated RF applications, although it is more general and can be used for other applications.

8.5.1 THEORY OF MULTISINE OFFSETTING

In order to achieve the purpose of distinct IM products by using offset multisine excitation, two approaches can be employed. First approach is to use incommensurate frequencies for the tones of the multisine excitation [51]. This choice of frequencies guarantees the distinction of IM products. However, the implementation of this excitation as well as the detection or capturing of the tones and generated IM products is very difficult because of being incommensurate. For example, to generate such an excitation, one may require having as many signal generators as the number of the excitation tones, since incommensurate frequencies cannot be generated with a single digital signal generator. The second approach for generating the desired offset multisine is using commensurate frequencies, but with a very small base frequency [50], as explained in the following. This approach requires only one digital signal generator.

A multisine with an odd number of tones M is considered. Note that choosing an odd number is just because the representation of tones used in this section is more compact in this way. Nevertheless, the same theory holds for an even number of tones. In a multisine excitation, the tone frequencies f_{m}'s may be chosen as

$$f_{\mathrm{m}} = F_{\mathrm{c}} + mF_{\mathrm{d}}, \quad m \in Z, \quad -\frac{M-1}{2} \leq m \leq \frac{M-1}{2} \tag{8.42}$$

where F_{c} is the center frequency of the multisine and F_{d} is the frequency spacing between two adjacent tones, and Z is the set of all integer numbers. In this section, we refer to this multisine as the conventional multisine. In Ref. [2] and in Section 8.3.1 it was shown that the IM products located around the fundamental frequency are the odd-order products and their frequencies are:

$$f_{\mathrm{IMD},2k+1} = F_{\mathrm{c}} + \left(\sum_{i=1}^{k+1} m_i - \sum_{j=1}^{k} n_j\right) F_{\mathrm{d}}, \quad m_i, n_j \in Z, \quad -\frac{M-1}{2} \leq m_i, n_j \leq \frac{M-1}{2} \tag{8.43}$$

Eq. (8.43) shows that different combinations of IM products, even from different orders, can have the same frequency locations [2]. Therefore, with this conventional multisine we are not able to distinguish between IM products in a straightforward way. Figure 8.12A and B demonstrate that IM3 products of different combinations can share the same frequency position for in-band and out-of-band IM, respectively.

The proposed approach is to offset the tone frequencies by different amounts, in such a way that each of the IM products has its own unique frequency, as illustrated in Figure 8.12C and D. The amount of offset is considered to be small. Therefore we shift the tone frequencies based on the frequency grid spacing F_{g} which is the frequency resolution of the simulation or measurement. Eq. (8.42) is then modified to:

$$\hat{f}_{\mathrm{m}} = F_{\mathrm{c}} + mF_{\mathrm{d}} + p_{\mathrm{m}}F_{\mathrm{g}}, \quad m, p_{\mathrm{m}} \in Z, \quad -\frac{M-1}{2} \leq m \leq \frac{M-1}{2} \tag{8.44}$$

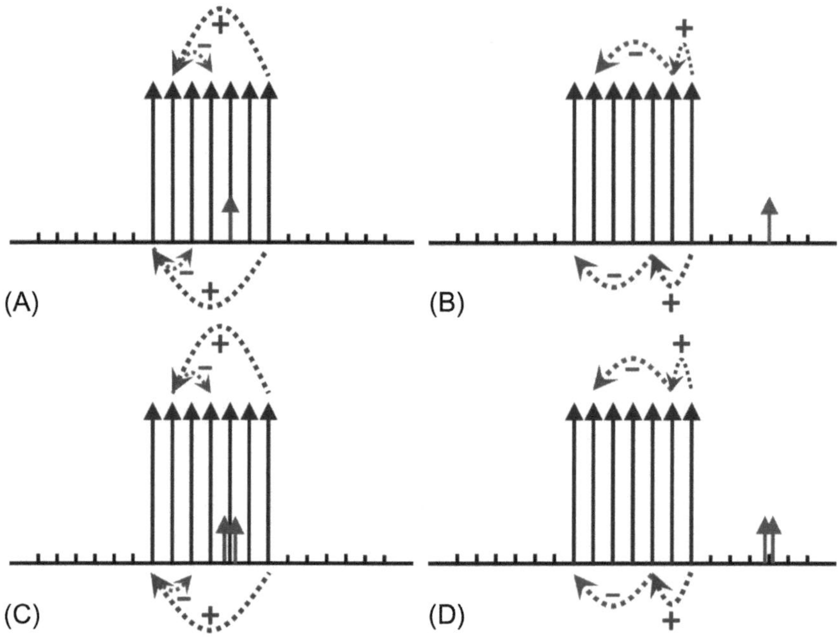

FIGURE 8.12

Formation of IM products *(vertical gray arrows)*: (A) In-band IM products resulting from conventional multisine excitation share the same location as those of the main tones. (B) Out-of-band IM products from different combination of tones resulting from conventional multisine excitation share the same location. (C) In-band IM products resulting from the offset multisine excitation and (D) out-of-band IM products resulting from the offset multisine excitation have distinct frequency locations. Here the *dotted arrows* connect frequencies that are combined to form third-order products.

in which the last term represents a small shift applied to each tone frequency. The hat sign above f_m indicates the shift in frequency, and Z is the set of all integer numbers. If we choose p_m values so that each $\hat{f}_m$ is unique with respect to other $\hat{f}_m$'s, we will achieve our target. Note that F_d is a multiple of the grid frequency F_g. Therefore, the tone frequencies in Eq. (8.44) are still commensurate, although they are now commensurate with a much smaller base frequency F_g, as opposed to the base frequency F_d in the conventional multisine setting. Eq. (8.44) states that each tone of the offset multisine is localized by coarse allocation (F_c+mF_d) followed by a small refinement. A computer code can be generated to find a set of p_m values as small as possible satisfying the uniqueness requirement.

Because of the finer frequency grid of the offset multisine (F_g) as compared to that of the conventional multisine (F_d), the period of the offset multisine is much longer than that of the conventional one. However, this disadvantage can be insignificant, because the period can still be comparable with the length of a typical real modulated signal used in testing.

There are several advantages for the offset multisine excitation. Three of them are seen with regard to in-band distortion, IM products, and waveform statistics. In real modulated, as well as conventional multisine waveforms, in-band distortion products are located at the same frequencies of the main signal. Therefore, they cannot be measured independently. However, with offset multisine excitation, due to the specific tailoring of the tone frequencies, the incoherent in-band distortion products are at distinct frequencies, as already shown in Figure 8.12C and also will be illustrated in Figures 8.13–8.15.

The frequency of each incoherent IM product is unique in the offset multisine excitation. This uniqueness can help in the characterization of phenomena in nonlinear circuits. For instance, one can easily assess the contribution of each nonlinearity order. Furthermore, as will be discussed in the following, memory effects can be well characterized. The offset multisine also has acceptable waveform statistics and dynamics, as explained in Ref. [50].

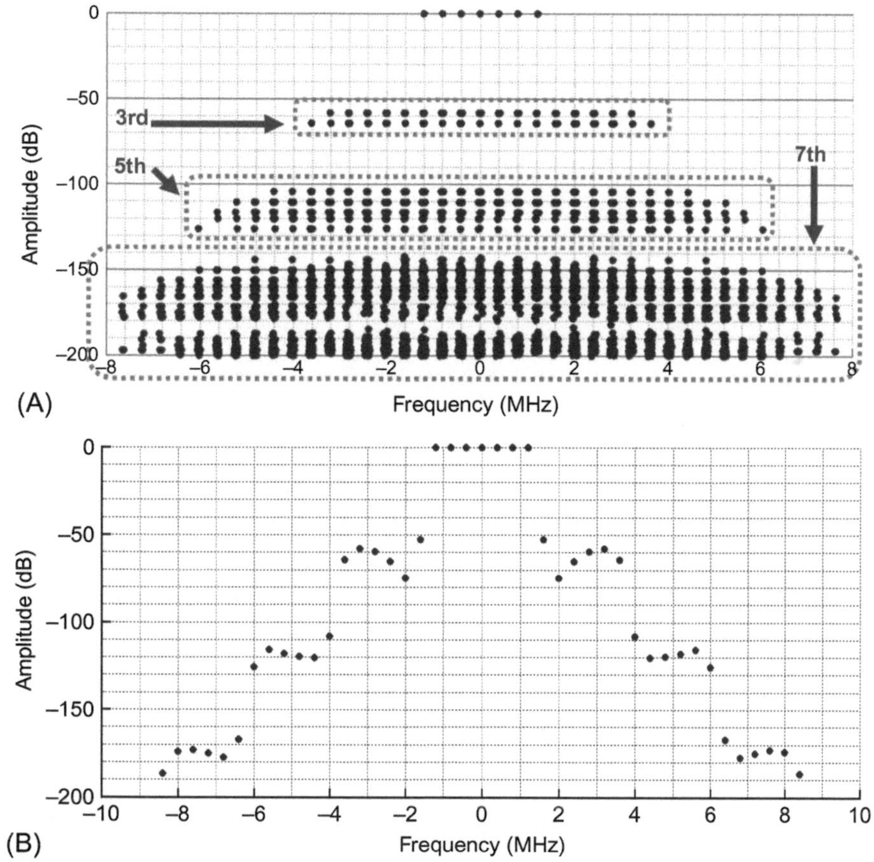

FIGURE 8.13

Output spectrum of Eq. (8.45) with (A) offset multisine excitation and (B) conventional multisine excitation. In (A) different nonlinearity orders of different degrees, even inside the main band, are easily distinguishable.

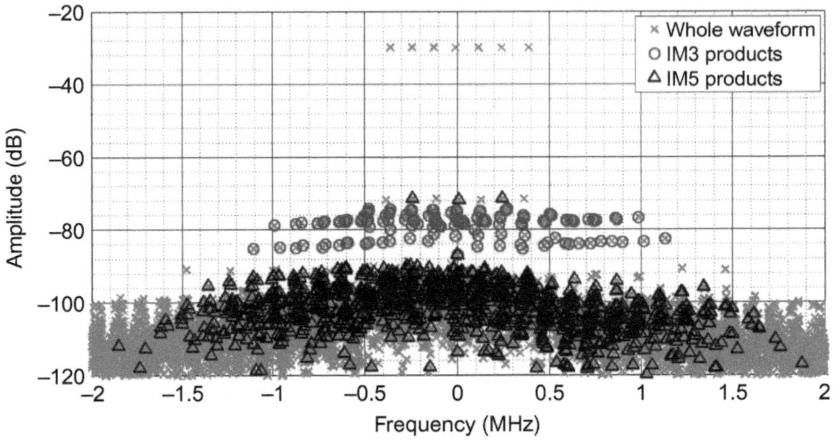

FIGURE 8.14

Measured baseband output spectrum of a PA at 21 dBm output power. The PA is excited with a seven-tone offset multisine with 720 kHz bandwidth. Different IM products from different nonlinearity orders are distinguishable in the offset multisine setting.

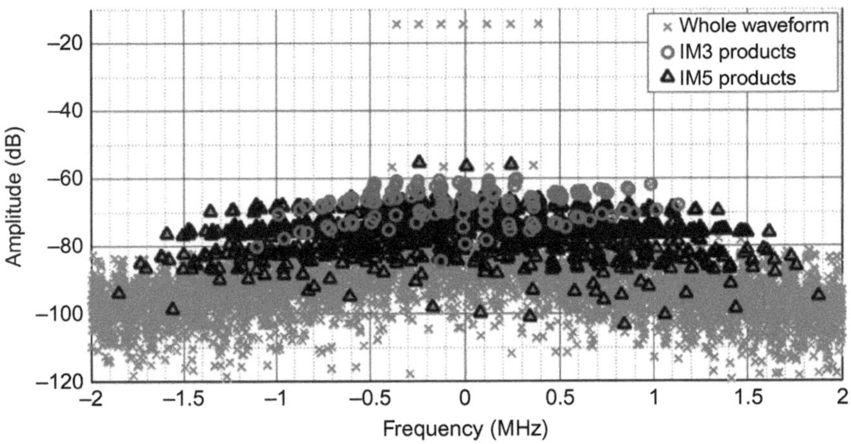

FIGURE 8.15

Measured baseband output spectrum of a PA at the output power of 24 dBm. The PA is excited with a seven-tone offset multisine with 720 kHz bandwidth.

8.5.2 SPECTRUM PLOTS WITH OFFSET MULTISINE EXCITATION

To illustrate the offset multisine method, we consider the following simulation example [50]. A seven-tone multisine with around 2.4 MHz bandwidth is selected. The set of frequency locations of the tones with respect to the center frequency is $\{-1200, -812, -422, -28, 378, 798\}$ kHz, chosen so that each IM product up to the fifth order has its own individual unique location. Since the aim of this section is to show the effect of offsetting the tones, the amplitude and phase manipulation of the tones are not considered here, and will be discussed in the subsequent section. The waveform is then used as an input to a nonlinear block with the characteristic function

$$y = 4(1/(1+e^{-x}) - 0.5) \tag{8.45}$$

The simulated output spectrum is plotted in Figure 8.13A. Four distinct bunches of IM products corresponding to the third-, fifth-, seventh-, and ninth-order IM products (denoted hereafter as IM3, IM5, IM7, and IM9, respectively) can be clearly seen. Interestingly, the in-band contributions of these nonlinearity orders can also be distinctly seen. These are the particular features of the offset multisine method that are absent in the conventional multisine approach whose respective spectrum is plotted in Figure 8.13B. Note that the power level difference of different IM orders changes with changing the signal power. Power levels of higher-order IM products increase faster than those of lower orders [2] as functions of the input power.

It can be seen in Figure 8.13A that each bunch of IM products has a different number of rows of frequency components. IM3 has two rows, IM5 has five rows and so forth. These numbers can be explained by examining frequency-domain Volterra kernels [21] as follows.

As explained in Section 8.4.3, IMD3 products correspond to terms of the type:

$$Y_3(\omega_1+\omega_2-\omega_3) = H_3(\omega_1, \omega_2, -\omega_3)X(\omega_1)X(\omega_2)X^*(\omega_3) \tag{8.46}$$

where H_3 is the frequency-domain Volterra kernel. There are four cases of different angular frequency combinations: (1) $\omega_1=\omega_2=\omega_3$, (2) $\omega_1=\omega_3$ or $\omega_2=\omega_3$, (3) $\omega_1\neq\omega_2$, $\omega_2\neq\omega_3$, $\omega_1\neq\omega_3$, and (4) $\omega_1=\omega_2\neq\omega_3$. Cases 1 and 2 lead to IM products which are exactly at the same frequency as the main large tones and are coherent with the tones [2]. Cases 3 and 4 generate products at distinct frequencies, and are incoherent with the tones [2]. However, the products in case 3 can be created in two ways: $\omega_1+\omega_2-\omega_3$ and $\omega_2+\omega_1-\omega_3$. Therefore the products from case 3 are two times larger in voltage than the products from case 4. This is why we see two rows of IM3 products with exactly 6 dB difference.

The same reasoning applies to the existence of five rows of IM5 products, as well as to the higher-order products. In the IM5 case, the levels of the five respective rows are weighted by 1, 2, 3, 6, and 12. Note that in the case of the fifth order, we have two sets of products that lie at the same locations of lower-order products. One set has the frequency combination of $(\omega_1, \omega_2, \omega_3, -\omega_2, -\omega_3)$, assuming that the ω values can be different or identical. The product of this combination lies at the location of the large main tones. The other set has the frequency combination of $(\omega_1, \omega_2, \omega_3, -\omega_3, -\omega_4)$ assuming that ω_4 is different from ω_1, ω_2, and ω_3. The product of this combination lies at the locations of the IM3 products. These products that are located at the frequencies of the lower-order terms normally are much smaller than them. We can estimate the values of these products by the most similar distinguishable IM5 products. Because the number of these special products is low as compared to the rest of IM products, this estimation will not affect the accuracy of the technique.

Now let us see what the spectrum plot can be in real applications. Figure 8.14 shows the spectrum for a cellular PA with the excited bandwidth of 720 kHz. In this plot, the IM3 and IM5 products are marked with circles and triangles, respectively, whereas the entire set of data points, including IM3 and IM5, are all marked with crosses. The plots clearly show that the two rows of IM3 for this PA in both the in-band and out-of-band regions are distorted as compared to those of Figure 8.13A: They are not perfectly flat anymore. This phenomenon is due to memory effects, as will be discussed in Sections 8.5.3 and 8.5.4.

Because the IM products are distinct from each other, by observing the output spectrum excited by the offset multisine, one can assess the amount of each nonlinearity order. Note that for IM analysis in this section nonlinearity orders up to five are considered and the conditions of interest are such that the higher orders are negligible. However, the approach can also be used for higher-order nonlinearity, provided that the offset multisine is modified to take into account the uniqueness of higher IM locations.

In Figure 8.14, the IM3 products are much larger than the IM5 products. Hence we can assume that those IM5 products that share the same frequencies as IM3 products do not have considerable impact on the IM3 levels. This means that as long as the IM3 products are substantially larger than the IM5 and other unwanted products, by changing the PA output power, the IM3 products pattern changes in level, not in shape. This fact can be easily verified, but not shown here. However, if we increase the power, the IM5 products start to affect the IM3 levels.

The IM5 products that share the same location as the lower IM3 products are the results of $\omega_1+\omega_1+\omega_3-\omega_3-\omega_2$ and all its combinations, where $\omega_1 \neq \omega_2$. The number of possible combinations is 11. This number is found in a way very similar to what is presented earlier in this section. Therefore, this specific accumulated IM5 level is at most 11 times higher than the level of the lowest IM5 row, which is itself composed of only one IM5 product. In other words, eleven IM5 products lie at the same frequency as that of each IM3 product. Thus, in order that the IM5 has impact on the IM3 products, the sum of eleven IM5 products should be comparable with the IM3 level. The best way to estimate this level is by using the level of the highest row of IM5, which is 12 times (21.6 dB) larger than the level of a single IM5 product (corresponding to the lowest IM5 row). Based on these numbers, the accumulated strength of the IM5 products that are impacting each IM3 product is, at most, around $20\log(12/11)=0.8$ dB lower than that of the highest IM5 row. So the highest IM5 row is a good measure to see how influential IM5 distortion is on IM3 products. All these discussions are under the assumption that the nonlinearity orders higher than five are negligible; otherwise we should extend our discussion to consider their effects as well.

Figure 8.15 illustrates the spectrum plots of the PA output at 24 dBm power. The shape of the IM3 pattern is clearly distorted, as opposed to Figure 8.14, due to large IM5 products whose power levels are now comparable with those of IM3 products: The highest IM5 level has even exceeded the lower IM3 level. Furthermore, the IM3 level is not increased by $(3\times 24-21)=9$ dB, again due to the IM5 impact. Note that the IM3 and IM5 products usually add up destructively.

Hereafter in this section, we focus on the cases where IM3 products are sufficiently larger than other unwanted products, including IM5. In this way, we can assume that, to a good approximation, the observed IM3 levels have resulted only from third-order nonlinearity, and not higher.

In both Figures 8.14 and 8.15, a row of seven products can be seen closely above the higher IM3 row. These are not the IM products, but the image of the multisine due to the imperfect image suppression of the IQ-modulator inside the ESG. These products can be suppressed by balancing the I and Q signals in ESG through the calibration process. However, we deliberately include the imbalance IQ case in Figures 8.14 and 8.15 to show another property of the offset multisine: if there is any image lying inside the band, they have frequency locations distinct from the main tones and its effect cannot be mistaken for other factors, such as modeling error in modeling application.

Figure 8.16 shows the IM3 spectrum of a PA at 23 dBm output power with 1.2 MHz bandwidth. IM3 products are marked based on $\omega_2 - \omega_3$ in Eq. (8.46). In this figure, Δm is the difference between the coefficients m of $\hat{f}_2$ and $\hat{f}_3$ in Eq. (8.44), where $\omega_2 = 2\pi\hat{f}_2$ and $\omega_3 = 2\pi\hat{f}_3$. As already explained, there are basically two rows of IM3 products. Each point in the upper one is the summation result of two combinations $\omega_1 + \omega_2 - \omega_3$ and $\omega_2 + \omega_1 - \omega_3$. That is why each point in the upper row has more than one marker at its location. However, the lower row consists of the

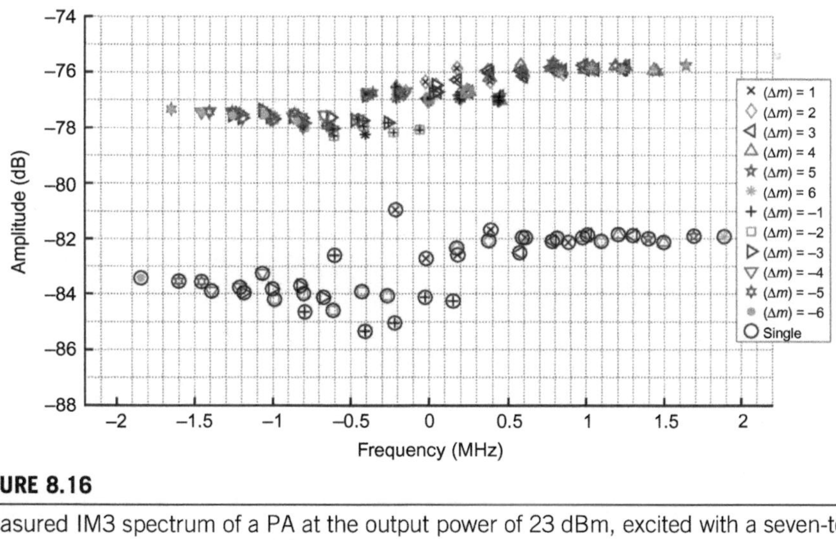

FIGURE 8.16

Measured IM3 spectrum of a PA at the output power of 23 dBm, excited with a seven-tone offset multisine with 1.2 MHz bandwidth. The IM3 products are marked separately based on their Δm values, where Δm is the difference $m_2 - m_3$. The coefficients m_2 and m_3 are the respective values of m in Eq. (8.44) for f_2 and f_3, from which the IM3 frequency is found by $f_{IM3} = f_1 + f_2 - f_3$.

points purely resulting from $\omega_1+\omega_1-\omega_2$ and no other combination. Henceforth, we call this row the "lower IM3". The points from the lower IM3 are circled in Figure 8.16. This purity makes the points of the lower IM3 suitable for our investigation. In fact, this row bears the information of the dependence of IM3 on the frequency difference $\omega_1-\omega_2$: considering the markers, the row shows that the IM3 level depends on this difference. To make this phenomenon clearer, we plot the lower IM3 profile in a different way in the following section.

8.5.3 IM3 PROFILE

In the next step, we sweep the multisine excitation bandwidth, gather IM3 data from all these measurements, and plot the lower IM3 data points versus frequency difference. The resulting plot is shown in Figure 8.17. In this figure, the horizontal axis is the frequency difference $(\omega_1-\omega_2)/2\pi$, which can also be negative. This is another way of demonstrating IM3 curves for memory effects as opposed to the typical one, in which the upper and lower IM3 curves are plotted on top of each other versus the absolute value of frequency difference.

Figure 8.17 clearly shows both asymmetry and unevenness of the IM3 profile, which are the consequences of slow (long-term) memory effects.

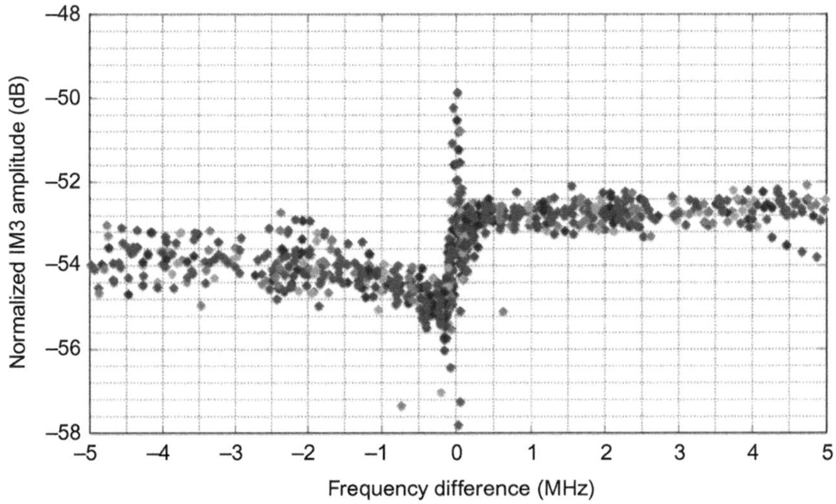

FIGURE 8.17

Measured IM3 profile of PA3 output at the output power of 23 dBm. Different colors in the plot indicate data points collected from different measurements.

It is well known that, assuming negligible nonlinearity order higher than three, the transfer function of the IM3 product at $2\omega_2-\omega_1$ is the summation products of three terms [52–54]. That is:

$$\begin{aligned} H_{\mathrm{IM3}}(\omega_2,\omega_2,-\omega_1) = & H_3(2\omega_2-\omega_1) \\ & + H_2(2\omega_2)\cdot H_1(-\omega_1) \\ & + H_2(\omega_2-\omega_1)\cdot H_1(\omega_2) \end{aligned} \tag{8.47}$$

where H_{IM3} is the IM3 transfer function, and H_1, H_2, and H_3 are the first, second, and third order transfer functions, respectively. Correspondingly, the IM3 product at $2\omega_1-\omega_2$ is:

$$\begin{aligned} H_{\mathrm{IM3}}(\omega_1,\omega_1,-\omega_2) = & H_3(2\omega_1-\omega_2) \\ & + H_2(2\omega_1)\cdot H_1(-\omega_2) \\ & + H_2(\omega_1-\omega_2)\cdot H_1(\omega_1) \end{aligned} \tag{8.48}$$

If the tone spacing between ω_2 and ω_1 is small compared to the system bandwidth, we can assume that $H_3(2\omega_2-\omega_1)\approx H_3(2\omega_1-\omega_2)$, $H_2(2\omega_2)\approx H_2(2\omega_1)$, and $H_1(\omega_1)\approx H_1(\omega_2)$.

However,

$$H_2(\omega_2-\omega_1) = H_2^*(\omega_1-\omega_2) \neq H_2(\omega_1-\omega_2) \tag{8.49}$$

where the asterisk denotes conjugation operation. These equations theoretically show that IM3 profile asymmetry happens if the imaginary part of the third term, as well as the imaginary part of the rest of equations (8.47) and (8.48), are not negligible [55].

Based on Eqs. (8.47) and (8.48), the unevenness of the IM3 profile also happens when $H_2(\omega)$ around DC is frequency dependent and is large compared to $H_2(2\omega)$, which is a likely situation due to bias circuits of PAs [53,54]. Note that second-order products around DC can modulate the PA bias due to the low-frequency response of the bias circuit.

Some spread can be seen in Figure 8.17, which is due to measurement accuracy limitation, as well as a small IM5 effect. However, the global shape is clear enough for the assessment of slow memory effects.

Note that with a real modulated waveform that has a continuous frequency spectrum, we are not able to distinguish the dependence of IM3 on frequency difference. In the continuous spectrum with the bandwidth *BW* we have every frequency difference between two components of the spectrum varying from zero to *BW*, and the IM products share the locations. Therefore, we cannot extract the IM3 profile with a continuous-spectrum signal. This drawback highlights one of the advantages of the offset multisine: while the statistics of the offset multisine are close to those of the real modulated waveform, the offset multisine provides some useful insight into IM behavior of the circuit which is not provided by real modulated waveforms. This advantage is reconfirmed in Ref. [50], where the offset multisine provides better assessment of model strength to predict IM behavior.

Similar IM3 profiles can be obtained by using two-tone excitations. However, since the waveform dynamics and statistics of the offset multisine can be designed to be more similar to the real modulated waveform [50] as compared to the two-tone waveform, the IM3 profile obtained by the offset multisine will be more reliable to emulate the real modulated case.

8.5.4 FOCUSED APPLICATION: MEMORY EFFECTS CHARACTERIZATION

As an interesting application of the offset multisine excitation, we focus on memory effects characterization. In the following, the methodology is applied to long-term memory effects investigation. The method can also be easily applied to short-term memory effects investigation, provided that we have a sufficiently wideband data-capturing equipment relative to the signal bandwidth.

We have already shown through Eqs. (8.47)–(8.49) that the frequency dependence of $H_2(.)$ around DC results in unevenness and asymmetry of the IM3 profile. One of the major factors determining $H_2(.)$ around DC is the bias circuit. This section investigates the effect of this factor on long-term memory and the IM3 profile.

Figure 8.18 shows the schematic of a GaN amplifier operating near 2 GHz, including its bias circuits. We show that we can efficiently inspect memory effects with the offset multisine technique, including changes induced by intentional manipulations of the bias networks. Figure 8.17 is the IM3 profile of the complete PA3 circuit. Figure 8.19 shows the IM3 profiles after two intentional changes to the bias network. Different brightness in the plots indicates data points collected from different measurements. In Figure 8.19A, the capacitor C15 is removed. This profile shows that C15 has a crucial role in low-frequency response of the circuit around DC. Due to nonlinearity of the transistor, the high-frequency components are mixed together and downconverted to frequencies around DC ($H_2(\omega_1 - \omega_2)$). If these downconverted products are not well suppressed in the bias line, they can modulate the DC supply of the transistor, be mixed with the RF large tones and upconverted back to the high frequencies around the main band (reflected in $H_2(\omega_1 - \omega_2)H_1(\omega_1)$). C15 suppresses

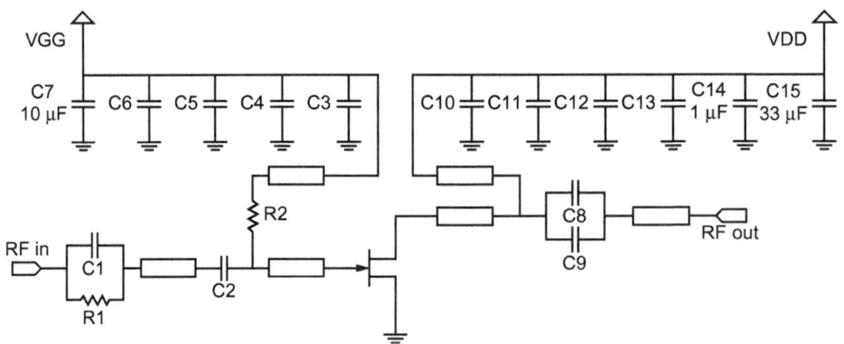

FIGURE 8.18

Simplified schematic of a PA module.

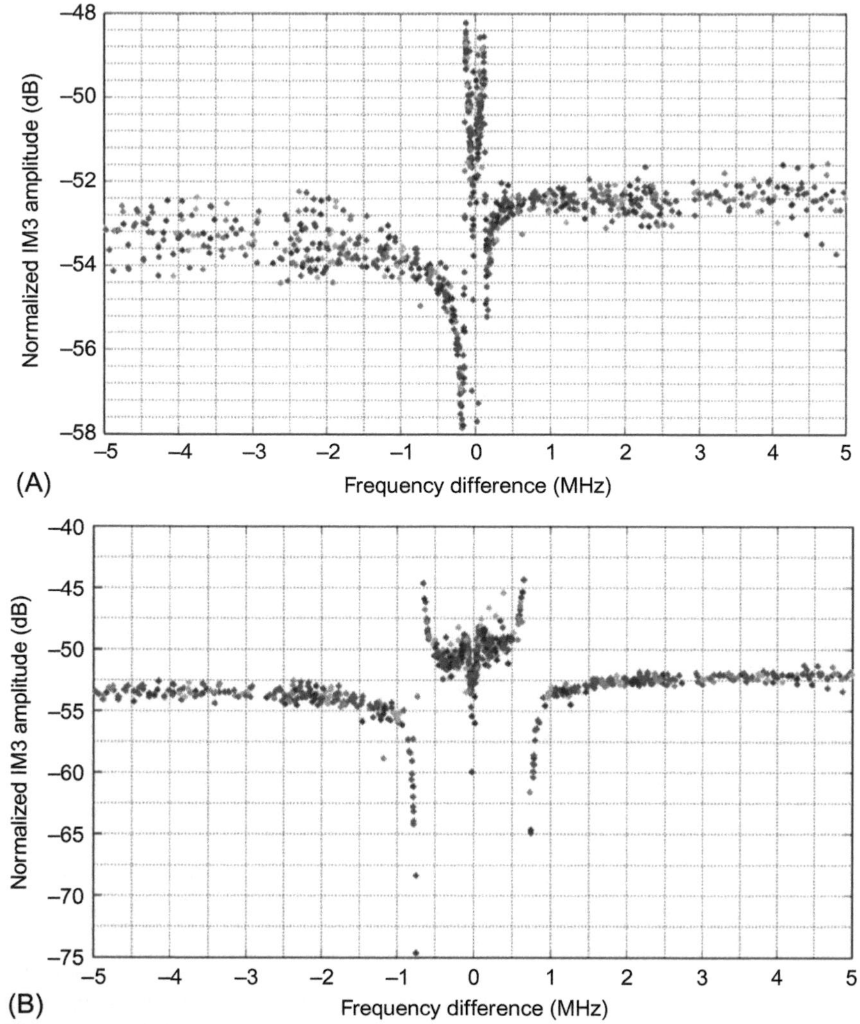

FIGURE 8.19

Measured IM3 profile of PA module of Figure 8.18 with (A) C15 removed, and (B) C7 removed. Different color in the plots indicate data points collected from different measurements.

the downconverted product by bypassing it to the ground. By removing C15, we allow this mixing mechanism to have its effect on the low-frequency IM3 profile. In other words, C15 reduces the amplitude of $H_2(\omega_1 - \omega_2)$. If C15 is removed or not large enough, $H_2(\omega_1 - \omega_2)$ is considerable and its frequency dependency highly affects the IM3 profile, leading to long-term memory.

Figure 8.19B is the IM3 profiles with removed C7. This figure highlights the importance of C7, a capacitor in the gate bias line, in memory effects and IM3 rejection.

In summary, the plots of Figure 8.19 provide information about the impact of bias line components on IM3 level variation and long-term memory effects. More demonstrations are provided in Ref. [50]. There are articles in the literature, for example [23,55–58], explain this phenomenon in more detail.

The offset multisine is much more suitable than the conventional multisine or modulated signals for investigation of long-term memory effects due to two main reasons: (1) in wideband applications, long-term memory effects show themselves in in-band frequency components and in others close to the main band. This localization is because of the fact that for long-term memory effects, the subtraction term in $\omega_1+(\omega_2-\omega_3)$ has a small value as compared to the tone frequencies. Because of this localization, we must use a method to measure in-band distortion in order to be able to properly characterize long-term memory effects; (2) in the conventional multisine or real modulated waveform, we can see only the accumulated effect of all IM3 products at each frequency, and we cannot separately study the individual products that lie at the same frequency. However, in the offset multisine method, this investigation is straightforward.

Because the offset multisine excitation discriminates different IM products, it is also a nice tool to evaluate the ability of models for predicting memory effects. This capability is elaborated in Ref. [50].

8.6 CHARACTERIZATION AND MODELING OF TRANSMITTER EMISSION INTO RECEIVE BAND

Most wireless transceivers use duplexers to share a common antenna between the transceiver and receiver. However, duplexers have limited isolation between transmitter and receiver ports. Therefore, the transmitter output emissions can leak into the receiver input. The transceiver emissions in the receive band can desensitize the receiver, and are commonly referred to as receive-band noise (RxBN), although a significant portion of them are deterministic.

In current cellular deployment scenarios, the receiver and PA are designed so that, along with duplexer isolation, the receiver sensitivity degradation is kept within acceptable limits — typically below 0.5 dB. However, future deployment of wider bandwidth and smaller frequency separation between transmit and receive bands, particularly in carrier aggregation mode, the emissions from the transmitter can significantly desensitize the receiver. These issues are further exacerbated by the need for lowering the supply voltage of the PAs, in average power tracking [59] or envelop tracking [60] modes, to improved efficiency, as discussed in Chapter 7.

In order to estimate the influence of RxBN on system performance, consider a typical case in handsets for the long-term evolution (LTE) standard. The receiver input noise leaking through the duplexer at the receiver input should be less than −184 dBm/Hz (10 dB below the noise floor of −174 dBm/Hz), to keep receiver desense below 0.5 dB. Considering 50 dB suppression by the duplexer, the generated RxBN at the PA output should be less than −134 dBm/Hz. A typical handset PA has a maximum output power around 24 dBm at the antenna, which is equivalent to power spectral density of −46 dBm/Hz for a 10 MHz LTE signal. Therefore, the RxBN should be at least 88 dBc, which is challenging for certain cellular bands. For example, as illustrated in Figure 8.20, 10 MHz LTE signal in band 13 with duplex spacing of 31 MHz, can easily fail the desired 88 dB suppression due to spectral regrowth caused by the nonlinearity in the PA.

There are increasing efforts to mitigate the RxBN in transceivers using different techniques, most of which are based on feed-forward cancelation [61–64]. These techniques are based on copying the generated RxBN in the RF path and subtracting it from the received signal, after gain and phase compensation, either in the analog or the digital domain. However, instead of copying the RxBN, one can use a model to estimate its value and use this estimation in the cancelation process [65–68]. Using the estimated RxBN is beneficial since it relaxes the specification requirements of the sampling circuit,[1] because of the robustness to unwanted noise. Furthermore, the RxBN models can be used in system-level simulations for fast analysis. In this section RxBN characterization and modeling approaches are explained [67].

RxBN modeling is different from the in-band modeling for EVM, since the frequencies of the input signal and the target signal are different. Also, unlike the modeling of adjacent-channel leakage power (ACLR), the RxBN is effectively a combined result of spectral regrowth caused by various high order nonlinearities in the PA. Moreover, the RxBN is practically too far from the Tx band to be estimated by a conventional behavioral model. Accordingly, the feedback path, which is usually narrow-band and does not include the receive band, used to sample the PA output for classical digital predistortion (DPD) cannot be deployed to sample the RxBN.

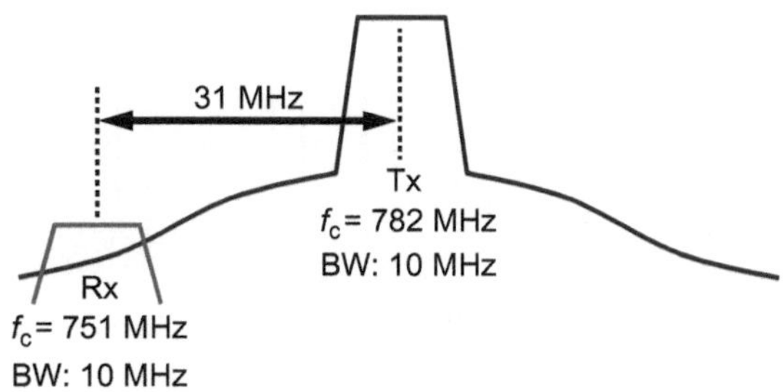

FIGURE 8.20

Frequency setting of a typical scenario of band 13 for LTE.

[1]The sampling circuit is a receiver that gets an attenuated sample of the PA output by using a coupler in the PA output path.

8.6.1 MEASURING THE DETERMINISTIC COMPONENTS OF RxBN

Figure 8.21 shows the measured spectrum of a cellular PA in the Rx band, including RxBN of the PA, at the average Tx output power of 24 dBm. The PA was excited by an 11-tone multisine in the Tx band with 10 MHz bandwidth. The roll-off at frequency offsets below around −40MHz and above −20MHz is due to the limited capturing bandwidth and internal filtering in the capturing instrument with 25 MHz bandwidth. The same holds for Figures 8.22 and 8.24.

The RxBN of the 11-tone excitation in Figure 8.21 has two types of frequency components: one part of the RxBN is a set of discrete tones. Considering their frequency spacing and locations we deduce that they are a deterministic noise component caused by nonlinearity. The other component is stochastic noise and manifests itself as continuous spectrum. Note that the observed stochastic RxBN is from the PA and is more than 10 dB above the noise floor of the instrument. The frequency scale in Figure 8.21 is referenced to the Tx center frequency. The observed slope in the continuous spectrum corresponds to a low roll-off of the duplexer, the matching network and the noise generated in the PA with increasing separation from the Tx band.

By using multisine excitation and carefully choosing appropriate phase relationship between the tones, a PAPR of 8.5 dB, comparable to LTE signal was achieved. As shown in Figure 8.21, RxBN has a major deterministic component evidenced by the tones that are well above the noise floor. We define the determinism quotient of the noise as:

$$D = P_{\mathrm{nd}}/P_{\mathrm{ns}} \tag{8.50}$$

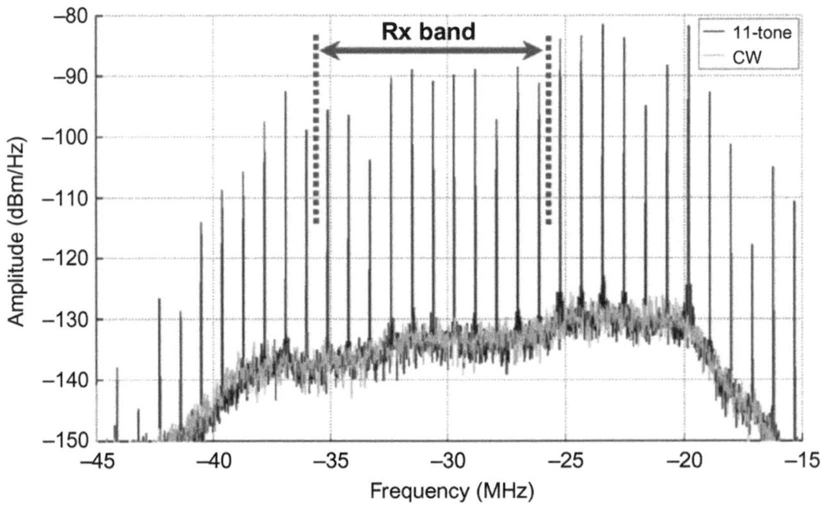

FIGURE 8.21

Measured RxBN spectrum of a PA excited by an 11-tone multisine and a CW signal with the same frequency and power. The frequency axis shows the offset from the Tx frequency.

where P_{nd} is the power of the deterministic component and P_{ns} is the power of the stochastic component of the RxBN. For the noise in Figure 8.21, the determinism quotient is calculated to be $D = 27.1$ dB (P_{nd} is more than a hundred times larger than P_{ns}).

The observed stochastic noise originates from random processes in the PA, and its power depends on the average power of the PA. The RxBN generated by a CW excitation is illustrated in Figure 8.21 against the discussed 11-tone multisine with the same frequency and average. Table 8.1 compares the calculated P_{ns} of measured data for 11-tone and CW excitations at different PA output powers. Figure 8.21 and Table 8.1 show that the stochastic noise power is mainly determined by the average power, because it is the same for both multisine and CW excitations with the same output power.

The multisine technique can distinguish between the deterministic and stochastic RxBN in real time. It uses the property of having discrete spectral contents with multisine excitation to discriminate between the deterministic and stochastic components of RxBN.

Measured RxBN with LTE excitation is shown in Figure 8.22. Unlike CW or multisine, LTE excitation results in continuous RxBN spectrum, where the deterministic and stochastic components are indistinguishable.

Another technique, capable of discriminating deterministic and stochastic RxBN even in continuous-spectrum cases, can be employed based on postprocessing the captured output signal. We call it the repetition-subtraction technique. In this approach, we capture more than one cycle of the signal, as illustrated in Figure 8.23. Assume the output RxBN is $y(t)$, the period of the input signal $x(t)$ is T and we have captured an output signal of length $2T$. Also assume that Y_1 is the output sequence in $t \in [0, T)$ and Y_2 is the sequence in $t \in [T, 2T)$. Due to the periodicity of $x(t)$, if the RxBN is all-deterministic then $Y_1 = Y_2$ and $Y_1 - Y_2 = 0$. Hence, if Y_2 is subtracted from Y_1, the result will be only the stochastic residual RxBN signal (with standard deviation increased by 3 dB) because the stochastic part is not periodic and it is assumed that the noise of the two periods are uncorrelated. This is a reasonable assumption as long as T is much larger than the noise variation time. Therefore, by this postprocessing technique we are able to quantitatively measure how much of the RxBN is deterministic and how much is stochastic.

The advantage of this approach is that it can be applied irrespective of the waveform modulation, and can thus be used even for continuous-spectrum signals. The

Table 8.1 Stochastic Part of Measured RxBN Power Density for Multisine (MS) and CW Excitations Versus PA Output Power (P_{out})

P_{out} (dBm)	18	20	22	24
P_{nd} (dBm/Hz), MS	−121.0	−115.9	−109.8	−106.8
P_{ns} (dBm/Hz), MS	−136.9	−135.8	−134.9	−133.9
P_{ns} (dBm/Hz), CW	−136.7	−135.7	−134.9	−134.0

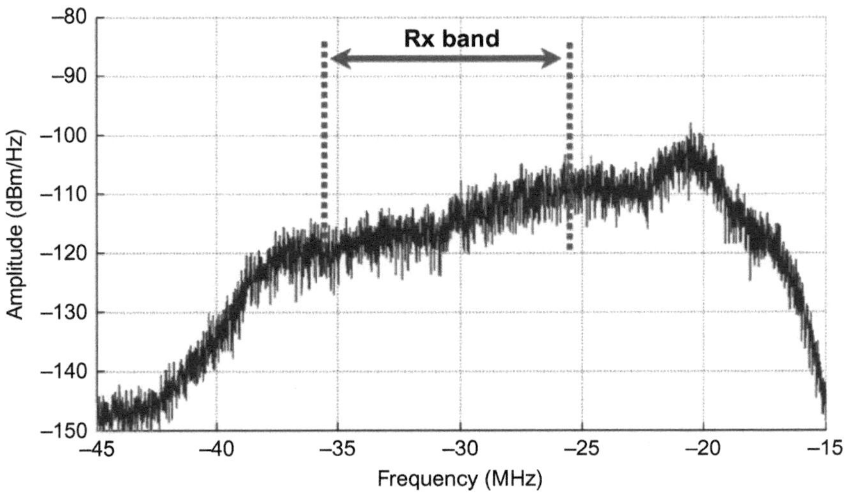

FIGURE 8.22

Measured RxBN spectrum of a PA excited by an LTE excitation with the same frequency and power as those of Figure 8.3. The frequency axis shows the offset from the Tx frequency.

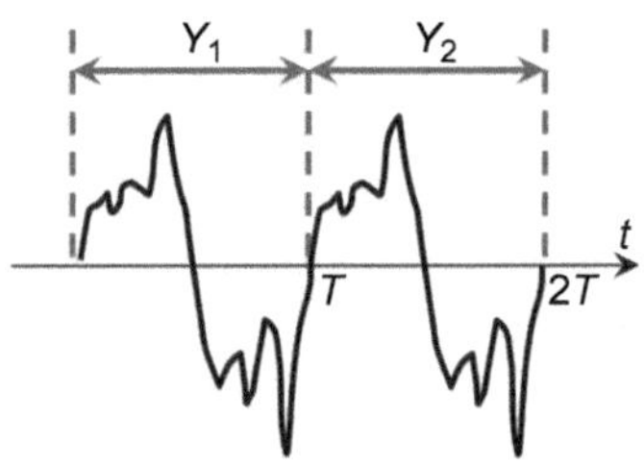

FIGURE 8.23

Repetition of signal, used for repetition-subtraction technique.

term "continuous-spectrum" has a special meaning in this discussion. The spectrum is discrete for periodic signals, including the signals which are the result of repetition. However, here we call the signal spectrum continuous if the frequency grid size of the repeating signal is much smaller than the signal bandwidth. In this case, the spectrum inside the signal band consists of a large number of nonzero components located at every frequency grid point, as if the spectrum is continuous.

Figure 8.24 shows the postprocessed stochastic noise against the total RxBN of different modulations, all measured at the average power around 24 dBm. Like Figure 8.21 and Table 8.1, this figure also shows that the power of the stochastic part of the RxBN is mainly determined by the average power and is almost independent of the signal dynamics.

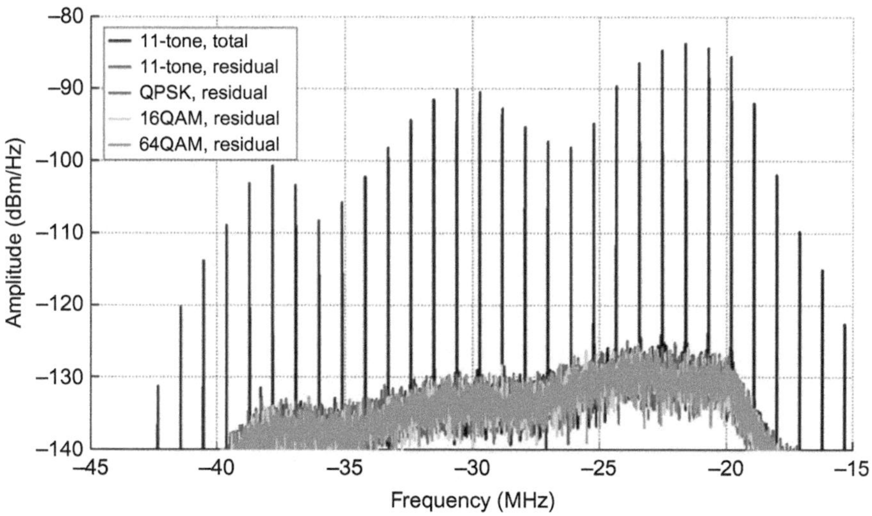

FIGURE 8.24

Total RxBN of measured data from a PA and its stochastic RxBN for different modulations, calculated with repetition-subtraction technique. The frequency axis shows the offset from the Tx frequency.

8.6.2 IDENTIFICATION OF NONLINEARITY ORDERS

Before starting to set the model parameters for identification, it is helpful to get a priori knowledge of the strength of nonlinearity orders. In this section, we demonstrate a method for the assessment of nonlinearity order strength contributing to the RxBN signal, by using only three tones as an excitation. For this characterization purpose the memory effects are ignored for the moment, as our primary goal is to assess the strength of the different nonlinearity orders qualitatively.

In this special three-tone offset multisine, the first and second tones lie at the same locations as those of the conventional multisine. However, the third tone has a small offset from its original location in the conventional case. Hence, the frequencies of the three tones are $f_1 = f_c - f_m$, $f_2 = f_c$, and $f_3 = f_c + f_m - \delta$, in which f_c is the center frequency, f_m is the major frequency spacing between the tones and δ is the offset of the third tone from its original location as compared to the conventional case. The tones have equal amplitudes. If a conventional multisine excitation is used, the deterministic RxBN will be a set of tones with f_m spacing. However, if the third tone is offset by δ there will be a number of tones scattered with δ spacing accumulated around the frequencies spaced by f_m. Figure 8.25A and B shows the RxBN of PA1 excited by a conventional and an offset three-tone multisine, respectively. In these figures $f_c = 782$ MHz, $f_m = 5$ MHz, and $\delta = 20$ kHz. The resulting deterministic RxBN consists of IM products caused by different nonlinearity orders. The IM products falling around the fundamental frequency, including those in the receive band, are caused by odd-order nonlinearities. In general, the frequencies of these IM products are:

$$f_{IM} = mf_1 + (1-(m-n))f_2 - nf_3 \tag{8.51}$$

where m and n are integer numbers. The resulting offset from the center frequency is $f_\Delta = -(m+n)f_m$ for the conventional multisine case and $-(m+n)f_m + n\delta$ for the offset

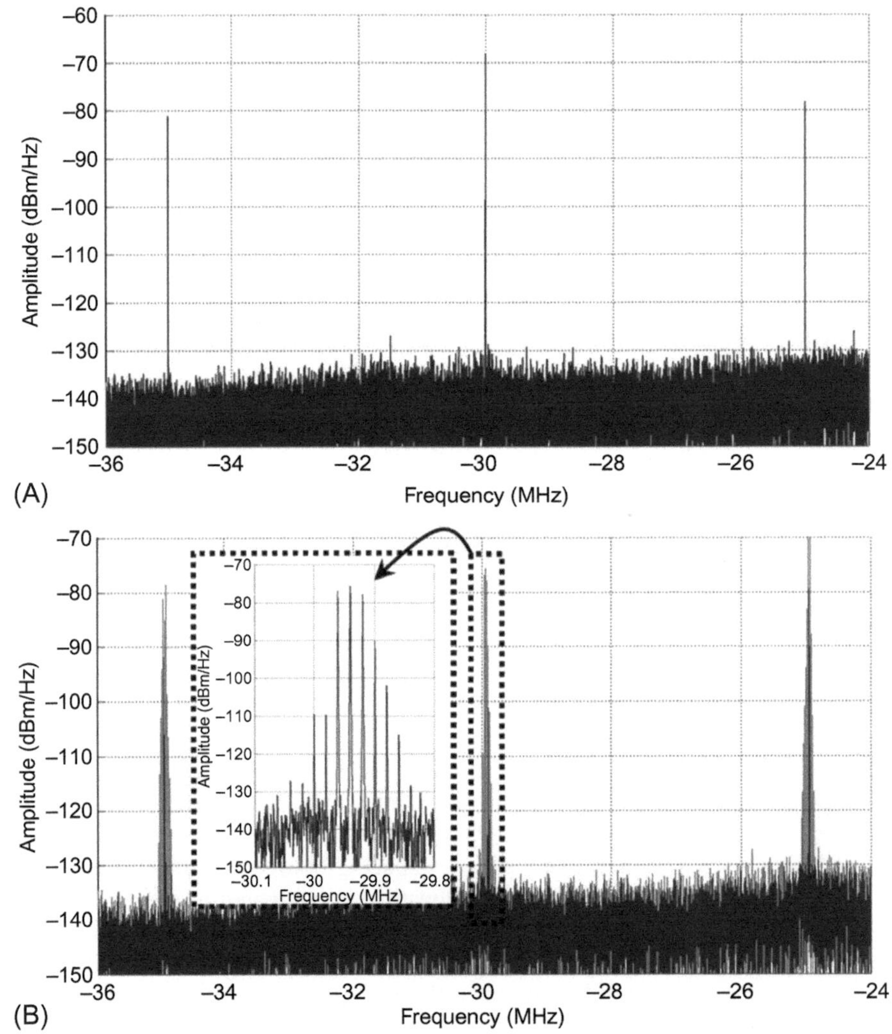

FIGURE 8.25

Measured PA output spectra excited by three-tone multisine with 10 MHz bandwidth: (A) conventional multisine, (B) offset multisine, and

(Continued)

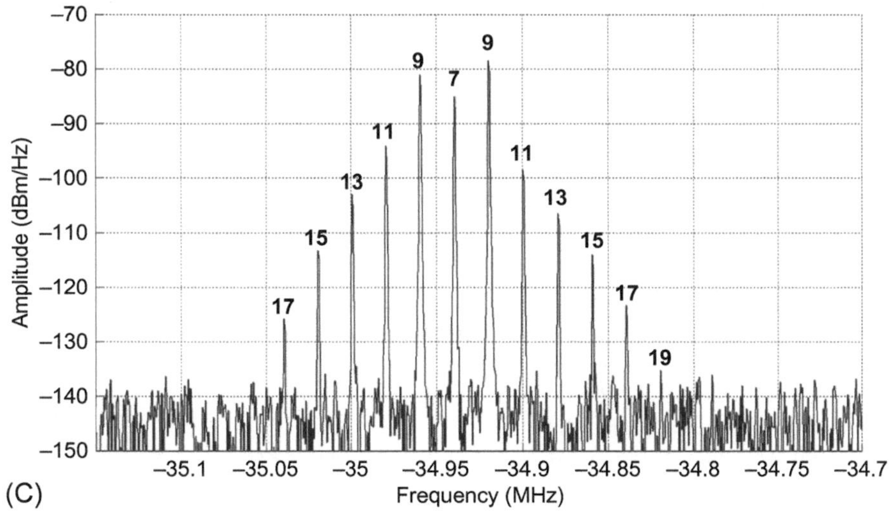

FIGURE 8.25, CONT'D (C) zoomed view of spectrum for offset multisine with reduced stochastic noise by averaging. The nonlinearity order corresponding to each tone is indicated above it. The frequency axis shows the offset from the Tx frequency.

multisine case. Therefore, the frequency difference between the two cases is $n\delta$. Hence, if we measure the deviation of each tone from $-(m+n)f_{\mathrm{m}}$, which is equal to $n\delta$, we can find the corresponding n-value. Having n and $f_{\Delta}=-(m+n)f_{\mathrm{m}}$, we can calculate the corresponding m. Finally, based on Eq. (8.51) the corresponding nonlinearity order becomes:

$$k=|m|+|n|+|1-(m-n)| \tag{8.52}$$

As can be seen in Figure 8.25C, the orders lower than seven do not contribute to the RxBN, because in this case, in which the bandwidth is 10 MHz and the duplex spacing is 31 MHz, the lower-order IM products are not far enough from the Tx band to lie in the Rx band.

Applying this technique to the spectrum of Figure 8.25B, the nonlinearity orders of the tones around -35 MHz offset from the transmit frequency are derived and illustrated in Figure 8.25C. Note that the tones assigned to lower nonlinearity orders also have contributions from higher orders with various phases, but not vice versa. Nevertheless, ignoring memory effects, we can assume that the contribution of, for example, the 17th order to the tones assigned to the 15th order is not larger than -10 dB, because the differences between the 15th and the 17th tones are about 10 dB. Therefore, the plot provides us qualitative assessment and comparison of the strength of nonlinearity orders, and helps to decide up to what order to consider for RxBN modeling. The offset multisine technique uses commensurate tone frequencies with a fine frequency grid. Therefore, this characterization technique can be used with the same setup that is used for the main experiment by only loading the waveform data of the offset multisine to the VSG. Thus, it is a fast and easy way to make a reliable initial guess of what orders have significant contributions to the RxBN.

8.6.3 MODELING OF DETERMINISTIC RxBN

In traditional Tx-band modeling that uses Volterra-based behavioral models, one tries to model the PA output around fundamental frequency based on the input. More precisely, the frequency components of the PA output in the Tx band and its vicinity are modeled by using a Volterra-based model. These models usually yield acceptable estimation accuracy within a bandwidth up to three or five times the input bandwidth. The frequency components beyond this bandwidth are usually not modeled well. This shortcoming makes RxBN estimation challenging, due to the separation of the Rx band from the Tx band.

Signal frequency components from the main band translate to receive band due to nonlinearity of the PA. The output of nonlinearity with the odd order $2k+1$ excited by a signal of bandwidth BW and with a center frequency f_c, is a signal with a frequency span from $f_c-(2k+1)BW/2$ to $f_c+(2k+1)BW/2$. If the nonlinearity order is large enough, the output signal spectrum can spread into the receive band frequency, causing RxBN generation.

The modeling approach used in most of the Volterra-based behavioral models for RF applications, including those in this section, is in baseband domain. To model the above-mentioned frequency translation mechanism, a modification to the conventional behavioral models is needed. For example, the MP model is very effective for main-band modeling, represented by:

$$y_n = \sum_{m=0}^{M-1} \sum_{k=k_0}^{K} b_{mk} |x_{n-m}|^k x_{n-m} \tag{8.53}$$

However, it is inefficient for RxBN modeling, because it has some excessive low-order terms that do not contribute to RxBN. Therefore, for RxBN modeling, the MP model is modified as:

$$\begin{aligned} y_n &= F\left\{ \sum_{m=0}^{M-1} \sum_{k=k_0}^{K} b_{mk} |x_{n-m}|^k x_{n-m} \right\} \\ &= \sum_{m=0}^{M-1} \sum_{k=k_0}^{K} b_{mk} \cdot F\left\{ |x_{n-m}|^k x_{n-m} \right\} \end{aligned} \tag{8.54}$$

where $F\{.\}$ is a linear time-invariant (LTI) filtering operator that keeps only the frequency components which lie inside the receive band. The parameter k_0 is the minimum nonlinearity order contributing to RxBN. In Ref. [67] it was shown that choosing an even value for k_0 as small as two can make considerable improvement

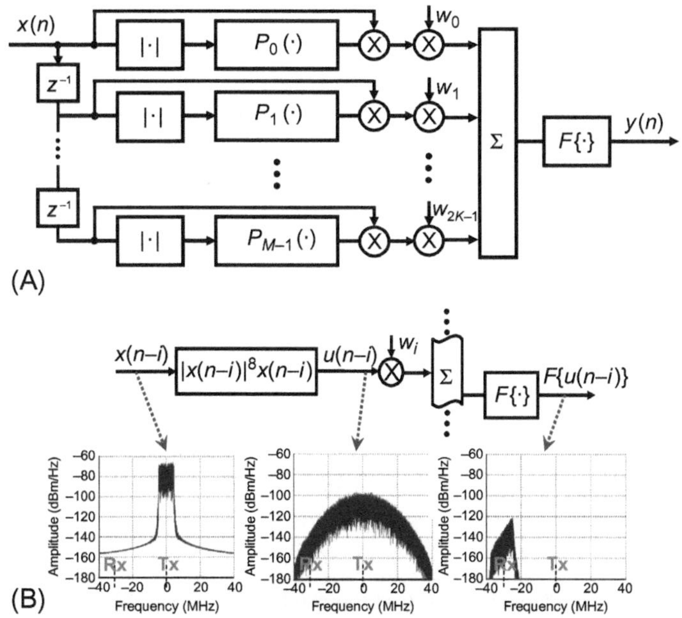

FIGURE 8.26

(A) Block diagram of the MP model used for RxBN modeling and (B) signal spectra at different points of the model path for the ninth-order nonlinearity.

in modeling performance. It was also shown that even orders can improve the RxBN modeling performance, like they do for main-band modeling.

The model is graphically illustrated in Figure 8.26A. In this figure, z^{-1} is the delay operator, $P_n(\cdot)$ is a polynomial of an even order $2K$ with the input encountering an n-tap delay, $F\{\cdot\}$ is the LTI band-pass filtering operator with the center frequency at f_{Rx} and w_n is the weight for $P_n\{\cdot\}$.

Each nonlinearity term in Eq. (8.54) gives rise to spectral reshape and regrowth of the signal that can lie within the Rx band. After filtering by the $F\{.\}$ operator, only the Rx band part of this signal remains and the rest will be removed. This mechanism is shown in Figure 8.26B, where the signal spectra at different points of the nonlinearity path in the model are represented.

Figure 8.27 shows an example of the RxBN modeling performance for the measured RxBN of a cellular PA excited by a 64-QAM modulation. This RxBN is the one whose stochastic part is shown in Figure 8.24. Comparing the residual error of Figure 8.27 with the stochastic noise of the 64-QAM case in Figure 8.24 confirms that the used RxBN model has been able to model the deterministic part acceptably, so that the residual error is close to the stochastic noise level.

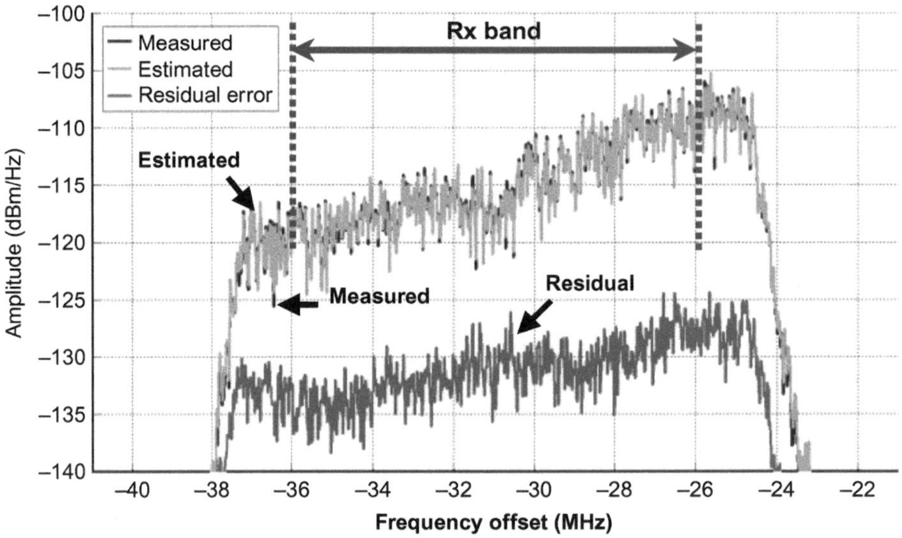

FIGURE 8.27

Spectra of measured and estimated RxBN modeled by MP with even orders included. The error signal is the difference between the measured and estimated signals. The frequency axis shows the offset from the Tx frequency.

8.7 FROM CHARACTERIZATION TO SYSTEM-LEVEL COMPENSATION

One of the profound applications of behavioral models is their use in predistortion. Predistortion of a circuit/system means adding a functional block prior to the non-ideal circuit, so that the overall response becomes as linear and memoryless as possible. The concept is shown in Figure 8.28 for a simple two-tone case. Without predistortion in Figure 8.28A, IM products are present at the PA output. By applying predistortion prior to the PA, a predistorted signal excites the PA. This signal has two IM products such that they cancel out the IM products generated by the PA. The result is a clean two-tone signal at the output, as shown in Figure 8.28B.

The predistortion block can be a software algorithm and/or a hardware circuit. The hardware predistorter can itself be an analog circuit or a digital one. DPD is increasingly of great interest [69–72] because of many reasons [73]: DPDs are flexible, accurate, reliable, and persistent against changes. Moreover, digital circuits usually consume less energy than analog circuits that have active devices. They are also appropriate for adaptive predistortion, to deal with the change of circuit characteristics over time. Some types of digital predistorters are explained in this chapter. They are based on digital signal processing techniques.

The DPD that is widely used is briefly illustrated in Figure 8.29. In this approach, two samples from the input and output of the nonideal circuit, the PA in this case, are

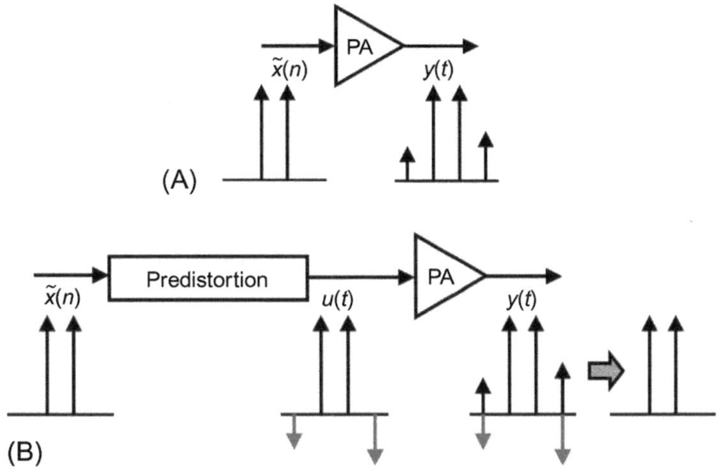

FIGURE 8.28

Demonstration of predistortion concept in two-tone case. The black small tones in (A), generated at the IM output due to IM, are canceled out in (B) by injecting the gray tones to the PA input. The gray tones are equal to the black ones in amplitude, but with opposite phase.

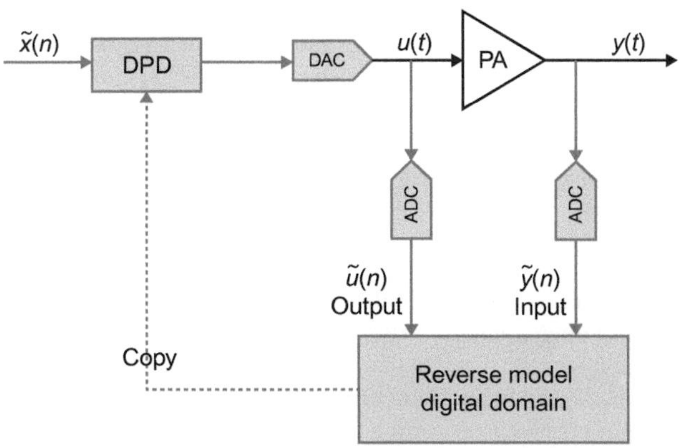

FIGURE 8.29

DPD approach based on indirect learning.

taken first. After conversion to baseband, producing $\tilde{u}(t)$ and $\tilde{y}(t)$, a behavioral model, characterized by one of the approaches explained in this chapter, is built with $\tilde{y}(t)$ as the input and $\tilde{u}(t)$ as the output. Learning such an inverse model is called indirect learning. The learned inverse model is then copied as the predistortion block, and will predistort the input signal to have desired output. In DPD, the predistortion block is in digital domain, hence before digital-to-analog conversion, as can be seen in Figure 8.29.

There is a huge amount of work on DPD in the literature. Most of these works differ in the behavioral model used in the reverse model. Different behavioral models have been applied in DPD. Old works and also the applications in which memory is not important usually use AM/AM and AM/PM models for DPD [74]. However, modern approaches mostly use Volterra-based models able to model memory effects [27]. Moreover, DPD approaches have also been used to compensate nonidealities in multiband applications, especially to remove cross-modulation effects [75,76]. DPD techniques were also used to compensate not only nonlinearity and memory effects, but also IQ imbalance [18–20].

REFERENCES

[1] S. Cripps, RF Power Amplifiers for Wireless Communications, second ed., Artech House, Norwood, MA, 2006.

[2] J. Pedro, N. Carvalho, Intermodulation Distortion in Microwave and Wireless Circuits, Artech House, Norwood, MA, 2002.

[3] W. Bosch, G. Gatti, Measurement and simulation of memory effects in predistortion linearizers, IEEE Trans. Microw. Theory Techn. 37 (12) (1989) 1885–1890.

[4] O. Shimbo, Effects of intermodulation, AM-PM conversion, and additive noise in multicarrier TWT systems, Proc. IEEE 59 (2) (1971) 230–238.

[5] P. Roblin, D. Root, J. Verspecht, Y. Ko, J.-P. Teyssier, New trends for the nonlinear measurement and modeling of high-power RF transistors and amplifiers with memory effects, IEEE Trans. Microw. Theory Techn. 60 (6) (2012) 1964–1978.

[6] J. Brinkhoff, Bandwidth-Dependent Intermodulation Distortion in FET Amplifiers, Macquarie University, Sydney, 2004 (Division of Information & Communication Sciences, Department of Electronics).

[7] J. Vuolevi, T. Rahkonen, J. Manninen, Measurement technique for characterizing memory effects in RF power amplifiers, IEEE Trans. Microw. Theory Techn. 49 (8) (2001) 1383–1389.

[8] T. Cunha, E. Lima, J. Pedro, Validation and physical interpretation of the power-amplifier polar Volterra model, IEEE Trans. Microw. Theory Techn. 58 (12) (2010) 4012–4021.

[9] S. Boumaiza, F. Ghannouchi, Thermal memory effects modeling and compensation in RF power amplifiers and predistortion linearizers, IEEE Trans. Microw. Theory Techn. 51 (12) (2003) 2427–2433.

[10] A. Santarelli, F. Filicori, G. Vannini, P. Rinaldi, 'Backgating' model including self-heating for low-frequency dispersive effects in III-V FETs, Electron. Lett. 34 (20) (1998) 1974–1976.

[11] K. Kunihiro, Y. Ohno, A large-signal equivalent circuit model for substrate-induced drain-lag phenomena in HJFETs, IEEE Trans. Electron Devices 43 (1996) 1336–1342.

[12] O. Jardel, F. De Groote, T. Reveyrand, J.-C. Jacquet, C. Charbonniaud, J.-P. Teyssier, D. Floriot, R. Quéré, An electrothermal model for AlGaN/GaN power HEMTs including trapping effects to improve large-signal simulation results on high VSWR, IEEE Trans. Microw. Theory Techn. 55 (12) (2007) 2660–2669.

[13] A. Raffo, V. Vadalà, D.M.M.-P. Schreurs, G. Crupi, G. Avolio, A. Caddemi, G. Vannini, Nonlinear dispersive modeling of electron devices oriented to GaN power amplifier design, IEEE Trans. Microw. Theory Techn. 58 (4) (2010) 710–718.
[14] A. Raffo, S. Di Falco, V. Vadalà, G. Vannini, Characterization of GaN HEMT low-frequency dispersion through a multiharmonic measurement system, IEEE Trans. Microw. Theory Techn. 58 (9) (2010) 2490–2496.
[15] J.B. King, T.J. Brazil, Nonlinear electrothermal GaN HEMT model applied to high-efficiency power amplifier design, IEEE Trans. Microw. Theory Techn. 61 (1) (2013) 444–454.
[16] J. Pedro, N. Carvalho, P. Lavrador, Modeling nonlinear behavior of band-pass memoryless and dynamic systems, in: IEEE MTT-S International Microwave Symposium Digest, vol. 3, 2003, pp. 2133–2136.
[17] B. Razavi, RF Microelectronics, Prentice Hall International Series in the Physical and Chemical Engineering Sciences Pearson Education, 2011.
[18] P. Zhan, K. Qin, S. Cai, Joint compensation model for memory power amplifier and frequency-dependent nonlinear IQ impairments, Electron. Lett. 47 (25) (2011) 1382–1384.
[19] H. Cao, A. Tehrani, C. Fager, T. Eriksson, H. Zirath, I/Q imbalance compensation using a nonlinear modeling approach, IEEE Trans. Microw. Theory Techn. 57 (3) (2009) 513–518.
[20] L. Anttila, P. Handel, M. Valkama, Joint mitigation of power amplifier and I/Q modulator impairments in broadband direct-conversion transmitters, IEEE Trans. Microw. Theory Techn. 58 (4) (2010) 730–739.
[21] D. Schreurs, M. O'Droma, A. Goacher, M. Gadringer, RF Power Amplifier Behavioral Modeling, Cambridge University Press, Cambridge, 2009.
[22] A.A.M. Saleh, Frequency-independent and frequency-dependent nonlinear models of TWT amplifiers, IEEE Trans. Commun. 29 (11) (1981) 1715–1720.
[23] P. Asbeck, H. Kobayashi, M. Iwamoto, G. Hanington, S. Nam, L. Larson, Augmented behavioral characterization for modeling the nonlinear response of power amplifiers, in: IEEE MTT-S International Microwave Symposium Digest, vol. 1, 2002, pp. 135–138.
[24] J. Demmel, Applied Numerical Linear Algebra, Society for Industrial and Applied Mathematics, Philadelphia, PA, 1997.
[25] D. Silveira, P. Gilabert, A. dos Santos, M. Gadringer, Analysis of variations of Volterra series models for RF power amplifiers, IEEE Microw. Wirel. Compon. Lett. 23 (8) (2013) 442–444.
[26] A. Zhu, J. Pedro, T.R. Cunha, Pruning the Volterra series for behavioral modeling of power amplifiers using physical knowledge, IEEE Trans. Microw. Theory Techn. 55 (5) (2007) 813–821.
[27] A. Tehrani, H. Cao, S. Afsardoost, T. Eriksson, M. Isaksson, C. Fager, A comparative analysis of the complexity/accuracy tradeoff in power amplifier behavioral models, IEEE Trans. Microw. Theory Techn. 58 (6) (2010) 1510–1520.
[28] J. Kim, K. Konstantinou, Digital predistortion of wideband signals based on power amplifier model with memory, Electron. Lett. 37 (23) (2001) 1417–1418.
[29] O. Hammi, F. Ghannouchi, B. Vassilakis, A compact envelope memory polynomial for RF transmitters modeling with application to baseband and RF-digital predistortion, IEEE Microw. Wirel. Compon. Lett. 18 (5) (2008) 359–361.

[30] A. Zhu, P. Draxler, J. Yan, T. Brazil, D. Kimball, P. Asbeck, Open loop digital predistorter for RF power amplifiers using dynamic deviation reduction-based Volterra series, IEEE Trans. Microw. Theory Techn. 56 (7) (2008) 1524–1534.
[31] D. Mirri, G. Iuculano, F. Filicori, G. Pasini, G. Vannini, G.P. Gualtieri, A modified Volterra series approach for nonlinear dynamic systems modeling, IEEE Trans. Circuits Syst. I Fundam. Theory Appl. 49 (8) (2002) 1118–1128.
[32] D. Morgan, Z. Ma, J. Kim, M. Zierdt, J. Pastalan, A generalized memory polynomial model for digital predistortion of RF power amplifiers, IEEE Trans. Signal Process. 54 (10) (2006) 3852–3860.
[33] G. Zhou, H. Qian, L. Ding, R. Raich, On the baseband representation of a bandpass nonlinearity, IEEE Trans. Signal Process. 53 (8) (2005) 2953–2957.
[34] L. Ding, G. Zhou, Effects of even-order nonlinear terms on power amplifier modeling and predistortion linearization, IEEE Trans. Veh. Technol. 53 (1) (2004) 156–162.
[35] E. Lima, T. Cunha, H. Teixeira, M. Pirola, J. Pedro, Base-band derived Volterra series for power amplifier modeling, in: IEEE MTT-S International Microwave Symposium Digest, Boston, MA, USA, 7–12 June 2009, 2009, pp. 1361–1364.
[36] J.P. Martins, P.M. Cabral, N.B. Carvalho, J.C. Pedro, A metric for the quantification of memory effects in power amplifiers, IEEE Trans. Microw. Theory Techn. 54 (12) (2006) 4432–4439.
[37] D. Schreurs, K.A. Remley, D.F. Williams, RF behavioural modelling from multisine measurements: influence of excitation type, in: European Microwave Conference, Munich, Germany, October 2003, 2003, pp. 1001–1014.
[38] D. Schreurs, K.A. Remley, Use of multisine signals for efficient behavioural modelling of RF circuits with short-memory effects, in: 61st Automatic RF Techniques Group (ARFTG) Conference, Philadelphia, PA, USA, June 2002, 2002, pp. 65–72.
[39] N.B. Carvalho, K.A. Remley, D. Schreurs, K.C. Gard, Multisine signals for wireless system test and design (Application notes). IEEE Microw. Mag. 9 (3) (2008) 122. 138.
[40] J. Verspecht, D. Root, Polyharmonic distortion modeling, IEEE Microw. Mag. 7 (6) (2006) 44–57.
[41] M. Myslinski, F. Verbeyst, M. Vanden Bossche, D. Schreurs, A method to select correct stimuli levels for S-functions behavioral model extraction, in: IEEE MTT-S International Microwave Symposium Digest, Anaheim, CA, USA, 23–28 May 2010, 2010, pp. 1170–1173.
[42] K. Kurokawa, Power waves and the scattering matrix, IEEE Trans. Microw. Theory Techn. 13 (2) (1965) 194–202.
[43] D. Root, J. Verspecht, D. Sharrit, J. Wood, A. Cognata, Broad-band poly-harmonic distortion (PHD) behavioral models from fast automated simulations and large-signal vectorial network measurements, IEEE Trans. Microw. Theory Techn. 53 (11) (2005) 3656–3664.
[44] J. Verspecht, J. Horn, L. Betts, D. Gunyan, R. Pollard, C. Gillease, D. Root, Extension of X-parameters to include long-term dynamic memory effects, in: IEEE MTT-S International Microwave Symposium Digest, Boston, MA, USA, 7–12 June 2009, 2009, pp. 741–744.
[45] J. Verspecht, J. Horn, D. Root, A simplified extension of X-parameters to describe memory effects for wideband modulated signals, in: 75th ARFTG Microwave Measurements Conference (ARFTG), May 2010, 2010, pp. 1–6.
[46] K. Vanhoenacker, T. Dobrowiecki, J. Schoukens, Design of multisine excitations to characterize the nonlinear distortions during FRF-measurements, IEEE Trans. Instrum. Meas. 50 (5) (2001) 1097–1102.

[47] G. Vandersteen, J. Schoukens, Detection of nonlinear distortions with multisine excitations in the case of nonideal behavior of the input signal, IEEE Trans. Instrum. Meas. 52 (3) (2003) 748–753.

[48] G. Vandersteen, G. Vandersteen, Y. Rolain, K. Vandermot, R. Pintelon, J. Schoukens, W. Van Moer, Quasi-analytical bit-error-rate analysis technique using best linear approximation modeling, IEEE Trans. Instrum. Meas. 58 (2) (2009) 475–482.

[49] M. Weiss, C. Evans, D. Rees, Identification of nonlinear cascade systems using paired multisine signals, IEEE Trans. Instrum. Meas. 47 (1) (1998) 332–336.

[50] S. Farsi, P. Draxler, H. Gheidi, B. Nauwelaers, P. Asbeck, D. Schreurs, Characterization of intermodulation and memory effects using offset multisine excitation, IEEE Trans. Microw. Theory Techn. 62 (3) (2014) 645–657.

[51] J.C. Pedro, N.B. Carvalho, Designing multisine excitations for nonlinear model testing, IEEE Trans. Microw. Theory Techn. 53 (1) (2005) 45–54.

[52] J.F. Sevic, K.L. Burger, M.B. Steer, A novel envelope-termination load-pull method for ACPR optimization of RF/microwave power amplifiers, in: IEEE MTT-S International Microwave Symposium Digest, vol. 2, 1998, pp. 723–726.

[53] J.P. Martins, N.B. Carvalho, J.C. Pedro, Intermodulation distortion of third-order nonlinear systems with memory under multisine excitations, IEEE Trans. Microw. Theory Techn. 55 (6) (2007) 1264–1271.

[54] N. Kim, V. Aparin, L.E. Larson, Analysis of IM asymmetry in MOSFET small-signal amplifiers, IEEE Trans. Circuits Syst. I Regul. Pap. 58 (4) (2011) 668–676.

[55] N.B. Carvalho, J.C. Pedro, A comprehensive explanation of distortion sideband asymmetries, IEEE Trans. Microw. Theory Techn. 50 (9) (2002) 2090–2101.

[56] J. Brinkhoff, A.E. Parker, Effect of baseband impedance on FET intermodulation, IEEE Trans. Microw. Theory Techn. 51 (3) (2003) 1045–1051.

[57] J. Brinkhoff, A.E. Parker, M. Leung, Baseband impedance and linearization of FET circuits, IEEE Trans. Microw. Theory Techn. 51 (12) (2003) 2523–2530.

[58] C.M. Olsen, S.L. Sweeney, K.A. Remley, Representing bias-network baseband characteristics when simulating intermodulation distortion, in: 76th Automatic RF Techniques Group (ARFTG) Conference, Clearwater Beach, FL, USA, December 2010, 2010, pp. 1–6.

[59] N. Cheng, J.P. Young, Challenges and requirements of multimode multiband power amplifiers for mobile applications, in: IEEE Compound Semiconductor Integrated Circuit Symposium (CSICS), Waikoloa, Hawaii, October 2011, 2011, pp. 1–4.

[60] J.J. Yan, C. Hsia, D.F. Kimball, P.M. Asbeck, Design of a 4-W envelope tracking power amplifier with more than one octave carrier bandwidth, IEEE J. Solid-State Circuits 47 (10) (2012) 2298–2308.

[61] T. O'Sullivan, R.A. York, B. Noren, P.M. Asbeck, Adaptive duplexer implemented using single-path and multipath feedforward techniques with BST phase shifters, IEEE Trans. Microw. Theory Techn. 53 (1) (2005) 106–114.

[62] A. Roussel, C.W.T. Nicholls, J.S. Wight, Frequency agile RF feedforward noise cancellation system, in: IEEE Radio and Wireless Symposium, Orlando, FL, USA, January 2008, 2008, pp. 109–112.

[63] E. Koukkari, P. Kurttio, A. Heiskanan, V.-M. Holappa, J. Peltokorpi, T. Tarvainen, S. Lanfranco, T. Kolmonen, H. Somerma, Small size receiver band self-interference cancellation amplifier for 4G transceivers, in: European Microwave Conference, Amsterdam, Netherlands, October 2012, 2012, pp. 1043–1046.

[64] M. Omer, R. Rimini, P. Heidmann, J.S. Kenney, A PA-noise cancellation technique for next generation highly integrated RF front-ends, in: IEEE Radio Frequency Integrated Circuits Symposium (RFIC), Montreal, QC, Canada, June 2012, 2012, pp. 471–474.

[65] A. Kiayani, L. Anttila, M. Valkama, Digital suppression of power amplifier spurious emissions at receiver band in FDD transceivers, IEEE Signal Process. Lett. 21 (1) (2014) 69–73.

[66] M. Omer, R. Rimini, P. Heidmann, J.S. Kenney, A compensation scheme to allow full duplex operation in the presence of highly nonlinear microwave components for 4 G systems, in: IEEE MTT-S International Microwave Symposium Digest, vol. 1, 2011.

[67] S. Farsi, H. Gheidi, H.-T. Dabag, P. Gudem, D. Schreurs, P. Asbeck, Modeling of deterministic output emissions of power amplifiers into adjacent receive bands, IEEE Trans. Microw. Theory Techn. 63 (4) (2015) 1250–1262.

[68] H. Gheidi, S. Farsi, Y. Liu, H.-T. Dabag, P. Gudem, P.M. Asbeck, Digital signal injection technique for cancellation of receive-band spurious emissions in FDD cellular transmitters, in: IEEE MTT-S International Microwave Symposium Digest, Phoenix, AZ, USA, 17–22 May 2015, 2015, pp. 1–4.

[69] A. Kheirkhahi, J. Yan, P. Asbeck, L. Larson, Improved envelope injection and termination (EIT) RF power amplifier with envelope equalization for mobile terminal applications, IEEE Trans. Microw. Theory Techn. 62 (2) (2014) 341–351.

[70] Y. Ma, Y. Yamao, Y. Akaiwa, K. Ishibashi, Wideband digital predistortion using spectral extrapolation of band-limited feedback signal, IEEE Trans. Circuits Syst. I Regul. Pap. 61 (7) (2014) 2088–2097.

[71] K.-F. Liang, J.-H. Chen, Y.-J. Chen, K.-F. Liang, J.-H. Chen, Y.-J. Chen, A quadratic-interpolated LUT-based digital predistortion technique for cellular power amplifiers, IEEE Trans. Circuits Syst. II Express Briefs 61 (3) (2014) 133–137.

[72] M. Rawat, K. Rawat, F. Ghannouchi, S. Bhattacharjee, H. Leung, Generalized rational functions for reduced-complexity behavioral modeling and digital predistortion of broadband wireless transmitters, IEEE Trans. Instrum. Meas. 63 (2) (2014) 485–498.

[73] S.N. Ba, Efficient digital baseband predistortion for modern wireless handsets (Thesis), Georgia Institute of Technology, Atlanta, GA, 2009.

[74] W. Wangmyong, M.D. Miller, J.S. Kenney, A hybrid digital/RF envelope predistortion linearization system for power amplifiers, IEEE Trans. Microw. Theory Techn. 53 (1) (2005) 229–237.

[75] S.A. Bassam, M. Helaoui, F.M. Ghannouchi, 2-D digital predistortion (2-D-DPD) architecture for concurrent dual-band transmitters, IEEE Trans. Microw. Theory Techn. 59 (10) (2011) 2547–2553.

[76] F.M. Ghannouchi, M. Younes, M. Rawat, Distortion and impairments mitigation and compensation of single- and multi-band wireless transmitters (invited), IET Microw. Antennas Propag. 7 (7) (2013) 518–534.

Index

Note: Page numbers followed by "*f*" indicate figures and "*t*" indicate tables.

E

F

G

H

I

K

L

M

N

Q

R

Kolibri 360 GmbH
Ferdinand-Jühlke-Straße 7
99095 Erfurt, Deutschland
produktsicherheit@kolibri360.de